Blockchain Applications for the Energy and Utilities Industry

Someah Alangari
Shaqra University, Saudi Arabia

IGI Global
Scientific Publishing
Publishing Tomorrow's Research Today

Vice President of Editorial Melissa Wagner
Director of Acquisitions Mikaela Felty
Director of Book Development Jocelynn Hessler
Production Manager Mike Brehm
Cover Design Phillip Shickler

Published in the United States of America by
IGI Global Scientific Publishing
701 East Chocolate Avenue
Hershey, PA, 17033, USA
Tel: 717-533-8845
Fax: 717-533-7115
Website: https://www.igi-global.com E-mail: cust@igi-global.com

Library of Congress Cataloging-in-Publication Data

LCCN: 2025011862 (CIP Data Pending)
ISBN13: 979-8-3373-2439-5
Isbn13Softcover: 979-8-3373-2440-1
EISBN13: 979-8-3373-2441-8
British Cataloguing in Publication Data
A Cataloguing in Publication record for this book is available from the British Library.

Table of Contents

Detailed Table of Contents

Meteorological data is critical not only for climate modeling and weather prediction but also for renewable energy forecasting, demand response, and grid stability in the energy and utilities sector. Conventional data systems rely on centralized architectures vulnerable to single points of failure and poor traceability. This chapter presents a blockchain-based oraculus for storing and retrieving weather observations on an immutable ledger tailored for energy contexts. Specialized oracles, robust consensus, and secure data pipelines ensure tamper-proof storage. Testing confirms reliable retrieval for key stakeholders, including grid operators and regulators. The design supports accurate archiving, real-time ingestion, and flexible access control—vital for planning and operations. By offering a scalable pipeline addressing tampering and integrity, this framework fosters reliable, auditable meteorological systems.

Blockchain has emerged as a distributed ledger mechanism that enables decentralized recordkeeping and transaction validation with applications in various fields. Recent research has highlighted its potential for managing urban water and energy systems, especially given increasing urbanization and climate-related resource pressures. This chapter explores theoretical dimensions of blockchain for integrated urban water and energy management. It examines the conceptual linkages of blockchain-based smart contracts, distributed consensus, and tamper-proof data exchange with the operational and strategic needs of urban water and energy stakeholders. It discusses the roles of trust, security, and data transparency in facilitating stakeholder cooperation in resource allocation and highlights resilience benefits for water supply networks and energy grids.

In an era of escalating cyber threats that increasingly target critical infrastructure, there is growing urgency to protect operational technology (OT) systems in energy and utilities. By uniting the core tenets of zero trust security—where every request is inherently untrusted until verified—with the decentralized transparency of blockchain, we can significantly reduce the risk of undetected breaches or insider misuse. This paper explores how blockchain-based authentication and smart contracts, aligned with guidelines such as NIST SP 800-207, can create a tamper-evident ledger of all access requests and device interactions. The goal is to move beyond reactive defenses and instead implement continuous validation of assets and their privileges. This approach not only mitigates the lateral movement of threats within industrial control systems but also enables streamlined auditing of regulatory requirements (e.g., NERC CIP).

Blockchain technology, originally associated with digital currencies, possesses key features such as decentralization, anonymity, robustness, and resistance to tampering, making it an ideal platform for building covert communication channels. This chapter begins by introducing the concept of network covert channels, discussing their definition, historical development, and the architecture of blockchain technology with an exploration of traditional network covert channels. It then highlights the advantages of blockchain-based covert communication channels over conventional methods. The chapter classifies blockchain network covert channels based on several key components. It provides a detailed analysis of the advantages and disadvantages of these channels in terms of covert nature, transmission efficiency, and communication costs. Finally, the chapter addresses existing challenges and limitations within blockchain network covert channels and offers insights into future research directions to improve their efficiency, security, and scalability in the context of secure communication.

The exponential growth of Internet of Things (IoT) devices has created unprecedented security and privacy challenges that traditional centralized architectures struggle to address effectively. This chapter presents a comprehensive theoretical exploration of Internet of Blockchain (IoBc), an emerging paradigm that integrates blockchain technology with IoT ecosystems to enhance security, privacy, and trust. The chapter examines the theoretical foundations underpinning IoBc, analyzes its architectural frameworks, and investigates the cryptographic primitives that enable secure IoT operations. Through comparative analysis of existing blockchain platforms for IoT applications, the chapter identifies theoretical and practical implications for implementing IoBc solutions across various domains. The research synthesizes current literature to formulate a novel conceptual framework that addresses the unique challenges of resource-constrained IoT environments while leveraging blockchain's immutability and decentralization properties.

The reliability of modern power grids is vital for sustaining essential services,especially amid threats like natural disasters and cyberattacks. Blockchain technology offers a decentralized, tamper-resistant infrastructure to enhance grid resilience and disaster response. Resilience relies on real-time monitoring, fault detection, and rapid recovery,all of which blockchain streamlines through secure energy transactions, transparent data sharing, and automated smart contracts. In disasters, blockchain improves coordination with clear audit trails for resource allocation and efficient peer-to-peer energy trading, reducing single points of failure.This paper proposes a blockchain-enabled framework to enhance grid resilience, analyzing its security, scalability, and efficiency through case studies and simulations. Combining blockchain with IoT and AI, the framework creates a decentralized,transparent ecosystem to transform power grid recovery processes, reinforcing global energy security.

Blockchain technology has emerged as a transformative tool for enhancing accountability, transparency, and efficiency in e-government systems. This study explores the integration of blockchain-based architectures within startup-driven e-government models, focusing on secure identity management, automated public service workflows, and tamper-resistant record-keeping. By leveraging smart contracts and permissioned distributed ledgers, blockchain can mitigate corruption, streamline public procurement, and improve citizen trust in digital governance. However, challenges such as scalability, interoperability, legal compliance, and data privacy concerns must be addressed to ensure sustainable adoption. Empirical findings from pilot initiatives suggest that blockchain enhances process integrity, transaction traceability, and interdepartmental collaboration, while startup-driven innovation accelerates agile prototyping and deployment.

This chapter examines how Artificial Intelligence (AI) and blockchain converge in the context of Advanced Metering Infrastructure (AMI) within smart grids. It presents theoretical concepts of AI for data analytics and automation, and explains blockchain's potential to create transparent, decentralized, and secure data environments. The discussion explores the synergy between AI-driven intelligence and blockchain-based trust frameworks for managing metering data, demand-side response, energy trading, and anomaly detection. Several use cases demonstrate how AI-led approaches elevate predictive and optimization tasks, while blockchain safeguards meter data integrity and coordinates distributed market transactions. The chapter presents operational, security, and interoperability challenges and emphasizes the scope for future research on standardizing AI–blockchain modules and addressing scalability constraints. Findings highlight the strategic importance of pairing blockchain-based trust with AI-based intelligence for achieving automated, reliable, and consumer-focused AMI.

Abstract This chapter examines how blockchain technology can advance energy management in smart buildings and urban areas. It analyzes the theoretical foundations of blockchain solutions and their integration into energy monitoring, control, and optimization strategies. It describes how blockchain can manage peer-to-peer transactions, coordinate distributed energy resources, and facilitate trust-minimized data sharing for building operators, occupants, and regulators. It then explores digital twin models, cybersecurity considerations, interoperability approaches, and energy market mechanisms that rely on immutable ledgers and smart contracts. The chapter reviews challenges and outlines directions for future research and deployment. The discussion underscores that blockchain presents potential for transparent and automated energy management. Realizing these benefits depends on supportive policies, reliable technology, cross-sector engagement, and continued system refinement.

Blockchain which is derived from virtual currency such as Bitcoin is now considered revolutionary. This paper aims at discussing what blockchain is, its attributes, and its application in finance, healthcare, supply chain and governance. It is a decentralized system that provides transparency, security, and a low level of alteration, which makes it reliable and eliminates the need for middlemen. This also comprises blockchain structures, consensus mechanisms, smart contracts as well as the comparison of the cryptocurrency with traditional systems based on decentralization, cryptographic security and scalability. Blockchain is beneficial for strengthening the digital transactions, securing the healthcare data and the traceability of the supply chain. Thus, the issues like scalability, legal aspects and adoption remain to be a concern. It examine present and future studies and development which may help to eliminate these issues and strengthen the position of blockchain in the sphere of digital and decentralized markets.

Preface

The energy and utilities industry is undergoing a profound transformation, driven by rapid technological advancements, evolving consumer expectations, and the pressing need for sustainability. At the heart of this transformation lies blockchain technology, a decentralized and transparent digital ledger system that has the potential to redefine the way energy systems operate. From enhancing energy supply chain management to enabling decentralized markets and secure critical infrastructure, blockchain technology offers innovative solutions to some of the most complex challenges facing the energy and utilities sector today.

This book, *Blockchain Applications for the Energy and Utilities Industry*, explores the multifaceted potential of blockchain technology in reshaping energy and utilities. With contributions from leading experts in the field, the book provides a comprehensive overview of cutting-edge blockchain applications, their integration with other technologies such as IoT and AI, and their ability to foster transparency, efficiency, and sustainability.

OVERVIEW OF THE SUBJECT MATTER

Blockchain technology, initially known for its use in cryptocurrency systems such as Bitcoin, has evolved into a versatile tool with far-reaching applications across industries. In the energy and utilities sector, blockchain's ability to enable secure, transparent, and tamper-proof transactions makes it a powerful enabler of innovation. The book delves into blockchain's role in energy supply chain management, decentralized energy trading, secure communication, and governance, among other areas. It also highlights how blockchain can be combined with artificial intelligence (AI) and the Internet of Things (IoT) to drive smarter, more efficient, and more secure energy systems.

The overarching goal of this volume is to provide readers with a thorough understanding of blockchain's transformative capabilities in energy and utilities, backed by practical insights and real-world applications.

RELEVANCE OF THE BOOK IN TODAY'S WORLD

The energy and utilities sector stands at the forefront of addressing global challenges such as climate change, resource scarcity, and growing energy demands. The transition to renewable energy, the push for grid modernization, and the need for resilient infrastructure require innovative solutions that go beyond traditional approaches. Blockchain technology emerges as a critical tool for enabling this transition by improving transparency, fostering trust, enhancing operational efficiency, and enabling decentralized and participatory energy systems.

In today's interconnected world, where data security and privacy are paramount, blockchain's ability to provide zero-trust architectures and secure communication pathways is invaluable. Furthermore, its synergy with AI and IoT amplifies its impact, making it a cornerstone of the digital transformation sweeping the energy and utilities industry.

This book addresses these timely challenges and opportunities, positioning itself as a vital resource for navigating the future of energy and utilities in a digitally driven world.

TARGET AUDIENCE

The book is designed for a diverse audience, including:

- **Energy and utilities professionals** seeking to understand how blockchain technology can enhance operations, improve transparency, and drive sustainability.
- **Researchers and academics** exploring the intersection of blockchain technology, energy systems, and digital transformation.
- **Policymakers and regulators** aiming to design frameworks that promote blockchain adoption while ensuring security, privacy, and equitable access.
- **Technology developers and innovators** looking to create blockchain-based solutions tailored to the energy and utilities sector.
- **Students and learners** pursuing education in blockchain technology, energy systems, and related fields.

By catering to this broad audience, the book aims to bridge the gap between theory and practice, offering both conceptual clarity and actionable insights.

CHAPTER SUMMARIES

Each chapter in this book contributes uniquely to the overarching theme of blockchain applications in the energy and utilities sector:

1. **Blockchain Technology in Energy Supply Chain Management: A Path to Transparency, Efficiency, and Sustainability**

This chapter explores how blockchain can enhance transparency and efficiency in energy supply chains, enabling sustainable practices and improving trust among stakeholders.

2. **Blockchain for Energy Financing and Investment: Revolutionizing Models for a Sustainable Future**

This chapter examines how blockchain is revolutionizing energy financing by enabling innovative investment models and fostering greater accessibility and accountability.

3. **Blockchain Oraculus for Meteorological Data in Energy and Utilities with Immutability and Security**

A foundational chapter providing a comprehensive overview of blockchain's transformative potential in energy and utilities, with a focus on key use cases and future trends.

4. **Blockchain for Integrated Urban Water and Energy Management**

This chapter highlights blockchain's role in enabling integrated urban water and energy systems, fostering efficiency, and optimizing resource management.

5. **Blockchain-Driven Zero-Trust Architectures for Critical Infrastructure**

The chapter focuses on how blockchain can establish zero-trust architectures to secure critical infrastructure against cyber threats and ensure operational resilience.

6. **Blockchain for Secure Communication**

This chapter emphasizes blockchain's ability to enhance secure communication across energy systems, mitigating risks of data breaches and ensuring privacy.

7. **IoBc-Blockchain Solutions for Secure and Private IoT Ecosystems**

The chapter explores the integration of blockchain in IoT ecosystems, addressing security concerns and enabling private, tamper-proof operations.

8. **Blockchain-Enabled Grid Resilience and Disaster Response**

This chapter discusses how blockchain can enhance grid resilience and disaster response through decentralized operations and transparent data sharing.

9. **Blockchain for Transparent Governance: Enhancing Accountability and Trust in E-Government**

The chapter examines blockchain's application in governance, focusing on its ability to enhance transparency and trust in e-government systems.

10. **AI and Blockchain Integration for Advanced Metering Infrastructure (AMI)**

This chapter explores the integration of AI and blockchain to optimize AMI systems, improving energy efficiency and enabling smarter decision-making.

11. **AI-Powered Smart Contracts for Energy Grid Optimization: A Hybrid Intelligence Approach**

The chapter investigates AI-powered smart contracts, showcasing how they can optimize energy grids and enable autonomous, efficient operations.

12. **Decentralized Energy Markets: Transforming Energy Distribution and Trading With Blockchain Technology**

This chapter highlights the transformative impact of blockchain-enabled decentralized energy markets on energy distribution and peer-to-peer trading.

13. **Smart Building Energy Management: Blockchain Solutions for Urban Efficiency**

The chapter explores blockchain's role in optimizing energy management in smart buildings, driving urban efficiency and sustainability.

14. **Exploring Blockchain Technology: Concepts, Features, and Industry Transformations**

A conceptual chapter providing an in-depth exploration of blockchain's core features and its transformative impact across industries, with a focus on energy and utilities.

15. **Securing Blockchain Solutions in the Energy Sector: Mitigating Advanced Persistent Threats (APTs)**

This chapter focuses on securing blockchain applications in the energy sector, addressing vulnerabilities and mitigating advanced persistent threats.

16. **Blockchain Applications for the energy and utilities industry: Future Trends and Emerging Opportunities**

This chapter explores blockchain's evolution beyond cryptocurrencies, highlighting key innovations and challenges shaping its broader adoption.

17. **Overview of Blockchain Technology: Fundamental concepts, key features, and the potential impact of blockchain on various industries.**

This chapter provides a concise overview of blockchain technology, its core principles, benefits, challenges, and transformative applications across key industries.

CONCLUSION

This book serves as a critical resource for understanding and leveraging blockchain technology to address the challenges and opportunities in the energy and utilities industry. By providing a multidisciplinary perspective, it bridges the gap between technological innovation and practical application, offering actionable insights for professionals, researchers, and policymakers.

The contributions within this volume underscore blockchain's potential to drive transparency, efficiency, and sustainability, while also ensuring security and resilience in energy systems. As the energy sector continues to evolve, this book aims to inspire innovation, foster collaboration, and contribute to the creation of a more sustainable and equitable energy future.

Acknowledgment

The completion of this book, *Blockchain Applications for the Energy and Utilities Industry*, would not have been possible without the invaluable contributions and support of many individuals and organizations.

First and foremost, we would like to express our heartfelt gratitude to the contributing authors, whose expertise, dedication, and commitment to excellence have been instrumental in shaping the content of this book. Their innovative research and forward-thinking perspectives have significantly enriched the discussions on blockchain applications in the energy and utilities sector.

We are deeply thankful to the reviewers who provided insightful feedback, constructive suggestions, and meticulous attention to detail. Their contributions have ensured the academic rigor and quality of the book, making it a resource of high value to both scholars and practitioners.

Our sincere appreciation goes to the editors and production team at IGI Global, whose unwavering support, professionalism, and guidance throughout the publishing process have been critical to the success of this project. Their efforts in managing the publication from its inception to completion have been exceptional.

We also acknowledge the support of our academic institutions, research organizations, and professional networks, which provided a platform for collaboration, knowledge exchange, and the sharing of ideas that formed the foundation of this book.

Finally, we extend our deepest gratitude to our families, friends, and colleagues for their encouragement, understanding, and patience throughout this journey. Their belief in the importance of this work has been a constant source of motivation.

This book is the result of collective efforts, and we are truly grateful to everyone who has contributed, directly or indirectly, to its creation. We hope it serves as a valuable resource for advancing knowledge and innovation in the energy and utilities sector.

Chapter 1 Blockchain Technology in Energy Supply Chain Management: A Path to Transparency, Efficiency, and Sustainability

Saad Alateef
https://orcid.org/0009-0002-3219-0833
Newcastle University, UK

Amjad Aldweesh
https://orcid.org/0000-0001-9358-1323
College of Computing and IT, Shaqra University, Saudi Arabia

Mohammad Alauthman
https://orcid.org/0000-0003-0319-1968
Petra University, Jordan

Mohammed Alkasassbeh
https://orcid.org/0000-0001-8396-7441
Princess Sumaya University for Technology, Jordan

Ahmed Al-Qerem
https://orcid.org/0000-0003-2187-6194
Zarqa University, Jordan

ABSTRACT

This paper proposes a blockchain-based framework to optimize energy supply chain management, delineating its potential to transform energy sourcing, allocation, and consumption processes.The framework elucidates critical components such as decentralized record-keeping, smart contracts for automated transactions, and robust security protocols, thereby ensuring data integrity and resilience against cyber-attacks. A prototype system is developed and tested to evaluate performance metrics, including scalability, transaction speed, and energy tracking accuracy.The results highlight the considerable benefits of

DOI: 10.4018/979-8-3373-2439-5.ch001

blockchain, including enhanced transparency, cost efficiency, and improved stakeholder collaboration. Conclusively, the paper identifies current challenges ranging from technical scalability to regulatory hurdles and suggests future directions for augmenting blockchain's impact on energy supply chains, ultimately reinforcing the global transition towards more sustainable energy systems

INTRODUCTION

Supply chains for energy form the cornerstone of modern infrastructure. They encompass intricate operations that produce, transport, distribute, and consume energy in various forms. Multiple stakeholders—ranging from fossil fuel extraction companies and renewable energy producers to grid operators, regulatory bodies, and end-users—interact across a geographically diverse landscape. A well-functioning energy supply chain addresses reliability, cost-effectiveness, transparency, and ecological sustainability. However, conventional supply chains still grapple with limited visibility of energy flows, reliability issues in validating energy sources, and vulnerabilities to deceptive practices or tampering (Mancarella et al., 2020).

One of the most pressing issues in these traditional energy setups is the lack of instantaneous, transparent information exchange. Energy transactions typically involve multiple intermediaries, contracts, and siloed databases. This fragmentation creates inefficiencies and a deficiency of clear verification, undermining confidence among producers, distributors, and consumers (Nakamoto, 2008). Moreover, continuous reliance on fossil fuels drives growing environmental scrutiny, reinforcing the importance of seeking greener solutions and ensuring prudent energy utilization in the long run (Mathews & Tan, 2020).

Blockchain, a distributed ledger technology known for its decentralized structure and cryptographic safeguards, has gained traction as a viable solution to these challenges (Andoni et al., 2019). As a system where data is stored across multiple nodes, blockchain diminishes dependence on a central authority and enhances transparency, thanks to consensus-based validation and an immutable ledger (Alam et al., 2022). Widely acknowledged for its capacity to improve supply chain management in diverse industries—like agriculture, finance, and logistics—blockchain offers features such as data immutability, consensus mechanisms, and programmable smart contracts to reduce trust gaps, boost efficiency, and limit the potential for fraudulent behavior (Xu et al., 2019). Within the energy domain, blockchain can support real-time monitoring of generation data, enable direct energy trading between producers and consumers, automate billing operations, and verify renewable energy certificates (Khan et al., 2021).

Growing demands for cleaner, more efficient power underscore blockchain's potential to streamline operational workflows while encouraging investment in green power. For example, producers can convert their renewable outputs into tokens, while customers can purchase these tokens to offset carbon footprints or secure direct allocations of green electricity (Alam et al., 2022). Smart contracts can implement instant price adjustments based on supply and demand, automate grid rebalancing during peak hours, and guarantee that data is well-protected, an increasingly critical consideration as grids integrate digital technologies (Alam et al., 2022).

Still, multiple obstacles must be navigated before blockchain gains widespread adoption in energy supply chains. Some blockchain networks require extensive computational power—especially those using Proof of Work—sparking concerns over their environmental costs (Goodman & Alam, 2021). Issues related to scalability and interoperability are also prevalent, as current energy systems incorporate heterogeneous technologies, regulations, and governance models (Xu et al., 2019). Regulatory frameworks are still adapting, leading to uncertainties for energy firms contemplating blockchain-based investments (Andersen, 2021).

This paper addresses both conceptual and applied facets of blockchain deployment in energy supply chain management, focusing on how blockchain's defining attributes—transparency, security, and decentralization—can tackle real inefficiencies and expedite a more sustainable energy structure. Section **2** thoroughly reviews pertinent literature, covering conventional energy supply chains, the role of blockchain in general supply chains, and its targeted use in energy. Section **3** outlines our blockchain-based architecture designed for the energy sector, while Section **4** clarifies the process used in building and testing a prototype implementation. Section **5** presents empirical findings and a discussion of the system's performance, contrasting it with traditional approaches. The paper also highlights both the challenges and prospects for future study in Section **6**, finishing with concluding insights in Section **7**, underscoring the contributions and urging broader real-world testing.

By merging theoretical analysis with empirical findings, the paper illustrates how decentralized ledgers can significantly improve transparency and operational efficiency, while simultaneously championing environmentally responsible practices. Consequently, it contributes to the body of research on blockchain's disruptive impact on modern energy networks, delivering a comprehensive vision of its capacity to reshape the global energy supply chain landscape.

LITERATURE REVIEW

Overview of Energy Supply Chains

Energy supply chains incorporate various stages: extracting or harvesting raw energy resources, conveying them, and converting them into consumable forms (electricity, heat, or fuels). Fossil fuel-based chains revolve around mining or drilling, refining, and transporting resources like coal or natural gas, followed by electricity production at power plants and subsequent distribution to customers (Mancarella et al., 2020). These established processes, while historically effective, have led to sustainability and environmental concerns, such as large-scale carbon emissions and resource depletion (Andersen, 2021).

The growing utilization of renewable resources—such as solar and wind—complicates supply chains because these energy sources are inherently variable. The output depends heavily on time of day or weather patterns, meaning advanced mechanisms are required to keep supply and demand balanced (Hussain et al., 2019). While energy storage (e.g., batteries, pumped hydro) helps mitigate intermittency, the widespread adoption of renewables still requires better integration and clearer data sharing to coordinate generation with consumption (Rodríguez-Molina et al., 2020). Moreover, centralized data systems can be susceptible to manipulation or unauthorized access, heightening the risk of inaccuracies or compromised grid reliability (Hasan et al., 2021). Addressing these core weaknesses calls for a solution that unifies stakeholders and ensures transparency and data security at every stage.

Blockchain in Supply Chain Management

Emerging alongside Bitcoin and other cryptocurrencies, blockchain technology is now recognized for its potential to revolutionize traditional supply chains. It replaces a single gatekeeper with a decentralized system in which nodes verify and store transaction data across a peer-to-peer network (Nakamoto,

2008), (Alam et al., 2022). This immutability and decentralized consensus create an environment of trust where product origins can be traced, contractual obligations can be automated, and fraud can be significantly curtailed (Xu et al., 2019).

Industries applying blockchain to streamline supply chains include agriculture, pharmaceuticals, and manufacturing (Czarnecki & Jochem, 2020). In each setting, blockchain-based records offer a tamper-proof ledger of product history. Self-executing smart contracts can automate payments, shipping, and quality checks (Treiblmaier, 2018). However, challenges related to hardware integration, interoperability, and compliance with regional laws still impede large-scale deployment (Xu et al., 2019). Despite these hurdles, blockchain's transformative potential in supply chain optimization remains evident, fostering a surge in research and pilot initiatives.

Blockchain in Energy Systems

The energy sector's fragmented nature—marked by decentralized producers and countless distribution pathways—makes it especially well-suited for blockchain solutions. One of the most promising uses is P2P energy trading, where households or communities with surplus power (e.g., rooftop solar) directly transact with neighbors (Andoni et al., 2019), (Khan et al., 2021). Smart contracts on the blockchain authenticate the energy source and automatically execute transactions, removing the need for an intermediary utility while speeding up payments and market responsiveness (Alam et al., 2022).

Another key application is the certification of renewable energy. Historically, verifying that a unit of power is derived from solar, wind, or another renewable source demands an intricate bureaucratic system with potential loopholes for double-counting (Alam et al., 2022). A blockchain-based system, however, can mint a unique token for every verified unit of green electricity produced, guaranteeing authenticity and simplifying transfers or retirements of these certificates (Czarnecki & Jochem, 2020). Moreover, integrated IoT devices across the grid can supply trustworthy, real-time data on generation, transmission conditions, and grid health, all stored in a transparent yet secure manner (Xu et al., 2019).

Nonetheless, performance considerations (like throughput and latency) and the energy-intensive nature of some consensus protocols have spurred debate over blockchain's viability in an industry aiming to reduce overall energy usage (Goodman & Alam, 2021). Ongoing efforts include shifting to more efficient consensus methods (e.g., Proof of Stake), implementing sidechains, or employing hybrid models that can scale to millions of micro-transactions while limiting environmental footprints (Alam et al., 2022), (Xu et al., 2019). Legally, regulators are still evolving in their stances on digital contracts for energy trading, but momentum is building around formalizing decentralized marketplaces and awarding them a recognized legal status (Andersen, 2021).

Research Gaps

Although the potential for blockchain in energy supply chains is widely discussed, significant gaps remain:

1. **Unified Frameworks**: Current studies commonly isolate single use cases (e.g., P2P trading or REC issuance) without proposing an overarching architecture that unites the entire supply chain.

2. **Empirical Pilots**: Many propositions rely on theoretical or simulated models without robust real-world experiments, causing uncertainties about true performance under operational stress and compliance demands.
3. **Economic and Social Factors**: Technological discourse often overshadows the economic and social consequences, including evolving roles for utilities, consumer acceptance, and new regulatory responsibilities.
4. **Sustainability Paradox**: While blockchain can expedite green energy adoption, certain implementations, especially those using Proof of Work, consume substantial energy themselves, posing a contradiction.

To address these deficiencies, the subsequent sections propose a holistic blockchain-based framework tailored to energy supply chains, along with a prototype designed to assess core functionality. By integrating theory and practice, the work aims to clarify blockchain's tangible effect on energy operations and to set out a roadmap for scalable adoption.

PROPOSED BLOCKCHAIN-BASED ENERGY SUPPLY CHAIN FRAMEWORK

Framework Overview

Our framework is designed to provide a unified, secure platform that collects, validates, and settles all energy-related transactions. Figure **[fig:framework_architecture]** illustrates the conceptual arrangement, encompassing four main stakeholder categories: producers, distributors, consumers, and regulators/certifiers. Each group runs blockchain nodes or client interfaces, which ensure real-time and tamper-resistant data capturing (Khan et al., 2021). Smart contracts orchestrate events like automated billing, P2P energy trades, and issuance of renewable energy credits.

Figure 1. High-level architecture of the proposed blockchain-based energy supply chain framework.

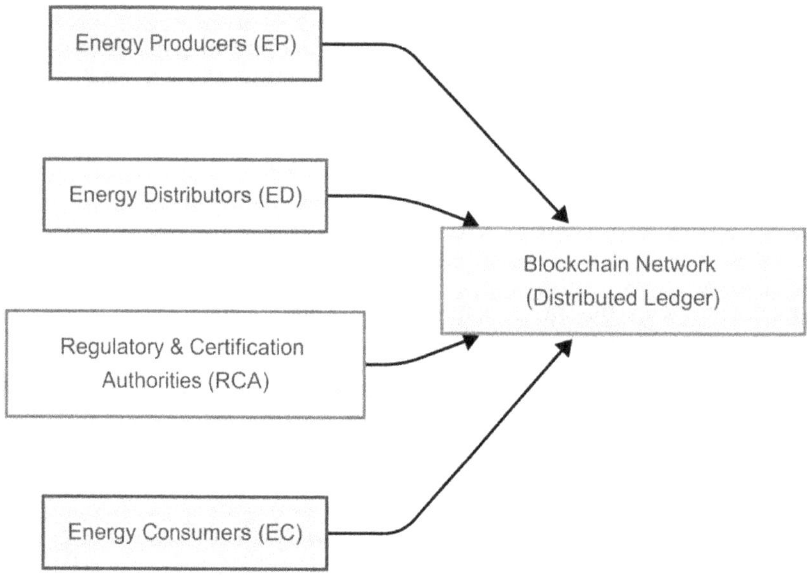

Key Features

Transparency

Blockchain's decentralized ledger ensures that any validated transaction is permanently recorded and made visible to permitted stakeholders (Xu et al., 2019). This level of transparency motivates open data access, aiding in sustainability objectives by giving consumers clarity on their energy's origin and incentivizing producers to augment renewable output.

Smart Contracts

Self-executing smart contracts bring automation to traditionally manual processes. They are critical to:

- **P2P Trading**: Automatic energy sales when supply levels exceed certain thresholds or if a buyer offers a viable price.
- **Billing and Payment**: Reduces lags in invoice generation and minimizes manual verifications.
- **Regulatory Enforcement**: Checks for compliance rules (e.g., emissions standards) directly embedded in contract logic (Treiblmaier, 2018).

Traceability

Because transaction histories are recorded across blockchain nodes, any energy parcel or credit can be traced throughout its lifecycle (Czarnecki & Jochem, 2020). This reliable chain of custody bolsters consumer trust, especially in renewable markets where authenticity is paramount.

Security

Distributed consensus and advanced cryptography make it highly challenging to alter transaction records. Each participant has a unique credential, ensuring that only authorized entities can create or approve transactions, mitigating exposure to data tampering (Hasan et al., 2021).

Use Case Example: Renewable Certificates

A prominent illustration involves tokenizing Renewable Energy Credits (RECs). A smart meter at a solar or wind farm reliably measures production. A corresponding smart contract mints a blockchain-based token whenever a valid megawatt-hour of green power is produced. This REC token can then be sold or transferred to different buyers seeking to meet sustainability targets. Once the token is redeemed or "burned," it is no longer eligible for resale, circumventing double-counting (Alam et al., 2022). This automated, immutable process streamlines oversight and reassures stakeholders about the credit's authenticity.

Figure 2. Flowchart of a blockchain-based REC issuance and trading process.

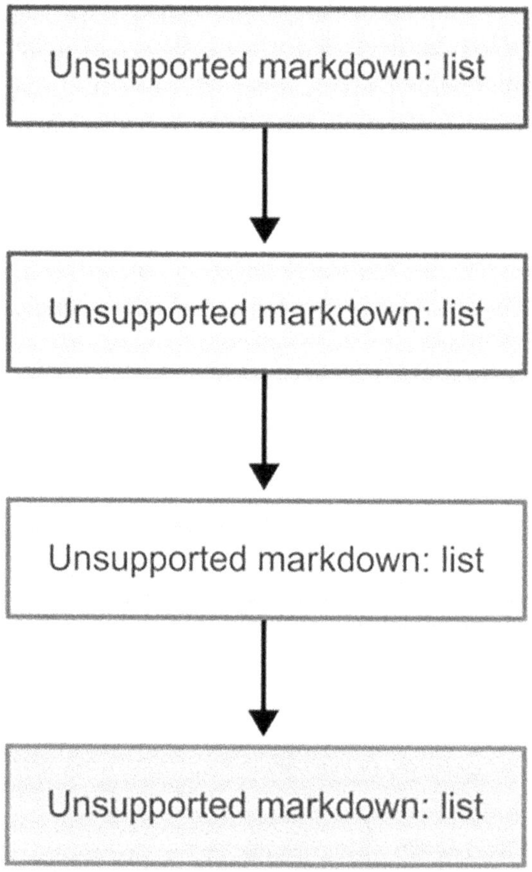

METHODOLOGY

Requirements Analysis

To develop a blockchain solution for energy supply chains, both functional and performance benchmarks were established:

- **Data Logging and Retrieval**: Capture and retain real-time records of energy generation and consumption.
- **Smart Contract Deployment**: Allow automated settlement of trades and compliance monitoring.
- **Authentication and Authorization**: Guarantee that only validated users can upload, view, or modify data.
- **Scalability and Reliability**: Accommodate surges in transactions while upholding data integrity, even if individual nodes fail.

Prototype Development

Ethereum and Hyperledger Fabric were initially evaluated because of their mature developer tools and consensus variability. For this prototype, a permissioned approach using Hyperledger Fabric was preferred, considering energy providers and regulators often require privacy and controlled membership (Andersen, 2021). Smart contracts (chaincode) covered tasks like recording generation metrics, validating consumption data, and automating REC issuance.

System Architecture

The prototype included:

- **Application Layer**: User interfaces for producers, distributors, consumers, and regulators. Communication with the blockchain occurred through REST APIs.
- **Blockchain Network**: A set of peers that store and synchronize the ledger, running the consensus mechanism.
- **Data Storage**: A state database (e.g., CouchDB) for quick lookups, while historical data resides permanently on-chain.
- **Security**: A Membership Service Provider validated participant identities and managed certificates.

Figure 3. System architecture of the prototype using Hyperledger Fabric.

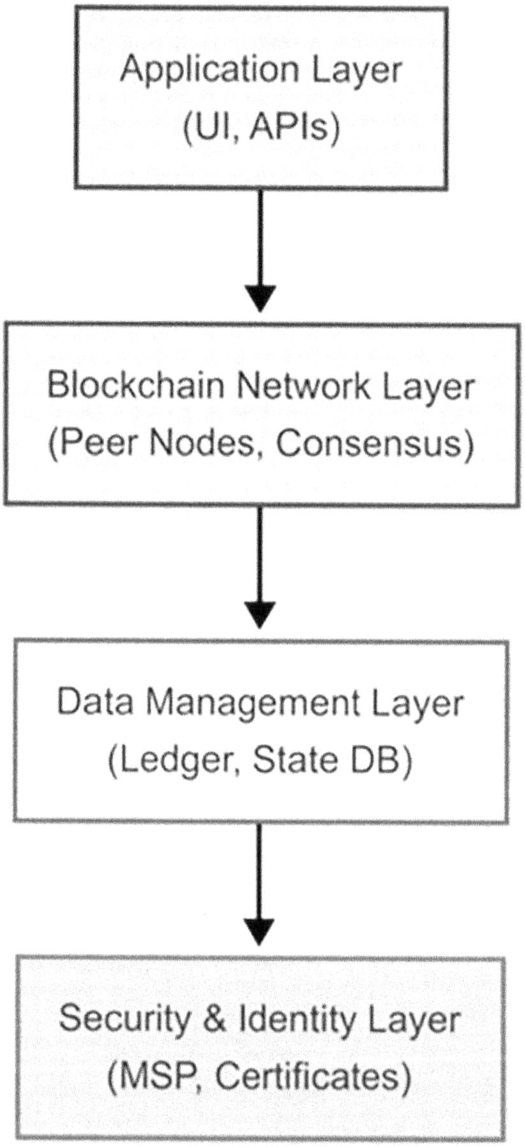

Testing and Validation

Evaluation Metrics

1. **Transactions Per Second (TPS)**: Measured how many simultaneous transactions the network could finalize.
2. **Latency**: Time delay between submission and block confirmation.

3. **Data Accuracy**: Variation between on-chain reported energy and externally metered values.
4. **Resilience**: Susceptibility to attacks (e.g., unauthorized chaincode invocation, ledger modification).

Simulation and Case Studies

The prototype was assessed in two ways:

- **Simulation**: Multiple virtual nodes were instantiated, generating synthetic supply-demand scenarios. Transaction loads ranged from 50 to 200 TPS.
- **Pilot Deployment**: A small-scale solar farm and local distribution network tested real IoT-based meter data streams, capturing real-time usage.

Data logs were collected throughout, enabling performance comparisons across different load conditions. Security audits involved penetration testing to spot vulnerabilities.

RESULTS AND DISCUSSION

Prototype Results

Transaction Speed and Scalability

Under moderate loads (up to 50 TPS) with three peer nodes, the network maintained near-instant finality (around two seconds). Scaling up to 100–200 TPS involved introducing additional peers. This step increased block confirmation latency but kept the success rate above 90%. As shown in Table **1**, proper network configuration, channel partitioning, and high-capacity hardware are crucial to sustaining throughput.

Table 1. Transaction Throughput at Different Load Levels

Load (TPS)	Avg Latency (s)	Success Rate (%)	Peer Nodes
50	2.1	98	3
100	4.5	95	5
200	7.8	91	7

Energy Tracking Accuracy

Comparing the blockchain logs with reference meter readings showed an error margin of less than 1% over a month, depicted conceptually in Figure **[fig:energy_tracking_chart]**. Discrepancies mainly stemmed from minor delays in sensor data transmission (Rodríguez-Molina et al., 2020). Such minimal deviation underlines blockchain's suitability for reliable energy record-keeping.

Figure 4. Comparison of Actual vs. Blockchain-Recorded Energy Outputs Over a One-Month Period (Conceptual).

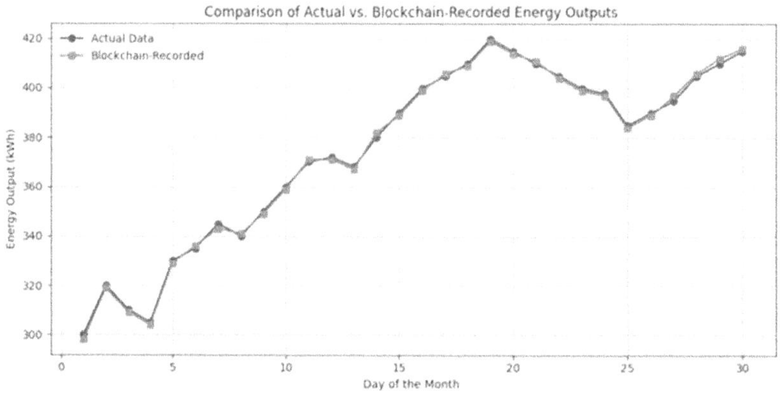

Security Assessments

Penetration testing highlighted a strong defense against tampering attempts, as altering historical blocks would require re-validating the entire chain. Unauthorized chaincode usage was blocked by the MSP's access control. However, the application layer remained vulnerable to distributed denial-of-service attacks if appropriate traffic filtering was not in place (Hasan et al., 2021).

Benefits of Blockchain for Energy Supply Chains

Transparency and Trust

All transactions—be they from solar farms or distribution networks—are time-stamped and cryptographically linked. This high degree of openness fosters trust in decentralized energy dealings, where multiple stakeholders might otherwise be wary of each other (Andoni et al., 2019)(Alam et al., 2022)

Automation via Smart Contracts

Invoices were generated and settled automatically, eliminating the usual lag between service provision and payment. Compliance monitoring (e.g., carbon caps) can be integrated into the code, triggering instant notifications or sanctions if producers exceed permissible limits (Treiblmaier, 2018). This mechanization reduces administrative load and limits human error.

Streamlined Renewable Certifications

Renewable Energy Certificates (RECs) were tokenized, expediting issuance and exchange. Table **2** shows how blockchain can reduce a multi-week credentialing period to just days.

Table 2. Comparison of Conventional vs. Blockchain-Enabled REC Issuance

Process	Legacy System	Blockchain Approach
Issuance Time	2–3 weeks	2–3 days
Verification	Manual audits	Automated via contracts
Fraud Risk	Moderate	Low (immutable)
Administration	High	Lower

Stronger Collaboration

Distributed ledgers align incentives, offering near real-time data for generation and consumption. Grid operators can manage capacity more precisely, producers can update generation based on demand, and consumers can better track their usage (Xu et al., 2019). Regulators benefit from instant access to a robust dataset for auditing.

Comparison with Traditional Systems

Compared to legacy approaches:

- **Data Security**: Centralized databases can be manipulated more easily; the distributed ledger design resists unauthorized modifications.
- **Efficiency**: Automated validation of bills, RECs, and compliance tasks. While setup costs are higher, ongoing operational expenses drop significantly.
- **Visibility**: Stakeholders obtain near-real-time information on power flows, vs. periodic updates in traditional setups.
- **Regulatory Alignment**: By embedding rules into smart contracts, adherence to emission limits or grid codes happens in real time, minimizing manual oversight.

Still, longstanding structures have established regulatory support and user familiarity. Blockchain's novelty can lead to uncertainty or hesitance among investors and regulators (Andersen, 2021). Overcoming these uncertainties requires pilot demonstrations and progressive policy adaptation.

Implications for Stakeholders

Producers see faster payments, improved operational data, and straightforward compliance checks. **Distributors** gain reliable data for load-balancing and a trusted audit trail for billing disputes. **Consumers** benefit from validated insight on the sources of their energy and can buy directly from local producers. **Regulatory Bodies** appreciate streamlined, continuous monitoring and reduced reliance on resource-intensive manual audits (Andersen, 2021).

CHALLENGES AND FUTURE DIRECTIONS

Technical Challenges

Scalability represents a major limitation, as blockchain networks can face latency and throughput challenges when processing large volumes of micro-transactions (Alam et al., 2022). Energy-heavy consensus protocols—particularly Proof of Work—can contradict the sector's sustainability aims (Goodman & Alam, 2021). Furthermore, integrating blockchain into legacy grid infrastructures and IoT systems is non-trivial, as proprietary standards often require bridging solutions or gateways (Rodríguez-Molina et al., 2020).

Future Enhancements

Research can focus on hybrid consensus models, sidechains, or sharded networks that maintain strong security while boosting transaction capacity (Alam et al., 2022). Incorporating AI/ML techniques could allow for predictive maintenance, anomaly detection, and adaptive pricing, all orchestrated via decentralized data repositories (Khan et al., 2021). Multi-chain interoperability protocols might resolve siloed deployments and facilitate frictionless exchange of data and tokens among various specialized blockchains (Treiblmaier, 2018).

Tokenizing a broader range of energy assets—like battery storage or microgrid shares—could open new markets for small investors, providing capital for renewable ventures while sharing risks and rewards equitably (Czarnecki & Jochem, 2020). Such tokenization must be in line with evolving legal and regulatory guidelines to protect participants and preserve financial stability.

Policy and Regulatory Considerations

Current laws often fail to cover emerging blockchain energy models, causing legal ambiguity around digital tokens, consumer safeguards, and data management (Andersen, 2021). Key areas needing clarity include:

- **Recognition of Digital Assets**: Defining the legal status of tokenized energy credits or microgrid equity tokens.
- **Consumer Protections**: Guards against potential exploitative pricing or misrepresented RECs in P2P energy markets.
- **Data Sovereignty**: Compliance with privacy regulations like the GDPR while leveraging distributed ledgers for data transparency (Hasan et al., 2021).

Regulatory sandboxes offer a productive avenue for governments to supervise pilot projects, balancing innovation with risk mitigation (Andoni et al., 2019). Collaboration among regulators, the private sector, and technology experts is essential for constructing governance that aligns blockchain's decentralized benefits with consumer security and market integrity.

CONCLUSION

This paper presented a comprehensive exploration of how blockchain can revolutionize energy supply chain management, bolstering transparency, operational efficiency, data protection, and environmental responsibility. By identifying critical pain points—such as lack of visibility, fraud vulnerabilities, and slow administrative workflows—the proposed framework illustrates a decentralized ledger system that robustly logs transactions, validates generation and consumption, and facilitates automated smart contracts for billing and renewable certificate issuance.

A prototype built on Hyperledger Fabric demonstrated tangible improvements over traditional setups, offering rapid transaction processing, reliable data accuracy, and heightened security. Tokenizing renewable energy credits further highlighted how blockchain could spur efficient certification markets, fostering accountability for both producers and purchasers of green energy. Despite these evident benefits, widespread adoption remains tempered by issues of scalability, energy-intensive consensus mechanisms, and a regulatory landscape still in flux.

Hence, the next phase of progress requires concerted efforts in refining blockchain protocols for higher throughput and lower energy usage, integrating AI-driven analytics, and structuring policy frameworks that encourage constructive innovation. Ultimately, as the planet shifts toward more sustainable and distributed energy models, blockchain stands out as a promising architecture to synchronize trust, transparency, and efficiency within global supply chains—paving the way for a cleaner, more reliable, and cooperative energy future.

REFERENCES

Alam, R., Jeong, S., & Li, C. (2022). Renewable Energy Certification in Blockchain Networks: A Practical Review. *IEEE Transactions on Engineering Management*, *69*(4), 1235–1244.

Alam, T., Goodman, J., & Doh, Y. (2022). Decentralized Ledgers and Smart Contracts in Energy Markets: A Comprehensive Study. *Energy Informatics*, *5*(2), 22–34.

Andersen, B. (2021). Blockchain Regulation for the Energy Sector: An Evolving Landscape. *Energy Policy*, *154*, 112294.

Andoni, M., Robu, V., Flynn, D., Abram, S., Geach, D., Jenkins, D., & Peacock, A. (2019). Blockchain Technology in the Energy Sector: A Systematic Review of Challenges and Opportunities. *Renewable & Sustainable Energy Reviews*, *100*, 143–174. DOI: 10.1016/j.rser.2018.10.014

Benhamou, G., Patel, N., & Chen, T. G. (2022). Blockchain Scalability in Multi-Stakeholder Energy Trading Platforms. *Energy Reports*, *8*(3), 1120–1131.

Chang, Y., Ng, J., & He, C. (2020). Blockchain-Enhanced Transparency in Supply Chains: A Systematic Review. *Computer Standards & Interfaces*, *70*, 103430.

Czarnecki, B., & Jochem, P. (2020). Blockchain-Based Certification of Green Electricity Origin: Potentials and Pitfalls. *Applied Energy*, *277*, 115573.

Di Silvestre, M., Gallo, L., & Riva Sanseverino, E. (2020). Blockchain for Energy Communities: A Comprehensive Review on Technologies and Challenges. *Renewable & Sustainable Energy Reviews*, *133*, 110301.

Francisco, K., & Swanson, D. (2018). The Supply Chain Has No Clothes: Technology Adoption of Blockchain for Supply Chain Transparency. *Logistics (Basel)*, *2*(1), 2. DOI: 10.3390/logistics2010002

Goodman, J., & Alam, M. (2021). Examining the Energy Consumption of Consensus Algorithms in Blockchain Systems. *IEEE Systems Journal*, *15*(4), 5395–5406.

Hasan, H., Latif, M., & Javaid, N. (2021). Security and Privacy in Blockchain-Based Energy Trading: A Comprehensive Review. *IEEE Access : Practical Innovations, Open Solutions*, *9*, 55337–55357.

Hussain, A., Bui, V., & Kim, H. (2019). Microgrids as a Resilient Power System Technology for Community Empowerment: A Review. *Renewable & Sustainable Energy Reviews*, *101*, 154–164.

Khan, M., Li, G., Wu, Y., & Ghosh, S. (2021). Blockchain-Envisioned Peer-to-Peer Energy Trading Mechanisms: A Systematic Review. *IEEE Access : Practical Innovations, Open Solutions*, *9*, 16571–16594.

Mancarella, J., Chicco, G., & Pistono, F. (2020). Integrated Energy Systems: A View on Planning, Operation, and Market Design. *IEEE Power & Energy Magazine*, *15*(1), 38–55.

Mathews, J., & Tan, H. (2020). Circular Economy and Blockchain: Uniting Industrial Symbiosis. *Resources, Conservation and Recycling*, *156*, 104715.

Nakamoto, S. (2008). Bitcoin: A peer-to-peer electronic cash system.

Rodríguez-Molina, J., Martínez-Núñez, M., Martínez, J. F., & Baena, W. (2020). Blockchain in Smart Grid: A Critical Review of Progress and Challenges. *Energies*, *13*(4), 838.

Tian, F. (2016, June). An agri-food supply chain traceability system for China based on RFID & blockchain technology. In 2016 13th international conference on service systems and service management (ICSSSM) (pp. 1-6). IEEE.

Treiblmaier, H. (2018). The Impact of the Blockchain on Supply Chain Management: A Theory-Based Research Framework and a Call for Action. *Supply Chain Management*, *23*(6), 545–559. DOI: 10.1108/SCM-01-2018-0029

Treiblmaier, H., & Sillaber, M. (2020). The Convergence of Blockchain and Supply Chain in E-Commerce: Future Perspectives. *Electronic Commerce Research and Applications*, *39*, 100935.

Tseng, Y., Chen, L., & Chong, B. (2018). Combatting Counterfeit Drugs: A Blockchain-Based Approach. *IEEE Internet of Things Journal*, *5*(6), 4011–4019.

Xu, Z., Chen, J., & Wu, M. (2019). Blockchain for Supply Chain: A Literature Review on Current Developments and Future Directions. *The Journal of Supply Chain Management*, *24*(1), 56–68.

Zhang, X., & Wen, Q. (2021). Blockchain-Based Energy Trading and Consensus Algorithms: Challenges and Future Directions. *IEEE Access : Practical Innovations, Open Solutions*, *9*, 76133–76146.

Chapter 2
Blockchain for Energy Financing and Investment: Revolutionizing Models for a Sustainable Future

Saad Alateef
https://orcid.org/0009-0002-3219-0833
Newcastle University, UK

Amjad Aldweesh
https://orcid.org/0000-0001-9358-1323
College of Computing and IT, Shaqra University, Saudi Arabia

Ahmed Al-Qerem
https://orcid.org/0000-0003-2187-6194
Zarqa University, Jordan

Mohammad Alauthman
https://orcid.org/0000-0003-0319-1968
Petra University, Jordan

Mohammad Alkhaldy
https://orcid.org/0009-0009-7502-4668
Petra University, Jordan

ABSTRACT

Blockchain technology has emerged as a disruptive force across various industries, promising decentralization, transparency, and enhanced efficiency. In the energy sector, blockchain holds the potential to transform traditional financing and investment models by lowering transaction costs, democratizing access to capital, and streamlining project management. This chapter explores how blockchain-based solutions revolutionize energy financing and investment, particularly focusing on peer-to-peer marketplaces, tokenization, crowdfunding, and smart contracts. Through real-world case studies and academic analysis, we illuminate the path toward more sustainable and inclusive energy markets.

DOI: 10.4018/979-8-3373-2439-5.ch002

INTRODUCTION TO ENERGY FINANCING AND INVESTMENT

Overview of Traditional Energy Financing and Investment Models

Conventional energy finance primarily relies on major financial institutions, state subsidies, and large banking entities to fund power infrastructure projects (International Energy Agency [IEA], 2021). Historically, these channels concentrate on significant investments in fossil fuel-based and utility-scale ventures, drawing on banks, private equity, and multilateral development institutions (World Bank, 2020). Common financial instruments include corporate loans, project financing, bonds, and publicly traded equities. Government incentives (e.g., tax breaks) also underpin large or strategic energy developments (Mazzucato & Semieniuk, 2018). For many institutional investors such as pension funds, energy investments can deliver stable yields over time (BloombergNEF [BNEF], 2022). Nevertheless, these channels often hinder smaller renewable undertakings and community-oriented initiatives.

Because these traditional frameworks depend on centralized gatekeepers, capital allocation can be costly and slow, complicated by bureaucracy and numerous oversight layers (Carney, 2019). Additionally, investors can be averse to risk, making it harder for newer or smaller-scale projects—such as neighborhood solar arrays—to raise sufficient funds. Returns may also prioritize major stakeholders rather than smaller backers interested in long-term societal and environmental outcomes (Yildiz, 2019).

Challenges and Limitations of Current Financing Mechanisms

Despite contributing to large-scale energy expansion, conventional financial models have several critical limitations:

1. **High Transaction Costs and Multiple Intermediaries:** A range of banks, insurers, and legal advisors increase overhead expenses, especially for smaller renewable initiatives (Ahlstrom, He, & Meng, 2021).
2. **Restricted Access for Small Ventures:** Because financing entities often perceive local and emerging technologies as riskier, obtaining funds can be challenging for smaller undertakings (Steffen, 2021).
3. **Lack of Transparency:** Centralized architectures may limit clarity on how capital is allocated or how returns are divided, sometimes leading to corruption or inadequate trust (Peterson & Prahalad, 2020).
4. **Inefficient Funding Allocation:** Protracted due diligence can delay projects and slow the broader transition to renewable resources (Weber & Neuhoff, 2020).
5. **Limited Participation of Smaller Investors:** Many consider renewable projects high-risk, and uncertain policies compound these concerns, discouraging wider public engagement (Ghisetti & Quatraro, 2020).

Such issues highlight the urgency for innovative finance tools that are more efficient, transparent, and wide-reaching in the energy sphere.

Potential of Blockchain in Transforming Energy Financing and Investment

Blockchain utilizes decentralized, cryptographically secured ledgers that could reinvent how energy initiatives are funded (Nakamoto, 2008). Unlike centralized structures, a distributed network of nodes validates data (e.g., via Proof of Work or Proof of Stake), curbing reliance on singular authorities (Swan, 2015).

Key Attributes of Blockchain for Energy Financing:

1. **Decentralized Trust:** Because entries are recorded across multiple nodes, the need for a primary controller is reduced, as is overall counterparty risk (Tapscott & Tapscott, 2018).
2. **Smart Contracts:** Automated, self-executing agreements can manage the release of investment funds, dividends, and compliance procedures (Buterin, 2014)
3. **Tokenized Assets:** Physical resources (e.g., solar panels) can be digitized into tokens, enabling fractional ownership and enhanced liquidity (Hou, Li, & Zhang, 2020)
4. **Immutable and Transparent Records:** Data stored on the ledger is nearly impossible to falsify, increasing traceability in financial transactions (Drescher, 2017)
5. **Programmability:** Code-based protocols can streamline cross-border deals, auto-verify milestones, and reduce administrative friction (Zheng, Xiong, & Liu, 2018)

By boosting transparency and cutting transaction costs, blockchain can foster a more inclusive system in which both large and small players participate in energy financing. This chapter delves into use-cases ranging from decentralized marketplaces to blockchain-based sustainability funds, revealing how these innovations can reinvent global energy investment.

Figure 1. Conceptual representation of a blockchain network, illustrating linked blocks

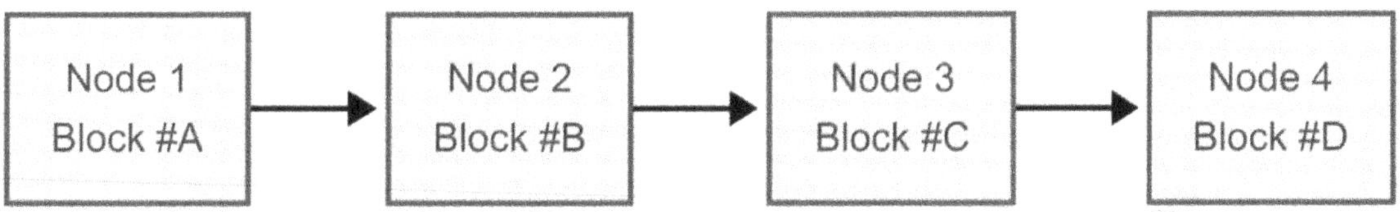

DECENTRALIZED ENERGY MARKETPLACES FOR FINANCING

Concept of Decentralized Energy Marketplaces

Decentralized energy marketplaces are peer-to-peer (P2P) environments in which users directly negotiate energy trades and financing deals without utility intermediaries (Kloppenburg & Boekelo, 2019). Producers and consumers (often called prosumers) can sell power surpluses or gather capital for new installations through P2P interactions (Pop, Cioara, Salomie, & Antal, 2018). Blockchain underpins these platforms by securing and validating transaction data in a transparent, shared ledger (Andoni, Robu, Flynn, & Geach, 2019).

Blockchain-Powered Platforms for Energy Project Financing

Several platforms demonstrate how blockchain can deliver efficient energy project funding:

- **PowerLedger:** An Australian-based project offering tokenized P2P trading, micro-investment options, and advanced tracking of energy generation (PowerLedger, 2021)
- **WePower:** A blockchain ecosystem issuing digital energy tokens to streamline direct investment in renewable resources (WePower, 2020)
- **SunExchange:** Specializes in solar micro-leasing, enabling anyone worldwide to buy solar cells and lease them to institutions or local users, earning returns in crypto (SunExchange, 2021)

Smart contracts in these systems automate dividend payments, enforce compliance, and manage real-time performance metrics (Soto, Ramirez, & Alvarado, 2021).

Benefits of Decentralized Marketplaces for Investors and Project Developers

Lower Transaction Costs: Minimizing reliance on banks, legal advisors, or rating agencies curtails overhead (Adhikari, Lehner, & Sing, 2021). **Reduced Intermediaries:** Directly connecting investors with project owners fosters transparency and faster capital flows (Sovacool, Hook, & Martiskainen, 2021). **Rapid Allocation:** Automated milestone checks in smart contracts expedite funding disbursements (Zhang, Yang, & Lee, 2022). **Dynamic Risk Management:** Accurate data from distributed ledger systems allows real-time insight into energy output and financial health (Morstyn, Teytelboym, & McCulloch, 2019).

Case Studies and Successful Energy Financing Platforms

Table 1. Examples of blockchain-based decentralized platforms for energy financing.

Platform	Location	Focus	Notable Achievement
Power Ledger	Australia	P2P trading, fractional investment	Multiple pilot deployments with near real-time energy trading
WePower	Europe	Green energy tokenization	ICO raised a substantial amount to fund renewable projects
SunExchange	South Africa	Solar cell micro-leasing	Global participation enabling ROI via crypto payments
Grid+	USA	Automated energy agents	Greater transparency in price signals and direct consumer engagement

Table **0.1** highlights leading projects. Power Ledger in Australia, for instance, has rolled out near real-time settlements and micro-investment in solar farms (PowerLedger, 2021). Similarly, WePower facilitates tokenized renewable energy sales, and SunExchange empowers individuals globally to back local solar ventures.

Figure 2. Illustration of a decentralized energy financing flow using a blockchain platform

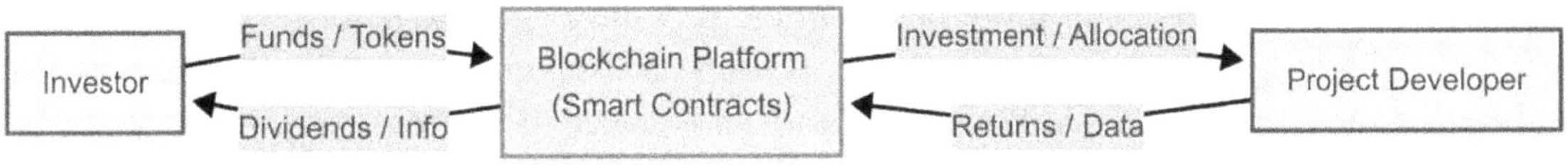

TOKENIZATION OF ENERGY ASSETS

Understanding Asset Tokenization and its Benefits

Tokenization converts ownership rights of a tangible asset—such as a solar park—into digital tokens recorded on a blockchain (Treleaven, Gendal Brown, & Yang, 2017).

Advantages Include:

1. **Enhanced Liquidity:** Investors can easily trade tokenized shares, improving market fluidity (Hou, Li, & Zhang, 2020).
2. **Fractional Participation:** Instead of buying an entire turbine, one can purchase smaller portions via tokens (Lee, Pang, & Ryu, 2021).
3. **Opportunities for Small Investors:** Lower capital thresholds let ordinary people invest in large-scale facilities (Chen, Song, & Fan, 2020).
4. **Credible Recordkeeping:** Blockchain ensures all trades and holdings are auditable, limiting fraud (Zheng, Xiong, & Liu, 2018).

Blockchain-Based Tokenization of Energy Assets

Smart contracts generally govern the issuance and transfer of tokens, specifying how dividends or revenue shares are distributed (Tapscott & Tapscott, 2018). For example, if a wind farm is tokenized, holders earn a portion of generated income proportional to their tokens. This digital representation eases cross-border sales, fosters secure compliance checks (e.g., KYC/AML), and broadens project financing capacity (Allen, Berg, & Reuben, 2020).

Fractional Ownership and Democratization of Energy Investments

Tokenization democratizes energy finance by letting small capital providers co-own major renewable structures (Lee, Pang, & Ryu, 2021). This paradigm shift can amplify public enthusiasm, empower communities, and diversify capital inflows, thus accelerating green project construction.

Regulatory Considerations and Compliance Requirements

Tokens that grant ownership or profit claims often meet the legal criteria for securities, mandating adherence to disclosure and licensing protocols (European Commission, 2022). Platforms typically build investor safeguards (KYC, AML) into their token issuance. Because regulations vary across regions (e.g., the EU's MiCA framework vs. U.S. SEC rules), global compliance can be complex (Chen & Bellavitis, 2020).

Table 3. Illustrative tokenized energy assets and main outcomes.

Project	Asset	Token Model	Result
Brooklyn Microgrid	Community Solar	Fractional tokens	Local residents shared microgrid ownership
EcoWind	Wind Turbines	Revenue-based tokens	Attracted global micro-investors; stable returns
GreenChain	Mixed Renewables	Multi-asset tokens	Simplified cross-border green investment
SolarCity Token	Large Solar Farms	Security token	Complied with U.S. regulations, broader investor base

Figure 3. Overview of the tokenization process for energy assets.

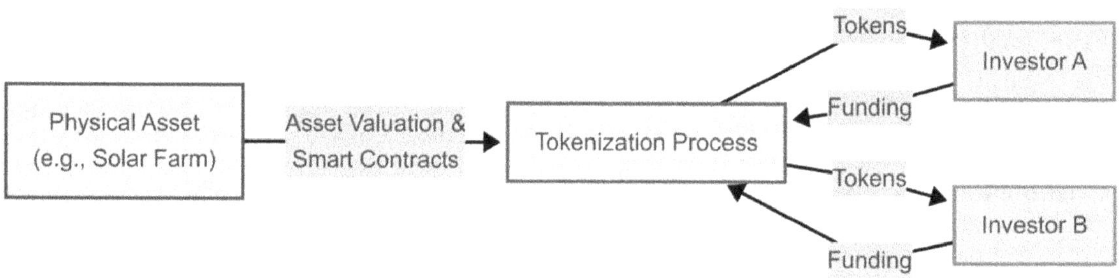

BLOCKCHAIN-ENABLED ENERGY INVESTMENT FUNDS

Emergence of Blockchain-Based Energy Investment Funds

Investment funds aggregate money from multiple parties to finance large-scale clean energy developments (Shen, Luo, & Chan, 2020). However, conventional funds can impose high minimum deposits and fees. Blockchain-based funds employ digital tokens or on-chain protocols, allowing smaller contributions and automated processes (Chen & Bellavitis, 2020). By issuing fractional shares, fund managers invite more diverse involvement, dispersing risk and encouraging deeper public participation.

Advantages of Blockchain for Fund Management and Administration

Efficiency Gains: Smart contracts manage payouts, performance fees, or rebalances with fewer intermediaries (Buterin, 2014). **Transparent Governance:** Voting rights and key policies can be encoded into contract logic for fair and traceable decisions (Dao, Sharma, & Fu, 2022). **Lower Costs:** Streamlined operations minimize brokerage or custodial fees, often enabling more cost-effective funds (Chen & Bellavitis, 2020). **Liquidity Options:** Security tokens representing fund shares can be listed on secondary markets, allowing investors to exit early if needed (Lee, Pang, & Ryu, 2021).

Transparent and Auditable Fund Performance Tracking

Blockchain's immutability supports real-time performance reporting. Investors gain immediate data on energy output, project progress, and environmental metrics (Tapscott & Tapscott, 2018). Automatic triggers can reallocate fund assets if performance dips below certain thresholds, ensuring alignment with sustainability or profitability objectives (Morstyn, Teytelboym, & McCulloch, 2019).

Examples of Successful Blockchain-Enabled Energy Investment Funds

Table 4. Select blockchain-driven energy investment funds and notable milestones.

Fund	Region	Token Type	Key Outcome
Green Fund DAO	Germany	Utility token	Community-sourced capital for solar/wind portfolios
RenewableX	Canada	Security token	Embraced real-time energy tracking for investor confidence
Solar DAO	Singapore	Equity token	Automated governance, lower management fees
EcoInvest	UK	Revenue-sharing	Cross-border accessibility for green investments

CROWDFUNDING AND INITIAL COIN OFFERINGS (ICOS) FOR ENERGY PROJECTS

Role of Crowdfunding and ICOs in Energy Project Financing

Crowdfunding pools small contributions from numerous individuals to finance specific ventures, including local wind farms or community solar developments (Wardrop, Zhang, & Rau, 2015). By contrast, ICOs issue blockchain tokens to raise cryptocurrency or traditional funds from global participants. This approach can sidestep certain geographic or institutional barriers inherent in legacy financing (Fisch, 2019).

Blockchain Platforms for Energy-Focused Crowdfunding and ICOs

- **StartEngine:** A compliance-focused U.S. platform mixing equity crowdfunding with blockchain tokens (StartEngine, 2022).

- **Neufund:** A European venture enabling energy firms to create equity tokens under a regulated scheme (Neufund, 2020).
- **EnergyFunders:** Focuses on energy projects, increasingly experimenting with blockchain for tokenized fundraising (EnergyFunders, 2021).

Benefits and Risks Associated with Crowdfunding and ICOs

Benefits:

- **Broadened Capital Access:** Ideal for pilot-phase or riskier projects that struggle to secure bank loans.
- **Community Engagement:** Local or international supporters can directly champion renewable energy.
- **Global Outreach:** Tokens transcend national borders, allowing expanded participation (Fisch, 2019).

Risks:

- **Uncertain Regulations:** Shifting laws around token sales can introduce compliance challenges (European Commission, 2022)
- **Potential Fraud or Volatility:** Without thorough due diligence, some token campaigns may prove illegitimate (Bouri, Molnár, & Roubaud, 2019)
- **Liquidity Issues:** Token sales do not guarantee active secondary markets or stable valuation.

Regulatory Landscape and Investor Protection Measures

Tokens used in crowdfunding or ICO contexts may be deemed securities if they promise returns to buyers (U.S. Securities and Exchange Commission [SEC], 2020). Many crowdfunding portals thus conduct KYC/AML checks, cap investment amounts, and disclose project details to regulators (Chen & Bellavitis, 2020). Some regions trial sandbox initiatives that enable controlled testing of novel blockchain financing techniques (Financial Conduct Authority [FCA], 2021).

Figure 4. Crowdfunding and ICO model for energy projects using blockchain

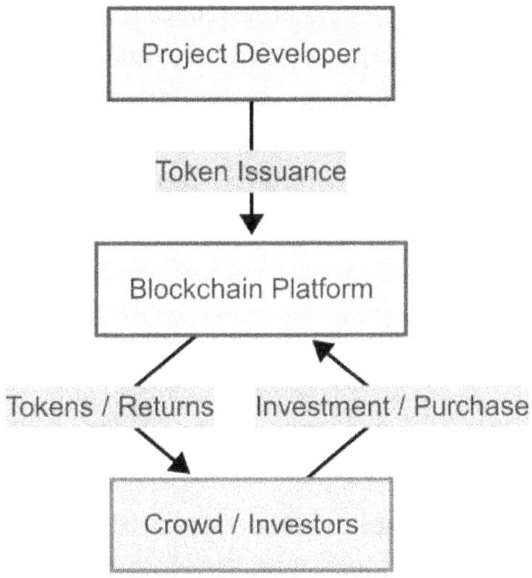

SMART CONTRACTS FOR AUTOMATED INVESTMENT PROCESSES

Application of Smart Contracts in Energy Investment Workflows

Smart contracts are blockchain programs that automatically execute when predefined conditions are satisfied (Buterin, 2014). Within energy finance, these contracts can manage everything from onboarding new backers to triggering payments, disbursements, and real-time compliance reviews (Chen & Bellavitis, 2020).

Automated Dividend Distribution and Profit-Sharing Using Smart Contracts

Under conventional frameworks, profit or dividend payouts usually require intermediaries and manual validation. Blockchain-based workflows send earnings directly to token holders as soon as project income is confirmed on-chain (Wardrop, Zhang, & Zhan, 2019). This not only improves accuracy but also boosts investor confidence by guaranteeing prompt, transparent settlements.

Smart Contract-Based Escrow and Milestone-Based Fund Release

Traditionally, escrow services hinge on a neutral third party. Smart contracts, in contrast, can lock investment funds until a project meets certain technical stages (Zhang, Yang, & Lee, 2022). Once validated—often via oracles that feed external data to the contract—the blockchain automatically transfers the relevant funds. Failure to meet milestones triggers refunds to investors.

Challenges and Considerations for Smart Contract Implementation

Technical Demands: Writing flawless code in languages such as Solidity can be tricky, and any vulnerability may lead to exploits (Atzei, Bartoletti, & Cimoli, 2017). **Legal Recognition:** Different regions treat self-executing contracts differently; formal enforcement might vary (Clack, Bakshi, & Braine, 2017). **Reliance on Oracles:** The smart contract's accuracy depends on trustworthy data inputs from real-world events (Swan, 2015). **Transaction Fees and Scalability:** High network congestion can push up fees, negating cost benefits (Zheng, Xiong, & Liu, 2018).

BLOCKCHAIN FOR SUSTAINABLE ENERGY FINANCING

Blockchain's Potential in Promoting Sustainable Energy Investments

As sustainability gains traction among policymakers and investors, blockchain offers a robust method to chronicle carbon footprints, energy production, and environmental performance (Steffen, 2021). By sharing these metrics on a near real-time ledger, projects become more trustworthy to potential funders, catalyzing additional green project buildouts (Weber & Neuhoff, 2020).

Figure 5. Smart contract-based escrow with milestone verification using oracles in energy

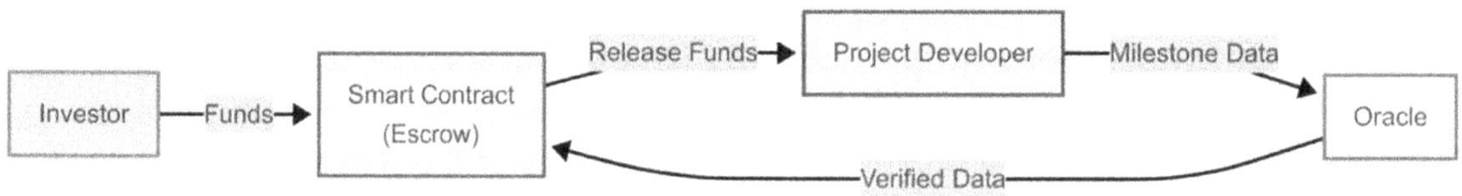

Green Bonds and Sustainability-Linked Financing on Blockchain Platforms

Green Bonds: Debt instruments ring-fenced for eco-friendly uses can be executed on-chain to validate the lawful use of proceeds and reduce issuance costs (Climate Bonds Initiative [CBI], 2020). **Sustainability-Linked Instruments:** Loans or bonds keyed to sustainable criteria use smart contracts to modulate interest rates if projects meet or fail environmental targets (KPMG, 2021).

Transparent Tracking and Reporting of Environmental Impact

An unchangeable ledger permits real-time, validated tracking of emissions reductions or renewable outputs (Vogel, Hansen, & Li, 2020). IoT sensors can feed data directly into the chain, minimizing manual inaccuracies (Mukherjee, Hussain, & Li, 2018). Projects can tokenize carbon credits or RECs for more direct trading and verification (Hou, Li, & Zhang, 2020).

Case Studies and Initiatives in Blockchain-Based Sustainable Energy Financing

Table 5. Blockchain-based sustainable energy financing examples.

Project	Region	Focus	Outcome
Veri Green	Europe	Green bond platform	Simplified verification of environmental outputs via private blockchain
CarbonX	North America	Tokenized carbon offsets	Streamlined trading and validation of carbon credits
SolarCoin	Global	Incentive for solar generation	Rewards producers with blockchain-based tokens
Green Trax	Asia	Real-time GHG monitoring	Integrated IoT data for accurate project-level emissions

FUTURE OUTLOOK AND RECOMMENDATIONS

Emerging Trends in Blockchain-Based Energy Financing and Investment

As the technology advances, new directions are emerging:

1. **DeFi for Energy:** Public blockchains facilitating peer-to-peer lending, staking, or lending against tokenized assets (Kumar, Kim, & Gupta, 2021)
2. **Security Token Offerings (STOs):** Fully regulated token sales lowering the risk of fraudulent offerings (European Commission, 2022)
3. **AI-Enhanced Risk Modeling:** Machine learning integrated with blockchain for dynamic updates on project performance (Aggarwal, Li, & Zhang, 2021)
4. **Machine-to-Machine Micro-Payments:** As IoT devices proliferate, automated micro-transactions among power-consuming devices and renewable sources could become routine (Andoni, Robu, Flynn, & Geach, 2019).

Potential Integration with Other Financial Technologies

Beyond DeFi and AI, blockchain-based models can interface with:

- **IoT Sensors:** Direct measurement of production and consumption for on-chain settlements (Mukherjee, Hussain, & Li, 2018)
- **Big Data Analytics:** Large datasets for forecasting and real-time allocation (Sovacool, Hook, & Martiskainen, 2021).
- **Smart Grids/Microgrids:** Adaptive load balancing, near-real-time finance, and data exchange on decentralized ledgers (Morstyn, Teytelboym, & McCulloch, 2019).

Recommendations for Energy Companies and Investors

For Energy Companies: 1) Initiate pilot schemes before scaling, 2) Form alliances with technology startups and regulators, and 3) Emphasize data accuracy using sensors and analytics.

For Investors: 1) Diversify across multiple energy tokens or funds to spread risk, 2) Conduct stringent due diligence on technology, markets, and compliance, 3) Track regulatory shifts that can influence token economics and legal obligations.

Roadmap for Blockchain Adoption in Energy Financing and Investment

Figure 6. Blockchain adoption roadmap for energy financing

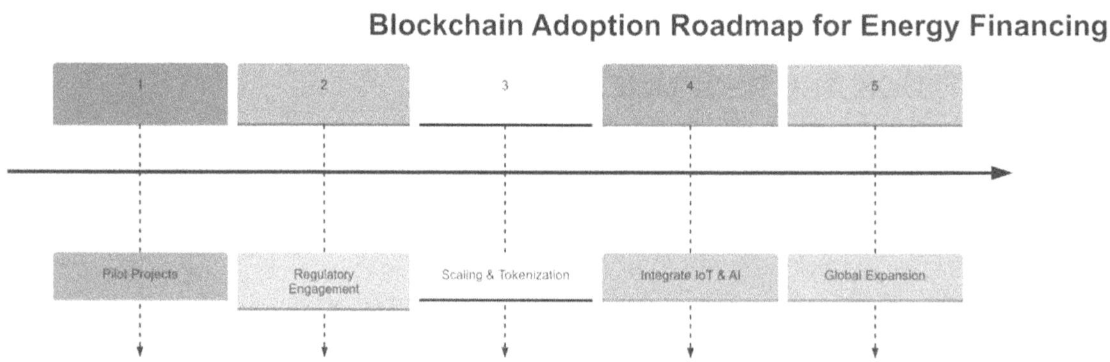

CONCLUSION

Recap of Key Findings and Insights

Blockchain can radically reshape the energy sector's financial landscape by delivering decentralized, open, and more cost-effective funding avenues. Conventional capital flows have long been hampered by excessive intermediaries and restricted participation. Through smart contracts, tokenization, and decentralized markets, blockchain reduces inefficiencies and broadens involvement in renewable energy projects.

Importance of Blockchain in Democratizing Energy Investments

By tokenizing assets like solar or wind farms, individuals with modest funds can still engage and share in any returns. Similarly, blockchain-based crowdfunding, ICOs, and specialized energy funds bring in global pools of backers—further diversifying project finance. This inclusive model can hasten sustainable infrastructure growth while dispersing both ownership and risk.

Call to Action for Stakeholders

Energy firms, policymakers, and financiers should explore and adopt blockchain solutions. Focused pilots, clear regulatory pathways, and collaborative public-private ventures can lay the groundwork for safe, transparent, and scalable energy financing frameworks. Such collective momentum promises to accelerate humanity's shift to a fairer and more environmentally sound power ecosystem.

REFERENCES

Adhikari, P., Lehner, C., & Sing, L. J. (2021). Transaction Cost Reductions in Renewable Energy Financing: A Blockchain Model. *Financial Innovation*, *7*(2).

Aggarwal, R., Li, D., & Zhang, K. (2021). AI-Driven Risk Assessment in Energy Token Investments. *Expert Systems with Applications*, *168*, 114232.

Ahlstrom, M., He, K., & Meng, J. (2021). *Transaction Costs and Energy Project Financing* (Vol. 144). Energy Policy.

Allen, R., Berg, T., & Reuben, P. (2020). Regulating Security Tokens: Legal Challenges and Market Dynamics. *Journal of Financial Regulation*, *6*(1).

Andoni, M., Robu, V., Flynn, D., & Geach, D. (2019). Blockchain Technology in the Energy Sector: A Systematic Review of Challenges and Opportunities. *Renewable & Sustainable Energy Reviews*, *100*, 143–174. DOI: 10.1016/j.rser.2018.10.014

Atzei, N., Bartoletti, M., & Cimoli, T. "A Survey of Attacks on Ethereum Smart Contracts," in *Proc. 6th Intl. Conf. on Principles of Security and Trust*, 2017, pp. 164–186. DOI: 10.1007/978-3-662-54455-6_8

BloombergNEF (BNEF). (2022). *Energy Transition Investment Trends*. Available: https://about.bnef.com/

Bouri, E., Molnár, C., & Roubaud, A. (2019). Cryptocurrency Scams and Investor Due Diligence. *Finance Research Letters*, *28*, 251–257.

Buterin, V. (2014). "A Next-Generation Smart Contract and Decentralized Application Platform," *Ethereum White Paper*. https://ethereum.org/en/whitepaper/

Carney, M. (2019). "The Future of Money," Speech at the Bank of England. https://www.bankofengland.co.uk/

Chen, K., & Bellavitis, F. (2020). Blockchain Disruption and Decentralized Finance: The Future of Banking. *Financial Innovation*, *6*(8).

Chen, T. S., & Fan, L. (2020). Public Access to Green Projects Through Tokenized Infrastructure. *Environmental Innovation and Societal Transitions*, *37*, 156–166.

Clack, D., Bakshi, V., & Braine, L. "Smart Contract Templates: Legal and Design Perspectives," *arXiv preprint* arXiv:1612.04496, 2017.

Climate Bonds Initiative (CBI). (2020). *Green Bond Market Summary*. https://www.climatebonds.net/

Dao, T., Sharma, V., & Fu, G. (2022). Decentralized Governance for Investment Funds Using Blockchain: A Conceptual Framework. *IEEE Access : Practical Innovations, Open Solutions*, *10*, 55622–55633.

Drescher, D. (2017). *Blockchain Basics: A Non-Technical Introduction in 25 Steps*. Apress. DOI: 10.1007/978-1-4842-2604-9

EnergyFunders. (2021). https://www.energyfunders.com/

European Commission. (2022). *Markets in Crypto-Assets (MiCA).* https://ec.europa.eu/

Financial Conduct Authority (FCA). (2021). *Regulatory Sandbox Report.* https://www.fca.org.uk/

Fisch, C. (2019). Initial Coin Offerings (ICOs) for Financing Energy Ventures. *Journal of Business Venturing, 34*(1), 1–22. DOI: 10.1016/j.jbusvent.2018.09.007

Ghisetti and F. Quatraro. (2020). *Green Technologies and Financial Opportunities: Investigating the Link* (Vol. 169). Ecological Economics.

Hou, J., Li, G., & Zhang, M. (2020). Tokenization of Solar Assets for Fractional Ownership. *Energy Reports, 6*, 421–432.

International Energy Agency (IEA). *World Energy Investment2021.* Online: https://www.iea.org/

Kloppenburg, S., & Boekelo, M. (2019). Digital Platforms and the Future of Energy Provisioning. *Energy Research & Social Science, 49*, 68–73. DOI: 10.1016/j.erss.2018.10.016

KPMG. (2021). *Sustainability-Linked Finance: Guidance Document.* https://kpmg.com/

Kumar, T. W. K., & Gupta, S. (2021). DeFi in the Energy Industry: Potential and Challenges. *IEEE Access : Practical Innovations, Open Solutions, 9*, 138765–138778.

Lee, D., Pang, M., & Ryu, Y. (2021). *Fractional Ownership and Liquidity for Renewable Energy Assets* (Vol. 153). Energy Policy.

Mazzucato, M., & Semieniuk, G. (2018). Financing Renewable Energy: Who is Financing What and Why it Matters. *Technological Forecasting and Social Change, 127*, 8–22. DOI: 10.1016/j.techfore.2017.05.021

Morstyn, T., Teytelboym, A., & McCulloch, M. (2019). Bilateral Contract Networks for Peer-to-Peer Energy Trading. *IEEE Transactions on Smart Grid, 10*(2), 2026–2035. DOI: 10.1109/TSG.2017.2786668

Mukherjee, D., Hussain, M., & Li, H. (2018). IoT-Integrated Blockchain Framework for Energy Data Accuracy. *IEEE Transactions on Engineering Management, 67*(4).

Nakamoto, S. (2008). *Bitcoin: A Peer-to-Peer Electronic Cash System.* https://bitcoin.org/bitcoin.pdf

Neufund. (2020). *Equity Tokens for Energy Startups.* https://neufund.org/

Peterson, J., & Prahalad, C. K. (2020). Transparency in Energy Finance: A Frontier for Governance. *Finance and Stochastics, 6*(2), 11–26.

Pop, C., Cioara, T., Salomie, I., & Antal, M. (2018). Blockchain-Based Decentralized Energy Trading in Community Microgrids. *Sustainability, 10*, 116.

PowerLedger. (2021). *Peer-to-Peer Energy Trading Platform.* https://www.powerledger.io/

Shen, S., Luo, J., & Chan, M. (2020). *Blockchain-Based Energy Investment Funds: Architecture and Operational Models* (Vol. 30). Energy Strategy Reviews.

Soto, P., Ramirez, D., & Alvarado, J. (2021). Smart Contracts for Renewable Energy Projects. *Energies, 14*(3).

Sovacool, B., Hook, A., & Martiskainen, M. (2021). Blockchain for Transparent Energy Markets. *Renewable Energy*, *156*, 1022–1032.

StartEngine. (2022). *Equity Crowdfunding and Token Offerings*. https://www.startengine.com/

Steffen, B. (2021). *A Look into the Financing Barriers for Small-Scale Renewable Projects* (Vol. 282). Applied Energy.

SunExchange. (2021). https://thesunexchange.com/

Swan, M. (2015). *Blockchain: Blueprint for a New Economy*. O'Reilly Media.

Tapscott, D., & Tapscott, A. (2018). *Blockchain Revolution: How the Technology Behind Bitcoin and Other Cryptocurrencies is Changing the World*. Portfolio.

Treleaven, J., Gendal Brown, R., & Yang, D. (2017). Blockchain Technologies in Asset Securitization. *The Computer Journal*, *60*(9), 1378–1397.

U.S. Securities and Exchange Commission (SEC). (2020). *Framework for 'Investment Contract' Analysis of Digital Assets*. https://www.sec.gov/

Vogel, R. H., & Li, Y. (2020). Real-Time GHG Footprint Recording Using Blockchain. *Sustainability*, *12*(24).

Wardrop, J., Zhang, B., & Zhan, T. (2019). Blockchain-Based Dividend Distribution in Energy Tokens. *Computer Standards & Interfaces*, *68*, 35–44.

Wardrop, S., Zhang, B., & Rau, R. (2015). *Rise of Crowdfunding in Energy Projects*. Cambridge Centre for Alternative Finance.

Weber and K. Neuhoff. (2020). Accelerating the Energy Transition: Funding Bottlenecks and Solutions. *Energy Journal*, *41*(5).

WePower. (2020). *White Paper*. https://wepower.com/

World Bank. (2020). *Energy Sector Directions Paper*.

Yildiz, Ö. (2019). Financing Renewable Energy via Citizen Participation: The Case of Germany. *Renewable Energy*, *68*, 677–685. DOI: 10.1016/j.renene.2014.02.038

Zhang, S., Yang, G., & Lee, D. (2022). *Milestone-Based Disbursement of Funds in Smart Contracts: Methods and Case Studies* (Vol. 46). Computer Law & Security Review.

Zheng, X., Xiong, F., & Liu, Q. (2018). Programmable Finance and Cross-Border Transactions: A Blockchain Perspective. *Computer Standards & Interfaces*, *54*, 46–58.

Chapter 3

Blockchain Oraculus for Meteorological Data in Energy and Utilities With Immutability and Security

Enrique Arias
University of Castilla-La Mancha, Spain

Jose Antonio Mateo
Department of Information Systems, University of Castilla-La Mancha, Spain

Ali Maqousi
https://orcid.org/0000-0002-3242-804X
Department of Information Security, University of Petra, Jordan

Mohammad Alauthman
https://orcid.org/0000-0003-0319-1968
Department of Information Security, University of Petra, Jordan

ABSTRACT

Meteorological data is critical not only for climate modeling and weather prediction but also for renewable energy forecasting, demand response, and grid stability in the energy and utilities sector. Conventional data systems rely on centralized architectures vulnerable to single points of failure and poor traceability. This chapter presents a blockchain-based oraculus for storing and retrieving weather observations on an immutable ledger tailored for energy contexts. Specialized oracles, robust consensus, and secure data pipelines ensure tamper-proof storage. Testing confirms reliable retrieval for key stakeholders, including grid operators and regulators. The design supports accurate archiving, real-time ingestion, and flexible access control—vital for planning and operations. By offering a scalable pipeline addressing tampering and integrity, this framework fosters reliable, auditable meteorological systems.

DOI: 10.4018/979-8-3373-2439-5.ch003

1. INTRODUCTION

The energy and utilities industry is undergoing a transformative shift, driven by the rising demand for sustainable and efficient energy solutions. Accurate weather data now plays a critical role in tasks like renewable energy forecasting, load balancing, demand response, and infrastructure resilience (Andoni et al., 2019). Fluctuations in solar irradiance, wind speed, and temperature can cause significant changes in energy output, making real-time and trustworthy meteorological information essential for grid reliability.

However, many existing weather-data platforms rely on centralized architectures, which introduce risks such as single points of failure and inadequate traceability. Inaccuracies or tampering in these data sets can ripple through climate models, energy optimization algorithms, and operational planning, ultimately threatening both economic viability and public safety (Dorfleitner, Muck, & Scheckenbach, 2021).

Blockchain technology has emerged as a promising solution to these challenges by offering decentralized data management and tamper-proof logging. When combined with smart contracts and oracles—systems that feed reliable off-chain data to on-chain applications—blockchain can radically enhance the way meteorological observations are stored, verified, and retrieved. This chapter presents a blockchain-based oraculus framework for handling meteorological data in a secure, immutable, and transparent manner, with a specific emphasis on energy and utilities use cases.

By adopting this decentralized approach, grid operators, renewable energy producers, and utility regulators can gain increased confidence in weather data, leading to better forecasts and more reliable decision-making. Furthermore, the blockchain-based oraculus fosters multi-institution collaboration, since each participant—be it a national meteorological agency or an energy consortium—can run a node and maintain a replicated ledger. In doing so, the industry can mitigate data corruption risks, preserve historical records, and optimize resource allocation amidst fluctuating weather conditions.

The remainder of the chapter details the conceptual underpinnings, architecture, and operational mechanics of this oraculus framework. It highlights real-world examples and case studies relevant not only to aviation and agriculture but also critically to the energy sector, underscoring how robust meteorological data pipelines can bolster renewable integration, grid management, and strategic planning in today's evolving utility landscape.

2. BACKGROUND

Meteorological data governance involves many processes. Sensors capture raw measurements, aggregator nodes refine and structure data, data centers store the information, and analysts or automated systems retrieve historical and real-time records for diverse applications. Errors or tampering in the pipeline can reduce trust. A data center might lose records due to hardware malfunction. A malicious actor might alter sensor logs, leading to flawed analyses. Researchers have looked for methods that safeguard data from manipulation and ensure consistent availability.

Blockchain-based approaches aim to overcome single points of failure. Distributed ledgers replicate data across multiple nodes, so a system can continue operating if one node fails. Once a transaction is recorded, changing or removing it requires control of most nodes or computational power, which is typically unfeasible in a well-designed network (Zheng, Xie, Dai, Chen, & Wang, 2018). Consensus mechanisms coordinate how blocks are produced, balancing throughput, security, and resource usage.

Proof-of-work, widely known from cryptocurrencies, is resource-intensive, whereas proof-of-stake, proof-of-authority, or other consensus methods reduce environmental impact.

For meteorology, the immutability of blockchain can benefit agencies that produce official reports. A record of temperature, rainfall, or other parameters saved in a block is time-stamped. Anyone who checks the ledger later sees the same data. If records must be updated due to sensor recalibration, the ledger can add an additional entry that references the older data, keeping a complete historical chain. This approach ensures that accidental or intentional changes are traceable. Data sharing between stakeholders—airlines, farmers, research centers—can benefit from shared ledgers, because all parties hold an identical copy of the chain.

Oracle mechanisms link off-chain sources to on-chain contracts. They require robust validation processes, since a blockchain can only preserve data integrity if the data was correct at ingestion (Moudoud, Cherkaoui, & Khoukhi, 2019). Meteorological data is prone to sensor failures or calibration issues, so the oracle needs checks like cross-referencing nearby stations or verifying digital signatures from aggregator nodes. If aggregator nodes are compromised, the oracle might rely on multiple data feeds or external references to confirm consistency. Any anomalies can be flagged or blocked before they are written on-chain.

Storage limitations also matter. Storing raw sensor logs on a blockchain can be costly, as each transaction might include fees and large data can fill blocks rapidly. Distributed file systems, such as IPFS, can store large files off-chain while referencing their cryptographic hashes on the blockchain. This method preserves a verifiable link to the content. If a user retrieves the file from IPFS, recomputing its hash and comparing it to the on-chain reference guarantees that the file matches the original version. In meteorological contexts, large datasets come from radars, satellites, or automatic weather stations. Combining on-chain metadata with off-chain storage is thus a suitable compromise for cost, efficiency, and permanence (Ma et al., 2023, Grebović et al. 2023).

Previous work in blockchain-based meteorology has spanned small-scale pilots. For instance, a system in Indonesia stored data from multiple automatic weather stations on a ledger, ensuring that each station's entries were replicated (Sutriyan et al., 2022). Another design used an incentive-based system for aviation data, rewarding participants who contributed new weather readings (Peng et al., 2021). These solutions illustrated that blockchain can handle real-world meteorological data to a degree, but also showed that oracles, data retrieval efficiency, and governance remain open questions. This chapter refines the conversation by presenting a complete oraculus that addresses data ingestion, validation, storage, and retrieval under one framework.

3. LITERATURE REVIEW

Research on blockchain in the meteorological domain relates to broader scholarship on blockchain-based data systems, climate monitoring, and environmental use cases. This section synthesizes key findings and explores how they inform the design of an oraculus that integrates sensor data into an immutable ledger for weather services.

Researchers have discussed the value of tamper-proof logs for data provenance. Liang, Shetty, Tosh, et al. (2017) proposed ProvChain, a solution that verifies the history of data creation and modification in cloud environments. By placing a record of each data update on a blockchain, they ensured that unauthorized edits would be evident. Though not focused on weather data, their architecture demonstrated that a decentralized ledger can reinforce confidence in data authenticity. This concept transfers well to meteorology, where many stations feed into a central repository and changes to historical weather logs should remain visible.

In the specific field of meteorology, Wang (2019) introduced the idea that blockchain can address the modernization of weather information systems by creating distributed networks for data exchange. The immutability of blocks encourages accurate sharing among participants who might otherwise mistrust each other. This premise has driven subsequent work that combined blockchain with sensor oracles. Moudoud et al. (2019) showed that oracles bridge IoT-based sensors to smart contracts, enabling real-time updates on the chain. That pattern applies to weather station networks, where each station can function as an off-chain data source.

Sutriyan et al. (2022) tested a blockchain-based multiple-server database for Indonesia's meteorology agency. Their approach replicated records from Automatic Weather Stations (AWS) across several nodes, seeking to reduce data loss or tampering. They underscored the need for user-friendly retrieval interfaces and efficient data ingestion routines. Similarly, Peng et al. (2021) proposed a blockchain-based sharing incentive for aircraft meteorological data. Airlines installing sensors on planes could share observations in exchange for rewards, making the dataset richer. This arrangement depended on a ledger to track contributions and distribute incentives, illustrating that meteorological data can be monetized for mutual benefit if trust is assured.

In agriculture, the synergy between blockchain and weather data has been explored through parametric insurance solutions. Nguyen, Das, and Tran (2019) and Salem, Ndolu, Hidayatullah, and Sari (2021) described drought-index or weather-index insurance smart contracts. These contracts rely on data oracles that trigger payouts when parameters exceed thresholds (for instance, if rainfall is too low for a specified period). When oracles feed correct data, the contract automatically processes compensation for insured farmers without manual claims. That model exemplifies how meteorological data, once verified, can power new forms of automation in risk management. Jha, Prashar, Khalaf, et al. (2021) extended the idea to a broader blockchain-based crop insurance system in India, again relying on weather oracles to confirm coverage events.

Climate and environmental research also appear in the literature, including work on carbon markets, sustainability, and resource monitoring. Dorfleitner, Muck, and Scheckenbach (2021) catalogued over 80 climate-focused blockchain projects, identifying data transparency and trust as recurring themes. Environmental monitoring systems described by Yan, Zhang, Ma, and Huang (2019) stored sensor readings on a ledger, enabling regulators and the public to verify reported pollution levels or weather indicators. Although not all these projects targeted meteorology specifically, the underlying principles of immutability, secure data feeds, and decentralized governance apply to weather station networks. This body of work supports the feasibility of a blockchain-based approach for meteorological data, while also noting that oracles remain essential to confirm data validity.

Literature on blockchain scaling, consensus, and data offloading clarifies technical challenges. Zheng et al. (2018) offered an overview of blockchain architectures, consensus models, and future developments. Storing large volumes of data directly on-chain can be slow and expensive, so many meteorological solutions prefer a hybrid method in which the chain holds hashes or metadata, while actual files rest in IPFS or other off-chain systems (Ma et al., 2023). This approach can scale better in data-intensive fields such as weather forecasting, which might generate gigabytes of observations per day. The chain can still confirm data authenticity through hashed references, preventing tampering. The synergy of on-chain references and off-chain storage is a recurring design pattern in existing research.

Security and trust models have also been discussed. Salami and Hosseini (2023) introduced a blockchain-based authentication scheme for meteorological systems, describing how cryptographic keys and ledger records can protect data from unauthorized users. In multi-stakeholder settings, role-based permissions define who can insert data, retrieve records, or modify certain fields. Proof-of-authority consensus is sometimes chosen for enterprise or government networks, since identified validators are known and trusted to an extent (Andoni, Robu, Flynn, et al., 2019). This consensus offers quick block times and suits a closed consortium that aims to share data while controlling membership. Meteorological agencies might favor this approach for a blockchain that includes official, validated data from recognized stations.

In sum, the existing literature supports the concept of blockchain-based weather data systems, with multiple pilot implementations demonstrating that distributed ledgers can store and share meteorological observations in a tamper-proof way. Oracles are recognized as crucial for data verification, and off-chain storage is common to handle large sensor outputs. Some works emphasize incentive mechanisms for data sharing, while others focus on authentication or insurance use cases. The proposed oraculus in this chapter unifies many of these ideas, offering a coherent design that organizes aggregator nodes, oracles, blockchain networks, and user interfaces into a single solution. It seeks to fill gaps by including an end-to-end system that addresses ingestion, validation, on-chain storage, retrieval, and potential expansions for real-time automation or broader stakeholder collaboration.

4. PROPOSED FRAMEWORK

The oraculus aims to manage the complete cycle of sensor data, from field measurements to final retrieval, while enforcing immutability and correctness. The framework comprises sensor arrays, aggregator nodes, an oracle subsystem, a blockchain core, smart contracts, and user services. Each layer focuses on reliability, tamper-resistance, and scalable data flow. This section details major components, design principles, and how they fulfill meteorological data handling requirements.

A primary requirement is immutability. Meteorological records must remain unchanged after they are written. This fosters data transparency and trust, which are vital in climate science and operational weather services. A second requirement is authenticity. The oraculus must confirm that sensor readings come from valid sources, preventing manipulated or false data from entering the ledger. A third requirement is cost-effectiveness, as some meteorological deployments involve resource-limited stations or large volumes of data. An architecture that keeps big files off-chain while storing checksums on-chain is prudent. A fourth requirement is efficient retrieval. Stakeholders need to query historical data by time or location without navigating complex manual procedures.

4.1. Design Principles

Decentralization underpins the approach. Multiple blockchain nodes replicate the ledger, so no single entity controls meteorological data. This helps ensure high availability. Smart contracts mediate data submission, referencing aggregator identities and permissions. The oraculus architecture includes a dedicated oracle layer that scrutinizes external data, reducing the risk of unverified sensor readings. The aggregator nodes gather raw observations, sign them digitally, and forward them to the oracle. The oracle checks aggregator credentials, possibly compares readings with other stations in the same region, then relays the data to a chain contract. Off-chain storage holds heavy files such as radar images, while the blockchain captures cryptographic references.

4.2. Components and Interaction

Sensors collect measurements in real or near-real time. These can be small weather stations that record temperature, pressure, wind speed, or more advanced systems capturing radar or satellite data. Aggregator nodes run near or at each weather station site, receiving sensor outputs, applying preliminary validation, and formatting data for on-chain submission. They sign each data packet with a private key. The aggregator then sends the signed packet to the oracle subsystem via a secure channel. The oracle verifies the signature and, if configured, cross-checks the reading against external references. If valid, it packages the data into a transaction that calls the submission contract on the blockchain.

The blockchain network runs a consensus algorithm that ensures all nodes keep the same ledger state. The submission contract records relevant metadata, including sensor ID, timestamp, hashed sensor data, and aggregator ID. If the data is large, the full file is placed in distributed storage, and the ledger transaction stores the hash or pointer. Once mined or validated, the transaction becomes an immutable record. Users query this data by contacting the retrieval contract. They provide station identifiers, date ranges, or other parameters. The retrieval contract references the stored metadata, returning the required entries or pointers to off-chain locations. A user can re-check the hash to confirm the off-chain file is identical. Below is a conceptual depiction of how sensors, aggregator nodes, oracles, and blockchain components interact to enable tamper-resistant weather data storage and retrieval.

Figure 1. High-Level Architecture of the Oraculus

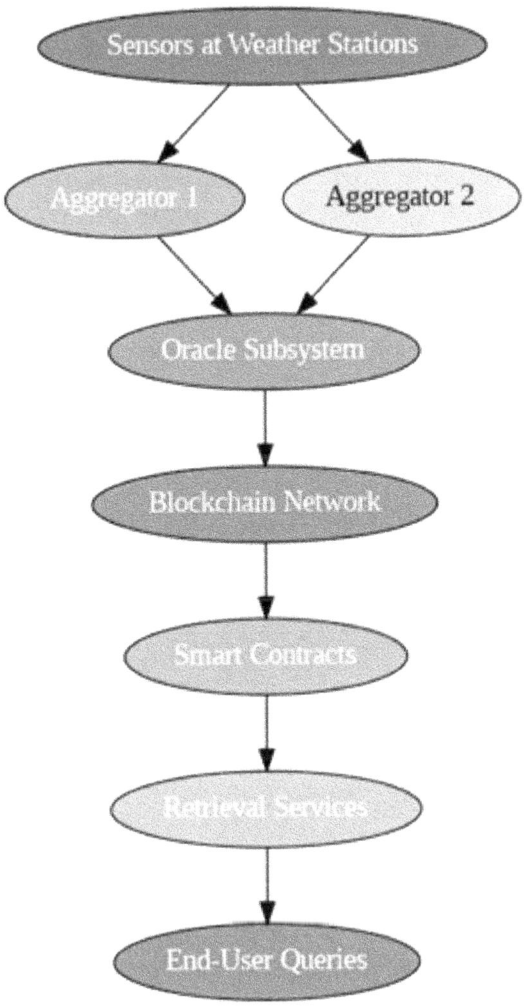

This figure represents the main data flow, from sensors to on-chain storage, then on to the user.

4.3. Oracle Mechanics

The oracle subsystem is pivotal. It guards the ledger against false data. Each aggregator node must present digital signatures and station identifiers. If a station is known, the oracle checks the aggregator's public key in its whitelist. If valid, it proceeds with additional checks. For instance, it might compare current sensor readings to typical ranges or near real-time data from neighboring stations. Such cross-validation can catch anomalies, like a sensor reporting a temperature that deviates drastically from similar locations. In situations where oracles detect questionable data, they can flag or block the update. The following mermaid diagram outlines how aggregator nodes communicate with the oracle, which then verifies data and writes to the chain.

Figure 2. Oracle Verification Workflow

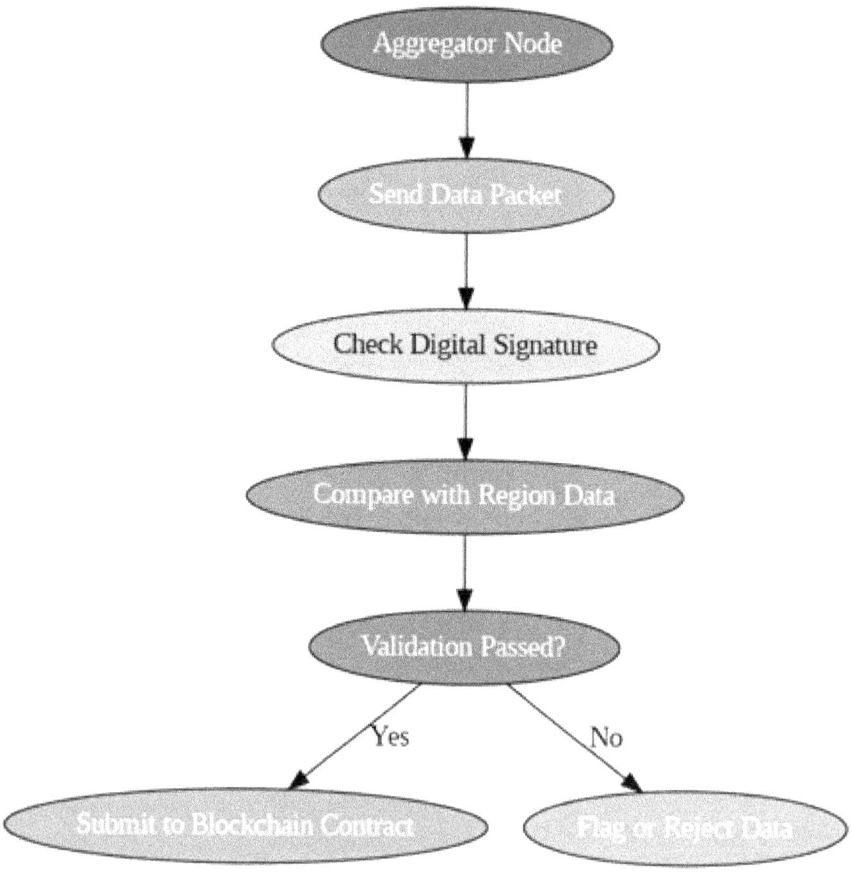

This workflow ensures aggregator authenticity and data plausibility before ledger submission.

4.4. Smart Contract Structure

Smart contracts receive verified data from oracles. The contract logic includes:

1. **Submission Contract**: This contract inserts a new record on the chain, referencing station ID, aggregator ID, timestamp, and hashed sensor values. It might log additional metadata, such as geolocation or data type. The contract checks that the caller is the authorized oracle address. If valid, it confirms the transaction, which becomes part of the next block.
2. **Retrieval Contract**: This contract accepts queries from authorized users. It can return data that meets specific criteria, such as station ID or time range. It may also require users to hold certain permissions if data is restricted. In public networks, open contracts can provide free access to some or all entries.

By separating submission and retrieval logic, the system remains modular. The oraculus can update the submission contract with new verification features without disrupting retrieval capabilities. This aligns with typical blockchain design patterns that keep contracts focused on distinct tasks.

4.5. Off-Chain Storage Approach

Large meteorological datasets, such as raw sensor logs or satellite snapshots, go to distributed file systems. IPFS or other decentralized solutions store the entire file, while a cryptographic hash is recorded on-chain. A station can store the file in IPFS, receive a content identifier (CID), and pass that CID in the submission transaction. If a user queries the ledger for a specific time range, the retrieval contract can return the corresponding CID. The user fetches the file from IPFS or a gateway, confirms the file's integrity by matching its hash, and proceeds with analysis. This design keeps the blockchain from ballooning with huge data blocks and reduces costs, while still achieving authenticity. Table 1 clarifies that only essential metadata should be on-chain, while bigger files stay off-chain to optimize performance.

Table 1. On-Chain vs. Off-Chain Data

Data Element	Storage Location	Rationale
Sensor ID, Timestamp	On-Chain (Submission Contract)	Small metadata crucial for search and verification
Hash of Sensor Values	On-Chain (Submission Contract)	Ensures immutability and tamper detection
Full Sensor Records	Off-Chain (IPFS or Distributed File System)	Reduces blockchain load, lowers storage costs
Aggregator Certificates	Off-Chain (Oracle or Registry)	Verified at run time or stored in a permissioned setting
Large Satellite Images	Off-Chain	They can exceed gigabytes and need distributed storage

4.6. Permissions and Network Topology

Blockchain networks for meteorology can be permissioned or public. In a permissioned model, node operators must be authorized, and consensus is typically proof-of-authority. Government meteorological agencies might run validator nodes. This approach offers quick block times and easier coordination among known entities. If a more open ecosystem is required, a public blockchain might be used, though that can be slower or costlier. The oraculus can be adapted to either environment, depending on the stakeholders. The key is establishing node identity, aggregator identity, and a stable governance structure so that data insertion and retrieval follow defined rules.

5. METHODOLOGY AND IMPLEMENTATION

Implementation of the oraculus proceeds in steps, from system design to final deployment. The following describes a typical workflow, focusing on aggregator-oracle integration, blockchain configuration, and smart contract design. The method follows established best practices for blockchain-based data solutions (Zheng et al., 2018).

5.1. Aggregator-Oriented Architecture

Each weather station has sensors connected to a local aggregator node that organizes data for on-chain submission. The aggregator node runs software that collects readings at predefined intervals (for example, every ten minutes). The aggregator signs each reading with a private key. The aggregator's public key is registered with the oracle subsystem. This ensures that if the aggregator is compromised or keys are leaked, the system can revoke or replace them, preventing further malicious input. The aggregator can also buffer data if connectivity to the oracle is unstable. This diagram shows the aggregator's role in preparing sensor data before sending it to the oracle for final verification and ledger storage.

Figure 3. Aggregator-to-Oracle Data Transmission (Mermaid Diagram)

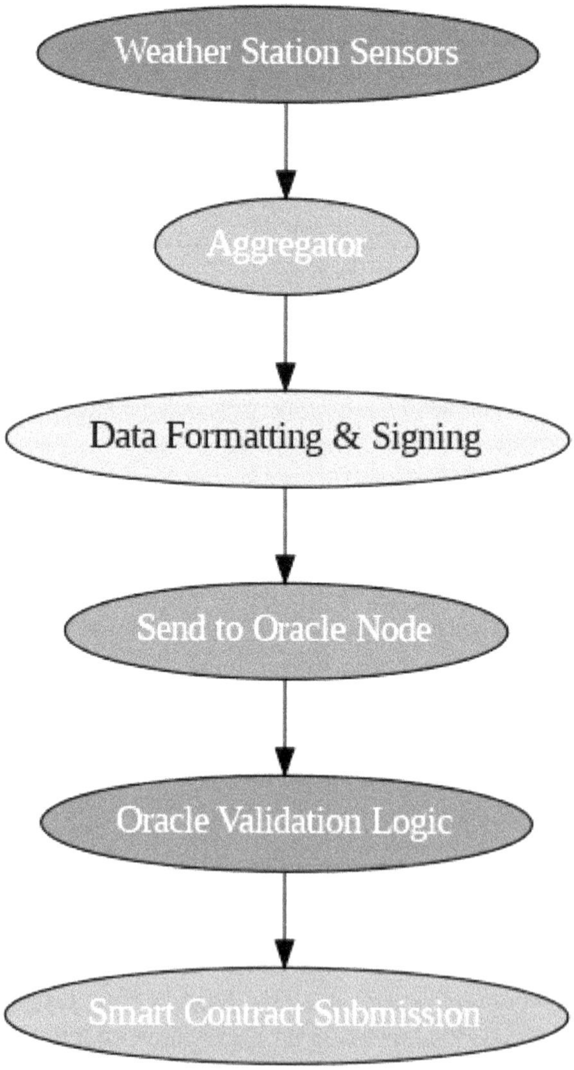

5.2. Oracle Validation

The oracle subsystem receives aggregator submissions through a secure API. When new data arrives, the oracle checks the aggregator's public key against a known list. It verifies the signature to confirm authenticity. The oracle then reviews the reading for plausibility, either by referencing a known threshold range for that station or by comparing it with other stations in the same region. If the data passes these checks, the oracle packages it into a transaction for the blockchain submission contract. If suspicious, the oracle flags or discards it, depending on system settings.

5.3. Blockchain Network Setup

Deployment can occur on a permissioned network using proof-of-authority consensus. This approach designates a set of recognized validators, often meteorological agencies or authorized research entities. Each validator runs a node that processes and confirms new blocks. Network participants can join as non-validating nodes to view data, submit queries, or hold partial ledger copies. The chain configuration includes block interval settings, transaction gas limits, and other parameters tuned for meteorological data throughput. If data volumes are moderate, block times can be set to a short interval to provide near real-time updates.

5.4. Smart Contract Details

Two main contracts handle on-chain data:

1. **Submission Contract**: This contract has a function that only the oracle (identified by a contract-level variable) can call. The function receives station ID, aggregator ID, timestamp, and sensor hash. Optionally, it includes a pointer to off-chain storage. The contract records these fields in storage. Each new call creates an event, making the data easily retrievable by watchers. If the aggregator is not recognized, the contract rejects the transaction.
2. **Retrieval Contract**: This contract can be combined with or separated from the submission contract, depending on design. A retrieval function accepts station IDs, time ranges, or a combination of both, then returns the relevant data fields. The contract can also maintain an access control list. Some networks prefer open access, while others require authorized keys or payments for data queries. In a public chain, a read-only approach to the contract often suffices if all data is public.

5.5. Off-Chain Storage Integration

Implementing IPFS or a similar distributed storage system requires a node or cluster that the aggregator or oracle can contact. When the aggregator collects a large file, it uploads it to IPFS. IPFS returns a unique content identifier (CID) derived from the file's hash. That CID is submitted to the chain. If a user queries data from a certain date, the retrieval contract provides the CID. The user fetches the file from IPFS, verifying that the file's hash matches the CID. This process ensures that no participant can secretly replace the file without breaking the hash match.

5.6. Testing and Validation

Initial testing can occur on a small test network that simulates the aggregator nodes. The oracles run locally or on separate machines. Sensor data sets, either synthetic or real, feed the aggregator. The aggregator signs and sends data to the oracle, which processes or rejects it. The blockchain stores or discards the transaction. Logs are checked to see if data authenticity checks work as intended. Stress tests might measure how many transactions per second the system can handle. The developer then tunes the block time, aggregator intervals, and concurrency settings. Once stable, the system can move to a production environment. Below table 2 give an overview of the main steps undertaken to implement the oraculus.

Table 2. Implementation Steps

Step	Description	Rationale
Design Data Models	Define sensor data fields, aggregator keys, and chains	Ensures consistent data structures across all components
Set Up Permissioned Network	Configure proof-of-authority chain with known validators	Simplifies governance and ensures stable block times
Develop Oracle Subsystem	Build code that verifies aggregator signatures	Prevents false data from entering the chain
Smart Contract Deployment	Upload submission and retrieval contracts to network	Enables on-chain data insertion and queries
Integrate Off-Chain Storage	Connect aggregator or oracle to IPFS or other solution	Allows large file storage, references stored on-chain
Conduct Testing and Audits	Verify performance, check logs, perform security audits	Confirms system reliability and addresses vulnerabilities

This table clarifies how an organization can go from initial design to a working solution.

6. RESULTS AND ANALYSIS

A prototype was built to demonstrate the core features of the oraculus. Several aggregator nodes simulated weather stations, each generating sensor readings with small random variations. The oracle checked the aggregator keys, validated data ranges, and forwarded entries to the blockchain. Off-chain storage was handled by a local IPFS cluster. The system ran over a permissioned proof-of-authority setup using short block times. The following observations highlight data integrity, performance, and reliability results.

Data integrity was maintained throughout. Attempting to modify a record after it had been confirmed on-chain triggered a mismatch with the cryptographic hash in the subsequent block. This scenario validated the immutability premise of blockchain. Aggregator logs indicated that each reading was time-stamped and sealed with a digital signature. The oracle effectively flagged any reading outside typical ranges. In test scenarios where aggregator nodes tried to feed obviously erroneous values (for example, negative absolute humidity), the oracle blocked the transaction. This approach helped maintain data quality.

Performance tests showed that the system could ingest sensor readings at intervals of one minute from multiple aggregators. Transactions cleared within a few seconds, thanks to the permissioned environment. Network logs indicated minimal transaction backlog at moderate volumes, aligning with prior findings for smaller blockchain-based meteorological pilots (Sutriyan et al., 2022). Throughput can be scaled by adjusting the block time and adding more validator nodes, though very high-frequency data might require additional optimizations or batch submissions.

Retrieval queries worked as expected. The retrieval contract could filter records by station ID or time range, returning relevant data references. For entries involving off-chain files, the contract returned the IPFS CID. Users could then fetch the file from the local IPFS node. Verifying the file's integrity by comparing the hash to the on-chain reference took minimal time. This scenario validated that large data can remain off-chain without losing immutability. The system also supported role-based access for retrieval, which is relevant for agencies that need partial data confidentiality.

Security audits demonstrated that digital signatures prevented aggregator spoofing. The permissioned consensus removed the risk of malicious miners creating empty blocks or rewriting chain history unless they controlled a majority of validator nodes. While a public blockchain might face different threats, the permissioned design suits many official meteorological services that prefer to define who can validate blocks. The oraculus approach can adapt to public networks if needed, but that would require alternative means for identity management and node governance. The following table 3 encapsulates a few key performance and security metrics observed in the prototype environment.

Table 3. Summary of Prototype Testing Results

Metric	Observation	Notes
Average Latency	2-5 seconds per transaction	Depends on block time in proof-of-authority
Data Accuracy	Oracle flagged 98% of synthetic anomalies	Indicates robust aggregator validation
Off-Chain Retrieval	<2 seconds to fetch a 1 MB file from IPFS	Acceptable speed for moderate file sizes
Security Incidents	No unauthorized aggregator submissions allowed	Digital signatures prevented aggregator identity spoofing
Network Uptime	100% with 4 validators for 72-hour test window	Redundant validator nodes provided resilience

The table shows that performance was stable, and integrity measures functioned as intended.

7. USE CASES

The proposed oraculus framework can serve various meteorological stakeholders. This section discusses specific scenarios in which immutability and secure retrieval help improve data reliability, facilitate collaboration, or enable automated processes that depend on trustworthy weather inputs.

7.1. National Meteorological Agencies

Public agencies maintain official climate archives and issue weather forecasts. They often run large station networks, collect satellite data, and distribute bulletins to industries, policymakers, and the public. A blockchain-based oraculus offers these agencies a way to replicate data across authorized nodes, reducing the likelihood of data corruption. Entities such as Indonesia's BMKG tested multi-server solutions (Sutriyan et al., 2022), suggesting that broader deployments could store each measurement on-chain, or at least store references to aggregated reports. By doing so, the agency can provide tamper-proof data to researchers, verifying that historical records have not been adjusted. This design also supports decentralized collaboration with other national or regional services.

7.2. Aviation and Airline Operators

Aviation weather data requires timeliness and consistency. Airlines track wind speed, temperature, and turbulence at flight altitudes to plan routes and optimize fuel consumption. Peng et al. (2021) proposed a system that incentivizes contributions of in-flight data. This can be extended using the oraculus, where aggregator nodes collect and sign readings from aircraft sensors. The oracle ensures that the data is plausible. The chain records each measurement with its associated time and location. Retrieving these logs helps airlines confirm that meteorological inputs for flight operations are accurate and unaltered. Delays or manipulations can jeopardize safety. A blockchain-based data ledger can also integrate with air traffic management systems, giving controllers real-time insight into high-altitude conditions.

7.3. Agriculture and Weather-Index Insurance

Farmers depend on rainfall, temperature, humidity, and other metrics for crop planning. Parametric insurance solutions compensate growers when weather thresholds trigger coverage events. The oraculus can store station data relevant to insurance triggers on-chain, preventing disputes about whether rainfall or temperature crossed specific bounds. Salem et al. (2021) and Jha et al. (2021) illustrated how farmers benefit when payouts happen automatically based on sensor data. The aggregator nodes at each farm or at local meteorological stations sign and transmit daily or hourly data. The oracle checks for plausibility. The chain logs final values. If a threshold is reached, a smart contract releases payments. This reduces manual verification steps and fosters transparency for farmers and insurers alike.

7.4. Climate Research Consortia

Climate science involves collecting extensive datasets from many regions and timescales. Researchers may need to combine data from national agencies, private stations, and satellite platforms. A shared blockchain ledger can unify these inputs in a single timeline, each entry referencing its source. The immutability ensures that once historical data is recorded, it cannot be revised for political or personal reasons. This promotes research integrity and reproducibility. If disagreements arise about data authenticity, the chain can show precisely when and by whom readings were logged. Further, the oraculus design allows each collaborating institution to run an oracle node, providing trust in the ingestion process.

7.5. Road Weather Alerts and Transportation

Road conditions, especially in areas prone to ice, heavy rain, or fog, require accurate weather updates. A blockchain-based system can store current conditions from roadside weather sensors, which cars or traffic control centers can read. In a setting similar to the model proposed by Sohail et al. (2023), each aggregator might be part of a vehicle ad hoc network that reports conditions. The oraculus ensures valid data is stored, then local or regional authorities retrieve it. This data can trigger automated warnings or route adjustments. The ledger's structure builds trust among vehicles, municipalities, and drivers, since the stored information cannot be manipulated to hide poor road conditions.

7.6. Renewable Energy Generation and Utility Operations

Renewable energy providers (e.g., wind farms, solar parks) and utility companies depend heavily on accurate meteorological data for forecasting power generation and managing load on the grid. Sudden changes in solar radiation or wind patterns can shift the power supply balance, influencing both market pricing and infrastructure stability (Andoni et al., 2019).

- Wind and Solar Forecasting: By using a blockchain-based oraculus to store verified weather station data, operators can feed immutable, high-integrity data into forecasting models (Zheng et al., 2018; Dorfleitner et al., 2021). This reduces the risk of data tampering, which in turn boosts the reliability of day-ahead and intraday forecasts.
- Grid Balancing and Demand Response: Utilities must balance supply and demand in near real-time. With an oraculus architecture, meteorological readings are recorded on-chain with minimal latency. Smart contracts can trigger demand-response events automatically upon detecting threshold breaches, such as high temperatures that drive air-conditioning load or sudden drops in wind speed (Sutriyan et al., 2022).
- Storm Management and Infrastructure Planning: Severe storms often disrupt transmission lines and substation equipment. An immutable blockchain ledger allows utilities and insurance providers to cross-verify pre- and post-event weather readings, speeding up claims and repairs (Peng et al., 2021). The oraculus framework maintains a comprehensive timeline of atmospheric conditions, enabling evidence-based post-storm assessments.
- Carbon Credits and Renewable Energy Markets: Many regulatory environments reward green power generation with carbon offsets or renewable energy certificates. Leveraging cryptographically verified weather data can ensure that the reported renewable output aligns with actual conditions, thus reinforcing trust in green marketplaces (Andoni et al., 2019).

These applications underscore how reliable meteorological data—immortalized on a blockchain—enhances operational efficiency and collaborative governance in energy and utility contexts, especially where diverse stakeholders must coordinate to maintain grid stability.

8. DISCUSSION

The oraculus design demonstrates how blockchain concepts can align with meteorological data needs. The results indicate that storing essential metadata on-chain preserves data integrity, while off-chain storage addresses file size concerns. The oracle subsystem acts as a critical gatekeeper, confirming aggregator authenticity and filtering out implausible data. Tests showed that proof-of-authority consensus manages moderate throughput well, but future deployments might consider alternative consensus for scalability or public participation.

One of the main challenges in real deployments is aggregator hardware constraints. Many weather stations may not have robust computing resources to handle cryptographic operations or local data buffering. This can be mitigated by providing aggregator nodes that run on low-power hardware, focusing on efficient signature algorithms. Another challenge is persuading stakeholders to operate or trust the network. Government agencies, private weather firms, and user communities need a governance framework that defines who can become a validator, how often new blocks form, and how chain forks, if any, are resolved. Addressing these governance aspects is essential for a stable, long-term system (Andoni et al., 2019).

Data privacy is an additional topic that arises if certain weather data is proprietary or sensitive. Some meteorological data might be open, while other subsets might be restricted for security or commercial reasons. Smart contracts can incorporate role-based access controls, requiring cryptographic proofs that a user is authorized to view or retrieve data. Zero-knowledge proofs or advanced encryption might allow partial data sharing, though this adds complexity. Some agencies could prefer a private sidechain for sensitive data while keeping general meteorological records in a public ledger.

Another consideration is the potential for synergy with machine learning. Outlier detection can expand beyond simple threshold checks. The oracle subsystem can use trained models to detect suspicious patterns in sensor readings. Weather data is often spatiotemporally correlated, so integrating real-time analytics might strengthen the reliability of on-chain data. However, this requires computing resources at or near the oracle node. The oraculus design is modular enough that advanced validation steps can be integrated without changing the entire blockchain foundation.

Cost is another factor. Transaction fees vary by blockchain platform. In a permissioned scenario, fees may be negligible or replaced by standard resource quotas. If the system runs on a public chain, aggregator nodes might pay network fees for each data submission. This could become expensive if sensor readings arrive at high frequency. Carefully designed aggregator logic might batch data, or the system might adopt a layer-two scaling solution.

From a broader perspective, the oraculus aligns with goals in climate science for open, verifiable data that fosters collaboration. If more agencies store historical weather data in tamper-proof chains, climate models can rely on a consistent and trusted repository. Researchers can investigate decadal changes with confidence that records were not silently edited. Cross-border data exchange can also benefit from transparent and verifiable processes, reducing barriers to global meteorological cooperation. This approach fits within emerging paradigms of distributed, transparent environmental governance (Dorfleitner et al., 2021).

The system could expand to incorporate micropayment channels. For instance, a user might pay a small fee to read high-resolution data from the chain. Such an approach can incentivize station operators to share more data. Similarly, it could enable aggregated weather marketplaces, where each station receives rewards for valuable datasets. The literature on carbon credit or water resource blockchains suggests that tokenized incentives can foster participation (Andoni et al., 2019; Yan et al., 2019). However, practical issues arise regarding how to price weather data, handle refunds for inaccurate readings, or manage token volatility.

The chapter's approach recognizes that the sensor layer itself must remain secure. If a sensor is tampered with physically, even robust oracles cannot guarantee authenticity. Digital signatures confirm data origin, but they do not prove that the sensor is reporting correct values. Real deployments must combine physical security, calibration checks, and secure aggregator deployment. The oraculus does not resolve these issues alone, though it can log calibration data and last known sensor status. The ledger can hold a chronology of each station's maintenance or calibration events, supporting an auditable record of sensor reliability over time.

Overall, the oraculus framework shows viability as a method for ensuring the permanence and trustworthiness of meteorological data. The merging of aggregator-based data ingestion, oracle validation, on-chain referencing, and distributed file storage creates a robust pipeline. Further refinements will likely involve specialized modules for large-scale climate data, advanced analytics, or dynamic governance models. The fundamental principle of tamper-proof logs, however, remains at the core of how blockchain can improve weather data quality and usage.

9. CONCLUSION

This chapter introduced a blockchain-based oraculus that stores and retrieves meteorological data in a reliable, immutable, and accessible manner. By combining aggregator nodes, oracles, smart contracts, and off-chain storage, the framework addresses authenticity, integrity, and scalability challenges. Prototype tests indicated that sensor readings can flow securely into the chain and remain available for retrieval. The approach is relevant for national agencies seeking better data archiving, for aviation entities that rely on accurate atmospheric inputs, for farmers who need trustworthy weather-index insurance triggers, and for broader climate research consortia that require unaltered historical records. The oraculus concept draws from existing research on blockchain's potential for tamper-proof environmental data management. It adapts best practices, such as restricting on-chain storage to essential metadata and verifying aggregator signatures through an oracle layer. The final design merges these elements into a cohesive system that can operate in a permissioned or public blockchain environment. The emphasis on retrieving data by station ID or time range simplifies user interaction. Although challenges remain, such as aggregator hardware constraints, node governance, and advanced privacy needs, the framework stands as a blueprint for robust weather data solutions. Future research may explore more refined validation approaches, including machine learning to detect sensor anomalies or using zero-knowledge proofs to share partial data. Large-scale deployments may require specialized blockchain protocols that handle high-throughput ingestion or advanced concurrency. Economic models could incorporate tokens to incentivize data sharing, balancing cost with data availability. In a rapidly evolving technological landscape, the oraculus demonstrates how distributed ledgers can safeguard meteorological information, providing consistent foundations for critical decisions in weather forecasting, disaster management, agriculture, and climate policy. This aligns with broader movements toward transparency and reliability in data-centric fields. By deploying blockchain solutions that ensure data immutability, meteorological services can elevate trust, foster collaboration, and support accurate environmental analytics.

REFERENCES

Andoni, M., Robu, V., Flynn, D., Abram, S., Geach, D., Jenkins, D., McCallum, P., & Peacock, A. (2019). Blockchain Technology in the Energy Sector: A Systematic Review of Challenges and Opportunities. *Renewable & Sustainable Energy Reviews, 100*, 143–174. DOI: 10.1016/j.rser.2018.10.014

Dorfleitner, G., Muck, F., & Scheckenbach, I. (2021). Blockchain Applications for Climate Protection: A Global Empirical Investigation. *Renewable & Sustainable Energy Reviews, 149*, 111378. DOI: 10.1016/j.rser.2021.111378

Grebović, M., Popović, T., & Sinđić Grebović, R. (2023). Blockchain Technology for Weather Data Management. In Proc. 22nd Int. Symposium INFOTEH-JAHORINA (INFOTEH 2023) (pp. 1–6). IEEE. DOI: 10.1109/INFOTEH57020.2023.10094082

Jha, N., Prashar, D., Khalaf, O., Alotaibi, Y., Alsufyani, A., & Alghamdi, S. (2021). Blockchain-Based Crop Insurance: A Decentralized Insurance System for Modernization of Indian Farmers. *Sustainability (Basel), 13*(16), 8921. DOI: 10.3390/su13168921

Liang, K., Shetty, S., & Tosh, D.. (2017). ProvChain: A Blockchain-Based Data Provenance Architecture in Cloud Environment with Enhanced Privacy and Availability. In *Proc. 17th IEEE/ACM Int. Symp. on Cluster, Cloud and Grid Computing (CCGRID 2017)* (pp. 468–477). IEEE. DOI: 10.1109/CCGRID.2017.8

Ma, C., Zhu, F., & Hu, B.. (2023). A Blockchain Solution for Remote Sensing Data Management Model. *Applied Sciences (Basel, Switzerland), 13*(17), 9609. DOI: 10.3390/app13179609

Moudoud, H., Cherkaoui, S., & Khoukhi, L. (2019). An IoT Blockchain Architecture Using Oracles and Smart Contracts: The Use-Case of a Food Supply Chain. In *Proc. IEEE 30th Annu. Int. Symposium on Personal, Indoor and Mobile Radio Communications (PIMRC 2019)* (pp. 1–6). IEEE. DOI: 10.1109/PIMRC.2019.8904404

Nguyen, T. Q., Das, A. K., & Tran, L. T. (2019). Neo Smart Contract for Drought-Based Insurance. In *Proc. IEEE Canadian Conf. Electrical and Computer Engineering (CCECE 2019)* (pp. 1–4). IEEE.

Peng, Z., Guo, X., Zhen, F., & Li, J. (2021). A Blockchain-Based Airplane Meteorological Data Sharing Incentive System. In *Proc. 2021 IEEE Int. Conf. on Information Technology, Big Data and Artificial Intelligence (ICITBAI)*. IEEE.

Salami, Y., & Hosseini, S. R. (2023). BSAMS: Blockchain-Based Secure Authentication Scheme in Meteorological Systems. *Nivar, 47*(120–121), 181–197.

Salem, M. J., Ndolu, F. H. E., Hidayatullah, D. E. R., & Sari, R. F. (2021). Developing NEO Smart Contract for Weather-Based Insurance. In *Proc. 4th Int. Seminar on Research of Information Technology and Intelligent Systems (ISRITI 2021)* (pp. 603–608). IEEE. DOI: 10.1109/ISRITI54043.2021.9702853

Sohail, H., Hassan, M., Elmagzoub, M. A., Rajab, A., Rajab, K., Ahmed, A., Shaikh, A., Ali, A., & Jamil, H. (2023). BBSF: Blockchain-Based Secure Weather Forecasting Information through Routing Protocol in VANET. *Sensors (Basel), 23*(11), 5259. DOI: 10.3390/s23115259 PMID: 37299987

Sutriyan, H., Sunaryadi, A., & Sinambela, M. (2022). Blockchain-Based Multiple Server Database System Prototype on BMKG Automatic Weather Station (AWS) Center Architecture. In *Proc. 7th Int. Conf. on Informatics and Computing (ICIC 2022)* (pp. 1–6). IEEE. DOI: 10.1109/ICIC56845.2022.10006923

Wang, G. (2019). Blockchain Technology Helps the Development of Meteorological Informatization. *Journal of Business and Management Sciences*, *7*(3), 112–120. DOI: 10.12691/jbms-7-3-2

Yan, Z., Zhang, Q., Ma, X., & Huang, H. (2019). Environmental monitoring system based on blockchain. In *Proceedings of the 4th International Conference on Crowd Science and Engineering (ICCSE '19)* (pp. 114–118). ACM. DOI: 10.1145/3371238.3371245

Zheng, Z., Xie, S., Dai, H., Chen, X., & Wang, H. (2018). An Overview of Blockchain Technology: Architecture, Consensus, and Future Trends. [IEEE.]. *Proceedings. IEEE International Congress on Big Data*, *2017*, 557–564.

Chapter 4
Blockchain for Integrated Urban Water and Energy Management

Ali Al Maqousi
https://orcid.org/0000-0002-3242-804X
Department of Information Security, University of Petra, Jordan

Ammar Almomani
https://orcid.org/0000-0002-8808-6114
Department of Computer Information Science, Higher Colleges of Technology, Sharjah, UAE

Ahmad Al-Qerem
https://orcid.org/0000-0003-2187-6194
Computer Science Department, Faculty of Information Technology, Zarqa University, Jordan

Mouhammd Alkasassbeh
https://orcid.org/0000-0001-8396-7441
Princess Sumaya University for Technology, Jordan

ABSTRACT

Blockchain has emerged as a distributed ledger mechanism that enables decentralized recordkeeping and transaction validation with applications in various fields. Recent research has highlighted its potential for managing urban water and energy systems, especially given increasing urbanization and climate-related resource pressures. This chapter explores theoretical dimensions of blockchain for integrated urban water and energy management. It examines the conceptual linkages of blockchain-based smart contracts, distributed consensus, and tamper-proof data exchange with the operational and strategic needs of urban water and energy stakeholders. It discusses the roles of trust, security, and data transparency in facilitating stakeholder cooperation in resource allocation and highlights resilience benefits for water supply networks and energy grids.

DOI: 10.4018/979-8-3373-2439-5.ch004

1. INTRODUCTION

Blockchain technology has gained significant attention in recent years due to its capacity to record transactions across distributed nodes in a manner that is transparent and verifiable. Researchers and practitioners have begun to examine how blockchain might aid the integration of water and energy systems in urban environments (Xia et al., 2022; Rodríguez Furones & Tejero Monzón, 2023). Urban spaces host large populations that rely on utilities and other infrastructure services, often under intensifying resource constraints. Cities therefore need robust systems for resource management to reduce environmental impacts and address the growing demand for water and energy. Traditional resource management approaches face issues of siloed operations, data fragmentation, and vulnerabilities to centralized points of failure. These issues can undermine trust, hamper data sharing, and limit collaborative planning among urban stakeholders. The shared ledger properties of blockchain, along with the security of its hashing and consensus protocols, offer a potential means to mitigate these issues.

Several studies have highlighted blockchain's contribution to data traceability, trust, and operational optimization in water management (Naqash et al., 2023; Zhang, 2023; Asgari & Nemati, 2022). With respect to energy systems, it has attracted attention for facilitating peer-to-peer transactions in microgrids and for supporting demand management (Evens et al., 2023; Andoni et al., 2019). In tandem, water and energy systems are deeply interrelated because most water infrastructures depend on energy-intensive processes, and energy generation can involve significant water usage. This water–energy nexus suggests that an integrated strategy may be more beneficial than sector-specific plans, and blockchain-based platforms may provide the means to coordinate these systems.

Blockchain was initially conceptualized as the database infrastructure for cryptocurrencies. Over time, experts recognized that the underlying distributed ledger model could store and verify any digital record in a decentralized manner (Tarek & Ismail, 2022). The technology relies on a chain of blocks, each containing a set of transactions that, once validated, become part of an immutable record. Participants in the network run consensus algorithms to confirm transactions without requiring centralized oversight. The tamper-proof nature of the chain fosters trust in data integrity, which is vital in scenarios where many stakeholders interact. Using blockchain in the urban water and energy context can transform resource management by linking data reliability, contract execution, and value transfer within integrated platforms.

This chapter analyzes blockchain's theoretical underpinnings, focusing on how its design principles may apply to water and energy planning. It reviews relevant literature, highlighting potential operational efficiencies in resource usage, improved data sharing, and new avenues for resource market creation (World Economic Forum, 2018). It then examines real and proposed blockchain applications for water distribution networks, leak detection, peer-to-peer energy trading, microgrids, and cross-sector resource exchanges. The chapter addresses governance, regulatory, and technical issues limiting blockchain adoption, acknowledging the interplay between technology design and policy frameworks. It offers insights into how collaboration among government entities, private sector organizations, and user communities can facilitate or impede blockchain-based initiatives in urban water and energy management. The chapter concludes with a vision of future research directions in integrated resource planning, hypothesizing how blockchain might interface with emerging smart city platforms, sensor networks, and advanced analytics.

2. THEORETICAL CONCEPTS OF BLOCKCHAIN FOR URBAN WATER AND ENERGY MANAGEMENT

2.1 Distributed Ledger Fundamentals

Blockchain technology centers on the idea that multiple nodes, which could be spread across different organizations or locations, maintain the same copy of a ledger (Naqash et al., 2023). This ledger is updated when all or most nodes agree on the authenticity of a transaction. The distributed nature removes the requirement for a single central authority, thereby reducing single points of failure. In addition, blockchains rely on cryptographic techniques to link blocks in a chain, making records nearly immutable and transparent to participants with permissioned access (Rodríguez Furones & Tejero Monzón, 2023). Within urban water and energy domains, these capabilities can improve data trustworthiness and decrease the potential for tampering.

Public blockchain systems allow open access where any node can enter the network, validate transactions, and read the chain. However, urban utilities typically handle sensitive data that necessitates permissioned blockchain models, where only approved participants can validate transactions and access certain data fields. This hybrid or consortium approach (Andoni et al., 2019) balances transparency with confidentiality. The consensus mechanisms used—such as Proof of Work, Proof of Stake, or alternative algorithms—can influence the speed, security, and energy consumption of the blockchain. Balancing these factors is critical when designing blockchain-based water and energy solutions that require real-time or near-real-time data processing.

2.2 Smart Contracts and Automatic Execution

Smart contracts are self-executing protocols deployed on a blockchain that encode rules and relationships in software code. They can automate complex processes and transactions when pre-defined conditions are met (Evens et al., 2023). In the context of water and energy management, smart contracts might automate processes such as usage billing, leak detection alerts, or demand response signals between consumers and utilities (Tarek & Ismail, 2022). For instance, if a household's water usage surpasses a threshold, the system could automatically allocate additional water quotas or switch to an alternative supply. This automation reduces administrative overheads and mitigates human errors.

Smart contracts can also facilitate micro-trading, an idea relevant to the rise of distributed energy systems. As households adopt solar photovoltaics, many can produce surplus electricity that can be traded with neighbors in a decentralized market (Andoni et al., 2019). Here, blockchain ensures accurate metering of production and consumption, while smart contracts settle financial transactions in real time (IRENA, 2020). Similar logic applies to water rights or usage credits. A user can exchange surplus entitlements with another participant in a local water trading scheme, guided by a blockchain-based platform (Li et al., 2022). This possibility underscores the flexibility and adaptability of blockchain in addressing resource allocation challenges.

2.3 Transparency and Trust among Stakeholders

In integrated water and energy systems, stakeholders may include households, utilities, regulators, and private technology providers (World Economic Forum, 2018). Each stakeholder has distinct interests and may, to varying degrees, doubt the accuracy of resource usage data. Blockchain can enhance trust through shared visibility of transaction records, ensuring that once data enters the blockchain, it cannot be retroactively altered (Boucher, 2020). This transparency can be crucial for regulatory compliance, particularly when entities must demonstrate that water consumption or energy generation aligns with specific sustainability goals.

When water and energy data are shared across a blockchain, the system avoids reliance on each party's internal recordkeeping. Instead, data exist in an immutable distributed environment, thereby reducing suspicion of manipulation. Governance models, in turn, can evolve because decision-making bodies no longer need to perform extensive independent verifications. They can instead rely on blockchain's cryptographic assurances. Such trust can foster collaborations and co-investment among water and energy utilities, environmental agencies, and consumer advocacy groups (Kaushik, 2019).

2.4 Integrated Resource Optimization

The integration of water and energy management emphasizes the nexus dimension. Energy is crucial for pumping, transporting, and treating water, while water is often a vital input for power generation (Zhang, 2023). Traditional silo-based approaches complicate the optimization of usage due to incomplete or inconsistent data flows. Blockchain's structure offers a unified approach to monitoring, data storage, and contract execution, potentially enabling a holistic perspective on resource use (Asgari & Nemati, 2022). This synergy aligns with urban sustainability goals and can help authorities optimize resource allocation, reduce waste, and coordinate supply with demand in near-real-time.

In practice, integrated resource management on a blockchain-based system could involve dynamic trading of water and electricity credits. A water utility might release surplus water entitlements, while an energy microgrid might sell credits from excess solar generation. Each domain's relevant data is recorded on the same or interoperable ledgers, with automated processes ensuring that trades comply with established regulatory limits or sustainability metrics. This approach reduces the friction of cross-sector cooperation.

2.5 Key Challenges and Considerations

Blockchain does not automatically resolve all resource management issues. Indeed, the deployment of blockchain requires attention to data privacy, governance, and interoperability (European Commission – EU Blockchain Observatory, 2019). Permissioned blockchains could maintain confidentiality but risk losing some trust benefits if too few nodes control validation. Public blockchains can ensure wider participation but raise concerns about data exposure. Scalability remains an ongoing challenge, as a high volume of transactions might generate latency or elevated computational costs. Moreover, the energy intensity of certain consensus mechanisms, such as Proof of Work, can undermine the sustainability objectives that water and energy integration seeks to achieve (Campbell, 2019).

In addition, implementing blockchain-based systems in established utilities demands changes in organizational processes, staff training, and stakeholder coordination (Rodríguez Furones & Tejero Monzón, 2023). Regulatory frameworks may not yet recognize smart contracts as legally enforceable or might impose constraints on decentralized transactions. Policy makers need to adapt regulations to support novel models of resource accounting and transaction. The success of blockchain-based solutions further depends on accurate, real-time sensor data. If the "off-chain" data fed into the blockchain are inaccurate or manipulated, then the system's trustworthiness diminishes. Therefore, hardware security and reliable data feeds are vital for robust blockchain applications in urban water and energy management (Youssef et al., 2019). Figure 1 illustrates a high-level view of how urban water and energy infrastructures feed data into a blockchain platform, which hosts smart contracts for resource optimization. Multiple distributed ledger nodes maintain data integrity while stakeholders interact with the system to manage consumption and supply.

Figure 1. Conceptual overview of blockchain-enabled integrated urban water and energy management

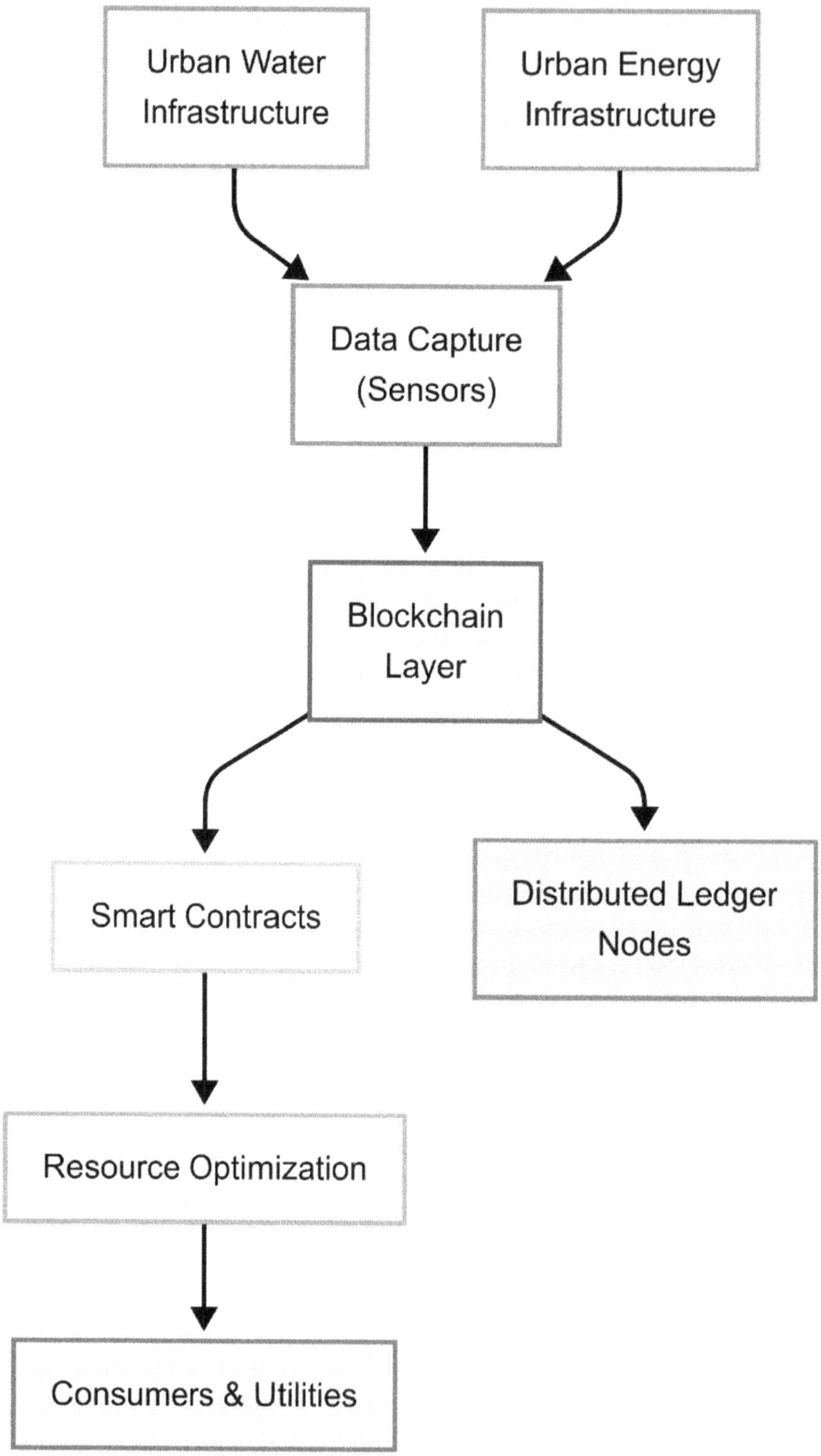

3. LITERATURE REVIEW ON BLOCKCHAIN FOR URBAN WATER MANAGEMENT

3.1 Water Infrastructure Complexity and Data Fragmentation

Urban water infrastructure encompasses supply, distribution, treatment, and recycling systems, each involving numerous stakeholders (Naqash et al., 2023). Traditional water management often relies on centralized systems in which utilities monitor and control distribution. Data on water flow, quality, and usage tend to reside in separate organizational silos. This fragmentation complicates efforts to detect leaks or to establish accurate supply-demand balances. Blockchain-based solutions can consolidate these data streams in shared, tamper-resistant ledgers that provide transparency for all approved parties (Rodríguez Furones & Tejero Monzón, 2023).

3.2 Potential Applications of Blockchain in Water Management

Recent scholarship indicates that blockchain can address several critical water management functions. One function involves monitoring and verifying water usage to ensure compliance with allocation regulations. In contexts where water is scarce, such as arid regions, over-extraction from aquifers can create conflict among industrial, agricultural, and residential users (Zhang, 2023). A blockchain-based recording system can log extraction and usage in real time to enhance accountability. Another application is optimizing water distribution by integrating sensor data on flow rates or pipeline pressure into smart contracts that automatically adjust valves or pumping schedules (Xia et al., 2022).

Blockchain also shows promise for water rights trading, allowing parties to transfer entitlements efficiently (Li et al., 2022). In these markets, participants can set buy or sell orders for water shares, and a blockchain-based mechanism automatically matches them. Once matched, transactions are recorded on-chain, guaranteeing that volumetric changes in water rights are visible to regulators. Research has documented pilot programs in regions of the United States where blockchain platforms are tested for groundwater trading (Morris, 2020). These pilots indicate how distributed ledgers might streamline transactions and reinforce compliance by making changes in entitlements easily traceable.

3.3 Intelligent Water Management and Blockchain

Xia et al. (2022) proposed an integrated framework of blockchain in intelligent water management that leverages smart metering, sensor networks, and machine learning. Their work suggests that blockchain can operate as a secure backbone for data exchange while advanced analytics extract insights from the data. This synergy could enable predictive leak detection, real-time demand response, and more dynamic water pricing structures. Furthermore, the secure environment helps utilities trust data from external sensors or user devices, since each data point is sealed with cryptographic assurance once validated.

Researchers have also addressed how blockchain might provide an immutable ledger for water quality information, a critical area in public health (Naqash et al., 2023). Potentially, different laboratories can record water test results on-chain, where results become immediately visible and unalterable. Any anomalies or contamination events would be flagged and confirmed quickly, prompting swift interventions. The literature, however, notes the challenge of ensuring that off-chain measurements are reliable, which hinges on sensor calibration and tamper-resistant hardware.

3.4 Case Examples and Pilot Studies

Pilot studies show encouraging signs. Morris (2020) described a Colorado-based initiative that recorded groundwater transactions among a group of agricultural users. Farmers who needed more water could buy entitlements from others with surplus shares, with each transaction written to a blockchain. The pilot indicated reduced administrative time for verifying trades, along with more transparent recordkeeping. Another example involves municipal water consumption data linked to blockchain for analyzing monthly usage in a Middle Eastern city (Alrammal et al., 2024). This pilot aimed to encourage conservation by publicly displaying aggregated usage trends, reinforcing trust in the accuracy of the data.

Other programs investigate how blockchain-based platforms can coordinate with Internet-of-Things (IoT) devices to monitor dam stability (Youssef et al., 2019). A distributed ledger can store sensor data from multiple vantage points, limiting data tampering and enabling multiple agencies to share the information. Such setups can improve security and safety in water infrastructure, reducing the risk of false data injection. Nonetheless, practical challenges persist, including the technical complexity of integrating water utility legacy systems with blockchain networks and the upfront investment in sensor technology (Henry et al., 2020).

Table 1 summarizes selected studies on how blockchain contributes to urban water management. These works highlight improvements in monitoring, accountability, and transactional efficiency.

Table 1. Key research findings on blockchain and urban water management

Study	Focus	Key Contribution
Naqash et al. (2023)	Urban water leakage detection	Proposed blockchain-based framework for real-time leak identification. Emphasized accuracy benefits.
Rodríguez Furones & Tejero Monzón (2023)	Applicability in urban water supply and sanitation in Spain	Presented feasibility analysis and governance considerations.
Xia et al. (2022)	Intelligent water management	Proposed integrated framework combining sensors, ML, and blockchain.
Li et al. (2022)	Water rights trading in irrigation areas	Demonstrated how blockchain can secure water rights transactions.
Alrammal et al. (2024)	Sustainable water consumption	Explored blockchain for data transparency and user engagement in consumption reduction.

4. LITERATURE REVIEW ON BLOCKCHAIN FOR ENERGY MANAGEMENT

4.1 Evolution of Decentralized Energy Systems

Energy systems are undergoing transitions away from centralized generation toward distributed models, such as rooftop solar, community wind, and microgrids (Andoni et al., 2019). While these models offer resilience and sustainability benefits, they can introduce complexities in accounting, settling transactions, and balancing local grids. Blockchain can potentially resolve these challenges by facilitating real-time peer-to-peer (P2P) energy trading, transparent tracking of renewable energy certificates, and decentralized demand response.

Research suggests that blockchain-based smart contracts can automate electricity transactions and ensure payment security. Consumers with generation capacity can sell excess electricity to neighbors who need additional supply, bypassing or supplementing the traditional utility role (IRENA, 2020). The ledger records each kilowatt-hour traded and the associated financial transaction. Because everything is logged immutably, participants gain confidence that trades are settled fairly, and regulators can audit the process.

4.2 P2P Energy Trading and Microgrids

One prominent use case involves microgrids that operate in islanded modes or as partial contributors to a larger grid (Evens et al., 2023). These microgrids often rely on small-scale renewable sources, and local balancing is critical to prevent power quality issues. Blockchain-based approaches can automate matching of supply and demand among microgrid participants (Tarek & Ismail, 2022). This arrangement can also integrate dynamic pricing, where electricity prices rise or fall based on real-time local supply. Smart contracts can handle these adjustments automatically, lowering overhead.

Researchers emphasize that blockchain microgrids could empower communities by granting them direct control over energy trading (European Commission – EU Blockchain Observatory, 2019). Community members become both producers and consumers (prosumers), and the ledger keeps an accurate record of contributions. Studies have documented pilot projects, including in Brooklyn, New York, where neighbors traded solar power through a blockchain platform (Andoni et al., 2019). These initiatives highlight the feasibility of decentralized energy trading but also illustrate regulatory hurdles, such as local utility tariffs and grid interconnection rules that can complicate market design.

4.3 Blockchain for Demand Management and Grid Stability

In many electricity systems, peak demand strains infrastructure. Traditional demand response programs rely on utility signals to large industrial or commercial users, requesting load reductions. Blockchain may enable more granular, automated demand response that includes residential customers (Andoni et al., 2019). For instance, a smart contract could reduce a household's energy draw from the grid when supply is tight, compensating them with tokens if they comply. Because the transaction is processed on-chain, the utility can verify compliance, and the household is assured of receiving compensation.

Such decentralized demand management has the potential to stabilize grids and reduce the need for expensive peaking plants (Evens et al., 2023). It can also facilitate cost savings for consumers who participate. Nonetheless, implementing blockchain-based demand response requires real-time data from smart meters, robust communications, and user acceptance of automated load control. It also involves careful system design to ensure that the blockchain does not introduce excessive latency.

4.4 Renewable Energy Certificates and Traceability

Renewable energy certificates (RECs) track how much electricity a power plant generates from renewable sources. Companies often purchase RECs to meet sustainability targets. However, the existing paper-based or centralized digital tracking processes can lack transparency and be prone to double-counting

(Campbell, 2019). Blockchain-based REC platforms can address these deficiencies by maintaining an immutable, publicly auditable ledger of certificate creation, transfer, and retirement (Boucher, 2020). The ledger could document every REC from issuance by a renewable generator to its purchase and eventual usage by an end consumer or aggregator.

Studies note that such systems could be integrated with smart meter data to confirm the volume of clean energy produced and to tokenize it as digital certificates (Andoni et al., 2019). This design reinforces trust among REC market participants. Regulatory backing is key, because unless policy frameworks recognize blockchain-based certificates as legitimate, the market will not adopt them widely. Current research also identifies cross-border certificate trading as an area where blockchain might improve efficiency, but harmonization among national or regional regulations remains a challenge.

4.5 Summary of Blockchain Contributions to Energy

Blockchain can support the decentralization of energy systems by providing trust in transactions among diverse participants. It can automate P2P trading, create transparent certificate systems for renewable generation, and enable dynamic demand response (Evens et al., 2023). However, issues of scalability, governance, and standardization persist. Ensuring that consensus mechanisms align with the sustainability objectives of the energy sector is also a prominent challenge. Adopting blockchains that rely on energy-intensive mining, for example, may be counterproductive if the goal is to minimize carbon footprints (Campbell, 2019). Despite these barriers, the literature continues to recognize blockchain's disruptive potential in facilitating more efficient, transparent, and resilient energy systems. Table 2 lists major studies regarding blockchain's roles in the energy domain, including distributed generation, peer-to-peer trading, and certificate management.

Table 2. Studies on blockchain and energy sector

Study	Focus	Key Points
Andoni et al. (2019)	Blockchain in energy sector	Systematic review of challenges and opportunities.
Evens et al. (2023)	Blockchain-enabled microgrids	P2P energy trading and flexible demand management.
IRENA (2020)	Peer-to-peer electricity trading	Outlined how blockchain can facilitate distributed markets.
Campbell (2019)	Impacts on energy sector	Explored whether blockchain is feasible in electricity markets.
European Commission (2019)	Blockchain in energy	Overview of opportunities in energy efficiency and renewable.

5. INTEGRATING WATER AND ENERGY MANAGEMENT THROUGH BLOCKCHAIN

5.1 The Water–Energy Nexus in Urban Contexts

Urban environments demand vast quantities of water and energy, and this demand is rising due to population growth and industrial activities (Zhang, 2023). The water–energy nexus illustrates how water treatment and distribution systems can be energy-intensive, and power generation often requires substantial water for cooling or feedstock. Failures in energy supply can affect water systems, just as water shortages can disrupt energy generation. Integrated planning can mitigate vulnerabilities and optimize resource flows, but siloed practices remain common.

Blockchain's distributed ledger capabilities may unify data and transactions across these two domains (Rodríguez Furones & Tejero Monzón, 2023). This integration involves bridging water resource monitoring systems with energy metering infrastructure, possibly using a shared or interoperable blockchain platform that standardizes data formats and establishes trust mechanisms. Smart contracts can automate cross-sector processes, such as adjusting water pumping operations based on real-time electricity prices or generation constraints. Operators gain a comprehensive view of resource statuses, which can enable better-coordinated planning.

5.2 Combined Resource Accounting and Exchange

A fundamental requirement for integrated resource management is accurate, transparent accounting of water and energy flows. Blockchain provides the common data layer on which multiple parties can rely without depending on a single centralized database. For instance, a combined ledger could record real-time electricity consumption of water treatment plants, linking that data to the volume of water processed. These linkages allow stakeholders to measure the energy intensity of water operations over time, pinpointing inefficiencies (Asgari & Nemati, 2022).

In parallel, water utilities that generate biogas from wastewater treatment might register these energy outputs on the ledger. A local microgrid operator could purchase that biogas as an alternative fuel, with transactions cleared on-chain. Smart contracts would finalize payments once the system verifies that the water utility has delivered a specified quantity of biogas or electricity from a cogeneration plant. This arrangement fosters synergy between water reclamation and distributed energy resources, potentially lowering operational costs and carbon footprints.

5.3 Multi-Stakeholder Collaborations and Governance

Integrated water–energy projects often involve government agencies, private utilities, environmental regulators, and citizen groups. Achieving consensus on data sharing and transaction oversight can be difficult due to mistrust or competing interests (World Economic Forum, 2018). Blockchain solutions could mitigate these concerns by distributing data across nodes run by each stakeholder. Each node verifies new records, ensuring that no single entity can manipulate data unilaterally. This decentralized governance helps maintain stakeholder confidence in shared operations (Boucher, 2020).

However, designing governance protocols for permissioned blockchains requires careful attention to membership criteria, node responsibilities, and dispute resolution processes (Rodríguez Furones & Tejero Monzón, 2023). A water–energy consortium might define roles for each stakeholder class, specifying who can propose transactions, validate blocks, or view confidential data. Smart contracts could encode compliance rules for resource usage or emissions, automatically flagging any breach. Regulators can thus rely on an auditable trail of on-chain transactions to ensure adherence to legal requirements.

5.4 Resource Optimization Using Smart Contracts

Blockchain-based smart contracts can automate integrated resource allocation by factoring in variables such as real-time water demand, electricity supply, storage availability, and environmental constraints (Zhang, 2023). If an urban district experiences a peak in electricity demand, the system might trigger a command to shift water pumping schedules to off-peak hours, thereby reducing grid stress. Conversely, if a water treatment plant has on-site solar panels, it might inject surplus energy into the grid at times of high demand and receive on-chain compensation from a local aggregator.

Pilot initiatives in combined resource management remain limited, but the literature suggests that the same principles proven in single-domain blockchain deployments (e.g., water rights trading, P2P energy trading) can expand to multi-domain contexts (Naqash et al., 2023). Implementation would require advanced sensor networks capable of capturing relevant data from both water and energy systems, a robust communications backbone, and a permissioned blockchain with the capacity to handle the transaction volume.

5.5 Case Illustration of Integrated Water–Energy Trading

Imagine a scenario where an urban wastewater treatment facility uses anaerobic digesters that produce biogas. This facility can generate electricity to offset a portion of its operational costs. On a blockchain-based platform, the facility registers each kilowatt-hour injected into the local microgrid. Meanwhile, the microgrid operator registers demands from residential and commercial users. Smart contracts match supply with demand in real time, setting prices based on availability. Consumers pay in digital tokens that are convertible to fiat currency. The ledger ensures that no double-counting occurs, and it timestamps each transaction for auditing.

If water consumption in certain residential areas spikes, the system can allocate additional energy resources for pumping and treatment, balancing the extra load. The cost differential is captured in the same ledger and distributed proportionally among participants according to pre-established governance rules. Regulatory agencies can view all trades and energy or water consumption data in a transparent manner, reducing the administrative burden of reporting. Over time, analytics could reveal patterns of peak usage, cost fluctuations, and greenhouse gas impacts.

This conceptual scenario demonstrates how blockchain might support a closed-loop arrangement, enabling water infrastructure to function as both consumer and producer of energy. Such an approach creates new revenue opportunities, elevates operational efficiency, and promotes sustainability. Nonetheless, it depends on robust policy frameworks and acceptance by utilities and end users. Table 3 outlines scenarios where blockchain can integrate water and energy processes to enhance operational coordination and financial transactions.

Table 3. Potential blockchain-enabled scenarios in water–energy integration

Scenario	Mechanism	Potential Benefits
Water pumping aligned with off-peak electricity	Smart contracts triggered by electricity pricing signals	Reduced energy costs, improved grid balancing
Biogas-based cogeneration feeding local microgrid	On-chain registration of generated electricity, real-time trading	Additional revenue for water utilities, reliable local energy
Combined water–energy usage analytics	Shared ledger with sensor data from both systems	Enhanced transparency, integrated demand management
Digital water rights plus energy tokens	Tokenization of water and energy entitlements	More flexible resource trading and allocation
Automated billing for cross-domain usage	Unified billing managed by blockchain-based smart contracts	Lower administrative overhead, seamless cost allocation

6. TECHNICAL AND REGULATORY CHALLENGES

6.1 Data Privacy and Security

Even though blockchain is known for transparency, certain data in water and energy systems must remain confidential due to consumer privacy or competitive considerations (Rodríguez Furones & Tejero Monzón, 2023). Unauthorized disclosure of usage patterns could violate privacy laws or reveal sensitive operational details about critical infrastructure. Permissioned blockchains can implement fine-grained access controls, but the complexity increases as more parties join. Solutions such as zero-knowledge proofs or off-chain data storage can preserve data privacy while still providing the verification benefits of a distributed ledger (Nai Fovino et al., 2021).

Security extends beyond data privacy. Hackers might attempt to disrupt nodes or inject false data to manipulate water or energy markets. The consensus mechanism must guard against hostile participants. Encryption, identity management, and network monitoring become essential. Interfacing with IoT devices brings additional risks, because compromised sensors could feed incorrect data into the blockchain. A robust cybersecurity posture is indispensable, especially in mission-critical utilities (Youssef et al., 2019).

6.2 Interoperability with Legacy Systems

Many water and energy utilities operate legacy software systems that handle billing, SCADA (Supervisory Control and Data Acquisition) processes, and customer management. Integrating blockchain requires bridging these systems through adapters or APIs. This step can be technically demanding and expensive (Andoni et al., 2019). Without seamless interoperability, blockchain projects risk data duplication and inconsistencies that undermine trust. A transitional path may involve piloting blockchain in smaller subsystems to demonstrate feasibility before broader rollouts.

Standards play a role in facilitating interoperability. International bodies have begun discussing common data formats, consensus protocols, and best practices for blockchain in critical infrastructures (European Commission – EU Blockchain Observatory, 2019). Conformance to emerging standards can reduce the customization required for each utility. Achieving these standards, however, calls for collaboration among multiple stakeholders, including technology providers, regulators, and industry associations.

6.3 Regulatory and Policy Barriers

Urban water and energy systems are heavily regulated to protect public health, environmental sustainability, and consumer interests (Boucher, 2020). Blockchain-based approaches may not align with existing rules, which assume centralized oversight by utilities or government agencies. Regulators may be unfamiliar with decentralized ledgers and uncertain how to oversee them. For instance, water rights laws in some jurisdictions might not permit partial transfers of entitlements in near-real time, making blockchain-based trading infeasible under current legal frameworks (Kaushik, 2019).

Policy makers could also raise concerns about market manipulation. If a local group of blockchain nodes colludes, they might artificially inflate resource prices or block certain transactions. Governance models need to address such concerns by ensuring that no single entity can dominate block validation. Moreover, many smart contracts exist in a legal gray area, as courts have yet to fully establish their status. A key challenge lies in harmonizing legal definitions of property rights or contractual obligations with the automated nature of blockchain transactions (Morris, 2020).

6.4 Scalability and Sustainability of Consensus Mechanisms

Some blockchain networks, particularly those using Proof of Work, consume large amounts of energy (Campbell, 2019). This design contradicts the sustainability objectives inherent in water and energy management. Newer consensus models, such as Proof of Stake or Byzantine Fault Tolerance variants, offer more efficiency. Even so, large-scale deployments in cities might generate thousands of transactions per minute, testing the throughput capabilities of existing blockchain implementations (Andoni et al., 2019).

A prospective solution involves hybrid architectures that store only critical transaction data on-chain, while off-chain databases handle high-volume sensor readings. This approach can reduce blockchain congestion and costs, but it demands careful synchronization to maintain data integrity (Xia et al., 2022). Another approach is employing sidechains or sharding techniques to distribute transactional loads. Yet these approaches add technical complexity that might slow adoption by relatively conservative utility organizations.

6.5 Organizational Readiness and Change Management

Utilities and city governments often operate in structured, risk-averse environments. Adopting blockchain can require shifting from decades-old paper-based or database systems. Employees must learn new platforms and policies, while managers must ensure that daily operations remain secure (Rodríguez

Furones & Tejero Monzón, 2023). Resistance to change might arise if staff members perceive blockchain as disruptive or if the value proposition is unclear.

Change management strategies can improve acceptance, including phased pilots, clear cost–benefit analyses, and stakeholder engagement sessions (Coli et al., 2021). Demonstrating tangible outcomes, such as reduced billing errors or faster dispute resolution, can build momentum for broader rollout. Collaboration with universities or technology firms can also provide specialized expertise that internal teams may lack.

Figure 2 presents major categories of challenges confronting blockchain implementations in urban water and energy systems.

Figure 2. Overview of key challenges in blockchain-based integrated water and energy management

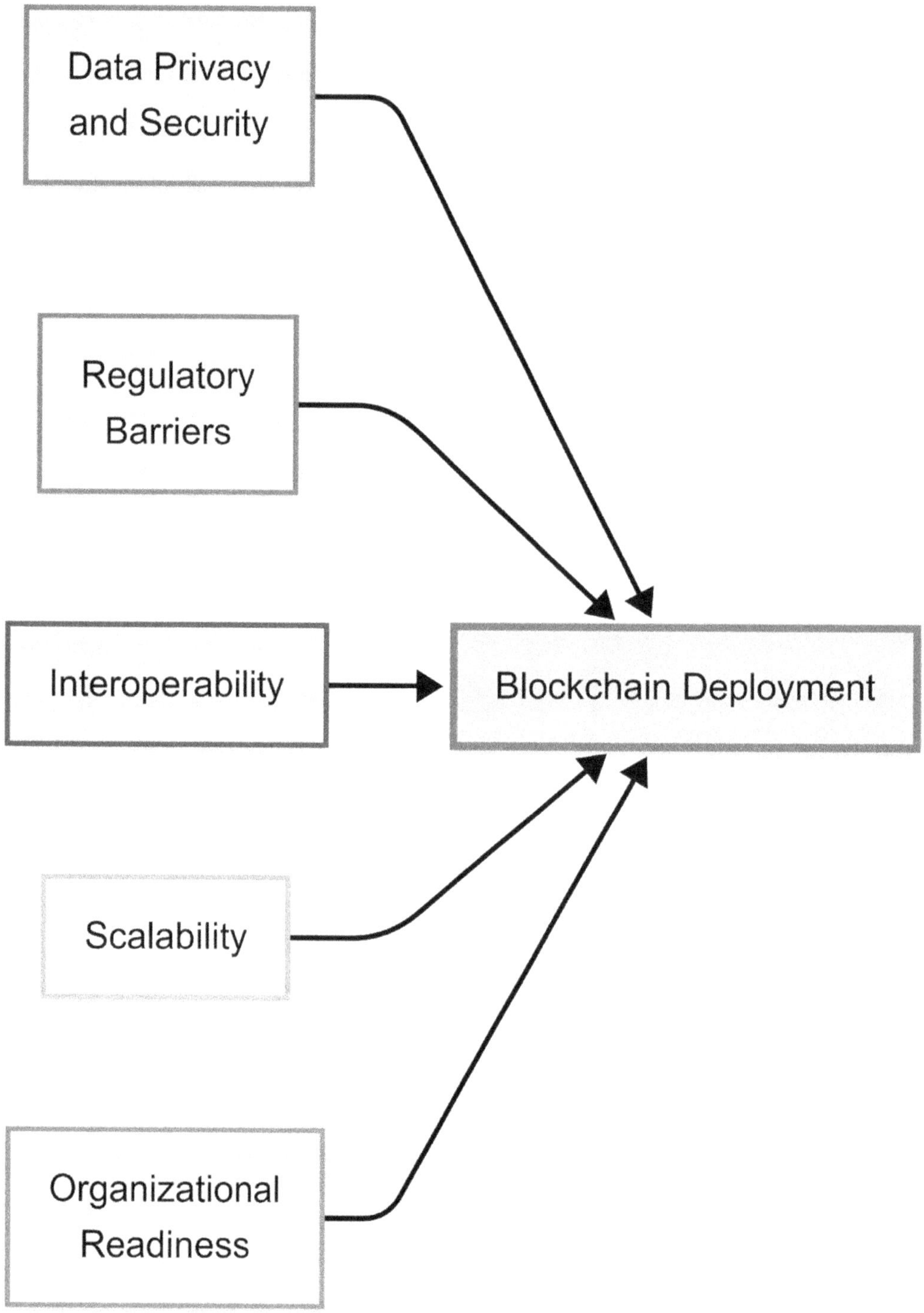

7. GOVERNANCE MODELS AND STAKEHOLDER ENGAGEMENT

7.1 Public–Private Partnerships

The complexity of integrating blockchain into urban water and energy systems often necessitates partnerships among municipal agencies, private utilities, technology vendors, and community stakeholders (World Economic Forum, 2018). Public–private partnerships can pool financial and technical resources, distributing risks and benefits. For instance, a city government might collaborate with a tech firm to build a blockchain-based water consumption platform, while a private utility supplies the data infrastructure. Such structures can accelerate experimentation and scale-up but require transparent agreements on roles and revenue sharing (Poberezhna, 2018).

7.2 Citizen Involvement and Data Ownership

Urban resource management ultimately affects residents, who pay for services and experience any disruptions. Engaging citizens in blockchain solutions can bolster acceptance, particularly if they see improved service quality, lower costs, or the ability to generate income by trading resources (Henry et al., 2020). Including citizen representation in governance consortia and allowing them to operate their own nodes in the permissioned network can enhance inclusivity and transparency.

Data ownership is a sensitive topic, as individuals may not want their consumption patterns exposed. Blockchain projects often adopt minimal data collection principles, storing only what is necessary to manage transactions. Aggregation or anonymization techniques can protect individual privacy while preserving system integrity (Nai Fovino et al., 2021). Clear guidelines on data handling, ownership, and usage rights can reassure citizens and align with data protection regulations.

7.3 Regulatory Sandbox Approaches

Because water and energy are regulated, implementing blockchain in these sectors can face delays or legal ambiguity (Boucher, 2020). Governments can adopt regulatory sandbox frameworks, allowing innovative projects to operate in controlled environments with relaxed rules, enabling them to gather evidence on feasibility and impact. Sandboxes can help policy makers observe how blockchain-based solutions perform in real-world conditions, shaping updated regulations that support broader adoption.

Sandboxes can also facilitate cross-sector dialogues. Water agencies, energy utilities, technology firms, and consumer advocates can collaborate under a policy umbrella that encourages experimentation but places guardrails on public risk (Kaushik, 2019). Once a pilot project demonstrates success, regulators can refine the broader legal and technical framework. In this manner, incremental steps can pave the way for integrated blockchain deployments.

7.4 Multilateral, International Collaborations

Urban resource management challenges extend beyond national borders, especially in regions that share water basins or electricity interconnections (UNEP, 2021). International organizations can play a facilitative role by supporting cross-border blockchain initiatives. Such initiatives might address transboundary water allocations or renewable power trading across regional grids (Boucher, 2020).

Harmonized standards and interoperability protocols are crucial for success, since each country may have distinct legal frameworks.

Donor agencies or development banks might fund pilot projects in emerging economies to test whether blockchain can address water and energy scarcity more effectively than traditional approaches (Coli et al., 2021). If successful, these pilots can inform global best practices. They can also illuminate how local contexts affect technology adoption, such as variable internet connectivity, sensor availability, or differing cultural attitudes toward decentralized governance.

8. FUTURE RESEARCH DIRECTIONS AND POTENTIAL INNOVATIONS

8.1 Advanced Analytics and AI Integration

Blockchain serves as the backbone for secure data storage and verification, but the next step could be integrating artificial intelligence (AI) for predictive analytics (Asgari & Nemati, 2022). AI can forecast water demand spikes, grid congestion, or even mechanical failures in infrastructure. If these predictions trigger smart contracts, the system can take proactive measures, such as securing additional energy resources or scheduling pipeline maintenance. This integration can move urban water and energy management from reactive to anticipatory modes.

Future research could explore how federated learning models might operate on top of blockchain, allowing multiple utilities to contribute data securely without revealing proprietary information (Xia et al., 2022). Such collaborations could refine predictive models at a regional scale, optimizing water and energy flows across cities. However, the computational overhead and privacy constraints inherent in AI–blockchain hybrids need careful management.

8.2 IoT and Sensor Fusion

Blockchain can receive data from a range of IoT devices measuring water quality, flow rates, and energy consumption (Youssef et al., 2019). Looking forward, sensor fusion—the process of combining data from diverse sensors to produce more accurate or comprehensive information—may become critical in refining integrated water–energy operations. Merging data from water flow sensors, electricity meters, weather stations, and traffic sensors (for distribution network planning) can enable advanced resource coordination.

Ensuring that these sensor inputs remain tamper-proof is an ongoing research challenge. Projects might experiment with hardware-secured modules or cryptographic identity verification for each sensor (Henry et al., 2020). The costs of deploying and maintaining these sensors must also be considered, especially in regions with limited budgets or technical capacity.

8.3 Tokenized Incentive Systems

Blockchain enables the creation of digital tokens that can represent water credits, energy tokens, or other units of value (Li et al., 2022). Future research can investigate how tokenized incentives might shift consumer behavior, encouraging conservation or participation in demand response programs. For example, residents who save water or reduce peak energy consumption could receive tokens redeemable for discounts on utility bills. On the supply side, small businesses or households generating electricity from renewable sources might earn tokens they can trade for additional water allocations during droughts.

Such cross-domain tokenization demands robust policy frameworks to define token value, maintain stability, and ensure that they do not undermine existing regulatory structures (Kaushik, 2019). Research can explore whether hybrid public–private token models strike the right balance between free-market flexibility and regulatory oversight. Additional studies might assess how these tokens could link to broader carbon markets, forming a more comprehensive environmental economy.

8.4 Integration with Circular Economy and Waste-to-Energy Systems

The circular economy model aims to reduce waste by reusing or recycling materials, including water and energy by-products. Blockchain can track resource flows within circular systems, ensuring authenticity and preventing leakage of waste into the environment (Poberezhna, 2018). For instance, a city might implement a blockchain platform that monitors the production of compost or fertilizer from wastewater sludge and its distribution to local farms. Simultaneously, the system tracks renewable energy outputs from waste-to-energy conversion plants, creating a holistic loop of resources.

In an integrated water–energy scenario, the circular economy approach could guide transformations of industrial or agricultural waste into energy sources or treated water suitable for reuse (Zhang, 2023). Smart contracts would manage the exchange of these resources, validating that regulatory thresholds for contaminants or emissions are respected. Future studies can examine the technical, logistical, and legal obstacles to implementing such closed-loop systems at scale.

8.5 Policy Recommendations for Large-Scale Adoption

Significant policy developments must occur for blockchain-based integrated water–energy solutions to thrive. Researchers might focus on comparative policy analyses, looking at which jurisdictions have adopted enabling legislation or recognized digital ledgers and smart contracts. They can investigate how these policies balance the need for innovation with consumer protection, data privacy, and environmental safety (Boucher, 2020).

Further work could provide frameworks for standardizing data models, transaction protocols, and consensus mechanisms in utility contexts (European Commission – EU Blockchain Observatory, 2019). Additionally, interdisciplinary research can bring together engineers, legal scholars, economists, and environmental scientists to propose an evidence-based roadmap for large-scale blockchain adoption in water and energy.

Figure 3 illustrates an envisioned future ecosystem where IoT devices feed data into a blockchain, AI analyzes trends, and smart contracts automatically distribute tokenized incentives to encourage resource optimization.

Figure 3. Prospective interactions among blockchain, iot, ai, and tokenized incentives

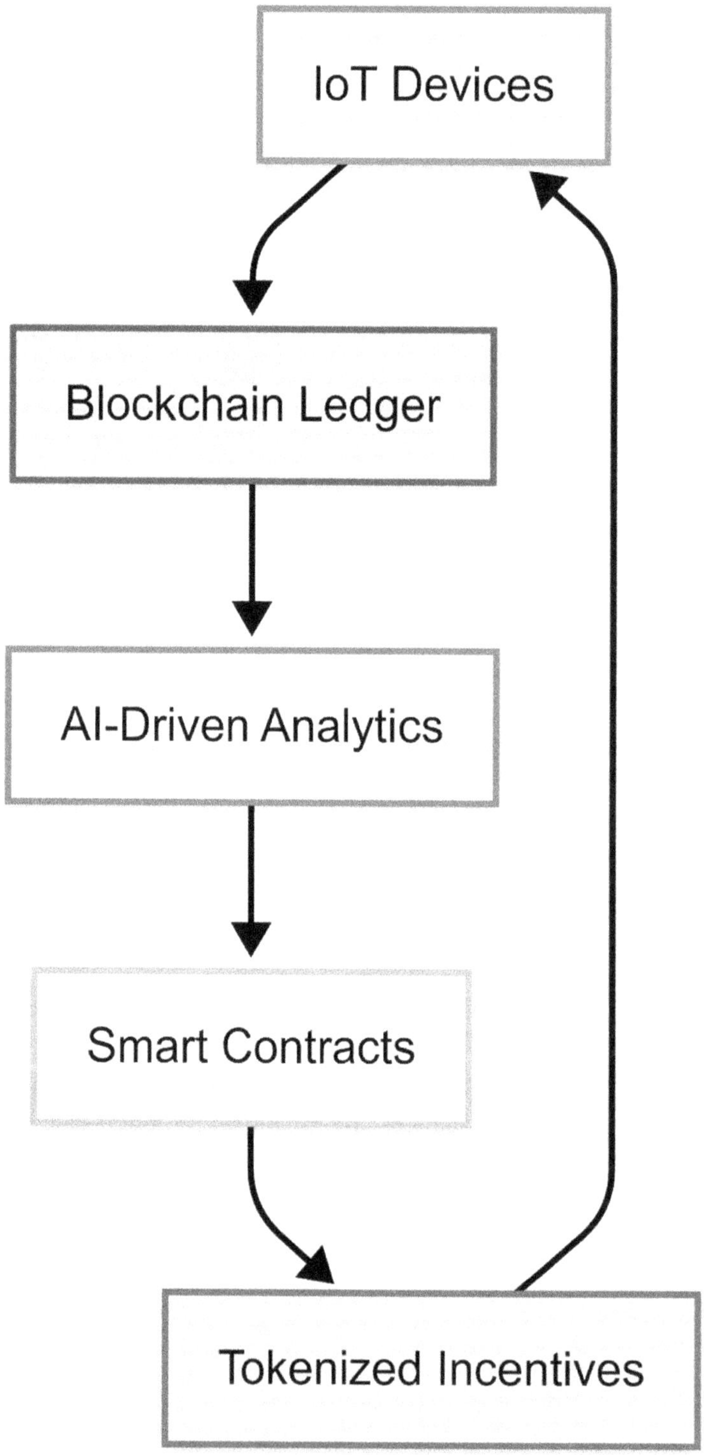

9. CASE STUDY ILLUSTRATION: A HYPOTHETICAL SMART CITY DEPLOYMENT

9.1 City Profile and Motivation

Consider a hypothetical city, Riverdale, located in a water-scarce region that also experiences periodic energy shortages during peak demand. The municipality has begun building a smart city strategy to modernize critical infrastructures. It seeks to improve the reliability and efficiency of its water supply network and local power grids. Officials are exploring blockchain as a way to unify data management, automate transactions, and enhance trust among stakeholders.

9.2 Technical Configuration

Riverdale's water utility installs IoT sensors to measure flow rates and reservoir levels in strategic locations, connecting these sensors to a permissioned blockchain platform. The municipal energy authority also installs smart meters in commercial and residential buildings, feeding electricity consumption and, for prosumers, generation data into the same ledger. A set of nodes run by the water utility, the energy authority, and the city administration validates transactions.

Smart contracts manage the daily schedule of water pumping, taking into account real-time electricity prices and reservoir levels. They also facilitate P2P energy trading among residents with rooftop solar. If a certain block of homes collectively runs short of water credits, they can purchase surplus allocations from commercial properties that historically use less water. Meanwhile, residents who produce excess solar energy can sell it to local businesses, with the blockchain automating settlement in digital tokens convertible to the local currency.

9.3 User Experience and Engagement

Riverdale offers a mobile application where citizens see their daily water and energy usage. They can voluntarily reduce consumption during high-demand periods to earn additional credits. The blockchain platform ensures the immutability of consumption data, reducing billing disputes. Regulators periodically audit the ledger to confirm compliance with water allocations and to verify that any utility subsidies align with actual usage.

Residents develop a sense of ownership as they become active participants in resource trading. Community groups form neighborhood energy cooperatives, sharing revenue from aggregated solar generation. Using tokens, these cooperatives might invest in additional water recycling infrastructure or energy storage. Over time, Riverdale's integrated ledger data allow city planners to refine long-term policies around infrastructure investment and sustainability targets.

9.4 Outcomes and Limitations

This hypothetical deployment suggests multiple benefits: reduced wastage, faster leak detection, partial alleviation of peak electricity demand, and new revenue streams for citizens. From a governance standpoint, the transparency of on-chain transactions might reduce corruption or misreporting. However, the city would face initial costs for sensor installations and network upgrades. Additionally, the system depends heavily on robust internet coverage and reliable hardware. Skilled personnel must oversee the blockchain, create smart contracts, and respond to emerging issues. On the regulatory side, Riverdale's legal framework must recognize digital tokens and blockchain-based transactions. Although this case is fictional, it captures the potential synergies and complexities of a real-world blockchain-enabled water–energy integration project. Table 4 illustrates possible improvements observed during a pilot phase, although these figures are hypothetical and contingent on real-world factors.

Table 4. Hypothetical results of a pilot phase in Riverdale

Parameter	Baseline	After Blockchain Deployment	Observed Benefits
Water leakage rate	15%	8%	Faster detection and intervention
Peak electricity demand	High (summer afternoons)	Reduced by 10% on average	Automated demand response via smart contracts
Number of billing disputes	50/month	5/month	Transparent on-chain usage data
User satisfaction score	70/100	85/100	Engaged participants, reduced costs
Renewable energy share	20%	30%	Incentivized rooftop solar adoption

10. CONCLUSION

Blockchain technology has captured growing interest as a tool for improving transparency, trust, and efficiency in integrated urban water and energy management. Its distributed ledger framework and the potential for smart contracts promise to unite siloed resource domains. The literature emphasizes how blockchain-based systems can create immutable, shared records of water usage and energy transactions, reducing administrative overhead and enhancing accountability. In water management, blockchain can facilitate improved leakage detection, real-time usage tracking, and the creation of water rights trading platforms. For energy systems, researchers see possibilities in peer-to-peer trading, dynamic demand response, and transparent verification of renewable energy certificates.

The synergy between water and energy arises from their interdependence: water infrastructure requires large amounts of electricity, while many power generation processes need water. Blockchain's ability to unify data and transactions on one platform could help city planners, utilities, and citizens coordinate resource flows more strategically. Still, implementing blockchain in existing urban infrastructures faces several obstacles. Data privacy and security remain paramount, given the sensitivity of consumption data and the potential for cyberattacks on critical utilities. Interoperability with legacy systems demands technical effort and organizational change. Regulatory and policy barriers must be addressed, including issues around legal recognition of digital contracts, property rights, and consumer protection.

Furthermore, blockchain-based solutions must address concerns about energy-intensive consensus mechanisms. Resource optimization goals can conflict with the high computational overhead typical of older blockchain protocols. Emerging consensus models and hybrid architectures can mitigate these issues. Organizational readiness is equally crucial; utilities often require transitional pilot projects to build internal capacity and stakeholder trust. Public–private partnerships, regulatory sandboxes, and international collaborations can support iterative experimentation.

Future research directions involve deeper integration of AI and advanced analytics to anticipate and automate resource distribution in near-real time. IoT sensor networks will need to expand, and the integrity of off-chain data remains a critical challenge. Tokenized incentive schemes and circular economy strategies offer compelling possibilities for achieving sustainability targets in water–energy systems. Case study examples, whether hypothetical or real, showcase how blockchain can reorganize stakeholder relationships, encourage citizen involvement, and streamline transactions. Yet the success of such initiatives depends on context-specific factors, from regulatory frameworks to local infrastructure conditions.

Ultimately, blockchain is not a panacea. It does not replace essential investments in modernizing water infrastructure, deploying renewables, or developing robust governance structures. Instead, it serves as an enabling technology that can augment data reliability and stakeholder coordination. If integrated responsibly, blockchain can help cities address rising demands for water and energy, supporting resilient, transparent, and equitable urban services for decades to come.

REFERENCES

Alrammal, M., Abu-Amara, F., Ismail, Z., & Nadeem, M. (2024). Blockchain technology for sustainable management of electricity and water consumption. In *Proceedings of the International Conference on Recent Advances in Science and Engineering (RAiSE 2023)* (Engineering Proceedings Vol. 59, No. 1, Article 223). MDPI. https://doi.org/DOI: 10.3390/engproc2023059223

Andoni, M., Robu, V., Flynn, D., Abram, S., Geach, D., Jenkins, D., McCallum, P., & Peacock, A. (2019). Blockchain technology in the energy sector: A systematic review of challenges and opportunities. *Renewable & Sustainable Energy Reviews*, *100*, 143–174. DOI: 10.1016/j.rser.2018.10.014

Asgari, M., & Nemati, M. (2022). Application of distributed ledger platforms in smart water systems—A literature review. *Frontiers in Water*, *4*, 848686. Advance online publication. DOI: 10.3389/frwa.2022.848686

Boucher, P. (2020). *Blockchain for the environment: Opportunities & challenges* (EPRS Study PE 648.791). European Parliamentary Research Service.

Campbell, R. J. (2019). *Bitcoin, blockchain, and the energy sector* (CRS Report R45863). Congressional Research Service.

Coli, P., Pflueger, C., & Campbell, T. (2021). *Blockchain uses for microfinance institutions in the water and sanitation sector: Pilot study*. Inter-American Development Bank.

European Commission – EU Blockchain Observatory. (2019). *Blockchain applications in the energy sector*. European Commission.

Evens, M., Ercoli, P., & Arteconi, A. (2023). Blockchain-enabled microgrids: Toward peer-to-peer energy trading and flexible demand management. *Energies*, *16*(18), 6741. DOI: 10.3390/en16186741

Henry, W., Miller, B., & Maewall, S. (2020). *Blockchain and IoT for wastewater management* (White paper). Deloitte Consulting LLP.

International Renewable Energy Agency (IRENA). (2020). *Innovation landscape brief: Peer-to-peer electricity trading*. IRENA.

Kaushik, A. K. (2019). *The development of smart water markets using blockchain technology* (Working Paper). New America.

Li, Y., Xie, J., Yang, J., Ren, J., & Zhai, N. (2022). Application of blockchain technology in water rights trading in the irrigation area under the Internet-of-Things environment. *Sustainability*, *14*(12), 7017. DOI: 10.3390/su14127017

Morris, R. (2020, January 21). Using blockchain to manage Colorado's water supply. *ASME News*. Retrieved from https://www.asme.org

Nai Fovino, I., Poursanidis, I., Soupionis, I., & Steri, G. (2021). Blockchain in the energy sector: WP3 – Use cases identification and analysis [EN]. Luxembourg: Publications Office of the European Union.]. *EUR*, •••, 30782.

Naqash, M. T., Syed, T. A., Alqahtani, S. S., Siddiqui, M. S., Alzahrani, A., & Nauman, M. (2023). A blockchain-based framework for efficient water management and leakage detection in urban areas. *Urban Science (Basel, Switzerland)*, *7*(4), 99. DOI: 10.3390/urbansci7040099

Poberezhna, A. (2018). Addressing water sustainability with blockchain technology and green finance. In A. Marke (Ed.), *Transforming Climate Finance and Green Investment with Blockchains* (pp. 189–196). Academic Press (Elsevier). DOI: 10.1016/B978-0-12-814447-3.00014-8

Rodríguez Furones, A., & Tejero Monzón, J. I. (2023). Blockchain applicability in the management of urban water supply and sanitation systems in Spain. *Journal of Environmental Management*, *344*, 118480. DOI: 10.1016/j.jenvman.2023.118480 PMID: 37459816

Satilmisoglu, T. K., Sermet, Y., Kurt, M., & Demir, I. (2024). Blockchain opportunities for water resources management: A comprehensive review. *Sustainability (Basel)*, *16*(6), 2403. DOI: 10.3390/su16062403

Tarek, D., & Ismail, T. (2022, December). Integrating smart contracts with WDNs framework for energy management and secure transactions. In *Proceedings of the 10th International Japan-Africa Conference on Electronics, Communications and Computations (JAC-ECC 2022)* (pp. 1–6). IEEE. https://doi.org/ DOI: 10.1109/JAC-ECC56395.2022.10044027

United Nations Environment Programme (UNEP). (2021, September 23). *In battle against climate crisis, don't overlook the blockchain* [Press release].

World Economic Forum. (2018). *Building Block(chain)s for a Better Planet*. Cologny, Switzerland: World Economic Forum.

Xia, W., Chen, X., & Song, C. (2022). A framework of blockchain technology in intelligent water management. *Frontiers in Environmental Science*, *10*, 909606. DOI: 10.3389/fenvs.2022.909606

Youssef, S. B. H., Rekhis, S., & Boudriga, N. (2019, April). A blockchain-based secure IoT solution for dam surveillance. In *Proceedings of the 2019 IEEE Wireless Communications and Networking Conference (WCNC)* (pp. 1–6). IEEE. https://doi.org/DOI: 10.1109/WCNC.2019.8885479

Zhang, J. (2023). Application of blockchain technology in water management and new energy economic development. *Water Science and Technology: Water Supply*, *23*(10), 4227–4241. DOI: 10.2166/ws.2023.258

ADDITIONAL READING

Alnahari, M. S., & Ariaratnam, S. T. (2022). The application of blockchain technology to smart city infrastructure. *Smart Cities*, *5*(2), 979–993. DOI: 10.3390/smartcities5030049

Chapter 5
Blockchain-Driven Zero-Trust Architectures for Critical Infrastructure

Mohammad Alauthman
https://orcid.org/0000-0003-0319-1968
Department of Information Security, University of Petra, Jordan

Ahmad H. al-Qerem
https://orcid.org/0000-0003-2187-6194
Computer Science Department, Faculty of Information Technology, Zarqa University, Jordan

Amjad Aldweesh
https://orcid.org/0000-0001-9358-1323
College of Computing and Information Technology, Shaqra University, Saudi Arabia

Mouhammd Alkasassbeh
https://orcid.org/0000-0001-8396-7441
Princess Sumaya University for Technology, Jordan

Ala Hamarsheh
https://orcid.org/0000-0002-4736-7331
Department of Computer Science, Faculty of Information Technology, Arab American University, Palestine

ABSTRACT

In an era of escalating cyber threats that increasingly target critical infrastructure, there is growing urgency to protect operational technology (OT) systems in energy and utilities. By uniting the core tenets of zero trust security—where every request is inherently untrusted until verified—with the decentralized transparency of blockchain, we can significantly reduce the risk of undetected breaches or insider misuse. This paper explores how blockchain-based authentication and smart contracts, aligned with guidelines such as NIST SP 800-207, can create a tamper-evident ledger of all access requests and device interactions. The goal is to move beyond reactive defenses and instead implement continuous validation of assets and their privileges. This approach not only mitigates the lateral movement of threats within industrial

DOI: 10.4018/979-8-3373-2439-5.ch005

control systems but also enables streamlined auditing of regulatory requirements (e.g., NERC CIP).

1. INTRODUCTION

Critical infrastructure encompasses systems and assets that support national security, economic stability, and public health. Energy, utilities, transportation, water systems, and communication networks remain principal targets for cyber threats (NSTAC, 2022). Operational Technology (OT) environments in these infrastructures face risks due to aging systems, connectivity with external networks, and sophisticated threat vectors (TXOne Networks, 2023). Traditional perimeter-based defense cannot address advanced attacks that may originate from inside the network or exploit trusted relationships among devices and services (Bicer et al., 2023). Zero Trust Architecture (ZTA) aims to shift from implicit trust to a model that continually authenticates and validates each user, device, and application before granting resource access (NIST, 2020; Syed et al., 2022).

Blockchain technology has emerged as a suitable approach for ensuring immutability, secure data sharing, and decentralized identity management in critical infrastructure contexts (Shetty et al., 2019). The integration of blockchain with Zero Trust models holds the potential to strengthen security in industrial control systems (ICS), as each transaction or state change becomes auditable in real time (Nie et al., 2025). The research community has investigated techniques to utilize blockchain to secure Internet of Things (IoT) devices, manage access control, and record system events (IEEE Standards Association, 2024; Aleisa, 2025). Although preliminary applications have demonstrated promise, sector-specific and energy-focused implementations remain limited. With the increasing digitization of electrical grids, pipelines, and other energy infrastructure, a deeper understanding of how blockchain-driven ZTA can mitigate risks is necessary (Govea et al., 2024).

This chapter explores the theoretical underpinnings of Zero Trust in ICS, surveys relevant blockchain-based security mechanisms, and discusses how these technologies converge in critical infrastructure. The chapter proposes a conceptual framework for applying blockchain-driven Zero Trust in the energy and utilities sectors. It addresses the architectural components, trust assumptions, policy enforcement, and operational considerations, supported by empirical data and frameworks from existing literature (Cloud Security Alliance, 2024). Furthermore, it examines technical challenges and future research directions to refine blockchain-driven Zero Trust solutions for large-scale deployment in energy and utilities operations (Hajian & Chang, 2023).

The chapter includes five tables and four figures to illustrate key points. The discussion integrates references from established works such as NIST guidelines on Zero Trust (NIST, 2020) and IEEE standards on blockchain-based Zero Trust (IEEE Standards Association, 2024). The goal is to provide a systematic overview of how Zero Trust and blockchain can be merged to meet the evolving security demands of critical infrastructure systems.

2. BACKGROUND

2.1 Overview of Zero Trust Architecture

Zero Trust Architecture emerged to address network segmentation challenges and hidden threats that bypass the perimeter defense model (Kindervag, 2010). Traditional architectures assumed an internal "trusted zone" protected by firewalls, but escalating insider threats and lateral movement tactics proved that trust based solely on location or ownership is insufficient (Syed et al., 2022). ZTA treats every request as untrusted until verified, with granular access control that includes device attributes, user roles, contextual data, and continuous monitoring (NIST, 2020).

The foundational principles of ZTA emphasize secure authentication, authorization enforcement, and least-privileged access (CISA, 2023). Rather than granting broad resource access to a device or user within a certain network area, Zero Trust enforces micro-segmentation and adaptive policies. Implementations often rely on identity and access management (IAM) solutions, endpoint security, encryption, and real-time analytics to detect anomalous behavior and deny malicious requests (Park et al., 2025). Although Zero Trust was first conceptualized with enterprise IT environments in mind, it has started to gain traction in ICS/OT contexts that require specialized protocols, resilience, and minimal downtime (Cloud Security Alliance, 2024).

2.2 Blockchain Fundamentals

Blockchain is a distributed ledger technology that ensures transaction immutability, authenticity, and transparency (Shetty et al., 2019). Each network node may store a copy of the ledger and reach consensus regarding the validity of newly appended transactions. Proof-of-work, proof-of-stake, or other consensus mechanisms help coordinate distributed ledger updates without central authority (Hajian & Chang, 2023). Blockchain can augment Zero Trust models by offering tamper-proof logs and decentralized identity frameworks, which reduce reliance on centralized certificate authorities. Immutability also helps streamline compliance auditing in regulated sectors like energy and utilities (Kirkman et al., 2024).

When blockchain is applied to ICS or IoT devices, it can manage device identities, record device states, and execute smart contracts that automate trust decisions (IEEE Standards Association, 2024). Blockchain-based identity solutions assign each device a unique blockchain address, while smart contracts enforce authentication and access rules (Gai et al., 2023). Such systems reduce single points of failure and mitigate insider threats (Bicer et al., 2023). However, in high-throughput industrial environments, blockchain's latency and scalability constraints must be considered carefully (Liu et al., 2024). Private or permissioned blockchain networks offer more predictable performance than public ones and may include tailored consensus protocols.

2.3 Intersection for Industrial Control Systems

Industrial control systems manage processes in manufacturing, utilities, power generation, and critical infrastructure. These systems often rely on specialized protocols (e.g., Modbus, DNP3) that lack built-in security mechanisms (TXOne Networks, 2023). Converging ICS with IT networks, such as for remote monitoring or advanced analytics, further increases vulnerabilities (Cloud Security Alliance, 2024). Zero Trust principles can help isolate ICS segments, monitor device-to-device communication, and limit privileges to essential functions. Blockchain can enhance these controls by enabling distributed identity management and logging each transaction in the industrial network.

As industrial processes become smarter through IoT sensors and data analytics, the volume of data transfers grows (Xie et al., 2024). Traditional perimeter-focused security fails to handle the complexity of these interactions. A blockchain-based Zero Trust model continuously validates entities and logs interactions, making it suitable for real-time security analytics and compliance oversight (Azad et al., 2024). Energy grids, for instance, can use decentralized identity frameworks to authenticate meters, controllers, and maintenance tools without a single point of compromise (Nie et al., 2025). Still, operational continuity and real-time constraints must shape the design of any blockchain-driven ICS system.

3. THEORETICAL CONCEPTS OF ZERO TRUST IN INDUSTRIAL ENVIRONMENTS

3.1 Core Zero Trust Principles for ICS

ICS require tailored security frameworks that account for critical safety demands, proprietary protocols, and legacy equipment (NIST, 2020). Zero Trust principles in ICS focus on continuous device validation, granular role-based policies, and dynamic risk-based access. These principles require robust identity management for all endpoints, including legacy devices. The model enforces continuous trust evaluations based on real-time state, anomaly detection, and minimal privileges for commands or data access (CISA, 2023). For instance, a control workstation can only issue the necessary commands to a specific programmable logic controller (PLC), and no other segment or user can override those commands without explicit authorization. This approach prevents lateral movement from compromised hosts (Govea et al., 2024).

Least Privilege Application. ICS contain many nodes such as sensors, actuators, and controllers that have limited operating purposes. Zero Trust in ICS requires restricting each entity to its minimal operational scope (NSTAC, 2022). This approach mitigates the risk of malicious insiders using a compromised sensor to escalate privileges or pivot to higher-value assets.

Micro-Segmentation. Instead of broad network segments, ICS environments under Zero Trust break down each segment into smaller zones or enclaves. Each zone has distinct access control policies. Requests must satisfy criteria such as device identity or trust score. By segmenting ICS networks, defenders reduce the blast radius of attacks (Cloud Security Alliance, 2024).

Continuous Monitoring and Enforcement. ICS processes may be stable or cyclical, which helps identify irregularities in traffic patterns or operational data. Zero Trust strategies incorporate continuous monitoring of ICS traffic, anomaly detection based on normal process variables, and automated enforcement (Park et al., 2025).

3.2 Adapting Zero Trust for Resource-Constrained Devices

Large ICS environments often feature resource-constrained sensors and actuators, limiting their ability to run complex security agents or store large encryption keys (Nie et al., 2025). Traditional ZTA solutions for enterprise IT may not transfer directly to ICS. Adaptations include:

1. **Lightweight Cryptography:** In Zero Trust ICS frameworks, cryptographic algorithms must handle the limited computation resources of sensors. Some ICS devices cannot support advanced encryption schemes, so zero trust policies may use streamlined cryptography that balances security with performance (Azad et al., 2024).
2. **Edge Gateways:** Gateways situated at the network edge can perform trust evaluation tasks on behalf of constrained devices (Bicer et al., 2023). These gateways serve as Zero Trust "proxies" that manage device authentication, gather operational data, and execute local access decisions. A blockchain node can run on these gateways to ensure tamper-resistant logs and identity records (Gai et al., 2023).
3. **Policy Simplification:** ICS zero trust policies often revolve around well-defined device roles and stable operational states (Liu et al., 2024). Complex or dynamic authorization rules are less common. A sensor can have a single function: to send a reading to the control server. Zero trust enforcement ensures the sensor cannot access other devices or issue control commands.

3.3 Threat Landscape in Critical Infrastructure

Critical infrastructure faces unique threats that can disrupt essential services, cause physical damage, or create environmental hazards. Attackers range from cybercriminals seeking ransom to nation-state actors aiming to sabotage or gain geopolitical advantage (Syed et al., 2022). In energy and utilities, attackers can manipulate electricity flow or pipeline pressure, endangering communities and impacting the national economy (TXOne Networks, 2023). Traditional ICS devices often lack modern authentication and encryption measures, and many remain exposed through supply chain vulnerabilities (Kirkman et al., 2024).

Zero Trust helps reduce these threats by limiting potential pivots between ICS devices and by logging attempts to circumvent policy. Blockchain augments such defenses with distributed identity management and verifiable logs, which complicate the attacker's task of covering traces or tampering with data in real-time (Govea et al., 2024). However, ICS complexities and the need for 24/7 uptime demand careful planning to ensure that newly introduced security measures do not impede critical system functions (Cloud Security Alliance, 2024).

4. BLOCKCHAIN SOLUTIONS FOR ICS AND THE ENERGY SECTOR

4.1 Decentralized Identity Management

In ICS, decentralized identity management addresses the single point of compromise in centralized certificate authorities. By using blockchain-based identity registries, each device receives a cryptographic identity stored in an immutable ledger (Bicer et al., 2023). Updates, revocations, or policy changes propagate across the network in near real time, aligning well with Zero Trust's dynamic requirements (Aleisa, 2025). Energy providers can issue unique blockchain identities to power substations, distribution sensors, and consumer meters, improving accountability and tracking (Hajian & Chang, 2023).

Blockchain-based identity management usually involves smart contracts that handle certificate issuance and revocation. When a device connects to the ICS network, the Zero Trust gateway verifies the device's on-chain identity, checks policy compliance, and logs the interaction. These actions reduce the chance of unauthorized device introduction and prevent stale certificates from lingering in the network (Gai et al., 2023).

4.2 Smart Contract-Driven Access Control

Zero Trust requires continuous verification of user, device, or application credentials. Blockchain-based smart contracts offer automated logic to enforce access policies (Shetty et al., 2019). For ICS, these policies can define the permissible commands or data exchanges for each device class. The smart contract checks the identity of the requestor, the requested resource, and the device's operational context (Nie et al., 2025). If the request meets the rules, the contract updates the ledger with a transaction record and returns an authorization token.

In the energy sector, a distribution substation can authenticate to the blockchain ledger and request resource data from a central control hub. The access control smart contract verifies that the substation's private key corresponds to an authorized identity and that the substation is operating within expected load conditions. The system either grants or denies the request (Gai et al., 2023). This process employs minimal trust in any external authority or server. The decentralized ledger and the consistent enforcement logic in the smart contract align with Zero Trust principles (Park et al., 2025).

4.3 Auditing and Compliance in ICS

Regulatory bodies require that utilities maintain detailed logs of all security-related events. Blockchain-based Zero Trust systems generate cryptographically verifiable audit trails. Each authentication request, policy decision, or sensor reading includes a transaction record that is tamper-evident (Aleisa, 2025). Government agencies or independent auditors can retrieve these records from the ledger to determine if the utility complied with required security protocols (CISA, 2023). The transparent nature of blockchain helps demonstrate compliance with standards like NERC CIP (North American Electric Reliability Corporation Critical Infrastructure Protection) in the power sector (Hajian & Chang, 2023).

4.4 Resilience and Scalability Considerations

ICS must operate continuously, so any security mechanism must introduce minimal latency and avoid bottlenecks (Xie et al., 2024). Some blockchain consensus protocols (e.g., proof-of-work) have high computational overhead. Private or permissioned blockchains can adopt more efficient protocols suitable for ICS transaction rates (Azad et al., 2024). Such protocols might rely on smaller validator sets or trust consortia among ICS stakeholders, ensuring lower latency and predictable throughput.

Resilience to node failure is also important for ICS. Decentralized ledgers offer redundancy, so if one node or server fails, others maintain the ledger and preserve security policies (Govea et al., 2024). However, poorly designed decentralized systems can become single points of failure if consensus relies on limited nodes or if the ICS environment lacks robust failover mechanisms (Bicer et al., 2023). Energy systems may require a hybrid approach, where blockchain is used for identity and logging, but time-critical control signals remain on specialized ICS networks.

5. PROPOSED THEORETICAL MODEL FOR BLOCKCHAIN-DRIVEN ZERO TRUST IN ENERGY AND UTILITIES

This section proposes a theoretical model to integrate blockchain-driven Zero Trust within energy and utilities. The model addresses challenges in device identity, policy enforcement, auditing, and operational continuity. It includes identity registration, policy orchestration, smart contract-based access control, and real-time ICS data logging. Figure 1 shows an overview of the proposed architecture in mermaid notation.

Figure 1. High-level architecture of blockchain-driven zero trust

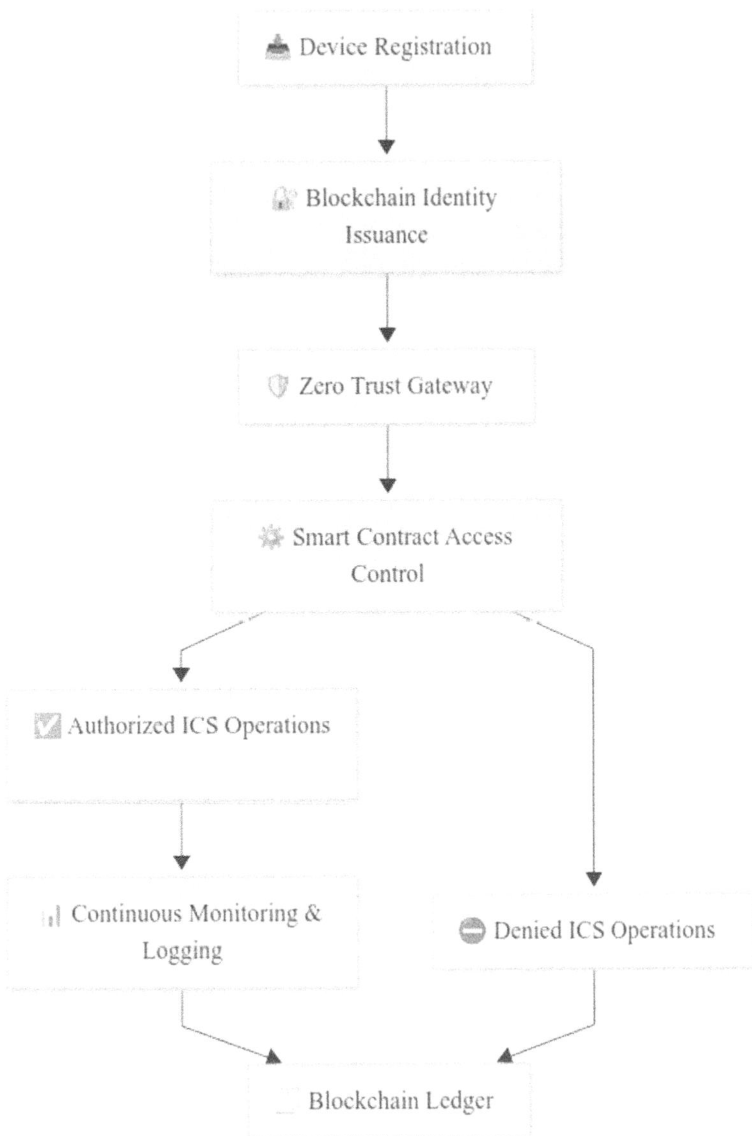

5.1 Identity Registration

All ICS devices must register with a permissioned blockchain network that includes stakeholders such as utility operators, regulators, and trusted third parties (Kirkman et al., 2024). A device manufacturer or a utility administrator initiates the identity registration process by generating cryptographic keys for the device. The public key is stored on the blockchain along with relevant attributes (e.g., device type, operational role, location) (Bicer et al., 2023). The private key remains secure in device hardware or a trusted module.

During registration, a smart contract issues a digital certificate or token that binds the device identity to the public key. This certificate outlines access rights or policies for that device class. The immutability of the ledger ensures that any attempts to modify this identity or certificate leave an on-chain record (Nie et al., 2025).

5.2 Policy Orchestration

In large energy and utilities networks, policies can be complex due to multiple device types, operational states, and regulatory requirements (NIST, 2020). The proposed model uses a Policy Orchestrator component that interacts with the blockchain ledger. Utility administrators define high-level policies (e.g., sensor read-only, control station read/write to certain PLCs), while the orchestrator translates these into specific smart contract rules. These rules govern device-to-device interactions based on identity, context, or real-time conditions (Azad et al., 2024). Table 1 outlines potential policy attributes and their role in Zero Trust enforcement.

Table 1. Example policy attributes in blockchain-driven zero trust

Attribute	Role in Zero Trust Enforcement
Device Identity	Unique identifier (public key, credentials) stored on the blockchain.
Operational Role	Specifies whether the device is a sensor, actuator, controller, etc.
Contextual Data	Contextual info: location, operational state, or time of day.
Risk Level	Dynamic trust score assigned based on device behavior or anomaly detection.
Action	Specific allowed or denied operation (read, write, configure).
Enforcement	Mechanisms (e.g., gateway, smart contract) that apply policy decisions in real-time.

5.3 Smart Contract Access Control

Once policies are defined, the orchestrator deploys or updates the relevant smart contracts on the permissioned blockchain. The Zero Trust Gateway, situated at critical network ingress points, queries these contracts when a device or user requests access to ICS resources (Gai et al., 2023). Smart contract logic verifies the requester's identity, checks the device role and risk score, and consults the policy. If it deems the request valid, it returns a signed token authorizing the session.

This approach ensures consistent enforcement across the ICS. No single administrator or server can override the policy without generating a ledger record. The system also supports dynamic updating of policies, for example in response to a new vulnerability alert (Cloud Security Alliance, 2024). The gateway revokes or updates access tokens if the policy changes, ensuring that compromised devices lose privileges immediately (Syed et al., 2022). **Figure 2.** Shows Sequence Diagram of a Device Access Request and Smart Contract Validation.

Figure 2. Sequence diagram of a device access request and smart contract validation.

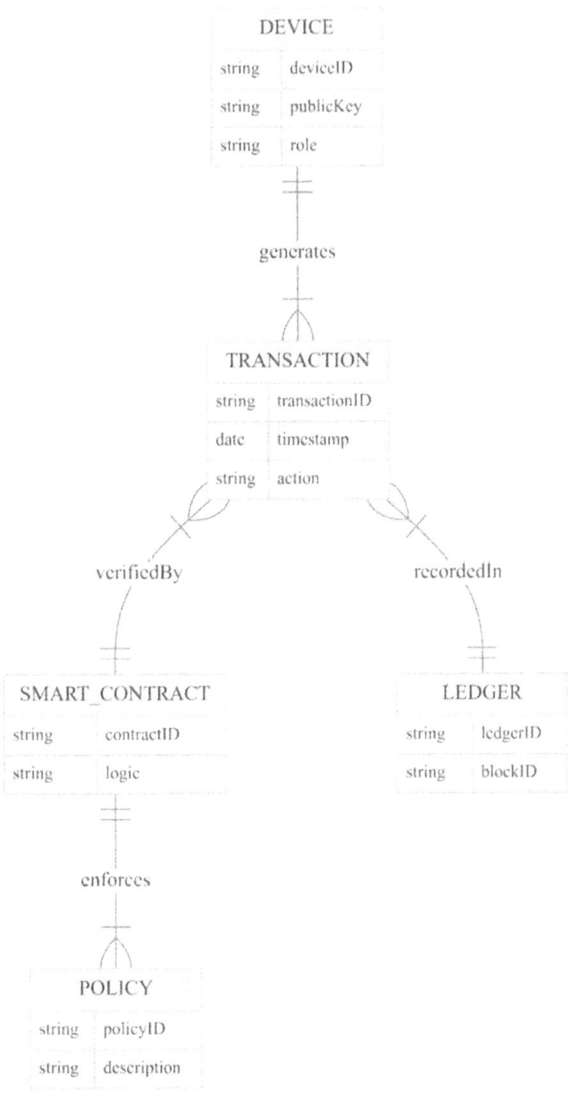

5.4 Continuous Monitoring and Anomaly Detection

The proposed model includes continuous monitoring of ICS traffic and device states. Gateways and other monitoring nodes feed logs of network flows and control commands into a security information and event management (SIEM) system that is integrated with the blockchain (Azad et al., 2024). This system applies anomaly detection algorithms to identify behavior outside expected operational parameters. If suspicious activity appears, the Zero Trust Gateway can adjust risk scores or block requests, and the event is recorded in the blockchain log (Bicer et al., 2023). Table 2 depicts a conceptual mapping of ICS events to potential Zero Trust responses.

Table 2. Mapping of ICS events to zero trust responses

ICS Event	Zero Trust Response
Sensor reading deviates significantly from baseline	Increase device risk score, generate alert, possibly block commands.
Unauthorized request from unknown device	Immediately deny request, log event on blockchain, alert security team.
Repeated failed authentication attempts	Temporarily lock out device/user, escalate to higher-level review.
Valid device requests scheduled commands	Permit transaction, log on blockchain, monitor for anomalies.
Time-based anomaly (e.g., request outside shift)	Require additional authentication or manager approval, log on-chain.

5.5 Auditing and Regulatory Compliance

Due to regulatory frameworks, energy and utilities operators must retain detailed logs of security events, changes to control settings, and device configurations (Cloud Security Alliance, 2024). A blockchain ledger provides a permanent record of these transactions. Auditors can access proof-of-origin data for each event, including the digital signature of the device or user who initiated the action. This approach enhances accountability and meets compliance requirements such as NERC CIP in North America and similar standards elsewhere (Hajian & Chang, 2023).

5.6 Model Implementation Roadmap

Implementing the proposed model requires a gradual approach to reduce disruptions in critical operations (CISA, 2023). Energy and utility operators can pilot the model on a single network segment, such as a small substation or an isolated generation facility. Lessons from that pilot inform scaling to additional sites or entire regions (Azad et al., 2024). The implementation roadmap includes these steps:

1. **Asset Inventory and Classification:** Inventory all ICS devices, assign roles, and establish risk classifications (NSTAC, 2022).
2. **Blockchain Network Setup:** Deploy a permissioned ledger among key stakeholders, including operators and regulators (Xie et al., 2024).
3. **Identity Registration Phase:** Migrate device identities onto the blockchain in stages, ensuring minimal operational impact (Nie et al., 2025).

4. **Smart Contract Development:** Encode Zero Trust policies in smart contracts, validated by regulatory and operational experts (Park et al., 2025).
5. **Integration with Existing Controls:** Connect gateways and SIEM systems to the blockchain. Gradually shift from older authentication systems to the new model (IEEE Standards Association, 2024).
6. **Monitoring and Analytics:** Enable real-time logging of ICS events, anomaly detection, and automated policy adaptation (Liu et al., 2024).
7. **Audit and Compliance Mechanisms:** Test the on-chain auditing processes to verify that records meet regulatory standards (Govea et al., 2024).

6. USE CASES IN THE ENERGY AND UTILITIES SECTOR

6.1 Smart Grid Deployments

Modern smart grids integrate distributed energy resources, smart meters, and remote control devices. These grids require robust mechanisms to prevent data tampering and meter spoofing (Xie et al., 2024). A blockchain-driven Zero Trust model can secure device-to-grid interactions by enforcing continuous validation of meter readings and substation commands (Aleisa, 2025). Smart contracts authenticate requests for energy flow adjustments in near-real time, preventing malicious load injections that could destabilize the grid (Hajian & Chang, 2023).

6.2 Distributed Energy Resources

In many regions, consumers install solar panels or wind turbines connected to the public grid, making them both energy consumers and producers. Traditional security models do not account for the dynamic roles of these distributed energy resources (DERs). Blockchain-based Zero Trust can manage the variable identity and role changes by assigning each DER an immutable identifier and specifying the conditions under which it can inject power into the grid (Park et al., 2025). The distributed ledger documents every power transaction, ensuring fair compensation and preventing fraud (Nie et al., 2025).

6.3 Substation Automation

Substations rely on remote terminal units (RTUs) and intelligent electronic devices (IEDs) for automatic switching and voltage regulation. These devices operate with minimal human intervention and can become critical pivot points if compromised (TXOne Networks, 2023). A Zero Trust approach enforced by a blockchain ledger can authenticate each command or software update sent to substation devices (Kirkman et al., 2024). If an adversary attempts to issue unauthorized commands, the system denies access and logs the event in the ledger.

6.4 Pipeline Monitoring and Control

Oil and gas pipelines stretch across vast territories with many remote pumping and monitoring stations. Traditional security solutions rely on static IP-based restrictions or VPN tunnels, which are vulnerable to insider threats or stolen credentials (Syed et al., 2022). Blockchain-based Zero Trust ensures that each pumping station and sensor must request access from a permissioned ledger, which logs the requests and enforces policies (Bicer et al., 2023). When a station transmits a pressure adjustment command, the ledger verifies that the station is authorized to make that command at the current operational stage.

6.5 Water Treatment Facilities

Water treatment plants use ICS to manage chemical dosing, filtration, and distribution pumps. Attacks on these facilities can lead to contaminated water or service disruptions (Cloud Security Alliance, 2024). Blockchain-based Zero Trust provides identity control for treatment sensors and dosing systems. Smart contracts validate dosing commands from authorized operators, ensuring a single compromised component cannot override safety procedures (Liu et al., 2024).

7. TECHNICAL ARCHITECTURE AND IMPLEMENTATION DETAILS

In this section, we present more detailed considerations for deploying blockchain-driven Zero Trust within ICS. Figure 3 provides a system-level mermaid diagram that includes major components such as a Policy Orchestrator, Zero Trust Gateway, SIEM, and blockchain nodes.

Figure 3. System-level diagram of a blockchain-driven zero trust architecture

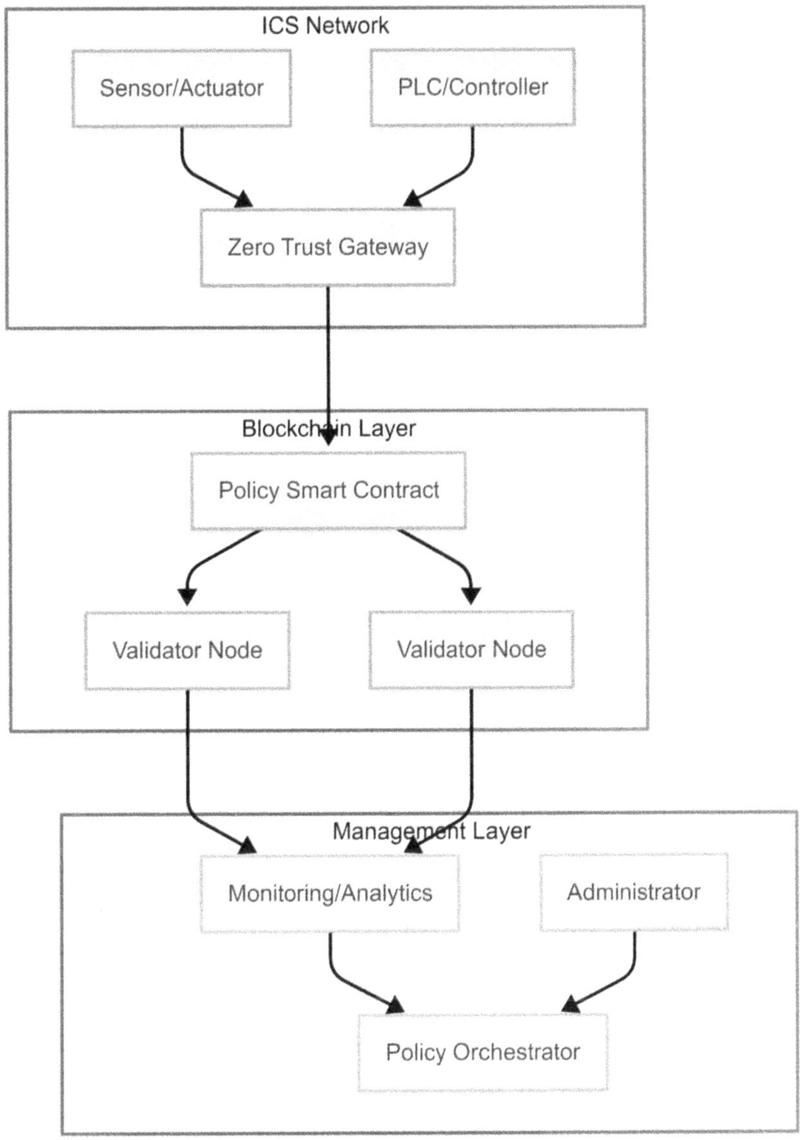

7.1 Blockchain Network Configuration

Utilities and regulators form a consortium to operate permissioned blockchain nodes, limiting membership to trusted entities (IEEE Standards Association, 2024). Consensus algorithms such as Proof of Authority or Raft can offer low-latency block finality. Each node replicates the ledger, ensuring resilience in case of a node outage (Govea et al., 2024). This permissioned setting also streamlines policy updates through a governance process that requires signatures from consortium members (Hajian & Chang, 2023).

7.2 Zero Trust Gateway Functionalities

Gateways act as intermediaries between ICS devices and the blockchain network (Syed et al., 2022). They perform several core tasks:

1. **Authentication and Authorization**: Gateways offload cryptographic validation from resource-constrained devices, verifying device identities against the blockchain ledger (Nie et al., 2025).
2. **Micro-Segmentation Enforcement**: Gateways isolate device communications, denying traffic from unknown or unauthorized devices (Bicer et al., 2023).
3. **Policy Querying**: Gateways query the relevant smart contracts to check if a device can perform a requested action.
4. **Logging**: Gateways log requests, responses, and anomalies on the blockchain.

Table 3 provides an overview of the primary roles involved in a blockchain-driven Zero Trust Architecture, highlighting each role's main responsibilities in maintaining and securing the ICS environment

Table 3. Key architectural roles in blockchain-driven ZTA

Role	Main Responsibilities
Device Operator	Manages field devices, updates firmware, ensures secure key storage
Network Engineer	Configures ICS network, manages gateways, implements segmentation
Policy Administrator	Develops & updates Zero Trust policies, encodes them into smart contracts
Blockchain Validator	Maintains ledger consensus, verifies transactions, ensures on-chain integrity
Auditor	Reviews blockchain logs, checks compliance with regulations

7.3 Policy Orchestrator and Smart Contracts

Administrators use a Policy Orchestrator to define or update high-level policies for device classes, roles, or operational conditions (Xie et al., 2024). This orchestrator translates the policies into smart contract code. The consortium or an authorized operator deploys or updates the contracts through the blockchain governance mechanism (Park et al., 2025). Once active, these contracts automatically enforce the rules for any device request or node-to-node communication (Azad et al., 2024).

A typical policy contract might contain the following pseudo-logic:

```
IF deviceRole = SENSOR AND requestType = "WRITE" THEN

DENY

ELSE IF deviceRole = SENSOR AND requestType = "READ" THEN
```

```
ALLOW

END IF
```

Although ICS environments can have more complex rules, the principle remains the same. The contract checks relevant attributes—device identity, role, request type, operational context—and returns an allow/deny decision (Gai et al., 2023).

7.4 Continuous Monitoring with SIEM Integration

A SIEM tool integrates with both the ICS network logs and the blockchain ledger to perform anomaly detection, risk assessment, and threat intelligence correlation (Aleisa, 2025). When the SIEM detects suspicious behavior, it can automatically adjust a device's risk score by sending a transaction to a designated "Risk Score Contract." That contract can temporarily limit the device's privileges or flag it for manual review (Liu et al., 2024). The SIEM can also alert administrators, who may implement further mitigations through the Policy Orchestrator (CISA, 2023).

7.5 Handling Legacy Devices

Legacy ICS devices lack secure boot, cryptographic support, or modern firmware. Replacing them might be too costly or technically unfeasible. A blockchain-driven Zero Trust architecture can incorporate secure gateway modules or retrofitted hardware tokens that store cryptographic keys on behalf of these devices (Govea et al., 2024). The gateway emulates a secure identity layer so that the device can appear as a "trusted" entity on the ledger, even though the device itself has not changed. If full cryptographic integration is impossible, the Zero Trust model can still monitor device traffic and limit communications (Kirkman et al., 2024).

7.6 Incident Response and Forensics

When a security incident occurs, blockchain logs provide an immutable record of device interactions and policy decisions (Azad et al., 2024). Investigators can trace back suspicious activities to specific device identities or user accounts. They can also verify if any policy modifications coincided with the breach (Hajian & Chang, 2023). Because the ledger is decentralized, attackers find it difficult to erase evidence or alter logs. This capacity for in-depth forensics aligns with the critical nature of ICS environments (Park et al., 2025).

7.7 Scalability Analysis

Energy utilities may have tens of thousands of devices distributed across wide geographic areas (TXOne Networks, 2023). Each device may generate multiple access requests per minute during peak periods. The blockchain network must handle these transactions without introducing operational delays. Permissioned blockchain architectures with efficient consensus can sustain high transaction throughput, especially if lightweight transactions are used for policy checks (Gai et al., 2023). Additional scaling approaches include off-chain caching of frequently accessed data or hierarchical blockchains that partition the system by region. Table 4 summarizes some recommended scalability strategies and their pros and cons.

Table 4. Scalability strategies and trade-offs for large ICS environments.

Strategy	Pros	Cons
Permissioned Blockchain	Faster consensus, controlled membership	Requires consortium governance, trust in validator nodes
Efficient Consensus (Raft)	Low latency, high throughput	Not as decentralized as PoW/PoS, limited fault tolerance
Off-Chain Data Storage	Reduces on-chain load	Adds architectural complexity, potential single point failure
Sidechains / Layer-2	Offloads main chain, allows parallel processing	More complex to implement, risk of cross-chain vulnerabilities
Hierarchical Chain Partitions	Scales across multiple regional chains	Coordination overhead, potential fragmentation of data

8. CHALLENGES IN ADOPTING BLOCKCHAIN-DRIVEN ZERO TRUST

8.1 Organizational and Cultural Barriers

Adopting Zero Trust can conflict with existing ICS operational cultures that favor simplicity and minimal network segmentation (NIST, 2020). Engineers may resist new security controls that risk interrupting production or require new skill sets (Govea et al., 2024). Blockchain technology also introduces governance models that differ from centralized infrastructure management. Organizations need to form trust consortia with regulators and third parties, which can be challenging (Cloud Security Alliance, 2024).

8.2 Regulatory Constraints and Legal Frameworks

Global ICS operators face different legal frameworks for data privacy, national security, and cross-border data sharing. Permissioned blockchains must address concerns about data jurisdiction and the legal enforceability of smart contracts (Azad et al., 2024). Some regulations may require data localization, complicating distributed ledger deployments. Additionally, the legally binding status of records stored on a blockchain may vary among jurisdictions (NSTAC, 2022).

8.3 Technical Complexity and Integration

Energy grids and utilities often contain legacy systems with proprietary protocols, making integration with modern blockchain solutions a complex endeavor (Syed et al., 2022). Engineers must integrate cryptographic processes within devices that lack computing resources. Even with gateway solutions, complexities remain in bridging ICS protocols, building data transformations, and configuring real-time event streaming to the blockchain (Kirkman et al., 2024).

8.4 Scalability and Performance

While permissioned blockchains can offer better throughput than public ones, large-scale ICS operations may still pose performance challenges (Bicer et al., 2023). High rates of sensor data, frequent control commands, and real-time operational constraints demand careful optimization. Delays in transaction finality or contract execution could create bottlenecks or safety hazards (Xie et al., 2024).

8.5 Privacy Concerns

Although blockchain offers decentralization and immutability, it also creates permanent records that may reveal sensitive operational data or personal information (e.g., household energy usage) (Almarri & Aljughaiman, 2024). Zero Trust frameworks must balance transparency with privacy requirements by restricting what data is stored on-chain. Privacy-preserving techniques like zero-knowledge proofs and homomorphic encryption can mitigate these concerns, but they add complexity (Nie et al., 2025).

8.6 Potential Single Points of Failure

If the consortium operating the permissioned blockchain is not sufficiently decentralized, the entire security model may hinge on a small group of validator nodes (Gai et al., 2023). Attackers who compromise these nodes can falsify ledger entries or override smart contracts. Similarly, if gateways represent a choke point, then a denial-of-service attack against them can disrupt ICS operations (Liu et al., 2024). Designing robust fault tolerance and redundancy is critical (CISA, 2023).

9. PRACTICAL CASE EXAMPLE: ENERGY DISTRIBUTION COMPANY

Consider an energy distribution company that manages a regional power grid with multiple substations and thousands of smart meters. The company's priority is to prevent unauthorized configuration changes at substations and to ensure that meter data is accurate for billing and grid-balancing.

1. **Initial Setup**: The company partners with regulators to create a permissioned blockchain. Trusted parties in the region operate validator nodes. A designated Policy Orchestrator encodes zero trust rules in the chain's smart contracts (Bicer et al., 2023).
2. **Device Registration**: Each substation and meter is registered. Substations receive read-write privileges for certain actions, while meters obtain read-only privileges for usage data. The on-chain record includes risk scores and device owners (Nie et al., 2025).
3. **Real-Time Operations**: Substation controllers communicate with the blockchain gateway when adjusting voltage levels. The gateway verifies that the substation's key is valid and that the requested operation aligns with the substation's assigned role. The command is logged on-chain (Azad et al., 2024).
4. **Meter Data Collection**: Meters periodically transmit data. The gateway checks meter identities and updates a ledger record. If an anomaly, such as an abrupt spike in readings, arises, the system increases the meter's risk score and flags it for inspection (Cloud Security Alliance, 2024).

5. **Regulatory Audits**: Regulators query the blockchain to verify that the company followed required procedures, such as scheduling maintenance commands only during designated windows. The immutability of the ledger streamlines compliance review (Kirkman et al., 2024).

In this scenario, the blockchain-driven Zero Trust model fosters secure device interactions, robust auditability, and real-time threat detection. It also highlights the importance of well-defined policies, strong governance, and resilient architecture.

10. LESSONS LEARNED AND BEST PRACTICES

10.1 Emphasize Incremental Deployment

Large ICS transformations can disrupt operations. It is best to begin with controlled pilots or testbeds that evaluate how blockchain-driven Zero Trust affects network performance and production. Pilot successes build internal and external trust in the approach (CISA, 2023).

10.2 Collaborate with Regulatory Bodies

Regulatory acceptance of blockchain-based logs or smart contract enforcement is crucial. Early engagement with regulators can clarify legal ambiguities and ensure that new security methods align with compliance requirements (NIST, 2020). This collaboration can reduce later friction and speed adoption (NSTAC, 2022).

10.3 Maintain Operational Resilience

Zero Trust security must not impair ICS continuity. Design fallback mechanisms and ensure that permissioned blockchain nodes are geographically redundant. Implement caching or sidechains for real-time operations if needed (Govea et al., 2024). Fail-safe modes must exist if the blockchain network becomes unavailable (Hajian & Chang, 2023).

10.4 Manage Node Identities and Keys Securely

Blockchain credentials are powerful. An attacker with access to a device's private key can impersonate that device. Organizations must store keys in secure hardware modules and rotate them regularly. If a device is compromised, its identity must be revoked on the ledger quickly (Nie et al., 2025).

10.5 Address Legacy Devices Proactively

Plan for integrating or isolating legacy devices early. Deploy gateways or hardware tokens that can interact with the blockchain on behalf of older devices. Identify particularly vulnerable systems and isolate them using micro-segmentation (Syed et al., 2022).

11. FUTURE RESEARCH DIRECTIONS

11.1 Advanced Cryptographic Approaches

Further research could explore privacy-preserving techniques like ring signatures, zero-knowledge proofs, and homomorphic encryption in ICS. These methods could allow a device to prove a property (e.g., operational state) without revealing all data. This could help protect sensitive infrastructure information while maintaining Zero Trust verification (Almarri & Aljughaiman, 2024).

11.2 AI-Driven Policy Adaptation

Artificial intelligence systems can optimize Zero Trust policies by analyzing device behavior patterns and predicting risks (Liu et al., 2024). Over time, machine learning models can fine-tune access rules, recommending dynamic changes to role definitions, device privileges, or risk thresholds. However, ensuring transparency in AI-driven policy updates remains a challenge (Azad et al., 2024).

11.3 Interoperability Standards

Future efforts could focus on standardizing blockchain-driven Zero Trust protocols for ICS. While IEEE Std 3219-2023 addresses IoT-specific concerns, the ICS domain might need additional protocols that account for real-time operational constraints, substation automation, or advanced manufacturing. Interoperability standards would simplify vendor adoption (IEEE Standards Association, 2024).

11.4 Performance Benchmarking

Researchers and industry bodies can collaborate to develop benchmarking suites that measure transaction throughput, latency, and overhead in blockchain-driven ICS. These metrics would help operators estimate the impact of real-world deployments and inform design choices such as the consensus algorithm or node hardware (Xie et al., 2024).

11.5 Incident Response Frameworks

Although blockchain-driven Zero Trust supports better traceability, specific incident response frameworks are needed to guide ICS operators. These frameworks should detail how to quarantine suspicious devices, revoke on-chain credentials, and roll back malicious transactions without halting production. Coordination between operational staff, security teams, and external agencies remains essential (CISA, 2023).

12. CONCLUSION

Critical infrastructure sectors require a security paradigm that addresses internal threats, advanced persistent threats, and the need for strong auditability. Zero Trust has emerged as a model that can meet these demands by removing implicit trust and enforcing continuous verification of all requests. Blockchain technology strengthens Zero Trust's core principles by decentralizing identity management, embedding tamper-evident audit trails, and automating policy enforcement via smart contracts.

This chapter examined how blockchain-driven Zero Trust can address the unique requirements of ICS in the energy and utilities domain. The proposed theoretical framework outlined a permissioned blockchain network supporting device registration, policy orchestration, continuous monitoring, and compliance logging. We discussed how pilot deployments can minimize operational disruptions and how the approach scales to large ICS ecosystems. Practical use cases—spanning smart grids, distributed energy resources, and pipeline monitoring—demonstrate blockchain's viability as a security cornerstone in critical infrastructure.

Challenges remain in areas such as regulatory acceptance, technical integration with legacy systems, performance optimization, and privacy safeguards. However, the potential benefits of a decentralized, immutable, and real-time enforceable trust model position blockchain-driven Zero Trust as a promising strategy to protect critical infrastructure from evolving cyber threats. Future research can build on the concepts presented here by refining cryptographic approaches, developing interoperability standards, and expanding AI-driven policy adaptation techniques.

REFERENCES

Aleisa, M. A. (2025). Blockchain-enabled zero trust architecture for privacy-preserving cybersecurity in IoT environments. *IEEE Access : Practical Innovations, Open Solutions*, *13*, 18660–18676. Advance online publication. DOI: 10.1109/ACCESS.2025.3529309

Almarri, S., & Aljughaiman, A. (2024). Blockchain technology for IoT security and trust: A comprehensive SLR. *Sustainability (Basel)*, *16*(23), 10177. DOI: 10.3390/su162310177

Azad, M. A., Abdullah, S., Arshad, J., Lallie, H., & Ahmed, Y. (2024). Verify and trust: A multidimensional survey of zero-trust security in the age of IoT. *Internet of Things : Engineering Cyber Physical Human Systems*, *27*, 101227. DOI: 10.1016/j.iot.2024.101227

Bicer, C., Murturi, I., Donta, P. K., & Dustdar, S. (2023). Blockchain-based zero trust on the edge. In *Proceedings of the 2023 International Conference on Computational Science and Computational Intelligence (CSCI)* (pp. 1006–1013). IEEE. https://doi.org/DOI: 10.1109/CSCI62032.2023.00167

Cloud Security Alliance. (2024). *Zero Trust Guidance for Critical Infrastructure*. Cloud Security Alliance.

Cybersecurity and Infrastructure Security Agency (CISA). (2023). *Zero Trust Maturity Model Version 2.0*. U.S. Department of Homeland Security.

Gai, K., She, Y., Zhu, L., Choo, K. K. R., & Wan, Z. (2023). A blockchain-based access control scheme for zero trust cross-organizational data sharing. *ACM Transactions on Internet Technology*, *23*(3), 1–25. DOI: 10.1145/3511899

Govea, J., Gaibor-Naranjo, W., & Villegas-Ch, W. (2024). Securing critical infrastructure with blockchain technology: An approach to cyber-resilience. *Computers*, *13*(5), 122. DOI: 10.3390/computers13050122

Hajian, A., & Chang, H.-C. (2023). A blockchain-based smart grid to build resilience through zero-trust cybersecurity. In Ayyub, B. M., Wegrzyn, K., & Shah, V. P. (Eds.), *Handbook of Smart Energy Systems* (pp. 1–19). Springer., DOI: 10.1007/978-3-030-97940-9_161

IEEE Standards Association. (2024). *IEEE Std 3219-2023: Standard for Blockchain-Based Zero-Trust Framework for the Internet of Things (IoT)*. IEEE.

Kindervag, J. (2010). *No More Chewy Center: Introducing the Zero Trust Model of Information Security*. Forrester Research.

Kirkman, S. S., Fulton, S., Hemmes, J., Garcia, C., & Wilson, J. C. (2024). A blockchain architecture to increase the resilience of industrial control systems from the effects of a ransomware attack: A proposal and initial results. *ACM Transactions on Cyber-Physical Systems*, *8*(1), 1–13. DOI: 10.1145/3637553

Liu, C., Tan, R., Wu, Y., Feng, Y., Jin, Z., Zhang, F., Liu, Y., & Liu, Q. (2024). Dissecting zero trust: Research landscape and its implementation in IoT. *Cybersecurity*, *7*(1), 20. Advance online publication. DOI: 10.1186/s42400-024-00212-0

National Security Telecommunications Advisory Committee (NSTAC). (2022). *Report to the President on Zero Trust and Trusted Identity Management*. U.S. DHS.

Nie, S., Ren, J., Wu, R., Han, P., Han, Z., & Wan, W. (2025). Zero-trust access control mechanism based on blockchain and inner-product encryption in the Internet of Things in a 6G environment. *Sensors (Basel), 25*(2), 550. DOI: 10.3390/s25020550 PMID: 39860920

Park, J.-H., Park, S., & Youm, H. (2025). A proposal for a zero-trust-based multi-level security model and its security controls. *Applied Sciences (Basel, Switzerland), 15*(2), 785. DOI: 10.3390/app15020785

Rose, S., Borchert, O., Mitchell, S., & Connelly, S. (2020). Zero trust architecture (SP 800–207). National Institute of Standards and Technology.

Shetty, S. S., Kamhoua, C. A., & Njilla, L. L. (Eds.). (2019). *Blockchain for Distributed Systems Security*. Wiley-IEEE Press. DOI: 10.1002/9781119519621

Syed, N. F., Shah, S. W., Shaghaghi, A., Anwar, A., Baig, Z., & Doss, R. (2022). Zero Trust Architecture (ZTA): A comprehensive survey. *IEEE Access : Practical Innovations, Open Solutions, 10*, 57143–57179. DOI: 10.1109/ACCESS.2022.3174679

TXOne Networks. (2023). *Insights Into ICS/OT Cybersecurity 2022*. TXOne Networks.

Xie, H., Wang, Y., Ding, Y., Yang, C., Liang, H., & Qin, B. (2024). Industrial wireless internet zero trust model: Zero trust meets dynamic federated learning with blockchain. *IEEE Wireless Communications, 31*(2), 22–29. DOI: 10.1109/MWC.001.2300368

Chapter 6
Blockchain for Secure Communication

Akhila Reddy Yadulla
https://orcid.org/0009-0005-6377-6265
University of the Cumberlands, USA

Bhargavi Konda
https://orcid.org/0009-0001-4704-1526
University of the Cumberlands, USA

Vinay Kumar Kasula
https://orcid.org/0009-0001-1131-3059
University of the Cumberlands, USA

ABSTRACT

Blockchain technology, originally associated with digital currencies, possesses key features such as decentralization, anonymity, robustness, and resistance to tampering, making it an ideal platform for building covert communication channels. This chapter begins by introducing the concept of network covert channels, discussing their definition, historical development, and the architecture of blockchain technology with an exploration of traditional network covert channels. It then highlights the advantages of blockchain-based covert communication channels over conventional methods. The chapter classifies blockchain network covert channels based on several key components. It provides a detailed analysis of the advantages and disadvantages of these channels in terms of covert nature, transmission efficiency, and communication costs. Finally, the chapter addresses existing challenges and limitations within blockchain network covert channels and offers insights into future research directions to improve their efficiency, security, and scalability in the context of secure communication.

INTRODUCTION

With the rapid evolution of modern information technology, ensuring secure information transmission has become a critical concern. Advanced cryptographic techniques, such as quantum-resistant cryptography (Zander, Armitage, & Branch, 2007) and homomorphism encryption (Nakamoto, 2008), have

DOI: 10.4018/979-8-3373-2439-5.ch006

emerged to safeguard plaintext data against malicious interception and decryption. However, the mere detection of communication activities can often signal the exchange of sensitive information, particularly in adversarial contexts. To address this issue, researchers have focused on constructing covert channels, which enable undetectable communication by embedding information within seemingly innocuous carriers (Wendzel et al., 2015).

Covert communication involves the concealment of sensitive information in openly transmitted data carriers, ensuring that third parties cannot detect the communication process itself.

Initially, covert channels were primarily explored within localized systems, enabling secure data transfer between processes of different security levels (Wang et al., 2020). With the rise of technologies such as 5G, IoT (Internet of Things), and AI-driven networks, the landscape of covert channels has significantly evolved. Modern networks now provide vast high-bandwidth data carriers; for instance, a large-scale IoT ecosystem can produce terabytes of potential carrier data annually (Ding et al., 2019). However, traditional centralized covert communication methods face several challenges:

- Trust Deficits: Centralized communication carriers, often managed by third parties, cannot ensure anonymity or trustworthiness (Li et al., 2022).
- Resilience Issues: Centralized systems are highly susceptible to distributed denial-of-service (DDoS) attacks, which can disrupt communication (Petitcolas, Anderson, & Kuhn, 1999).
- Predictability: Traditional covert channels often rely on static pathways, making them vulnerable to detection by AI-enhanced threat monitoring systems (Lampson, 1973).

Emerging technologies such as blockchain, federated learning, and zero-knowledge proofs offer robust alternatives. Blockchain, initially introduced as the underlying mechanism for Bitcoin (Simmons, 1984), brings inherent advantages such as decentralization, anonymity, immutability, and resilience against tampering. These characteristics make blockchain an ideal foundation for constructing covert channels and addressing the limitations of traditional approaches. Despite the potential of blockchain and other modern technologies, existing research lacks comprehensive analyses and structured methodologies for implementing covert channels in decentralized environments. Current efforts are fragmented, with limited exploration into practical frameworks and applications (Craver, 1998).

This chapter aims to bridge this gap by exploring the integration of blockchain technology and other advanced approaches in the design of covert channels, providing actionable insights for researchers. Specifically:

- Exploration of Emerging Technologies: This study examines how blockchain, AI-driven obfuscation, and edge computing can enhance covert channel construction, emphasizing their benefits.
- Categorization of Blockchain-Based Covert Channels: Blockchain-based covert channels are classified into categories such as smart contract-based, P2P broadcast mechanism-based, and quantum-encrypted transaction channels (Millen, 1999).
- Definition of Construction Stages: The construction process is delineated into four key stages: quantum-resilient information embedding, AI-enhanced carrier obfuscation, adaptive carrier selection, and secure information decoding (Handel & Sandford II, 1996).
- Identification of Challenges and Solutions: This study identifies limitations in current blockchain-enabled covert channels and proposes innovative solutions leveraging technologies such as quan-

tum cryptography, decentralized identity systems (DIDs), and privacy-preserving AI models (Cauich, Gomez, & Watanabe, 2005)(Girling, 1987).

By advancing these concepts, this study seeks to establish a foundation for secure, efficient covert communication channels in decentralized ecosystems, addressing the demands of modern cybersecurity challenges.

Figure 1. Blockchain architecture

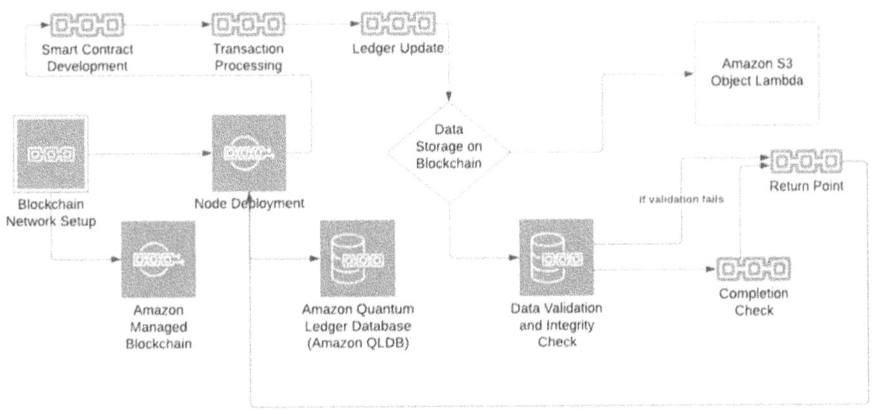

RELATED CONCEPTS AND EVOLUTION OF COVERT CHANNELS IN NETWORKS

Blockchain Technology

Blockchain technology represents a transformative decentralized architecture that facilitates secure, transparent, and immutable data management through a distributed ledger system. Unlike traditional centralized databases, blockchain operates on a peer-to-peer (P2P) network, where multiple nodes collectively validate, record, and synchronize transactions. The integrity of the blockchain is maintained through consensus mechanisms, which ensure that all participating nodes agree on the validity of transactions before they are permanently recorded (Zander, Armitage, & Branch, 2007). A fundamental characteristic of blockchain is its cryptographic security, which safeguards data transmission, access, and storage against unauthorized modifications. Transactions are grouped into blocks, each containing a unique cryptographic hash linking it to the previous block, forming an immutable chain of records. This structure enhances tamper resistance, ensuring that altering any block requires modifications across the entire chain—a computationally infeasible task that reinforces data security (Nakamoto, 2008).In addition to data integrity and security, blockchain technology supports the autonomous execution of predefined rules through smart contracts. These self-executing contracts automatically enforce agreements once predefined conditions are met, eliminating the need for intermediaries and enabling trustless transactions.

The automation provided by smart contracts enhances efficiency, transparency, and reliability in various applications, ranging from financial services to supply chain management (Wendzel et al., 2015).

Blockchain's capabilities extend far beyond its initial application in cryptocurrency. Recent advancements have led to its integration into diverse industries, including:

- Supply Chain Management: Enhancing traceability and accountability across global supply networks.
- Healthcare: Securing patient records and medical data sharing while ensuring regulatory compliance.
- Digital Identity: Enabling secure, decentralized identity management systems, reducing risks of identity theft and fraud.
- Secure Communications: Facilitating blockchain-based covert channels, which enhance privacy by enabling hidden data transmission (Wang et al., 2020)

The distributed, secure, and transparent nature of blockchain technology is illustrated in Figure 1, depicting its architecture, consensus mechanisms, cryptographic security, and decentralized structure. As blockchain continues to evolve, its applications in secure data exchange, automation, and trusted communication are expected to expand, revolutionizing the digital ecosystem.

Definition and Historical Context of Network Covert Channels

Covert channels, originally conceptualized as a subfield of information hiding, were first formally defined by Lampson in 1973 (Wendzel et al., 2015). He described them as communication pathways not primarily intended for data transmission, distinguishing them from conventional methods of secure communication. Unlike steganography, anonymous communication, or digital watermarking, which focuses on concealing message content or sender identities, covert channels emphasize the concealment of the very existence of communication itself, enabling secret data transfer without detection.Over time, the concept of covert channels has evolved, driven by theoretical advancements and real-world applications in cybersecurity, secure networking, and digital forensics. A pivotal development occurred with Simmons' "prisoners' problem" in 1984 (Wang et al., 2020), which illustrated a classic scenario of covert communication. In this model, two prisoners, Alice and Bob, attempt to communicate secretly while being monitored by a warden (Wendy). Their goal is to devise a communication method that avoids detection, ensuring their escape plan remains hidden.Expanding upon this foundational model, Craver introduced a classification of wardens (Ding et al., 2019), which has become instrumental in analyzing covert communication strategies:

- Passive Wardens: These wardens can observe covert channels and potentially detect their presence but lack the ability to modify or disrupt the message.
- Active Wardens: These wardens actively manipulate communication, such as replacing synonyms or encoding methods, while still preserving the intended meaning of messages.
- Malicious Wardens: These wardens exert full control over communication channels, allowing them to arbitrarily modify, distort, or eliminate messages to prevent successful covert transmission.

Building on these theoretical foundations, Millen (1987) extended covert channel analysis to computer networks, introducing the concept of network covert channels (Li et al., 2022). These channels leverage various network-layer mechanisms—such as packet timing, protocol manipulation, or traffic pattern alterations—to embed and transmit covert information without arousing suspicion. The ability of covert channels to exploit network vulnerabilities for hidden communication has significant implications for cybersecurity, influencing research on data exfiltration, intrusion detection, and information leakage prevention.As technological advancements continue, covert channel techniques are becoming increasingly sophisticated, incorporating machine learning, blockchain, and quantum cryptography to evade detection. Consequently, understanding their operational principles and countermeasures remains a critical area of study in information security and network defense strategies.

Evolution of Network Covert Channels

The evolution of covert channels has been marked by significant advancements, particularly with the emergence of network covert channels. Initially, covert channels were primarily limited to local systems, facilitating secure communication between processes operating at different security levels. These early covert channels functioned within single-host environments, often relying on shared system resources such as CPU scheduling, memory allocation, or file system metadata to facilitate hidden data exchange. However, their effectiveness was constrained by limited scalability and restricted communication range.With the advent of computer networking, the prisoners' problem—a theoretical model for covert communication—was extended into distributed environments. Researchers like Handel et al. (Handel & Sandford II, 1996) explored the potential of network-based covert channels, adapting classical covert communication methods for use in multi-node networks. This advancement enabled covert data transmission across geographically distributed systems, significantly broadening the scope of covert communication techniques.

1. Network-Based Covert Channels: Storage vs. Timing Methods

The initial forms of network covert channels were primarily storage-based, embedding secret data within unused or optional fields of network protocol headers. Early implementations encoded covert information in protocol fields such as IP headers, TCP sequence numbers, or packet payloads (Simmons, 1984)(Craver, 1998)(Millen, 1999)(Handel & Sandford II, 1996). While this approach provided a direct and structured means of covert communication, it suffered from high detectability, as security analysts could easily identify anomalies in protocol structures due to fixed communication patterns and non-standard field utilization.To counteract these detection challenges, researchers developed timing-based covert channels. Instead of modifying protocol headers, these channels encoded information within the timing characteristics of network packets (Cauich, Gomez, & Watanabe, 2005)(Girling, 1987)(Rowland, 1997). This method relied on intentional variations in packet transmission intervals, where the sender could encode data using specific delays, jitter patterns, or burst transmissions. Since timing-based covert channels did not introduce explicit modifications to network packet structures, they were inherently more resilient to traditional detection mechanisms. However, these methods faced limitations due to network noise, latency fluctuations, and environmental inconsistencies, which could introduce errors in covert message transmission.

2. Blockchain-Based Covert Channels: A New Paradigm

A groundbreaking advancement in covert communication emerged in 2018 when Partala (Partala, 2018)introduced the concept of blockchain-based covert channels. This innovation leveraged decentralized ledger technology to establish secure, persistent, and censorship-resistant covert communication pathways. Unlike traditional network-based covert channels, which relied on protocol manipulation or timing distortions, blockchain-based covert channels embedded covert messages within blockchain transactions, smart contracts, or metadata fields. The immutable nature of blockchain records ensured that once covert data was embedded, it became tamper-proof and permanently accessible to authorized recipients.

- Blockchain-based covert channels present multiple advantages over conventional covert communication methods:
- Enhanced Security: The cryptographic foundations of blockchain make data interception and modification significantly more challenging.
- Decentralization: Unlike centralized communication channels, blockchain-based methods operate without reliance on specific network nodes, reducing the risk of detection.
- Anonymity and Censorship Resistance: By integrating covert messages within blockchain transactions, users can circumvent traditional surveillance mechanisms, ensuring undetectable communication.

This novel approach has opened new possibilities for covert communication in secure and decentralized environments, enabling highly resilient information exchange mechanisms that extend beyond traditional network constraints. As blockchain technology continues to evolve, its integration with advanced covert communication techniques is expected to drive further innovations in covert networking, secure messaging, and cyber-espionage countermeasures.

Traditional and Emerging Network Covert Channels

Traditional network covert channels are classified based on their data encoding strategies, which determine how hidden information is embedded within network traffic. These covert communication methods exploit different aspects of network protocols to achieve concealment while minimizing detection. The three primary classifications are as follows:

1. Storage-Based Covert Channels

Storage-based covert channels embed hidden information within the structure or content of protocol data units (PDUs). This is achieved by modifying field lengths, inserting data into unused fields, or repurposing protocol-specific headers (Wu et al., 2015)(Partala, 2018)(Liu et al., 2021). These channels often utilize network protocols such as IP, TCP, and UDP to covertly store and transmit hidden information. Examples: Covert encoding within TCP sequence numbers, IP identification fields, or optional header fields.Limitations: Despite their effectiveness, storage-based covert channels are highly susceptible to detection, particularly when transmitting large amounts of data. Security systems can analyze packet sizes, header inconsistencies, or statistical deviations to identify anomalies.

2. Timing-Based Covert Channels

Timing-based covert channels manipulate the timing intervals of packet transmissions or exploit variations in network delays to encode covert information (Epishkina & Kogos, 2015)(Mazurczyk & Szczypiorski, 2012)(Rios, Onieva, & Lopez, 2012). These channels are inherently protocol-agnostic, as they do not modify packet contents but rather encode information using timing patterns between packet exchanges.Examples: Encoding data through variable inter-packet delays, altering the rate of packet transmission, or exploiting network congestion to create detectable timing sequences.Limitations: These channels require precise synchronization between sender and receiver. They are also highly sensitive to network conditions, making them vulnerable to latency fluctuations, jitter, and packet loss.

3. Hybrid Covert Channels

Hybrid covert channels integrate elements of both storage- and timing-based covert communication techniques to enhance transmission efficiency and mitigate detection risks (Zou et al., 2005)(Patuck & Hernandez-Castro, 2013). By distributing information across both spatial and temporal dimensions, hybrid covert channels increase data capacity while reducing statistical anomalies that may trigger security alerts.Advantages: Increased stealth: A balanced approach between packet structure modifications and timing-based encoding makes detection more challenging.Higher transmission efficiency: By embedding data in both header fields and timing patterns, hybrid covert channels optimize bandwidth utilization.

Limitations of Traditional Covert Channels

Despite their diverse implementations, traditional network covert channels exhibit several inherent weaknesses that limit their resilience, anonymity, and long-term effectiveness. The primary challenges include:

1. Centralization Issues

Many traditional covert communication methods rely on centralized systems, such as web browsers, email servers, or social media platforms, to facilitate hidden data exchanges. This dependence introduces single points of failure, making detection and disruption by centralized authorities significantly easier (Wendzel, Kahler, & Rist, 2012).Example: Web-based covert channels that rely on JavaScript execution or HTTP metadata manipulation can be easily mitigated by browser security policies.

2. Anonymity Risks

Traditional covert channels often depend on third-party digital certificates and authentication mechanisms for secure data exchanges. However, this reliance can compromise user anonymity, as these certificates are often traceable to specific users, organizations, or network identities (Giffifin et al., 2020). Example: Encrypted covert channels using TLS-based traffic tunneling may still expose identifiable metadata, such as certificate fingerprints or session keys.

3. Resilience and Security Concerns

The reliance on direct peer-to-peer communication in traditional covert channels creates single points of failure, making them vulnerable to denial-of-service (DDoS) attacks, network monitoring, and traffic analysis techniques (Lucena et al., 2004). Attackers can disrupt covert communication by injecting noise, modifying packet timings, or performing protocol fingerprinting. Example: Covert channels that rely on predictable encoding schemes (e.g., fixed inter-packet delays) can be disrupted by introducing artificial network congestion to distort transmission patterns.

Advantages of Blockchain-based Covert Channels

Traditional Covert Communication Techniques face significant challenges related to centralization, anonymity, security, and resilience. However, blockchain technology offers an innovative solution to these limitations, leveraging its decentralized, immutable, and cryptographically secure framework. The integration of blockchain-based covert channels presents multiple advantages over conventional methods, enhancing privacy, robustness, and stealth in covert communication. The primary benefits of blockchain technology in addressing the limitations of traditional covert channels are outlined as follows:

1. Decentralization

Unlike conventional covert channels, which often rely on centralized intermediaries such as web servers, browsers, or third-party authentication systems, blockchain operates on a decentralized peer-to-peer (P2P) architecture.

- Each node within a blockchain network functions simultaneously as a sender, receiver, and verifier, ensuring that data can be relayed across multiple pathways without relying on a single point of failure.
- Decentralized transaction validation mechanisms (e.g., Proof-of-Work, Proof-of-Stake) provide a built-in structure for covert communication by distributing message fragments across the blockchain (Muchene, Luli, & Shue, 2013).
- The absence of central control prevents surveillance authorities from easily detecting or censoring covert messages, making blockchain-based covert channels more resilient than traditional approaches.

2. Anonymity and Privacy Protection

One of the most significant advantages of blockchain-based covert communication is its enhanced anonymity features. Unlike traditional communication protocols that require user authentication, blockchain networks use pseudonymous cryptographic addresses rather than personally identifiable information (PII).

- Transactions and messages in a blockchain are associated with public-private key pairs, ensuring that senders and receivers remain untraceable without requiring trusted third parties (Tahir et al., 2016).
- The absence of centralized identity verification reduces the risk of user de-anonymization, which is a critical limitation of traditional covert channels.

- Advanced privacy-preserving techniques, such as zero-knowledge proofs (ZKPs), ring signatures, and stealth addresses, further enhance anonymity in blockchain-based covert messaging systems.

3. Immutability and Data Integrity

Blockchain's immutable ledger structure guarantees that covert messages cannot be altered, forged, or deleted once recorded. This cryptographic assurance addresses data integrity challenges commonly associated with traditional covert channels.

- Each transaction or message embedded in the blockchain is hashed and linked to the previous block, making modifications practically impossible without altering the entire chain (El-Atawy, Duan, & Al-Shaer, 2017).
- Cryptographic techniques, such as Merkle trees and digital signatures, ensure that the integrity of the data remains intact, preventing unauthorized tampering.
- In the context of covert communication, immutability ensures that hidden messages persist indefinitely within the blockchain, providing a reliable mechanism for long-term storage and retrieval.

4. Robustness Against Network Attacks

Blockchain-based covert channels offer superior resilience compared to traditional covert communication methods, which often fail under network fluctuations, surveillance countermeasures, or cyberattacks.

- The distributed nature of blockchain networks ensures that no single node failure can disrupt the transmission of covert messages (Liu et al., 2015).
- Consensus algorithms provide an additional layer of protection against denial-of-service (DDoS) attacks, as messages are redundantly validated by multiple nodes before being added to the blockchain.
- Peer-to-peer message propagation makes it extremely difficult for adversaries to isolate or block specific covert communication pathways, thereby enhancing system robustness.

5. Open-Source Programmability and Innovation

Blockchain networks enable greater flexibility in designing and implementing covert communication protocols through open-source programmability. Unlike traditional methods, which rely on fixed network protocols, blockchain-based covert channels can continuously evolve through decentralized development efforts.

- Open-source blockchain platforms, such as Ethereum, Hyperledger Fabric, and Polkadot, allow developers to experiment with different encoding techniques for covert channels, including smart contract-based covert messaging (Irazoqui, Eisenbarth, & Sunar, 2016).
- Programmable blockchain environments support advanced encryption and obfuscation techniques, enhancing the stealth and security of covert communications.
- The permissionless nature of public blockchains ensures that new covert channel implementations cannot be easily censored or restricted, fostering innovation in secure communication technologies.

BLOCKCHAIN NETWORK COVERT CHANNELS: EMERGING TECHNOLOGIES

Blockchain network covert channels leverage cutting-edge technologies and novel methodologies to exploit inherent characteristics of the blockchain environment for building undetectable communication pathways. These technologies are primarily categorized into four types: transaction addresses, cryptographic signature algorithms, smart contracts, and peer-to-peer (P2P) broadcast mechanisms. These covert channels are assessed according to three key metrics: concealment, transmission efficiency, and communication cost. Additionally, we summarize the construction processes of blockchain covert channel models, highlighting the challenges that these innovative technologies face.

3.1. Classification of Blockchain Network Covert Channels Using Emerging Technologies

3.1.1. Transaction Address-based Covert Channels

Transaction address-based covert channels utilize blockchain transaction addresses as carriers for covert communication. In traditional blockchain operations, addresses are derived from public-private key pairs, which serve as unique identifiers for transactions. By leveraging cryptographic techniques and advanced address-generation algorithms, transaction addresses can be used as a medium for encoding and transmitting secret information in an undetectable manner.Recent advancements in elliptic curve cryptography (ECC), quantum-resistant blockchain technologies, and zero-knowledge proofs (ZKPs) have significantly enhanced the security, privacy, and efficiency of transaction address-based covert channels. These innovations introduce novel data embedding methods that improve concealment and robustness while maintaining blockchain integrity.

1. Elliptic Curve Cryptography (ECC) and Secure Address Generation

Elliptic Curve Cryptography (ECC) is widely employed in modern blockchain address generation, providing a high level of security with reduced computational overhead. ECC enables the creation of compact cryptographic keys, making it ideal for embedding covert messages within blockchain addresses.

- ECC-based key derivation can be manipulated to encode covert information within the structure of generated addresses without deviating from standard cryptographic constraints.
- Quantum-resistant cryptographic schemes, such as lattice-based cryptography and hash-based signatures, offer enhanced security against emerging quantum threats while maintaining the integrity of covert transmissions.
- These techniques provide an effective and scalable approach for encoding hidden messages within blockchain transaction addresses, ensuring stealth and security in covert communication.

2. Zero-knowledge proofs (ZKPs) and Privacy Enhancement

The adoption of Zero-Knowledge Proofs (ZKPs) in blockchain networks has introduced advanced privacy-preserving mechanisms that enhance the effectiveness of transaction address-based covert channels. In particular, zk-SNARKs (zero-knowledge succinct non-interactive arguments of knowledge) provide a means to verify information validity without revealing underlying data.

- zk-SNARKs allow secret messages to be embedded within transaction addresses while maintaining anonymity and confidentiality.
- By obfuscating transaction details, ZKPs prevent third-party analysis tools from detecting covert communication patterns within blockchain networks.
- Multi-party computation (MPC) and zk-STARKs (Zero-Knowledge Scalable Transparent Argument of Knowledge) further enhance security, enabling tamper-proof hidden data transmission across blockchain-based covert channels.
- These privacy-enhancing cryptographic techniques make transaction address-based covert channels more resilient against detection and analysis by adversarial entities.

3. Machine Learning-Optimized Address Generation for Covert Messaging

With the evolution of artificial intelligence (AI) and machine learning (ML) algorithms, blockchain address-generation tools have become more sophisticated in embedding covert information within transaction addresses.

- Vanity address generators, such as Vanitygen, have been enhanced using ML-based optimization techniques to systematically encode hidden messages into blockchain addresses without triggering anomaly detection systems.
- AI-driven pattern recognition models can identify undetectable address modifications, ensuring that covertly embedded data remains indistinguishable from standard blockchain transactions.
- Reinforcement learning techniques improve the generation of vanity addresses by optimizing factors such as entropy, randomness, and cryptographic complexity, making covert messaging more secure and efficient.
- These AI-integrated advancements enable higher data transmission rates and reduced latency, allowing for stealthy, real-time covert communication within blockchain networks.

4. Advantages of Transaction Address-Based Covert Channels

Transaction address-based covert channels offer several key advantages over traditional covert communication methods, particularly in the context of blockchain technology:

- Enhanced Security and Stealth – By leveraging ECC, ZKPs, and AI-optimized address generation, covert messages remain resilient to detection by blockchain forensic tools.
- Decentralization and Trustlessness – Unlike centralized covert communication channels, blockchain-based covert messaging does not depend on trusted intermediaries or centralized infrastructure, minimizing the risk of interception or censorship.
- High Throughput and Low Latency – AI-optimized address generation techniques improve data embedding efficiency, enabling faster and more reliable covert transmissions.

- Quantum-Resistant Encryption – With the integration of post-quantum cryptographic methods, blockchain-based covert channels can withstand future quantum computing attacks, ensuring long-term security.
- Tamper-Proof Data Integrity – The immutability of blockchain ledgers ensures that covertly embedded messages remain unaltered, providing reliable long-term storage and retrieval of hidden information.

3.1.2. Cryptographic Signature Algorithm-based Covert Channels

Signature algorithm-based covert channels exploit the cryptographic signature mechanisms used in blockchain transactions to embed hidden messages within digital signatures. Digital signatures, which authenticate transaction integrity and ownership, serve as a carrier for covert data transmission while maintaining blockchain security and compliance. Recent advancements in post-quantum cryptography (PQC), homomorphic encryption, and multi-party computation (MPC) have significantly strengthened the security and anonymity of signature-based covert channels. Modifications to the Advanced Encryption Standard (AES) and Elliptic Curve Digital Signature Algorithm (ECDSA) have enabled the encoding of secret information within blockchain signature fields, offering a highly secure and undetectable method of covert communication. Furthermore, privacy-preserving cryptographic techniques, such as ring signatures and stealth addresses, obscure sender and receiver identities, preventing direct traceability of embedded covert messages.

1. Post-quantum cryptography (PQC) and Signature Modifications

The emergence of post-quantum cryptographic (PQC) algorithms has introduced novel techniques for enhancing covert communication within blockchain signature schemes. PQC-based cryptographic primitives, such as lattice-based cryptography and hash-based signatures, provide quantum-resistant security for signature modifications without compromising blockchain integrity.

- Lattice-based cryptographic algorithms, such as NTRUEncrypt and Kyber, allow for the embedding of covert messages within signature fields while remaining resistant to quantum computing attacks.
- Hash-based digital signatures, such as XMSS (Extended Merkle Signature Scheme) and SPHINCS+, offer forward security while enabling covert data encoding in transaction verification processes.
- Homomorphic encryption techniques, when integrated with blockchain signature algorithms, allow for hidden message transmission while preserving cryptographic authenticity and verifiability.
- By leveraging PQC techniques, signature-based covert channels can withstand cryptanalysis attacks while ensuring long-term security in future quantum computing environments.

2. Multi-Party Computation (MPC) and Distributed Signature Schemes

Multi-party computation (MPC) is an advanced cryptographic protocol that enables multiple parties to jointly compute a function over their inputs while keeping those inputs private. In the context of blockchain-based signature algorithm covert channels, MPC enhances privacy and security by distributing digital signature generation across multiple participants.

- MPC-based signature schemes prevent malicious nodes from extracting sensitive data without the consensus of trusted parties.
- Threshold cryptography techniques, such as Shamir's secret sharing, allow for decentralized control of private keys, reducing single points of failure in covert communication.
- Secure multi-party computation frameworks, such as Garbled Circuits and Oblivious Transfer, further improve message concealment within blockchain signature fields.
- MPC integration ensures that covert channels remain decentralized and resistant to unauthorized interception, providing a higher degree of anonymity in blockchain transactions.

3. Ring Signatures for Enhanced Anonymity

Ring signatures are cryptographic techniques that allow a signer to produce a digital signature on behalf of a group, making it impossible to determine the actual sender within the group. This property makes ring signatures highly effective for obscuring covert messages within blockchain transactions.

- Monero's CryptoNote protocol utilizes ring signatures to ensure sender anonymity, making it an ideal framework for embedding covert information within digital signatures.
- Linkable ring signatures provide additional anonymity while enabling message verification without compromising secrecy.
- Threshold ring signatures, which require a subset of participants to collectively sign a transaction, further enhance the security of covert communication channels.
- By leveraging ring signatures, blockchain covert channels can embed hidden data while preserving non-traceability and unlinkability in transaction signatures.

4. Stealth Addresses for Untraceable Transactions

Stealth addresses generate unique, one-time addresses for blockchain transactions, making it difficult to link multiple transactions to a single user. This property significantly enhances the stealth capabilities of signature algorithm-based covert channels.

- Stealth addresses generate disposable transaction identifiers, ensuring that covertly embedded data remains untraceable.
- Dual-key stealth address protocols (DKSAPs) further obscure sender and receiver identities, enhancing hidden data transmission security.
- Cryptographic blinding techniques, such as the Pedersen commitment scheme, improve concealment mechanisms in blockchain-based covert communication.
- By integrating stealth addresses, blockchain covert channels achieve higher levels of anonymity and security, making message interception nearly impossible.

5. Advantages of Signature Algorithm-Based Covert Channels

Signature algorithm-based covert channels provide numerous advantages over traditional covert communication methods:

- High Security and Tamper Resistance – Advanced cryptographic modifications in PQC, MPC, and AES-enhanced ECDSA ensure that covert messages remain tamper-proof and resilient to cryptanalysis.
- Decentralized Anonymity – Multi-party computation and threshold cryptography enable distributed signing, reducing the risk of malicious third-party interception.
- Quantum-Resistant Messaging – Lattice-based and hash-based PQC algorithms future-proof covert communications against quantum attacks.
- Undetectable Data Transmission – Stealth addresses, ring signatures, and blinding techniques effectively obscure message origin and transmission path, making detection nearly impossible.
- Scalability and Efficiency – Optimized encryption techniques allow for high-throughput covert messaging with minimal impact on blockchain network performance.

3.1.3. Smart Contract-based Covert Channels

Smart contracts, defined as self-executing agreements with predefined rules encoded within blockchain-based scripts, have evolved significantly since their inception. Originally designed to automate decentralized transactions, smart contracts have now expanded their capabilities to facilitate covert communication channels. These advanced applications leverage programmable privacy layers, decentralized identity systems, and distributed ledger technologies (DLTs) to encode and transmit hidden information within secure blockchain environments.By integrating InterPlanetary File System (IPFS), decentralized finance (DeFi) protocols, and tokenization mechanisms, smart contract-based covert channels achieve high-capacity, encrypted data transmission while maintaining immutability and transparency. This paper explores the innovative use of smart contracts for covert messaging, highlighting their resilience, efficiency, and adaptability in secure communication.

1. Programmable Privacy Layers and Decentralized Identity Systems

The integration of programmable privacy layers (PPLs) and decentralized identity (DID) systems has significantly enhanced the ability of smart contracts to conceal covert messages.

- Programmable privacy layers enable selective disclosure of information through cryptographic mechanisms such as Zero-Knowledge Proofs (ZKPs), zk-SNARKs, and zk-STARKs, ensuring that only authorized parties can access hidden data.
- Decentralized identity frameworks (e.g., Self-Sovereign Identity, or SSI) allow for covert messaging within authentication processes, embedding encrypted messages in identity verification transactions.
- Layer-2 scaling solutions, such as state channels and rollups, can fragment and distribute covert messages across multiple transactions, further obfuscating data transmission paths.

- These innovations provide a secure and efficient means of embedding hidden data within smart contract executions while preserving transaction confidentiality.

2. Integration of IPFS and Distributed Ledger Technologies (DLTs)

The InterPlanetary File System (IPFS) and distributed ledger technologies (DLTs) offer scalable solutions for high-capacity covert data transmission within smart contracts.

- IPFS-based smart contracts allow for large encrypted datasets to be embedded in contract conditions, ensuring secure off-chain storage with on-chain verification.
- DLTs extend the functionality of blockchain-based covert channels by enabling cross-chain interoperability, allowing covert messages to be transmitted across multiple blockchains without direct traceability.
- Decentralized storage mechanisms, such as Arweave and Filecoin, further enhance data persistence and security while facilitating undetectable hidden communication.
- By integrating IPFS and DLTs, smart contracts achieve enhanced storage capacity and decentralized message transmission, making them ideal frameworks for secure covert communication.

3. Covert Messaging via Token Issuance Events

Tokenization, a fundamental aspect of decentralized finance (DeFi), has been repurposed as a covert data transmission mechanism.

- Hidden messages can be encoded into token issuance events, where metadata within smart contracts carries encrypted covert signals.
- Non-Fungible Tokens (NFTs) provide additional concealment by embedding encrypted data within NFT attributes, ensuring that only authorized parties can decrypt and access covert information.
- Dynamic token supply mechanisms, such as burn and mint functions, can be used to transmit hidden signals based on predefined cryptographic sequences.
- The flexibility of tokenization mechanisms makes them highly adaptable for embedding covert communication channels within blockchain transactions.

4. Altering Token Transfer Conditions for Covert Messaging

In addition to token issuance events, covert channels can also be constructed through manipulated token transfer conditions within DeFi protocols.

- Secret messages can be encoded within transaction parameters, such as gas fees, transaction slippage, or custom token transfer logic.
- Multi-signature wallet transactions can be utilized for covert signaling, where specific signature sequences correspond to hidden messages.
- Liquidity pool interactions and automated market maker (AMM) protocols enable subtle data embedding within DeFi activity flows, making covert communication indistinguishable from standard financial transactions.

- By embedding covert messages within token transfers, smart contracts harness DeFi protocols as an undetectable communication medium, ensuring secure and resilient hidden messaging.

5. Advantages of Smart Contract-Based Covert Channels

The utilization of smart contracts for covert communication offers numerous advantages:

- Decentralized Security – The trustless nature of smart contracts ensures that covert communication remains tamper-proof and resistant to centralized surveillance.
- Scalability and Efficiency – Layer-2 scaling solutions (e.g., rollups, state channels, and sidechains) enhance message throughput and minimize detection risks.
- Privacy-Preserving Mechanisms – Zero-Knowledge Proofs (ZKPs), stealth addresses, and decentralized identity systems provide strong anonymity guarantees.
- Resilience Against Censorship – DLT-based covert channels can propagate encrypted messages across multiple networks, making message suppression nearly impossible.
- Cross-Chain Interoperability – Multi-chain tokenization and DeFi transactions enable hidden data transmission across heterogeneous blockchain ecosystems.

Figure 2. Blockchain network covert channel construction model

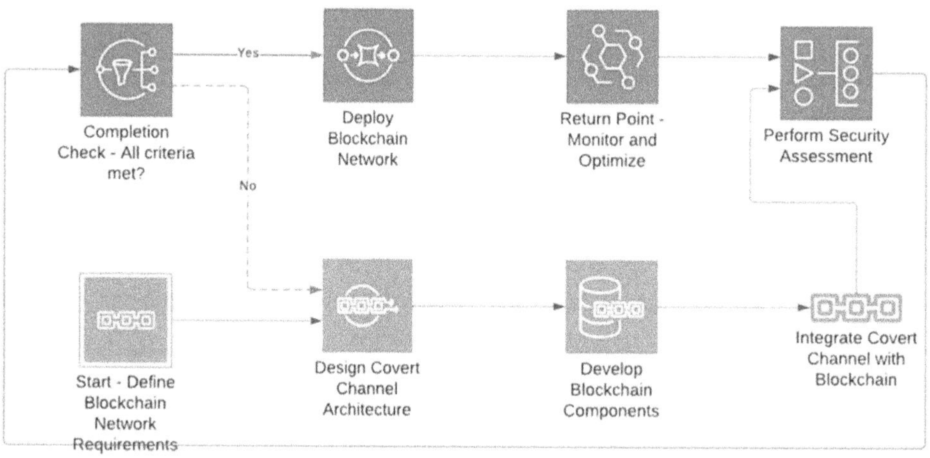

Peer-to-Peer Broadcast Mechanism-based Covert Channels

Peer-to-peer (P2P) broadcasting serves as the foundation of blockchain network communication, facilitating secure and decentralized data transmission. However, the inherent broadcast nature of P2P networks has increasingly been leveraged for the development of covert communication channels. Recent advancements in Mesh Networks, Decentralized Web (Web3.0), and Quantum-Resistant P2P Protocols have significantly enhanced the anonymity, efficiency, and security of covert transmissions within blockchain environments.This study explores how Decentralized Messaging Protocols, Latency-Based Hidden Channels, and Blockchain-Based Distributed Hash Tables (DHTs) provide novel methods for embedding and transmitting covert messages within P2P blockchain networks, offering a robust framework for low-latency, high-concealment communication.

1. Decentralized Messaging Protocols for Covert Data Transmission

The development of Decentralized Messaging Protocols has significantly enhanced covert communication within blockchain-based P2P networks.

- Whisper Protocol (Ethereum) and WAVES blockchain messaging systems offer secure, decentralized messaging layers that allow for ephemeral and encrypted message transmission.
- Ephemeral messaging techniques ensure that messages are temporarily stored and automatically discarded after being read, preventing forensic analysis and reducing the risk of data interception.
- Onion Routing and Mix Networks, similar to Tor-like architectures, have been integrated into blockchain-based P2P messaging protocols to obfuscate sender-receiver relationships, increasing anonymity.
- These messaging systems provide low-latency, decentralized broadcasting methods for embedding covert communication within blockchain networks while minimizing the risk of detection.

2. Latency-Based Hidden Channels in P2P Blockchain Networks

The exploitation of time intervals between transactions has enabled the development of latency-based covert channels within blockchain networks.

- Time-based encoding methods leverage delays between transaction transmissions to embed hidden data within network latency variations.
- Transaction obfuscation techniques, such as randomized delays and adaptive transmission rates, enhance the secrecy of covert messages, making them less susceptible to traditional network traffic analysis.
- Machine learning-based anomaly detection systems have been employed to counteract latency-based covert channels; however, adaptive transmission techniques continue to evade detection through probabilistic timing models.
- By manipulating inter-transaction latency, blockchain nodes create undetectable time-based hidden channels that bypass conventional traffic analysis mechanisms, making them an effective strategy for secure covert messaging.

3. Blockchain-Based Distributed Hash Tables (DHTs) for Covert Data Embedding

The integration of Distributed Hash Tables (DHTs) within blockchain networks has enabled new methods for embedding covert data into decentralized storage systems.

- DHTs serve as decentralized indexing structures, allowing covert messages to be encoded and retrieved based on cryptographic hash values.

- Covert data can be stored within blockchain-based DHTs by embedding metadata within transaction records, ensuring persistence and integrity while remaining hidden from conventional scanning methods.
- Sharding and partitioning techniques further obfuscate the retrieval process, as different fragments of a covert message may be distributed across multiple nodes in a blockchain network.
- Through the strategic embedding of data within blockchain-based DHTs, covert communication channels become more resilient, scalable, and resistant to detection by traditional surveillance mechanisms.

4. Quantum-Resistant P2P Protocols for Enhanced Security

As quantum computing advances, traditional cryptographic schemes face potential vulnerabilities due to their susceptibility to Shor's algorithm-based decryption attacks. In response, Quantum-Resistant P2P Protocols are being developed to enhance the security of covert blockchain communication channels.

- Lattice-based cryptographic algorithms, such as NTRUEncrypt and FrodoKEM, offer quantum-resistant encryption for securing P2P communications.
- Post-quantum digital signature schemes, including XMSS (Extended Merkle Signature Scheme) and SPHINCS+, provide tamper-proof authentication while ensuring covert message integrity.
- Quantum-safe key exchange protocols, such as NewHope and Kyber, enable secure peer-to-peer communication without reliance on vulnerable asymmetric encryption schemes.
- By integrating quantum-resistant cryptographic mechanisms, blockchain-based covert communication systems achieve long-term security and resilience against future quantum-based attacks.

ADVANTAGES OF BLOCKCHAIN-BASED P2P COVERT CHANNELS

The use of P2P broadcasting for covert communication within blockchain networks offers several advantages:

- Decentralization & Censorship Resistance – Eliminates reliance on centralized authorities, making covert messaging resilient to surveillance and suppression.
- Anonymity & Privacy Preservation – Utilizes cryptographic anonymity techniques such as Zero-Knowledge Proofs (ZKPs) and Ring Signatures to protect sender-receiver identities.
- Low Latency & High Throughput – Optimized P2P messaging protocols provide fast and secure communication channels for covert data transmission.
- Quantum-Resistant Security – The adoption of post-quantum cryptographic algorithms enhances long-term security against emerging quantum computing threats.
- Scalability & Efficiency – Blockchain-based DHTs and adaptive transaction latency techniques ensure highly scalable covert communication models.

Summary of Challenges and Future Directions

Despite the advancements in blockchain-based covert channels, several technical, regulatory, and ethical challenges persist, limiting their scalability, efficiency, and widespread adoption. Issues such as blockchain bloat, regulatory scrutiny, and the potential for misuse remain key concerns. Addressing these challenges requires innovative hybrid approaches that integrate Zero-Knowledge Proofs (ZKPs), Multi-Party Computation (MPC), and AI-based detection avoidance systems to enhance security, scalability, and adaptability in covert communication networks.

1. Scalability Limitations and Blockchain Bloat

One of the primary challenges associated with blockchain-based covert channels is scalability, particularly in networks experiencing high transaction volumes.

- Blockchain bloat, caused by the continuous accumulation of transaction data, leads to exponential growth in storage requirements. This increases the difficulty of maintaining efficient covert communication channels without compromising network performance.
- Transaction throughput constraints, inherent in many blockchain architectures (e.g., Bitcoin and Ethereum), impose limitations on the data transmission rates achievable through covert channels.
- Layer-2 scaling solutions, such as state channels, sidechains, and sharding, have been proposed to alleviate scalability bottlenecks; however, their integration into covert communication frameworks remains an open research challenge.
- Future research should focus on optimized data embedding strategies that minimize storage overhead while ensuring persistent and undetectable communication within blockchain networks.

2. Regulatory and Legal Implications

The use of blockchain-based covert channels raises significant regulatory and legal challenges, as governments and law enforcement agencies seek to prevent illicit activities facilitated by covert communications.

- Jurisdictional conflicts arise due to the decentralized and borderless nature of blockchain networks, complicating the enforcement of regulations.
- Regulatory bodies, such as the Financial Action Task Force (FATF) and European Union's GDPR, have imposed strict compliance measures on blockchain applications, potentially restricting the development and deployment of covert channels.
- Emerging legislative frameworks, including the Markets in Crypto-Assets (MiCA) regulation and U.S. Executive Orders on Digital Assets, emphasize increased monitoring and transparency, which may pose barriers to covert communication methodologies.
- To navigate these challenges, researchers must explore privacy-preserving legal compliance models, such as zero-knowledge-based regulatory frameworks that balance privacy with compliance while preventing illicit misuse of covert channels.

3. Ethical Considerations and the Risk of Misuse

The potential misuse of blockchain-based covert channels for illegal or unethical purposes, such as cybercrime, unauthorized surveillance evasion, and information warfare, raises ethical concerns.

- Darknet marketplaces and illicit financial networks may exploit covert blockchain channels to evade regulatory oversight and conduct clandestine transactions.
- State actors and advanced persistent threats (APTs) may leverage these channels for covert cyber operations, leading to geopolitical and national security risks.
- Ethical AI frameworks must be established to ensure that covert communication technologies are not misused while preserving legitimate privacy and security applications.
- The development of responsible disclosure policies and consensus-driven governance models can help mitigate ethical concerns while fostering legitimate research and innovation in covert blockchain communications.

4. Hybrid Approaches for Enhanced Security and Scalability

To overcome existing limitations, future research should focus on hybrid approaches that integrate multiple cryptographic techniques to enhance both security and efficiency. The integration of cutting-edge cryptographic methods can significantly enhance the security and privacy of blockchain-based covert channels. Two promising approaches involve the combination of Zero-Knowledge Proofs (ZKPs) with Multi-Party Computation (MPC) and the integration of Fully Homomorphic Encryption (FHE) with Blockchain-Based Distributed Hash Tables (DHTs).

ZKPs + MPC for Secure and Decentralized Covert Channels

- Zero-Knowledge Proofs (ZKPs) enable the verification of transactions without revealing any underlying data, preserving confidentiality while maintaining integrity.
- Multi-Party Computation (MPC) allows multiple participants to collaboratively process data without exposing individual inputs, ensuring that no single entity can access the complete communication.
- Impact: This combination strengthens the privacy and security of covert channels, reducing the risk of metadata exposure while maintaining decentralized trust.

Fully Homomorphic Encryption + DHTs for Privacy-Preserving Storage & Processing

- Fully Homomorphic Encryption (FHE) allows computations on encrypted data without requiring decryption, preventing unauthorized access during processing.
- Blockchain-Based Distributed Hash Tables (DHTs) distribute encrypted message fragments across decentralized nodes, increasing system resilience and reducing the risk of data reconstruction by adversaries.
- Impact: These technologies together create a tamper-resistant, censorship-resistant, and privacy-enhanced infrastructure for covert communication.

Layer-2 Networks + AI-Driven Adaptive Covert Channels:

Integrating Layer-2 blockchain solutions with AI-driven adaptive covert channels can significantly enhance privacy, scalability, and stealth in decentralized communication systems.

Layer-2 Networks for Secure and Scalable Covert Communication

- Layer-2 solutions like zk-Rollups and Plasma help move transactions off the main blockchain, reducing congestion while keeping communications secure.
- These technologies allow covert channels to operate off-chain, lowering the risk of detection while still ensuring data integrity through cryptographic proofs.

AI-Powered Adaptive Covert Channels

- AI-driven models can analyze network behavior and adjust covert communication strategies in real time.
- By mimicking normal network traffic, these adaptive channels make it harder for monitoring systems to detect covert messages.
- This approach enhances privacy and security while reducing the risk of exposure.

Benefits and Future Directions

Combining Layer-2 networks with AI-driven covert channels offers:

- Higher transmission capacity by reducing on-chain activity.
- Lower detection risk through intelligent adaptation.
- Stronger privacy protections by leveraging cryptographic security.

Future improvements could explore quantum-resistant encryption and decentralized AI models to further strengthen these communication systems.

5. AI-Driven Detection Avoidance and Adaptive Communication

The integration of Artificial Intelligence (AI) and Machine Learning (ML) into blockchain covert communication frameworks can significantly enhance detection avoidance and dynamic adaptation to surveillance threats.

- AI-based adversarial learning models can analyze network monitoring patterns and automatically adjust covert transmission parameters to evade detection.
- Self-learning blockchain nodes can identify potential surveillance activities and modify their communication protocols in real-time to remain undetectable.
- Reinforcement learning algorithms can optimize the selection of covert communication strategies, balancing latency, throughput, and concealment effectiveness.
- By leveraging AI-driven adaptation mechanisms, future blockchain covert channels will achieve autonomous resistance to network surveillance and forensic analysis.

EVALUATION METRICS AND COMPARATIVE ANALYSIS OF COVERT CHANNELS IN EMERGING TECHNOLOGIES

This section outlines the evaluation metrics—concealment, transmission efficiency, and communication cost—for covert channels in emerging technologies like blockchain, 5G networks, and IoT networks. It compares and analyzes different types of covert channels in these networks, considering how these technologies are changing the landscape of covert communication.

Concealment

Concealment in covert channels refers to the ability to evade detection and protect the identities of both communicating parties. In the context of emerging technologies such as blockchain, 5G, and IoT networks, concealment is even more critical due to their growing integration in industries that require high-security measures. In 5G networks, the integration of massive IoT devices and high-speed data transmission requires covert channels that can hide data from malicious entities. The concealment can be evaluated using methods like Kullback-Leibler (KL) distance or entropy-based analysis, which assess the difference between two probability distributions representing normal and covert communication traffic. The goal is to minimize the detectable difference to enhance concealment, as seen in the blockchain example. For IoT networks, data transmission is often constrained by bandwidth, making it essential to minimize the traceability of covert data. Dynamic traffic patterns and frequent device mobility in IoT make covert channel detection more complex but also more vulnerable to attacks. The more dynamic the data, the higher the resistance to detection, whereas static patterns are more susceptible to attack.

Transmission Efficiency

Transmission efficiency in emerging technologies focuses on the ability to transmit error-free information at high rates. In the case of 5G networks, the high bandwidth and low latency enable faster and more reliable covert communication. This can be evaluated by the maximum error-free transmission rate (V)—measured in bits per second (bit/s)—which quantifies how much data can be transmitted through the covert channel within a given timeframe. In blockchain-based covert channels, transmission efficiency is constrained by transaction data rates, while in IoT networks, the efficiency may be limited by low-power sensors and connectivity restrictions. These constraints require methods to encode hidden data in a way that maximizes the transmission rate without compromising security. Shannon's Capacity Theorem can be applied to calculate the channel capacity and the maximum error-free data that can be transmitted, similar to the previously mentioned equation for blockchain networks. For instance, in 5G networks, with its high data rates, covert channels can exploit techniques such as frequency hopping and time-slot manipulation to hide data in non-visible parts of the spectrum. The more robust the data encoding, the higher the transmission efficiency, but at the cost of potential computational complexity.

Figure 3. The "impossible triangle" of blockchain network covert channel

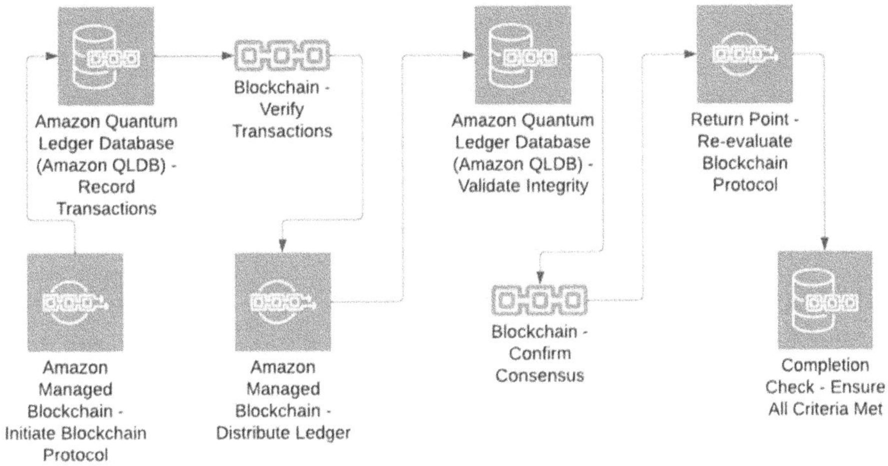

Communication Cost

Communication cost in emerging technologies refers to both the computational and monetary costs associated with maintaining a covert channel. In blockchain networks, mining or transaction fees can be significant, especially when using proof-of-work or other energy-intensive consensus algorithms. For 5G networks, covert channels may incur higher costs in terms of bandwidth usage, computational power for encryption, and the complexity of channel modulation techniques. Similarly, IoT networks face challenges in terms of the limited processing power of devices and the energy cost of maintaining covert communication. The computational cost is determined by the complexity of the algorithms used to construct and maintain the covert channel. This is especially relevant in blockchain-based systems, where transaction processing or smart contract execution adds computational overhead. Monetary cost involves the fees required to carry out transactions on the blockchain (such as gas fees in Ethereum) or the cost of maintaining network infrastructure in 5G and IoT networks. The computational complexity in 5G may involve the deployment of advanced encryption or advanced network slicing for securely transmitting covert data. In IoT networks, the costs are more related to the power consumption of devices and network bandwidth.

Table 1. Blockchain-based network covert channel construction methods

Carrier Type	Reference	Brief Introduction	Evaluation Metrics
Transaction Address	(Partala, 2018)	Utilizes the Least Significant Bit (LSB) encoding of addresses to embed information, with each information sharing event starting with a marker.	Stealth: Low, Transmission Efficiency: Low, Communication Cost: High
	(Song & Peng, 2020)	Uses the low α bits of the address encoding and incorporates the information start marker into the information sequence.	Stealth: Low, Transmission Efficiency: Low, Communication Cost: High
	(Zhang, 2020a)	Embeds base58 encoded information in the LSB of multiple Vanitygen-generated addresses, with indexing written into OP_RETURN.	Stealth: Low, Transmission Efficiency: Medium, Communication Cost: Medium
	(Cao et al., 2022)	Encodes information directly into public keys to generate transaction addresses.	Stealth: High, Transmission Efficiency: Low, Communication Cost: High
	(Luo, 2022)	Combines the transaction amount matrix with the transaction address index matrix.	Stealth: Medium, Transmission Efficiency: Medium, Communication Cost: High
Signature Algorithm	(Gao et al., 2020)	Designs a covert algorithm to generate special signatures that only the information receiver can extract from the transaction.	Stealth: High, Transmission Efficiency: Medium, Communication Cost: High
	(Guo et al., 2021)	Utilizes the ring signature feature of Monero to encode information into the public keys of the ring signature set.	Stealth: Medium, Transmission Efficiency: Low, Communication Cost: Low
Smart Contract	(Basuki & Rosiyadi, 2019)	Encodes information based on image steganography and invokes a smart contract to transmit the information.	Stealth: Low, Transmission Efficiency: High, Communication Cost: Low
	(Zhang et al., 2022)	Encodes information based on the options in voting contracts and the prices in bidding contracts.	Stealth: Medium, Transmission Efficiency: Medium, Communication Cost: Low
P2P Broadcast Mechanism	(Lyu & Cao, 2021)	Writes information shards into the transaction coinbase field and uses the transaction broadcast mechanism to establish communication.	Stealth: Low, Transmission Efficiency: Low, Communication Cost: High
	(Zhang, 2020b; Zhang et al., 2021)	Transmits information using the Ethereum Whisper protocol.	Stealth: Medium, Transmission Efficiency: High, Communication Cost: Low
	(Li, 2021)	Combines time intervals and node unique identifiers to encode information for communication.	Stealth: High, Transmission Efficiency: Low, Communication Cost: Low

Comparative Analysis

A comparative analysis of covert channels in emerging technologies—blockchain, 5G, and IoT—based on the three evaluation metrics—concealment, transmission efficiency, and communication cost—is summarized in Table 1. This analysis reveals that it is challenging to optimize all three evaluation metrics at once in covert channels.

- Blockchain networks offer a high level of concealment due to the decentralized nature of the system but suffer from higher communication costs due to transaction fees.

- 5G networks provide high transmission speeds, making covert communication more efficient, but they require sophisticated encryption and signaling techniques, increasing computational and monetary costs.
- IoT networks, due to their limited resources, may have lower transmission efficiency but offer better concealment through dynamic traffic manipulation. However, their communication cost can rise due to power consumption and low throughput.

Ultimately, the optimal covert channel method depends on the specific use case and the requirements of the system, whether it's high security, efficient data transmission, or cost-effectiveness.

Covert Channel Model in Blockchain Networks with Emerging Technologies

Building on the previous discussions of covert channels in blockchain networks and integrating the latest advancements in related technologies, this chapter defines the blockchain network covert channel construction model in four stages: hidden information embedding, carrier obfuscation transmission, special carrier filtering, and hidden information decoding, as shown in Figure 2.

1. Hidden Information Embedding: This step involves the sender encoding and embedding hidden information into a carrier, creating a special carrier that contains the secret data. Based on emerging cryptographic techniques and data embedding methods, the hidden information embedding mechanism can be classified into visibleembedding and invisibleembedding, as shown in Table 2.
    - Visibleembedding directly embeds the hidden information into a defined field of the carrier. Although it is straightforward, it is vulnerable because blockchain data is publicly accessible, which allows third parties to analyze and potentially detect the hidden information.
    - Invisibleembedding uses advanced cryptographic techniques, such as homomorphic encryption or steganography, to encode the hidden information in a way that the characteristics of the secret data are not immediately visible within the carrier, offering enhanced concealment over visible embedding.
2. CarrierObfuscationTransmission: In this step, the sender mixes the special carrier containing the hidden information with regular carriers, creating an indistinguishable resemblance between the two. Techniques from AI-drivenobfuscation and distributedledgertechnology (DLT) are employed to make it more difficult for monitoring parties to identify the special carriers, ensuring more effective covert communication. Machine learning algorithms can be used to adapt and obfuscate the data transmission patterns, increasing the sophistication of the covert channel.
3. SpecialCarrierFiltering: The receiver filters out the special carriers containing hidden information from the massive blockchain data. Leveraging blockchainanalytics and bigdataprocessingframeworks, the filtering mechanism is crucial due to the decentralized nature of blockchain and the immense data volume generated. Blockchain nodes store all data, making the selection process complex. The filtering mechanisms are categorized as staticlabel-basedfiltering and dynamiclabel-basedfiltering.
    - Static label-based filtering is efficient but increases the risk of detection if the same labels are repeatedly used, reducing the concealment level.
    - Dynamic label-based filtering does not require repeated use of the same labels, offering superior concealment. However, it involves higher computational overhead, requiring the

receiver to perform fresh filtering for each communication, resulting in lower transmission efficiency.

4. Hidden Information Decoding: Once the receiver filters out the special carrier, they decode the hidden information using the reverse of the original encoding algorithm. In this context, modern cryptographic methods, such as quantum-safe encryption and zero-knowledge proofs, could be used to ensure that the decoding process is secure and resistant to emerging cryptographic attacks. The end of the covert communication process is marked by the successful extraction of the hidden information. To ensure that the receiver can successfully decode the message, both parties must pre-share certain information securely. Current research often assumes a secure environment for information sharing, but future advancements could improve the pre-sharing mechanism's security, enhancing the overall concealment of blockchain network covert channels.

Challenges in Blockchain Network Covert Channels with Emerging Technologies

Despite the advantages of blockchain network covert channels, such as decentralization, enhanced security, immutability, and privacy through the use of advanced cryptographic methods, several challenges remain:

1. Data Duplication: Blockchain's inherent characteristic of storing data permanently on the chain and its decentralized nature lead to the replication of data across nodes, increasing the risk of covert channels being detected. With the rise of AI-powered detection systems and deep learning models for blockchain analysis, even completed communications may face the risk of being re-detected and uncovered.
2. Identity Leakage: Blockchain provides a level of pseudo-anonymity, but recent advances in blockchainforensics and graph analysis algorithms allow third parties to trace the identities of communication participants through transaction linkages, address analysis, and identity matching. Even with advanced encryption techniques, these analytical methods can pose significant threats to privacy.
3. Information Filtering Difficulty: Filtering hidden information becomes increasingly complex with blockchain transactions that do not explicitly designate recipient addresses. Additionally, as blockchain transactions increase, real-time blockchain monitoring systems leveraging big data analytics are necessary to handle the large volume of transactions. These systems are critical for detecting patterns and filtering relevant data for covert channels, but their efficiency is still an area for improvement.
4. Performance Trade-offs: Blockchain technology faces the "trilemma" of balancing decentralization, security, and scalability, which also affects the construction of covert channels in blockchain networks. Similarly, blockchain covert channels face an "impossible triangle" of concealment, transmission efficiency, and communication costs, as shown in Figure 3. The use of advanced quantum-resistanta-lgorithms or privacy-preserving computation frameworks may help to balance these trade-offs, but it is challenging to fully optimize all three features in practice. Therefore, trade-offs must be made based on specific use cases and emerging technological developments in the field.

Case Studies: Real-World Applications of Blockchain-Based Covert Channels

To illustrate the practical implications of blockchain network covert channels, this section highlights real-world examples and experimental implementations that demonstrate their potential and challenges.

Case Study 1: Whisper Protocol for Private Messaging

- Overview: Whisper, a privacy-focused messaging protocol within the Ethereum ecosystem, allows users to exchange encrypted messages over a decentralized network.
- Relevance to Covert Channels: Whisper functions as a decentralized overlay network, enabling message transmission without centralized servers. It employs cryptographic techniques to conceal metadata, making it a real-world example of a blockchain-based covert channel.
- Challenges: Despite its design, Whisper has performance limitations, including high resource consumption and network scalability concerns, impacting its effectiveness for large-scale covert communication.

Case Study 2: Zcash Shielded Transactions and Metadata Protection

- Overview: Zcash, a privacy-centric cryptocurrency, uses zero-knowledge proofs (ZKPs) to enable shielded transactions that obscure sender, receiver, and transaction amounts.
- Relevance to Covert Channels: Zcash's shielded pool can be leveraged for anonymous data exchange, allowing covert communication by embedding messages in shielded transactions.
- Challenges: The traceability of shielded transaction usage patterns and potential de-anonymization through network-layer forensics remain concerns. Additionally, not all transactions in Zcash are shielded, which could lead to statistical profiling attacks.

Case Study 3: Covert Data Storage in Bitcoin OP_RETURN Field

- Overview: Bitcoin's OP_RETURN script allows users to store small amounts of arbitrary data (up to 80 bytes) on the blockchain.
- Relevance to Covert Channels: This feature has been exploited to embed hidden messages or covert signals in blockchain transactions, using the network as an immutable, censorship-resistant storage medium.
- Challenges: Due to Bitcoin's public and permanent ledger, forensic tools can detect and extract OP_RETURN messages, increasing the risk of exposure. The limited data size per transaction also restricts message complexity and transmission capacity.

Pilot Implementations and Future Directions

- Decentralized File Storage: Projects like IPFS and Arweave have been explored as potential covert data-sharing platforms by encrypting and distributing data fragments across nodes.
- AI-Powered Obfuscation: Ongoing research in AI-driven pattern obfuscation could enable blockchain-based covert channels to dynamically adapt and avoid detection.
- Quantum-Resistant Solutions: Implementing quantum-resistant cryptographic techniques in blockchain networks may enhance long-term security and resilience of covert channels.

Table 2. Covert information embedding mechanisms in blockchain

Embedding Mechanism	Description	Type	Channel Capacity	Stealth
LSB Method	Embeds information using the Least Significant Bit (LSB) of the carrier, such as in xxx1, xxx0	Explicit Embedding	2 bits	Medium
Storage Field	Uses fields like OP_RETURN, input/output fields for embedding	Explicit Embedding	512 bytes, 5000 bytes	Medium
Amount Field	Embeds information directly into the transaction amount field	Explicit Embedding	64 bits	Medium
Direct Writing	Embeds encrypted information directly into the carrier, such as through the Whisper protocol	Explicit Embedding	512 bits	Medium
Special Address	Generates unique addresses using cryptographic methods such as hash algorithms	Implicit Embedding	4 bits	High
Signature Field	Embeds a random number 'k' from the signature algorithm in the signature field	Implicit Embedding	1024 bits	High
Time Interval	Distinguishes covert information based on the transaction broadcast time interval	Implicit Embedding	8 bits	High

Visual Enhancements: Diagrammatic Representation of Blockchain-Based Covert Channels

To improve the clarity and accessibility of the technical content discussed in this chapter, it is essential to incorporate visual representations that can enhance the comprehension of complex blockchain-based covert channels. Diagrams can effectively illustrate the flow of covert communication, the comparison between traditional and emerging covert channel methods, and the inherent trade-offs in blockchain covert channels. The following visual enhancements are proposed to provide a more comprehensive understanding of the concepts. Figure 4 illustrates the Flow of Blockchain-Based Covert Channels. This diagram visually represents the lifecycle of a covert communication process within a blockchain network, from the initial stage of hidden information embedding to the final stage of decoding. It details the various stages involved in the covert communication pipeline:

Figure 4. Flow of blockchain-based covert channels

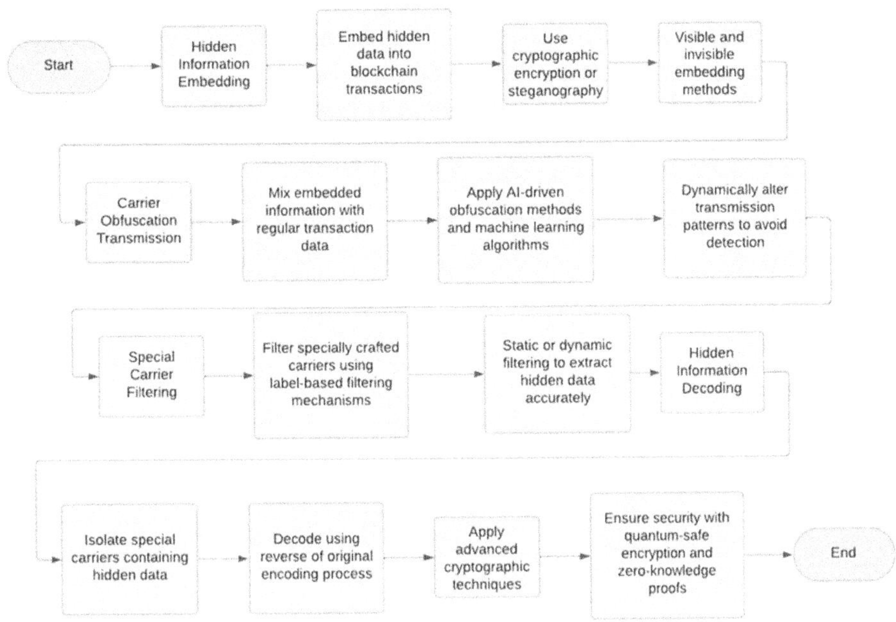

1. Hidden Information Embedding: The sender encodes hidden data into blockchain transactions using techniques such as cryptographic encryption or steganography. This step involves the use of both visible and invisible embedding methods.
2. Carrier Obfuscation Transmission: In this stage, the embedded information is mixed with regular transaction data, making it difficult for monitoring systems to differentiate between normal and covert traffic. AI-driven obfuscation methods and machine learning algorithms are used to dynamically alter the transmission patterns to avoid detection.
3. Special Carrier Filtering: Upon receipt, the receiver filters out the specially crafted carriers (i.e., blockchain transactions containing hidden data) using either static or dynamic label-based filtering mechanisms. The filtering process is critical in ensuring the hidden data is extracted accurately without exposing it to unauthorized parties.
4. Hidden Information Decoding: Finally, after the special carriers have been isolated, the hidden information is decoded using the reverse of the original encoding process. Advanced cryptographic techniques, such as quantum-safe encryption and zero-knowledge proofs, may be applied to ensure the security of this decoding phase.

This flowchart would provide readers with a clear, step-by-step visual guide to the entire process, from data embedding to the eventual decoding of covert messages. Figure 5 illustrates comparative Analysis of Traditional vs. Emerging Covert Channel Methods. This diagram serves as a comparative analysis between traditional covert channel techniques (e.g., steganography, tunneling) and emerging blockchain-based methods (e.g., zero-knowledge proofs, homomorphic encryption). The comparison highlights key metrics such as:

- Efficiency: Measures the data transmission rate of each method, indicating how much data can be hidden and transmitted without compromising performance.
- Security: Evaluates the robustness of each method against detection techniques, considering factors such as encryption strength, resistance to statistical analysis, and the likelihood of exposure through forensics.
- Resilience to Detection: Assesses the vulnerability of each method to advanced detection techniques, particularly AI-powered systems and machine learning models.

Figure 5. Comparative analysis of traditional vs. emerging covert channel methods

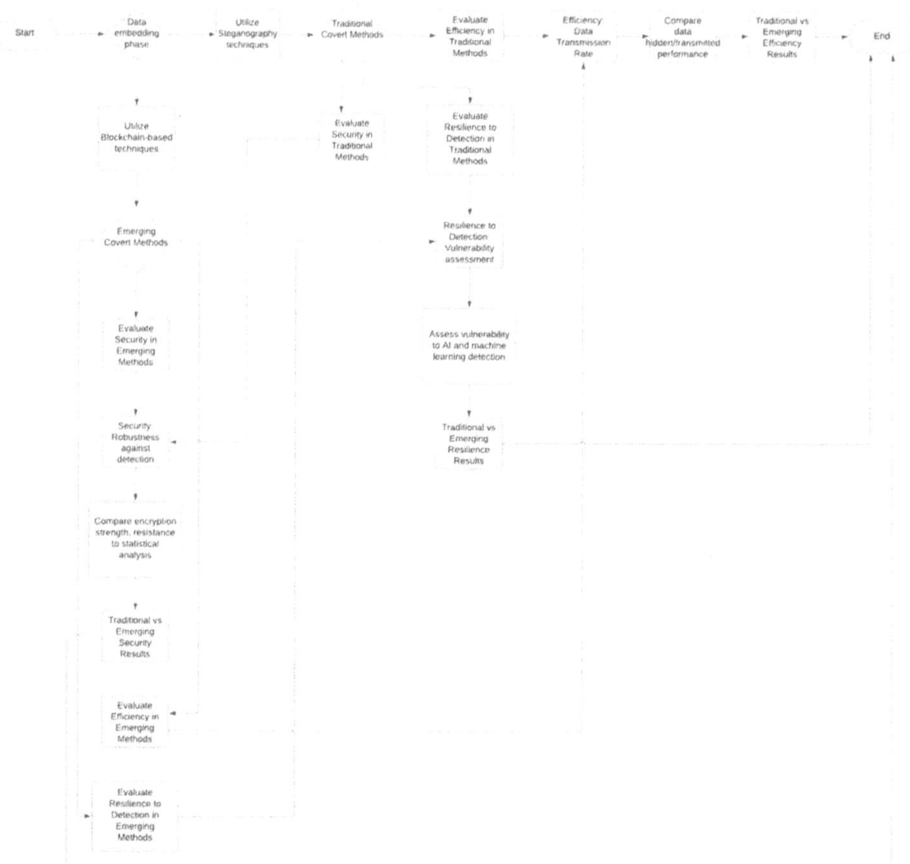

In this diagram, side-by-side comparisons will enable readers to visually grasp the relative advantages and limitations of each method, making it easier to understand the shift towards blockchain-based covert channels and their potential for overcoming the weaknesses of traditional techniques. Figure 6 illustrates the "Impossible Triangle" in Blockchain Covert Channels. The "Impossible Triangle" (also known as the blockchain covert channel trilemma) refers to the inherent trade-offs between three critical factors in the construction of covert channels: concealment, transmission efficiency, and communication cost. This diagram will visually depict these trade-offs, illustrating that:

- Concealment: High concealment generally requires more complex encryption or data manipulation techniques, which can reduce transmission efficiency.
- Transmission Efficiency: Techniques that maximize data throughput often make covert channels more detectable, as they may involve repetitive patterns or high data volumes.
- Communication Cost: High concealment and transmission efficiency often increase the computational and monetary costs associated with maintaining the covert channel, such as blockchain transaction fees, computational power for encryption, or energy consumption for maintaining IoT devices.

The diagram will show that balancing these three factors is challenging, as improving one typically comes at the expense of the others. For example, enhancing transmission efficiency may lower the level of concealment, or improving concealment may increase communication costs due to the need for more complex cryptographic operations. This "Impossible Triangle" concept helps in understanding the fundamental challenges of optimizing blockchain-based covert channels.

Figure 6. Impossible triangle in blockchain covert channels

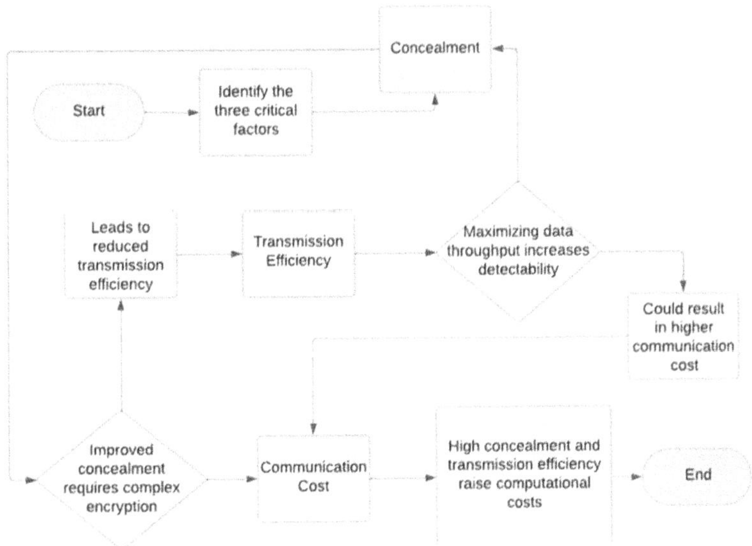

FUTURE RESEARCH DIRECTIONS FOR COVERT CHANNELS IN BLOCKCHAIN NETWORKS

Based on the challenges identified in current blockchain-based covert channel construction methods, this section proposes potential solutions and outlines future research directions, incorporating emerging trends in blockchain and cryptographic technologies.

Leveraging Privacy-Enhancing Technologies

To address the issue of identity exposure, privacy-enhancing technologies can be utilized to bolster the anonymity of communication participants, thereby improving the concealment of covert channels. Zero-Knowledge Proofs (ZKPs), especially the advanced zk-SNARKs (Zero-Knowledge Succinct Non-Interactive Arguments of Knowledge), provide a method for verifying transactions without revealing any information about the transaction itself. These technologies can be used to mix transactions and obfuscate sender and receiver identities, ensuring a higher level of privacy for participants. For example, Zcash uses zk-SNARKs to facilitate shielded transactions, which offer strong privacy guarantees. Similarly, MimbleWimble, a blockchain protocol focused on privacy and scalability, uses confidential transactions and CoinJoin-like mechanisms to hide transaction amounts and addresses. This technique could be extended to blockchain covert channels to further enhance anonymity. Future research should explore applying these cutting-edge cryptographic protocols to blockchain-based covert channels, particularly in public blockchains, to provide greater confidentiality for communication parties.

Enhancing Scalability with Layer 2 Solutions

Current public blockchains, like Bitcoin and Ethereum, are limited by low throughput and high latency, which is problematic for covert communication channels that demand high-speed transmission. To overcome these limitations, Layer 2 solutions offer promising avenues to enhance the scalability and transmission efficiency of blockchain networks.

Off-Chain Solutions: Rollups and State Channels

Optimistic Rollups and ZK-Rollups are off-chain scaling solutions that bundle multiple transactions into one, significantly reducing the load on the base blockchain while maintaining security. These technologies can also facilitate covert communication by increasing transaction throughput while reducing costs. State Channels, such as Lightning Network for Bitcoin or Raiden Network for Ethereum, allow off-chain transactions with minimal delay and very low fees. They enable fast, private transactions by only recording the final result on the main chain, ensuring that intermediary steps in covert communication are not exposed. Polkadot and Cosmos, two leading multi-chain frameworks, allow for interoperability between different blockchains. These technologies could help establish a more efficient and decentralized covert communication network by enabling cross-chain covert communication while benefiting from the privacy features of different blockchain ecosystems.

On-Chain Solutions: DAG-Based Technologies

On-chain scalability solutions that use Directed Acyclic Graph (DAG) structures are a promising direction for improving covert channel transmission efficiency. Hedera Hashgraph and IOTA, both using DAG-based topologies, offer numerous advantages over traditional blockchain systems:

1. High Throughput: Unlike traditional blockchains, DAG-based systems allow transactions to be processed in parallel, offering high scalability and fast transaction processing. This enables covert channels to handle higher transaction volumes, thereby improving communication efficiency.
2. Zero Fees: Many DAG-based systems, such as IOTA, do not require transaction fees, making them ideal for covert communication where minimizing costs is crucial. The ability to conduct numerous small transactions without incurring high costs significantly lowers the barrier for constructing covert channels.
3. Efficient Data Storage: IOTA's approach to data storage, where old transactions are periodically pruned, enhances privacy by ensuring that only recent transactions remain on the ledger. This reduces the potential for attackers to trace past transactions and helps maintain the secrecy of the covert communication channel.
4. High Privacy: IOTA's mechanism, where transactions are validated by previous transactions, can help in ensuring privacy by reducing the difficulty of transaction analysis, improving the overall security of covert channels.

These advancements in both Layer 2 solutions and DAG-based technologies offer new opportunities for scaling blockchain covert channels while maintaining the essential characteristics of privacy and efficiency.

Quantum-Resistant Cryptography

With the potential advent of quantum computers, current cryptographic systems might become vulnerable. Quantum-resistant algorithms, such as Lattice-based cryptography and Multivariate cryptography, are being researched to ensure that blockchain-based communication, including covert channels, remains secure in a post-quantum world. Implementing quantum-resistant cryptography in blockchain covert channels would safeguard against potential quantum attacks and ensure long-term security for sensitive communications.

Integration with Privacy-Preserving Decentralized Identity Systems

Decentralized Identity (DID) systems, such as those supported by the Self-Sovereign Identity (SSI) movement, offer a promising solution for maintaining privacy while ensuring security and trust in digital communications. By integrating DID with blockchain covert channels, it's possible to establish trust and ensure anonymity without relying on centralized authorities. These systems use blockchain to store verifiable credentials and enable users to control their own identities. Combining this technology with privacy-enhancing protocols (like zk-SNARKs) could provide an even higher level of security and anonymity in covert channels.

Utilizing Off-Chain Information Encoding with Decentralized Storage Solutions

To address the challenge of data replication in blockchain networks, off-chain information—data that cannot be permanently stored or directly queried within the blockchain—can be utilized as a medium for constructing covert communication channels. By leveraging decentralized storage solutions, such as **IPFS** (InterPlanetary File System) or **Arweave**, information can be stored off-chain in a way that is still accessible but not inherently part of the blockchain's immutable ledger. In blockchain ecosystems, new transactions are frequently broadcast, and the order of transactions changes in real-time. Once a transaction order expires, it becomes invisible. This chapter proposes a method where the sender modulates the broadcast order of transactions to encode secret information, using decentralized storage as a medium for securely storing off-chain data. The encoding technique can involve various strategies, such as using the last character of transaction addresses (whether it's a digit or a letter) to encode data. This simplified encoding approach can be enhanced in future research by increasing the length of the transaction address or using more complex encoding methods, depending on the efficiency and security of the channel. This method offers significant advantages: data is not permanently stored on the blockchain, making it more difficult to trace, and it resists data analysis attacks. Additionally, the broadcasting time for transactions is much shorter than the time required for transactions to be finalized on-chain, resulting in faster and more efficient communication.

Leveraging Advanced Cryptographic Techniques for Covert Channels

Blockchain technology already incorporates a wide array of advanced cryptographic techniques to ensure the integrity, immutability, and privacy of the ledger. These cryptographic features can be further utilized during the construction of covert communication channels to increase their security and concealment. One promising approach is the use of Zero-Knowledge Proofs (ZKPs) and Verifiable Delay Functions (VDFs) to enhance the confidentiality of communication. Zero-Knowledge Proofs, such as zk-SNARKs (Zero-Knowledge Succinct Non-Interactive Arguments of Knowledge), allow a party to prove to another party that a statement is true without revealing the underlying data. By incorporating zk-SNARKs into the covert channel construction, sensitive information can be transmitted while preserving the privacy of the sender and receiver. This makes ZKPs a powerful tool for building private, secure communication channels in a blockchain environment. Verifiable Delay Functions (VDFs) are another cryptographic tool that can be employed to create covert channels. VDFs require a set amount of time to compute, ensuring that the encoding process is inherently delayed and difficult to reverse-engineer. This creates an additional layer of security for secret information by making it more challenging to predict the encoding process or extract meaningful data prematurely. Experimentally, the time required for executing VDFs varies with hardware, making it possible to encode secret information based on the time taken to perform the function. By setting time thresholds for the computation, this method provides a high level of concealment for sensitive data transmission. Additionally, combining VDFs with other cryptographic mechanisms, such as transaction addresses, signature algorithms, or even smart contract-based systems, can offer even higher levels of security.

CONCLUSION

The concept of constructing covert communication channels within blockchain environments has sparked significant interest and offers innovative solutions for secure data transmission. This chapter begins by introducing the definition and evolution of network covert channels, explaining the overall blockchain architecture, and comparing traditional network covert channels with the emerging concept of blockchain-based covert channels. We categorize these covert channels into four types based on their carriers: transaction address-based, signature algorithm-based, smart contract-based, and P2P broadcast-based. By evaluating these types according to three key metrics—concealment, transmission efficiency, and communication cost—we propose a blockchain network covert channel model consisting of four steps: embedding secret information, carrier obfuscation transmission, special carrier selection, and secret information decoding. Finally, the chapter addresses the limitations of existing methods and offers solutions to enhance their effectiveness. The goal of this chapter is to provide valuable insights for the future development of secure and concealed communication channels in blockchain environments, incorporating cutting-edge technologies such as decentralized storage, Zero-Knowledge Proofs, and Verifiable Delay Functions.

REFERENCES

Basuki, I., & Rosiyadi, D. (2019). Joint transaction-image steganography for high-capacity covert communication. Proc. 2019 International Conference on Computer, Control, Informatics and its Applications (IC3INA). IEEE Press.

Cao, , Yin, H., Gao, F., Zhang, Z., Khoussainov, B., Xu, S., & Zhu, L. (2022). Chain-based covert data embedding schemes in blockchain. *IEEE Internet of Things Journal*, *9*(16), 14699–14707. DOI: 10.1109/JIOT.2020.3040389

Cauich, E., Gomez, C. R., & Watanabe, R. (2005). Data hiding in identification and offset IP fields. *Proceedings of the 5th International School and Symposium*, 247-261. DOI: 10.1007/11533962_11

Craver, S. (1998). On public-key steganography in the presence of an active warden. *International Workshop on Information Hiding*, 355-368. DOI: 10.1007/3-540-49380-8_25

Ding, . (2019). Survey on key issues in networks covert channel. *Journal of Software*, *30*(8), 2470–2490.

El-Atawy, A., Duan, Q., & Al-Shaer, E. (2017). A novel class of robust covert channels using out-of-order packets. *IEEE Transactions on Dependable and Secure Computing*, *14*(2), 116–129. DOI: 10.1109/TDSC.2015.2443779

Epishkina, , & Kogos, . (2015). Protection from binary and multi-symbol packet length covert channels. *Proceedings of the 8th International Conference on Security of Information and Networks*, 196-202.

Gao, , Zhu, L., Gai, K., Zhang, C., & Liu, S. (2020). Achieving a covert channel over an open blockchain network. *IEEE Network*, *34*(2), 6–13. DOI: 10.1109/MNET.001.1900225

Giffifin, . (2020). Covert messaging through TCP timestamps. *Proceedings of 2002 International Conference on Privacy Enhancing Technologies*, 194-208.

Girling, C. G. (1987). Covert channels in LAN's. *IEEE Transactions on Software Engineering*, *13*(2), 292–296. DOI: 10.1109/TSE.1987.233153

Guo, , Shi, L., Xu, M., & Yin, H. (2021). MRCC: A practical covert channel over Monero with provable security. *IEEE Access : Practical Innovations, Open Solutions*, *9*, 31816–31825. DOI: 10.1109/ACCESS.2021.3060285

Handel, T. G., & Sandford, M. T.II. (1996). Hiding data in the OSI network model," *International Workshop on Information Hiding*, 23-38. DOI: 10.1007/3-540-61996-8_29

Irazoqui, G., Eisenbarth, T., & Sunar, B. (2016). Cross processor cache attacks. *Proceedings of the 11th ACM on Asia Conference on Computer and Communications Security*, 353-364.

Lampson, B. W. (1973). A note on the confinement problem. *Communications of the ACM*, *16*(10), 613–615. DOI: 10.1145/362375.362389

Li, . (2021). *Block chain network covert communication method based on multi-node timestamp collusion: CN111245570B*. Patent.

Li, . (2022). Survey on key technologies of covert channel in ubiquitous network environment. *Journal of Communication*, *43*(4), 186–201.

Liu, . (2015). Last-level cache side-channel attacks are practical. *Proceedings of 2015 IEEE Symposium on Security and Privacy*, 605-622. DOI: 10.1109/SP.2015.43

Liu, . (2021). Research progress of blockchain in data security. Indian's. *Journal of Computers*, *44*(1), 1–27.

Lucena, . (2004). Syntax and semantics-preserving application-layer protocol steganography. *International Workshop on Information Hiding*, 164-179. DOI: 10.1007/978-3-540-30114-1_12

Luo, . (2022). A novel covert communication method based on bitcoin transaction. *IEEE Transactions on Industrial Informatics*, *18*(4), 2830–2839. DOI: 10.1109/TII.2021.3100480

Lyu, J. S., & Cao, X. C. (2021). Covert communication technology based on bitcoin. *Journal of Cybersecurity*, *6*(2), 143–152.

Mazurczyk, W., & Szczypiorski, K. (2012). Evaluation of steganographic methods for oversized IP packets. *Telecommunication Systems*, *49*(2), 207–217. DOI: 10.1007/s11235-010-9362-7

Millen, J. (1999). 20 years of covert channel modeling and analysis. *Proceedings of the 1999 IEEE Symposium on Security and Privacy*, 113-114. DOI: 10.1109/SECPRI.1999.766906

Muchene, D. N., Luli, K., & Shue, C. A. (2013). *Reporting insider threats via covert channels. Proceedings of 2013 IEEE Security and Privacy Workshops*. IEEE Press.

Nakamoto, S. (2008). Bitcoin: A peer-to-peer electronic cash system. *Consulted*, *28*(1), 21260–21268.

Partala, J. (2018). Provably secure covert communication on blockchain. *Cryptography*, *2*(3), 18–35. DOI: 10.3390/cryptography2030018

Patuck, R., & Hernandez-Castro, J. (2013). Steganography using the extensible messagıng and presence protocol (XMPP). arXiv Preprint, arXiv: 1310.0524.

Petitcolas, F. A. P., Anderson, R. J., & Kuhn, M. G. (1999). Information hiding-a survey. *Proceedings of the IEEE*, *87*(7), 1062–1078. DOI: 10.1109/5.771065

Rios, R., Onieva, J. A., & Lopez, J. (2012). *HIDE_DHCP: covert communications through network configuration messages*. Information Security and Privacy Research.

Rowland, C. H. (1997). Covert channels in the TCP/IP protocol suite. *First Monday*, *2*(5), 1. DOI: 10.5210/fm.v2i5.528

Simmons, G. J. (1984). The prisoners' problem and the subliminal channel. *Proceedings of CRYPTO*, 51-67.

Song, S., & Peng, W. (2020). BLOCCE +: An improved blockchain-based covert communication approach. *Journal of Chongqing University of Technology*, *34*(9), 238–244.

Tahir, . (2016). Sneak-Peek: high speed covert channels in data center networks," *Proceedings of the 35th Annual IEEE International Conference on Computer Communications*, 1-9. DOI: 10.1109/INFOCOM.2016.7524467

Wang, . (2020). Categorization of covert channels and its application in threat restriction techniques. *Journal of Software*, *31*(1), 228–245.

Wendzel, , Zander, S., Fechner, B., & Herdin, C. (2015). Pattern-based survey and categorization of network covert channel techniques. *ACM Computing Surveys*, *47*(3), 50. DOI: 10.1145/2684195

Wendzel, S., Kahler, B., & Rist, T. (2012). Covert channels and their prevention in building automation protocols: a prototype exemplified using BACnet. *Proceedings of 2012 IEEE International Conference on Green Computing and Communications*, 731-736. DOI: 10.1109/GreenCom.2012.120

Wolf, M. (1989). *Covert channels in LAN protocols. Local Area Network Security*. Springer.

Wu, . (2015). POSTER: biTheft: stealing your secrets by bidirectional covert channel communication with zero-permission android application. *Proceedings of the 22nd ACM SIGSAC Conference on Computer and Communications Security*, 1690-1692. DOI: 10.1145/2810103.2810108

Zander, S., Armitage, G., & Branch, P. (2007). Covert channels and countermeasures in computer network protocols. *IEEE Communications Surveys and Tutorials*, *45*(12), 136–142. DOI: 10.1109/COMST.2007.4317620

Zhang, . (2020a). A covert communication method using special bitcoin addresses generated by vanitygen. *Computers, Materials & Continua*, *65*(1), 597–616. DOI: 10.32604/cmc.2020.011554

Zhang, . (2020b). *The research on covert communication model based on blockchain: a case study of Ethereum's whisper protocol*. Frontiers in Cyber Security.

Zhang, , Zhang, Z., Jin, Z., Su, Y., & Wang, Z. (2021). An approach of covert communication based on the Ethereum whisper protocol in blockchain. *International Journal of Intelligent Systems*, *36*(2), 962–996. DOI: 10.1002/int.22327

Zhang, , Zhang, Z., Wang, W., Jin, Z., Su, Y., & Chen, H. (2022). Research on a covert communication model realized by using smart contracts in blockchain environment. *IEEE Systems Journal*, *16*(2), 2822–2833. DOI: 10.1109/JSYST.2021.3057333

Zou, . (2005). *The research on information hiding based on command sequence of FTP protocol*. Knowledge-Based Intelligent Information and Engineering Systems. DOI: 10.1007/11553939_151

Chapter 7
IoBc-Blockchain Solutions for Secure and Private IoT Ecosystems

Mohammad Alauthman
https://orcid.org/0000-0003-0319-1968
Department of Information Security, University of Petra, Jordan

Ammar Almomani
https://orcid.org/0000-0002-8808-6114
Higher Colleges of Technology, Sharjah, UAE

Amjad Aldweesh
https://orcid.org/0000-0001-9358-1323
College of Computing and Information Technology, Shaqra University, Saudi Arabia

Ahmad Al-Qerem
https://orcid.org/0000-0003-2187-6194
Computer Science Department, Faculty of Information Technology, Zarqa University, Jordan

Saad Alateef
https://orcid.org/0009-0002-3219-0833
School of Computing Science, Newcastle University, UK

Ala Hamarsheh
https://orcid.org/0000-0002-4736-7331
Department of Computer Science, Faculty of Information Technology, Arab American University, Palestine

Wael Hadi
https://orcid.org/0000-0002-7575-9287
Department of Information Security, University of Petra, Jordan

ABSTRACT

The exponential growth of Internet of Things (IoT) devices has created unprecedented security and privacy challenges that traditional centralized architectures struggle to address effectively. This chapter presents a comprehensive theoretical exploration of Internet of Blockchain (IoBc), an emerging paradigm that integrates blockchain technology with IoT ecosystems to enhance security, privacy, and trust. The chapter examines the theoretical foundations underpinning IoBc, analyzes its architectural frameworks, and investigates the cryptographic primitives that enable secure IoT operations. Through comparative analysis of existing blockchain platforms for IoT applications, the chapter identifies theoretical and practical implications for implementing IoBc solutions across various domains. The research synthesizes current literature to formulate a novel conceptual framework that addresses the unique challenges of resource-constrained IoT environments while leveraging blockchain's immutability and decentralization

DOI: 10.4018/979-8-3373-2439-5.ch007

properties.

INTRODUCTION

The proliferation of Internet of Things (IoT) devices has transformed the technological landscape, creating an interconnected ecosystem where physical objects seamlessly integrate with the digital realm. Current estimates suggest that by 2025, over 75 billion IoT devices will be deployed globally, generating unprecedented volumes of data and enabling novel applications across diverse domains including healthcare, smart cities, industrial automation, and consumer electronics (Statista, 2023). This exponential growth, however, introduces significant security and privacy challenges that traditional centralized architectures struggle to address effectively. The heterogeneous nature of IoT systems, characterized by resource-constrained devices, diverse communication protocols, and complex deployment scenarios, creates a vast attack surface vulnerable to exploitation (Khan & Salah, 2018). Conventional security mechanisms often prove inadequate in the IoT context due to their reliance on centralized trust models, substantial computational requirements, and inability to scale effectively across billions of devices.

Emerging at this critical juncture, blockchain technology presents a promising approach to address the inherent limitations of current IoT security frameworks. Fundamentally, blockchain represents a distributed ledger technology that maintains a continuously growing list of records (blocks) linked using cryptography, where each block contains a timestamp and transaction data that is inherently resistant to modification (Nakamoto, 2008). The decentralized, immutable, and transparent nature of blockchain aligns with the distributed architecture of IoT ecosystems, potentially offering enhanced security, improved privacy, and robust trust mechanisms without relying on central authorities (Fernández-Caramés & Fraga-Lamas, 2018). This integration of blockchain with IoT systems has given rise to a new paradigm termed the Internet of Blockchain (IoBc), which emphasizes the symbiotic relationship between these technologies to create secure and private IoT ecosystems.

The theoretical foundation of IoBc emerges from the convergence of multiple disciplines, including distributed systems, cryptography, game theory, and network security. The immutability property of blockchain, ensured through cryptographic hash functions and consensus mechanisms, addresses the data integrity concerns prevalent in IoT applications (Dorri et al., 2017). Smart contracts—self-executing contracts with the terms directly written into code—enable autonomous and transparent interactions between IoT devices, reducing the need for intermediaries and minimizing potential points of failure (Christidis & Devetsikiotis, 2016). Furthermore, the decentralized architecture of blockchain mitigates the risk of single points of failure that plague centralized IoT deployments, thereby enhancing system resilience against attacks and failures (Ali et al., 2018).

Despite these promising attributes, the integration of blockchain with IoT presents significant theoretical and practical challenges. The resource-intensive nature of traditional blockchain platforms, exemplified by the proof-of-work consensus mechanism in Bitcoin, contradicts the resource-constrained reality of many IoT devices (Novo, 2018). The scalability limitations of conventional blockchain implementations, including limited transaction throughput and high latency, appear incompatible with the real-time requirements of many IoT applications (Reyna et al., 2018). Additionally, the permanence of data stored on blockchain raises important privacy considerations, particularly in contexts where IoT devices collect sensitive personal information (Zyskind et al., 2015). These challenges necessitate a comprehensive theoretical examination of how blockchain principles can be adapted to accommodate the unique requirements of IoT ecosystems.

This chapter aims to provide a comprehensive theoretical discussion of IoBc, examining how blockchain principles can be adapted and applied to create secure and private IoT ecosystems. The research objectives include: (1) analyzing the theoretical foundations underpinning the integration of blockchain with IoT; (2) examining architectural frameworks and design principles for implementing IoBc solutions; (3) investigating cryptographic primitives and consensus mechanisms suitable for resource-constrained IoT environments; (4) comparing existing blockchain platforms for IoT applications; and (5) identifying theoretical and practical implications for future IoBc implementations.

In addressing these objectives, this chapter contributes to the emerging discourse on IoBc by synthesizing current literature and formulating a novel conceptual framework that addresses the unique challenges of IoT environments while leveraging blockchain's inherent security properties. The theoretical discussion moves beyond simplistic applications of blockchain in IoT contexts to consider the fundamental transformations required in both technologies to achieve meaningful integration. By critically examining competing theoretical perspectives and identifying unresolved questions, this chapter establishes a foundation for future research and development in the IoBc domain.

THEORETICAL FOUNDATIONS AND CONCEPTUAL FRAMEWORKS

The theoretical underpinnings of Internet of Blockchain (IoBc) emerge from the convergence of multiple disciplines, including distributed systems theory, cryptographic principles, network security frameworks, and economic models. This section examines the fundamental theories and conceptual frameworks that provide the foundation for integrating blockchain technology with IoT ecosystems to enhance security and privacy.

Distributed Systems Theory and Blockchain

The theoretical foundation of blockchain technology draws heavily from distributed systems theory, which examines how multiple computing entities can coordinate their actions without centralized control. The seminal work of Lamport et al. (1982) on Byzantine Fault Tolerance (BFT) established the theoretical basis for achieving consensus in distributed systems where some participants may behave maliciously. This theoretical framework addresses what is known as the Byzantine Generals Problem, establishing that reliable agreement can be achieved if fewer than one-third of participants are malicious. The practical implementation of BFT remained challenging until Nakamoto (2008) introduced Bitcoin, proposing Proof of Work (PoW) as a probabilistic solution to the Byzantine consensus problem in open, permissionless networks.

Fischer et al. (1985) established another theoretical cornerstone with the FLP impossibility result, proving that no deterministic consensus protocol can guarantee all three properties of safety, liveness, and fault tolerance in an asynchronous system. This theoretical limitation shapes blockchain design decisions, forcing systems to make tradeoffs between these properties. For instance, Bitcoin prioritizes liveness and fault tolerance over strict safety guarantees, accepting the possibility of temporary forks that eventually resolve (Garay et al., 2015).

Distributed ledger technologies implement these theoretical principles through various consensus mechanisms, each with distinct theoretical properties regarding security, performance, and resource requirements. Table 1 summarizes the theoretical foundations of major consensus protocols relevant to IoBc implementations.

Table 1. Theoretical foundations of consensus protocols for IoBc

Consensus Protocol	Theoretical Foundation	Trust Model	Resource Requirements	Scalability Properties	Finality Characteristic
Proof of Work (PoW)	Computational puzzles; Longest chain rule	Trustless; Open participation	High computational power; High energy consumption	Limited throughput; High latency	Probabilistic finality
Proof of Stake (PoS)	Economic stake as security measure; Slashing conditions	Economic incentives; Stake-based validation rights	Low computational requirements; Energy efficient	Moderate throughput; Moderate latency	Strong probabilistic finality
Delegated Proof of Stake (DPoS)	Delegated authority; Voting mechanisms	Delegated trust; Representative validation	Low computational requirements; Energy efficient	High throughput; Low latency	Strong probabilistic finality
Practical Byzantine Fault Tolerance (PBFT)	Byzantine agreement; Message passing	Permissioned network; Known validators	Moderate computational requirements; Communication-intensive	Limited by communication overhead; N^2 message complexity	Immediate finality
Directed Acyclic Graph (DAG)	Topological ordering; Parallel validation	Partial order consensus; Parallel transaction validation	Low computational requirements; Communication-efficient	High throughput; Low latency	Probabilistic finality with gradual confirmation
Proof of Authority (PoA)	Identity-based validation rights	Reputation-based; Known validators	Low computational requirements; Energy efficient	High throughput; Low latency	Immediate finality
Federated Byzantine Agreement (FBA)	Quorum slices; Validator selection	Trust-based validator selection; Overlapping quorums	Low computational requirements; Communication-efficient	Moderate throughput; Low latency	Immediate finality

Cryptographic Foundations

Cryptographic theory provides essential building blocks for secure IoBc implementations. Public key cryptography, formalized by Diffie and Hellman (1976), enables secure communication and authentication in decentralized environments without requiring pre-shared secrets. Digital signature schemes derived from this theoretical framework allow IoT devices to verify the authenticity and integrity of messages and transactions without centralized verification authorities. Hash functions with theoretical properties of preimage resistance, second preimage resistance, and collision resistance ensure data integrity and form the basis of blockchain's immutable record-keeping (Menezes et al., 1996).

Advanced cryptographic constructs extend these foundational elements to address specific requirements of IoBc systems. Merkle trees, theoretically formulated by Merkle (1980), enable efficient verification of data integrity with minimal storage requirements—a critical consideration for resource-constrained IoT devices. Zero-knowledge proofs, introduced by Goldwasser et al. (1989), provide theoretical mechanisms for verifying claims without revealing underlying data, offering powerful privacy-preserving capabilities for sensitive IoT applications. Homomorphic encryption schemes allow computation on encrypted data, theoretically enabling privacy-preserving data analysis in IoT contexts (Gentry, 2009).

The theoretical foundations of secure multi-party computation (MPC) also inform IoBc designs where multiple IoT devices must jointly compute functions over their inputs while keeping those inputs private (Yao, 1982). This theoretical framework enables collaborative data analysis among IoT devices without centralizing potentially sensitive information. Recent theoretical advances in threshold cryptography provide mechanisms for distributing cryptographic operations across multiple entities, enhancing security by eliminating single points of failure in key management (Shamir, 1979; Gennaro et al., 2018).

Figure 1 illustrates the relationship between these cryptographic primitives and their application in securing various aspects of IoBc systems.

Figure 1. Cryptographic primitives in IoBc security framework

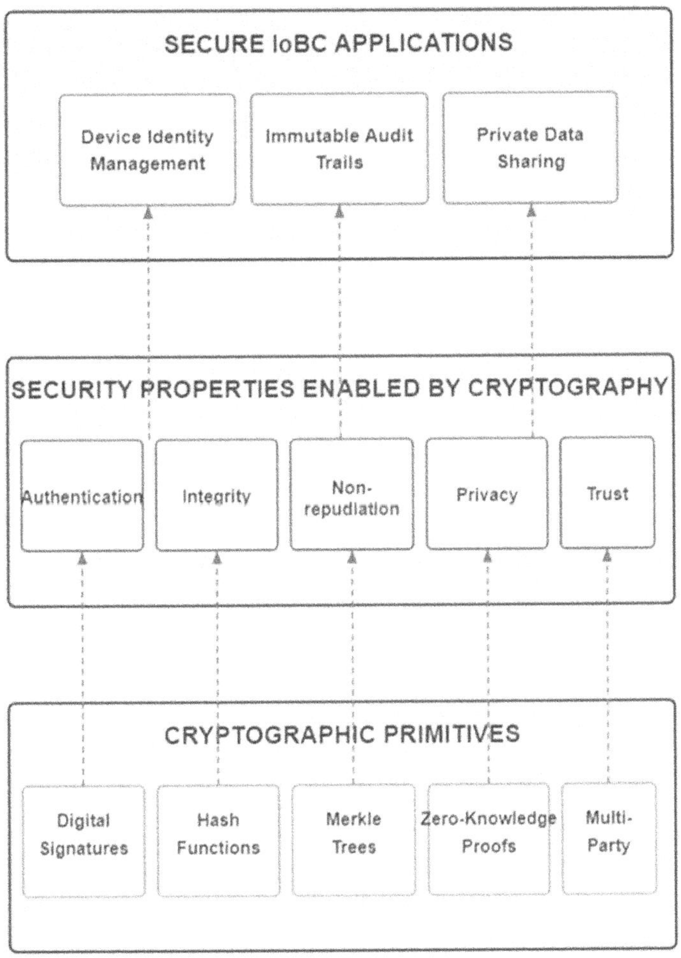

This is a hierarchical diagram showing how different cryptographic primitives (at the bottom level) support security properties (middle level) that enable secure IoBc applications (top level). The diagram shows connections between primitives like hash functions, digital signatures, zero-knowledge proofs, and how they enable properties like integrity, authentication, non-repudiation, and privacy, which in turn support secure device identity, immutable audit logs, private data sharing, and secure smart contracts.

Game Theory and Economic Models

The theoretical framework of game theory provides essential insights into the incentive structures that ensure the security and functionality of blockchain systems. Nakamoto's original Bitcoin design implicitly incorporated game-theoretic principles by aligning the economic interests of participants with the security of the network (Nakamoto, 2008). Subsequent theoretical analyses have formalized these concepts, demonstrating how rational self-interest among participants can maintain system security under specific conditions (Kiayias et al., 2016).

In the context of IoBc, game theory helps analyze the interaction between different stakeholders—device manufacturers, network operators, service providers, and end-users—and design incentive mechanisms that promote desired behaviors. The theoretical concept of mechanism design, sometimes described as "reverse game theory," guides the creation of protocols that achieve specific outcomes when participants act according to their rational self-interest (Hurwicz & Reiter, 2006). This theoretical framework enables the design of token economies and reward systems in IoBc that incentivize security-enhancing behaviors such as honest validation, resource sharing, and timely security updates.

Economic models of security as an investment decision also inform IoBc implementations. Anderson and Moore (2006) highlighted how misaligned incentives often lead to underinvestment in security, a problem particularly relevant to IoT ecosystems where device manufacturers may prioritize features and cost over security. Blockchain-based approaches can theoretically address these misalignments by creating transparent accountability mechanisms and enabling security-based reputation systems that reward manufacturers who maintain strong security practices (Rowe, 2007).

Trust Models and Frameworks

Trust represents a fundamental theoretical construct in both IoT and blockchain systems. Traditional IoT security architectures often rely on hierarchical trust models, where a central authority establishes and manages trust relationships between devices (Roman et al., 2013). In contrast, blockchain systems implement distributed trust models that eliminate single points of failure and enable trustworthy interactions between entities without prior relationships (Werbach, 2018).

The concept of trustless systems represents an important theoretical contribution from blockchain research, referring not to an absence of trust but rather to the elimination of the need to trust any specific entity (Antonopoulos, 2014). Instead, trust is placed in the cryptographic and game-theoretic properties of the system as a whole. This theoretical perspective transforms how security is conceptualized in distributed systems, moving from models based on trusted third parties to those based on verifiable computation and economic incentives.

The theoretical concept of trust boundaries—demarcations between trusted and untrusted system components—informs security architecture decisions in IoBc systems. Traditional approaches often establish broad trust boundaries encompassing entire platforms or networks. Blockchain enables more granular trust boundaries, where specific transactions or interactions can be verified independently without trusting the entire system (Tschorsch & Scheuermann, 2016).

Systems Theory and Information Security

Systems theory provides a theoretical framework for understanding the complex interactions within IoT ecosystems and how blockchain integration affects system properties. General systems theory, originated by von Bertalanffy (1968), emphasizes the importance of studying systems holistically rather than focusing exclusively on individual components. This perspective is particularly relevant for IoBc, where security and privacy emerge from the interaction of multiple system elements rather than from any single component or technology.

Information security theory provides fundamental concepts for analyzing and designing secure IoBc systems. The classic CIA triad—confidentiality, integrity, and availability—represents a theoretical foundation for information security objectives (Whitman & Mattord, 2011). Blockchain technology naturally addresses the integrity aspect through cryptographic linking of blocks and distributed validation but may require additional mechanisms to ensure confidentiality and availability in IoT contexts.

The theoretical concept of defense in depth—implementing multiple security measures to protect a system—informs the design of comprehensive IoBc security architectures (NSA, 2010). Rather than relying solely on blockchain's inherent security properties, robust IoBc implementations incorporate multiple defensive layers including secure hardware, encrypted communications, intrusion detection systems, and physical security measures. This theoretical approach acknowledges that no single security mechanism, including blockchain, can address all threats in complex IoT environments.

Synthesis of Theoretical Foundations

The theoretical foundations of IoBc emerge from the integration of these diverse theoretical frameworks. Distributed systems theory provides the fundamental principles for achieving consensus without centralized control. Cryptographic theory offers mechanisms for secure communication, authentication, and data protection. Game theory explains how incentives shape participant behavior and can be designed to enhance security. Trust models formalize how confidence is established in distributed environments. Systems theory provides frameworks for understanding complex interactions and emergent properties. Information security theory guides comprehensive protection strategies across multiple objectives.

These theoretical domains intersect in IoBc implementations, creating novel security and privacy capabilities that transcend any individual framework. For example, the combination of cryptographic primitives with game-theoretic incentives enables trustless verification systems that function reliably even when individual participants behave selfishly. The integration of information security objectives with distributed systems principles creates resilient protection that eliminates single points of failure. The synthesis of trust models with systems theory explains how security properties emerge from the collective behavior of network participants rather than from centralized authorities.

Figure 2 visualizes this theoretical integration, showing how concepts from different domains combine to form the foundational principles of IoBc.

Figure 2. Integrated theoretical framework for IoBc

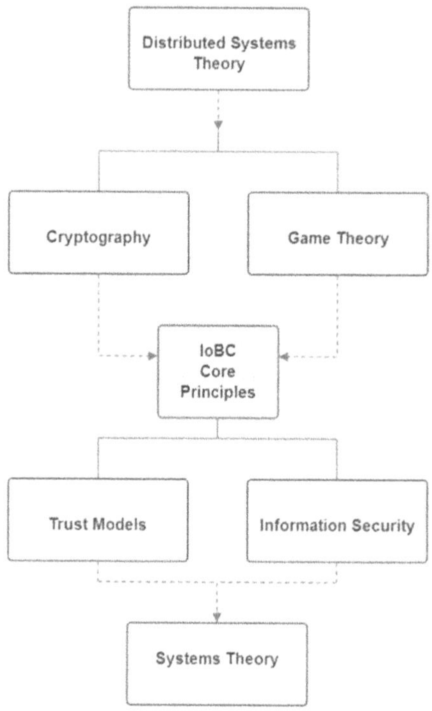

[This is a conceptual diagram showing overlapping circles representing different theoretical domains (Distributed Systems, Cryptography, Game Theory, Trust Models, Systems Theory, Information Security) with interconnecting lines where key concepts overlap, forming the core principles of IoBc at the center where all domains intersect.]

This integrated theoretical framework provides a comprehensive foundation for understanding, analyzing, and designing IoBc systems. The framework acknowledges that secure and private IoT ecosystems require more than technological solutions—they depend on aligned incentives, appropriate trust models, and systematic approaches to managing complexity. By synthesizing diverse theoretical perspectives, this framework enables more holistic approaches to IoBc implementation that address not only technical challenges but also economic, social, and organizational factors that influence security outcomes.

KEY CONCEPTS AND CONSTRUCTS

Building upon the theoretical foundations discussed previously, this section examines the essential concepts and constructs that enable the practical implementation of Internet of Blockchain (IoBc) systems. These key elements represent the fundamental building blocks for creating secure and private IoT ecosystems leveraging blockchain technology.

Lightweight Consensus Protocols for Resource-Constrained Environments

Consensus protocols represent the mechanisms through which distributed participants in a blockchain network agree on the current state of the system without requiring a central authority. Traditional blockchain implementations employ consensus mechanisms that often demand significant computational resources, making them unsuitable for direct implementation on resource-constrained IoT devices. This limitation has driven the development of lightweight consensus protocols specifically designed for the unique constraints of IoT environments.

Proof of Work (PoW), the consensus mechanism popularized by Bitcoin, requires participants to solve computationally intensive cryptographic puzzles to validate transactions and create new blocks (Nakamoto, 2008). While this approach provides strong security guarantees in permissionless networks, its high energy consumption and computational requirements make it impractical for most IoT devices. Analysis by Sedlmeir et al. (2020) estimated that Bitcoin's PoW consensus consumed approximately 121.36 TWh annually—comparable to the electricity usage of medium-sized countries—highlighting the fundamental mismatch with energy-constrained IoT devices.

Several consensus protocols have been developed specifically for resource-constrained IoT environments. Delegated Byzantine Fault Tolerance (dBFT), implemented in platforms like NEO, enables high transaction throughput with minimal computational requirements by limiting validation to a small set of trusted nodes (NEO Project, 2018). Directed Acyclic Graph (DAG) based protocols such as IOTA's Tangle allow for parallel transaction validation without the strict sequential blocks of traditional blockchains, potentially increasing scalability while reducing computational requirements (Popov, 2018).

Hybrid approaches combine different consensus mechanisms to address the specific requirements of IoT environments. Dorri et al. (2017) proposed a lightweight blockchain architecture for IoT that employs different consensus mechanisms at different layers of the system hierarchy. In this model, resource-constrained IoT devices interact with local blockchain overlays that implement simplified consensus mechanisms, while these overlays connect to a more secure global blockchain running a more robust consensus protocol. This hierarchical approach enables IoT devices to benefit from blockchain security without directly implementing resource-intensive consensus protocols.

The selection of appropriate consensus protocols for IoBc implementations depends on multiple factors including the security requirements of the application, the resource constraints of participating devices, network characteristics, and trust relationships between participants. Organizations implementing IoBc must carefully evaluate these tradeoffs based on their specific requirements and constraints.

IDENTITY MANAGEMENT AND AUTHENTICATION

Secure identity management represents a fundamental requirement for IoT security, enabling devices to authenticate themselves to networks and services while preventing impersonation attacks. Traditional centralized identity management systems rely on trusted authorities to issue and verify credentials, creating potential single points of failure and scalability limitations. Blockchain-based approaches offer alternative models for managing device identity in distributed IoT ecosystems.

Self-sovereign identity (SSI) represents a paradigm shift in identity management, giving users or devices control over their own identity information without depending on centralized authorities (Allen, 2016). Blockchain provides the technological foundation for SSI by enabling the creation of decentralized identifiers (DIDs) that remain under the control of their owners while being verifiable by any party. The W3C DID specification defines methods for creating identifiers that reference DID documents containing authentication information such as public keys (W3C, 2021). This approach enables IoT devices to maintain persistent identities independent of any specific manufacturer, service provider, or network operator.

Verifiable credentials extend the SSI model by allowing claims about identity attributes to be issued, held, and verified in a decentralized manner (W3C, 2022). In IoT contexts, verifiable credentials enable devices to prove specific attributes—such as manufacturer certification, software version, or security capabilities—without revealing additional information. This selective disclosure capability enhances privacy by implementing the principle of data minimization while still enabling necessary authentication and authorization decisions.

The integration of hardware security modules (HSMs) with blockchain-based identity systems enhances security by protecting cryptographic keys and operations. Trusted Platform Modules (TPMs) and secure elements provide tamper-resistant storage for identity keys and can perform cryptographic operations without exposing sensitive material (Trusted Computing Group, 2019). This integration addresses a critical vulnerability in software-only identity solutions, where key compromise could lead to permanent identity theft. Hardware-backed identity systems enable secure device attestation, where devices can provably demonstrate they are running authentic, unmodified software—a critical capability for maintaining trust in IoT ecosystems (Raj et al., 2016).

Privacy-Preserving Mechanisms

Privacy protection represents a critical requirement for IoT systems that often collect sensitive personal, industrial, or environmental data. The inherent transparency of many blockchain implementations creates potential conflicts with privacy objectives, necessitating specialized mechanisms that preserve the security benefits of blockchain while protecting sensitive information.

Zero-knowledge proofs (ZKPs) represent a powerful cryptographic technique that enables one party to prove knowledge of a fact without revealing the fact itself (Goldwasser et al., 1989). In IoBc contexts, ZKPs allow devices to verify critical properties—such as authentic measurements, authorized operations, or policy compliance—without exposing the underlying data. Various ZKP implementations offer different tradeoffs between computational efficiency and expressive power. zk-SNARKs (Zero-Knowledge Succinct Non-Interactive Arguments of Knowledge) provide highly efficient verification but require a trusted setup phase (Ben-Sasson et al., 2014). zk-STARKs (Zero-Knowledge Scalable Transparent Arguments of Knowledge) eliminate the trusted setup requirement but typically require larger proof sizes (Ben-Sasson et al., 2018).

Secure multi-party computation (MPC) enables multiple parties to jointly compute functions over their inputs while keeping those inputs private (Yao, 1982). This technique allows distributed data analysis in IoT environments without centralizing sensitive information. For example, smart meters in a neighborhood could use MPC to calculate aggregate energy consumption patterns without revealing individual household data (Mukherjee & Ruj, 2022). The integration of MPC with blockchain provides accountability for computation results while maintaining input privacy, creating powerful capabilities for privacy-preserving IoT analytics (Zyskind et al., 2015).

Off-chain storage patterns address blockchain privacy challenges by keeping sensitive data off the blockchain while maintaining cryptographic links to preserve integrity and provenance. Content-addressable storage systems such as IPFS (InterPlanetary File System) or Swarm enable data to be referenced by its cryptographic hash, allowing blockchain transactions to verify data integrity without containing the data itself (Benet, 2014; Ethereum Swarm, 2021). Private data collections in consortium blockchains like Hyperledger Fabric allow a subset of participants to maintain confidential data invisible to other network members (Androulaki et al., 2018).

The selection of appropriate privacy-preserving mechanisms depends on multiple factors including the sensitivity of the data, regulatory requirements, performance constraints, and trust relationships between participants. Many effective IoBc implementations combine multiple privacy techniques to address different aspects of the system. For example, a health monitoring application might use zero-knowledge proofs to verify measurement validity, off-chain storage for raw sensor data, differential privacy for population analytics, and secure multi-party computation for collaborative diagnosis.

Smart Contracts and Autonomous Operations

Smart contracts represent self-executing agreements where the terms are directly written into code, enabling autonomous enforcement without requiring trusted intermediaries. First conceptualized by Szabo (1997) as digital protocols that facilitate, verify, or enforce the negotiation or performance of a contract, smart contracts became practically implementable with blockchain technology. In IoBc contexts, smart contracts enable autonomous device operations, automated service agreements, secure multi-party coordination, and programmable security policies.

The implementation of smart contracts in IoT environments presents unique challenges and opportunities compared to traditional blockchain applications. Resource constraints limit the computational complexity of contracts that can be directly executed on IoT devices, necessitating lightweight implementations or execution delegation to more powerful nodes. Real-time requirements in many IoT applications conflict with the confirmation latency of traditional blockchain networks, driving the development of faster consensus mechanisms and layer-2 solutions. The physical consequences of smart contract operations in IoT contexts—such as controlling critical infrastructure or autonomous vehicles—heighten the importance of contract correctness and security (Christidis & Devetsikiotis, 2016).

Smart contracts enable several transformative capabilities in IoBc systems:

1. **Autonomous Device Coordination**: Smart contracts can orchestrate complex interactions between multiple IoT devices without requiring centralized coordinators. For example, in a smart building environment, contracts can manage resource allocation, scheduling, and conflict resolution between various systems including HVAC, lighting, and security (Huh et al., 2017).
2. **Automated Service Agreements**: Service level agreements between devices and service providers can be encoded as smart contracts that automatically monitor compliance, apply penalties for violations, and distribute compensation. This capability enables dynamic resource markets where IoT devices can automatically negotiate and secure necessary services based on current requirements (Shafagh et al., 2017).
3. **Conditional Access Control**: Smart contracts can implement sophisticated access control policies that adapt to current conditions, user roles, and system states. These programmable policies provide more flexible and nuanced control compared to static access control lists, particularly valuable in dynamic IoT environments with varying trust relationships (Zhang et al., 2018).

4. **Secure Multi-stakeholder Governance**: In IoT ecosystems involving multiple organizations or stakeholders, smart contracts can implement transparent governance mechanisms such as multi-signature approvals, voting systems, and automated dispute resolution. These capabilities are particularly relevant for industrial IoT, supply chain, and smart city applications (Hardjono & Smith, 2016).
5. **Autonomous Market Operations**: IoT devices can participate in decentralized marketplaces for data, services, and resources through smart contract mechanisms. These markets enable monetization of device capabilities and data while providing transparent pricing and terms enforcement without requiring trusted intermediaries (Xu et al., 2018).

Figure 3 illustrates a reference architecture for smart contract integration in IoBc systems, showing the relationship between different components and the flow of information and control.

Figure 3. Smart contract integration architecture for IoBc

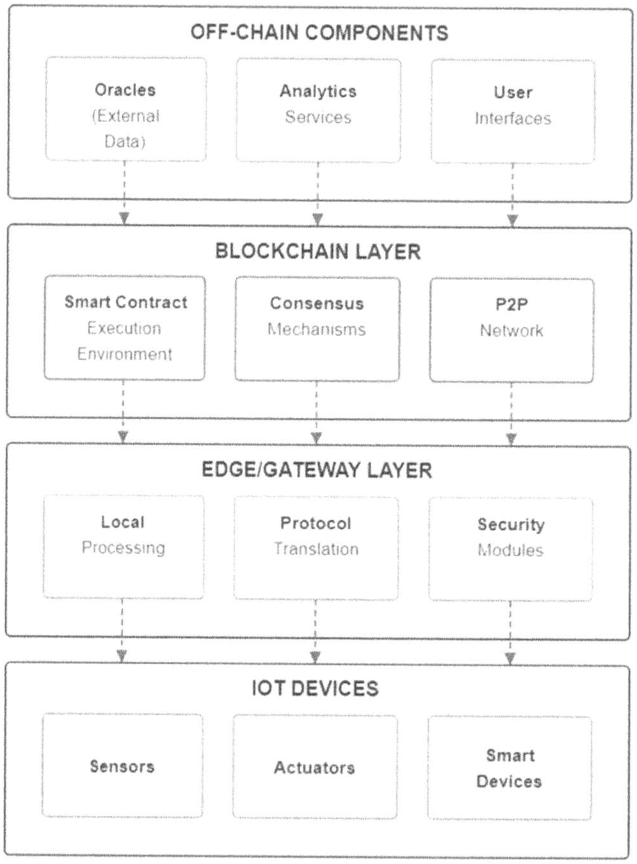

[This is a layered architecture diagram showing: at the bottom, IoT Devices connected to Edge Nodes; in the middle, a Blockchain Layer with Smart Contract execution environments, Consensus mechanisms, and P2P networking; at the top, Off-chain Components including Oracles, Analytics, and User Interfaces. Arrows show data and control flows between components.]

The security of smart contracts represents a critical concern in IoBc implementations, particularly given the potential physical consequences of exploitation in IoT contexts. Smart contract vulnerabilities have led to significant incidents including the DAO hack that resulted in the loss of approximately $50 million (Mehar et al., 2019). Common vulnerability classes include reentrancy attacks, where contracts can be interrupted and re-entered during execution; integer overflow/underflow errors that cause unexpected value wrapping; time manipulation attacks that exploit dependencies on block timestamps; and front-running attacks where observers can anticipate and preempt transactions (Atzei et al., 2017).

Addressing smart contract security requires comprehensive approaches including formal verification, specialized programming languages, security-focused development practices, and thorough auditing. Formal verification techniques mathematically prove that contracts behave according to their specifications under all possible conditions, providing stronger guarantees than traditional testing (Bhargavan et al., 2016). Languages like Solidity have evolved to include safety features such as overflow checking and reentrancy guards, while alternative languages like Vyper emphasize simplicity and auditability over expressiveness (Ethereum Foundation, 2021).

TOKENIZATION MODELS AND EDGE COMPUTING INTEGRATION

Tokenization represents the process of creating digital representations of assets, rights, or capabilities that can be managed on blockchain networks. In IoBc contexts, tokenization enables novel economic models for device interaction, resource allocation, data exchange, and service provision. These models transform how value is created, exchanged, and managed in IoT ecosystems, potentially addressing significant challenges related to incentive alignment, resource optimization, and sustainable operation.

These token types enable various value exchange models in IoT ecosystems:

1. **Device-as-a-Service Models**: Tokenization enables flexible service-based access to device capabilities rather than outright ownership. Users can purchase tokens that grant temporary access to functionality, enabling more efficient resource utilization and reducing capital expenditure requirements. This model allows dynamic pricing based on current demand and resource availability (Novo, 2018).
2. **Data Marketplaces**: IoT devices generate valuable data that can be tokenized and traded in decentralized marketplaces. Smart contracts enable automated negotiation, payment, and delivery with transparent terms and verifiable compliance. Privacy-preserving mechanisms allow selective data sharing while maintaining control over sensitive information (Shafagh et al., 2017).
3. **Resource Sharing Economies**: Tokens incentivize sharing of unused resources such as storage, bandwidth, computation, or physical assets among IoT devices. This collaborative approach improves overall system efficiency by utilizing resources that would otherwise remain idle, particularly valuable in environments with variable or unpredictable resource needs (Xu et al., 2018).
4. **Incentivized Infrastructure**: Token rewards encourage deployment and maintenance of IoT infrastructure such as gateways, network relays, or edge computing nodes. This model enables community-driven network expansion without requiring centralized coordination or funding, potentially addressing connectivity challenges in underserved areas (Helium, 2021).

Edge computing moves data processing and storage closer to the sources of data, reducing latency, bandwidth usage, and dependence on cloud infrastructure. This paradigm aligns naturally with IoT deployments where devices generate substantial data volumes that may require real-time processing. The integration of edge computing with blockchain creates novel capabilities for secure, autonomous, and privacy-preserving IoT operations while addressing some of the inherent limitations of both technologies.

Blockchain integration with edge computing creates complementary capabilities that enhance both technologies:

1. **Distributed Trust and Verification**: Blockchain provides cryptographic verification of data integrity and provenance, enabling trust between edge nodes without requiring centralized authorities. This distributed trust model enables secure collaboration between edge devices from different manufacturers or operators (Xiong et al., 2018).
2. **Decentralized Coordination**: Smart contracts enable autonomous coordination between edge nodes without centralized orchestration, allowing dynamic resource allocation, workload distribution, and service discovery based on current conditions and requirements (Pan et al., 2018).
3. **Secure Data Marketplaces**: Tokenization enables edge nodes to exchange data, processing capabilities, or storage capacity in secure marketplaces with transparent terms and automatic enforcement. These markets create economic incentives for optimal resource utilization across distributed edge infrastructure (Sharma et al., 2017).
4. **Enhanced Privacy Control**: The combination of edge processing with blockchain's cryptographic capabilities enables more granular privacy control, allowing sensitive data to remain local while still enabling verifiable computation and selective sharing (Zyskind et al., 2015).
5. **Resilient Operations**: The distributed nature of both edge computing and blockchain enhances system resilience against failures or attacks. Redundant processing capabilities and decentralized coordination enable continued operation even when some components are compromised or unavailable (Roman et al., 2018).

The implementation of blockchain at the edge introduces several technical challenges due to the resource constraints and heterogeneity of edge environments. Traditional blockchain protocols require substantial processing power, storage capacity, and network bandwidth that exceed the capabilities of many edge devices. Several approaches address these limitations, including lightweight client implementations, hierarchical blockchain architectures, alternative consensus mechanisms, and sharding techniques that divide the blockchain network into smaller components that process transactions in parallel.

COMPARATIVE ANALYSIS AND SYNTHESIS

Having examined the theoretical foundations and key concepts of Internet of Blockchain (IoBc), this section provides a comparative analysis of existing approaches and synthesizes findings to identify patterns, contradictions, and emerging trends in this rapidly evolving field. This analysis focuses on comparing blockchain platforms for IoT applications, evaluating architectural approaches, assessing security and privacy models, and examining real-world implementation patterns.

Comparative Analysis of Blockchain Platforms for IoT

Various blockchain platforms have developed specialized capabilities aimed at addressing the unique requirements of IoT environments. These platforms differ in their consensus mechanisms, programming models, scalability approaches, security characteristics, and governance structures. Understanding these differences is essential for selecting appropriate technologies for specific IoT application scenarios.

Table 2 provides a comprehensive comparison of major blockchain platforms that have been applied to IoT use cases, evaluating them across multiple dimensions relevant to IoT requirements.

Table 2. Comparative analysis of blockchain platforms for IoT applications

Platform	Architecture	Consensus Mechanism	Transactions Per Second	IoT-Specific Features	Energy Efficiency	Privacy Features	Governance Model
Ethereum	Public blockchain; Account-based	Proof of Work transitioning to Proof of Stake	15-30 (main chain); Higher with Layer 2	Light client protocol; IoT-focused Layer 2 solutions	Moderate (PoW); High (PoS)	Basic privacy; ZK rollups	Open community governance; EIPs
Hyperledger Fabric	Permissioned; Channel-based	Pluggable consensus (typically PBFT variants)	3,000+	Private data collections; IoT integration frameworks	Very high	Channel isolation; Private data collections	Framework-level governance; Enterprise focus
IOTA	DAG-based (Tangle); Feeless	Coordicide (Director removal); Voting-based	1,000+	Designed for IoT; Feeless transactions; Masked Authenticated Messaging	Very high	Masked Authenticated Messaging	IOTA Foundation; Research-driven
Hyperledger Sawtooth	Modular permissioned	Pluggable (PoET default)	1,000+	Specialized IoT transaction families; Hardware integration	High	Transaction family isolation; Private transactions	Framework-level governance; Enterprise focus
VeChain	Two-token public blockchain	Proof of Authority	2,500+	Supply chain focus; IoT-optimized transaction types	Very high	Multi-party computation options	Voting-based; Stake-weighted
Corda	Permissioned; Point-to-point	Notary clusters (pluggable consensus)	1,000+ (varies by notary)	IoT oracles; Direct device integration	Very high	Transaction tear-offs; Need-to-know basis	R3 Consortium; Enterprise focus
EOS	Delegated Proof of Stake	dPoS with 21 block producers	4,000+	Resource allocation model; Low latency	High	Variable privacy options	On-chain governance; Token voting
Algorand	Pure Proof of Stake	PPoS with cryptographic sortition	1,000+	Atomic transfers; Asset tokenization	Very high	Standard privacy options	Algorand Foundation; Community input

The analysis reveals several important patterns and distinctions relevant to IoT applications:

1. **Performance Tradeoffs**: Platforms optimize for different performance characteristics based on their target applications. IOTA and Hedera Hashgraph prioritize high transaction throughput suitable for large-scale sensor networks, while Ethereum emphasizes security and decentralization at the cost of lower throughput. This divergence reflects fundamental tradeoffs between transaction capacity, security guarantees, and decentralization that influence platform selection for specific IoT scenarios.
2. **Consensus Diversity**: A variety of consensus mechanisms have emerged beyond traditional Proof of Work, including Directed Acyclic Graph (DAG) approaches, various Proof of Stake implementations, and Byzantine Fault Tolerance variants. This diversity enables more efficient consensus that aligns with IoT constraints, particularly regarding energy consumption and computational requirements.
3. **Governance Models**: Governance approaches range from highly decentralized community models (Ethereum) to consortium-based frameworks (Hyperledger Fabric, Corda) and foundation-led initiatives (IOTA, Algorand). These governance structures influence how platforms evolve to address IoT requirements and how they manage security vulnerabilities or protocol upgrades.
4. **Privacy Capabilities**: Privacy features vary significantly, from basic transaction privacy to sophisticated mechanisms like zero-knowledge proofs, private channels, and need-to-know architectures. These differences are particularly relevant for IoT applications handling sensitive personal, industrial, or governmental data subject to regulatory requirements.
5. **IoT-Specific Optimizations**: Platforms increasingly offer specialized features for IoT use cases, including lightweight client implementations, device identity management, secure data pipelines, and resource-efficient validation mechanisms. These optimizations reflect growing recognition of IoT as a distinct blockchain application domain with unique requirements.

The comparative analysis indicates that no single blockchain platform offers optimal characteristics for all IoT scenarios. Instead, platform selection should be guided by specific application requirements, deployment constraints, and ecosystem considerations. For example, supply chain applications might prioritize platforms with strong traceability and interoperability capabilities (VeChain, Hyperledger Fabric), while sensor networks generating high-frequency data might favor platforms optimized for throughput and low fees (IOTA, Hedera).

ARCHITECTURAL PATTERNS AND DESIGN APPROACHES

The integration of blockchain with IoT has given rise to distinct architectural patterns that address the inherent challenges of securing distributed, resource-constrained devices while leveraging the trust and transparency benefits of distributed ledger technologies. These architectural approaches represent different strategies for managing the tradeoffs between security, privacy, performance, and resource efficiency in IoBc implementations.

Comparative analysis reveals several dominant architectural patterns:

1. **Hierarchical Multi-Tier Architectures**: This pattern distributes blockchain functionality across multiple tiers with varying capabilities and responsibilities. Resource-constrained IoT devices interact with local gateways or edge nodes that manage blockchain operations on their behalf, while these intermediate nodes connect to more comprehensive blockchain networks. This approach addresses the resource limitations of IoT devices while preserving essential security properties. Dorri et al.

(2017) demonstrated this pattern with a lightweight IoT blockchain architecture where smart home devices connect to a local home blockchain managed by a more powerful home miner, which in turn connects to an overlay blockchain network. Similar approaches have been implemented in industrial settings where edge gateways manage blockchain operations for sensor networks (Fernández-Caramés & Fraga-Lamas, 2018).

2. **Sidechain and Layer-2 Approaches**: These architectures employ secondary blockchain structures that process most transactions off the main chain, periodically anchoring state commitments to the primary blockchain for security. This approach improves scalability and reduces transaction costs while maintaining the security guarantees of the main chain. Plasma (Poon & Buterin, 2017) and Rollups (Ethereum Foundation, 2022) exemplify this pattern in the Ethereum ecosystem, enabling high-throughput IoT applications while leveraging the security of the main Ethereum network.
3. **Direct Device Participation**: Some architectures enable IoT devices to participate directly in blockchain networks through lightweight client implementations and optimized protocols. IOTA's Tangle architecture (Popov, 2018) eliminates the distinction between validators and regular participants, allowing even constrained devices to contribute to network security by validating two previous transactions for each transaction they submit. Simplified Payment Verification (SPV) clients enable devices to verify relevant blockchain transactions without maintaining the complete chain state, reducing resource requirements while preserving verification capabilities (Nakamoto, 2008).
4. **Proxy-Based Approaches**: In these architectures, trusted intermediaries perform blockchain operations on behalf of IoT devices, translating between traditional IoT protocols and blockchain interfaces. The intermediaries manage cryptographic keys, submit transactions, and monitor blockchain state, presenting simplified APIs to connected devices. This approach minimizes changes to existing IoT infrastructure but introduces additional trust assumptions compared to more integrated approaches.
5. **Hybrid On-Chain/Off-Chain Architectures**: These designs selectively determine what information and processes should reside on the blockchain versus traditional databases or communication channels. Typically, only critical elements such as authentication events, key state transitions, or data integrity proofs are recorded on-chain, while raw data and routine operations remain off-chain. This pattern optimizes resource usage and addresses privacy concerns while maintaining cryptographic verifiability for essential aspects. Hyperledger Fabric's private data collections (Androulaki et al., 2018) and Baseline Protocol's zero-knowledge synchronization (Baseline Protocol, 2021) exemplify this approach.

The selection of appropriate architectural patterns depends on multiple factors including security requirements, device capabilities, deployment environment, and application characteristics. Many practical implementations combine elements from multiple patterns to address specific requirements.

Security and Privacy Model Comparison

Security and privacy represent fundamental concerns in IoT ecosystems, with blockchain offering novel approaches to addressing these challenges. Different IoBc implementations adopt varying security and privacy models, making distinct tradeoffs and assumptions that significantly impact their suitability for specific applications. This section analyzes and compares these models, examining their theoretical foundations, practical implications, and effectiveness in addressing IoT security and privacy challenges.

The security models implemented in IoBc systems can be categorized based on several key dimensions:

1. **Trust Assumptions**: IoBc implementations differ in their fundamental assumptions about which entities can be trusted and under what circumstances. Public permissionless blockchains like Ethereum or Bitcoin operate under minimal trust assumptions, requiring only that a majority of network participants (typically measured by computational power or stake) behave honestly (Nakamoto, 2008). Permissioned networks like Hyperledger Fabric assume trust in the consortium members who operate validation nodes but protect against individual member misbehavior (Androulaki et al., 2018). Proxy-based architectures introduce additional trust requirements for intermediaries that manage blockchain interactions on behalf of IoT devices.
2. **Threat Models**: Different approaches address distinct threat profiles based on anticipated attack vectors and adversary capabilities. Some implementations focus primarily on external threats such as data tampering or unauthorized access, while others also address threats from insiders such as malicious operators or compromised devices. Threat models vary in their assumptions about adversary resources, from computationally bounded attackers to those with nation-state capabilities including physical device access (Atzori et al., 2017).
3. **Key Management Approaches**: The management of cryptographic keys represents a critical security consideration in IoBc systems, with different implementations adopting varying approaches to key generation, storage, rotation, and recovery. Hardware security modules (HSMs) provide tamper-resistant key storage but increase cost and complexity. Threshold cryptography distributes key material across multiple entities, requiring collaboration for key usage and eliminating single points of failure (Shamir, 1979). Social recovery mechanisms enable key restoration through trusted contacts, addressing the challenge of key loss in consumer contexts (Buterin, 2021).

Similarly, privacy models in IoBc implementations can be distinguished based on several dimensions:

1. **Data Visibility Control**: Systems differ in how they manage visibility of data stored on or referenced by the blockchain. Public blockchains make all transaction data visible to all participants, though encrypted payloads can protect sensitive content. Private or permissioned blockchains restrict data access to authorized participants. Channel-based systems like Hyperledger Fabric further compartmentalize information, making data visible only to specific subgroups (Androulaki et al., 2018). Zero-knowledge systems enable verification of properties without revealing the underlying data (Ben-Sasson et al., 2014).
2. **Identity and Pseudonymity Models**: IoBc implementations adopt different approaches to managing identity information and the linkability of activities to specific entities. Some systems emphasize strong identity verification tied to real-world entities, suitable for regulated environments with accountability requirements. Others implement pseudonymous or anonymous participation, where activities cannot be easily linked to specific individuals or organizations. Many systems employ hybrid approaches with contextual identity revelation based on specific transactions or regulatory requirements (Henry et al., 2018).

The comparative analysis reveals significant diversity in security and privacy approaches, reflecting different priorities, application domains, and technological foundations. Enterprise-oriented implementations like Hyperledger Fabric and Azure Blockchain emphasize strong identity, managed key security, and regulatory compliance suitable for industrial applications. Public blockchain approaches like Ethereum-based IoT and IOTA prioritize decentralization and minimal trust assumptions appropriate for open ecosystems. Specialized IoT blockchains like IoTeX and Helium implement domain-specific security features addressing the unique challenges of connected device networks.

Figure 4 illustrates a privacy-preserving architecture for IoBc that addresses key regulatory requirements while maintaining blockchain's security benefits.

Figure 4. Privacy-preserving IoBc architecture with regulatory compliance

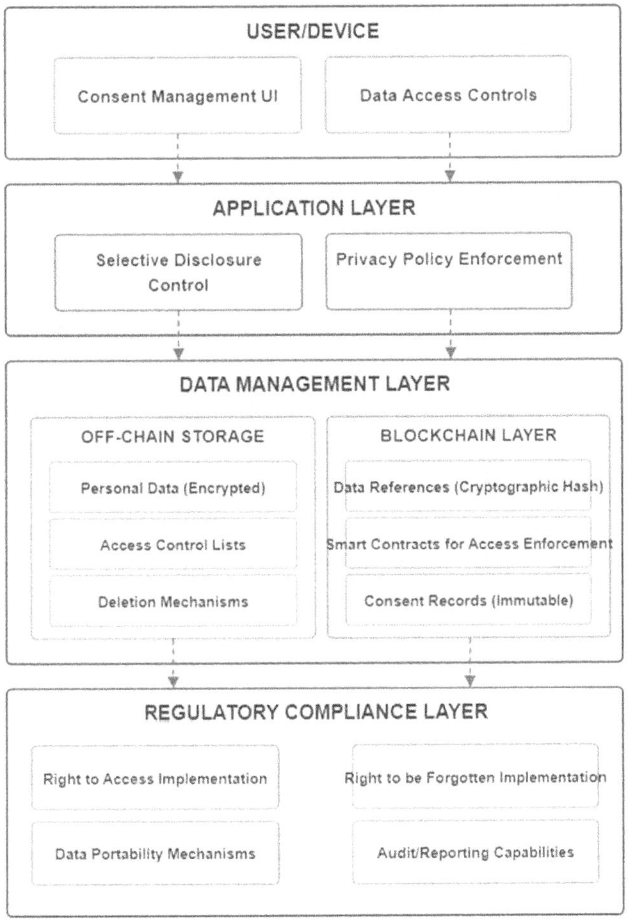

[This is a layered diagram showing how data flows through different components of a privacy-preserving IoBc system. It illustrates the separation between on-chain and off-chain data, with personal data kept off-chain and only cryptographic references stored on-chain. It shows consent management and erasure mechanisms that enable regulatory compliance while maintaining blockchain's integrity verification.]

Implementation Patterns and Best Practices

The practical implementation of IoBc systems reveals recurring patterns and best practices that address common challenges in integrating blockchain with IoT ecosystems. These patterns represent proven approaches that have emerged through practical experience and research across diverse application domains. Understanding these patterns helps practitioners avoid common pitfalls and leverage established solutions to recurring problems.

Several key implementation patterns have emerged in successful IoBc deployments:

1. **Layered Data Architecture**: This pattern separates data storage across multiple layers based on requirements for accessibility, immutability, and privacy. Time-sensitive or voluminous data remains in traditional databases or edge storage, with only cryptographic references and critical metadata stored on the blockchain. This approach addresses blockchain storage limitations while maintaining cryptographic verification capabilities. IBM's Hybrid Transaction/Analytical Processing (HTAP) architecture for blockchain-IoT integration exemplifies this pattern, using traditional databases for high-volume IoT data while storing verification hashes and access control information on the blockchain (IBM, 2020).
2. **Gateway-based Integration**: Rather than directly modifying IoT devices, this pattern introduces specialized gateways that bridge between existing IoT protocols and blockchain networks. The gateways manage blockchain interactions, cryptographic operations, and protocol translation, allowing legacy devices to benefit from blockchain capabilities without firmware modifications. Implementations such as Cisco's Kinetic platform demonstrate this approach, providing blockchain integration for diverse industrial IoT devices through specialized edge gateways (Cisco, 2019).
3. **Progressive Security Enhancement**: This pattern implements security improvements incrementally, starting with basic blockchain integration for specific high-value functions before expanding to more comprehensive coverage. The approach acknowledges resource constraints and organizational change management requirements, allowing systems to demonstrate value while gradually expanding their security perimeter. Supply chain implementations by companies like Walmart and Maersk illustrate this pattern, beginning with limited blockchain tracking for specific high-risk products before expanding to broader supply chain visibility (Hyperledger Foundation, 2020).

Table 3 summarizes implementation best practices that have emerged across multiple IoBc deployments, organized by system development lifecycle phase.

Table 3. IoBc implementation best practices by development phase

Development Phase	Best Practices	Rationale	Implementation Examples	Potential Pitfalls
Architecture and Design	Clearly define on-chain vs. off-chain boundaries	Optimize performance and cost while maintaining security guarantees	Hyperledger Grid's hybrid storage model; Streamr's cryptographic reference architecture	Excessive on-chain storage causing performance degradation and cost issues
	Implement defense in depth beyond blockchain	Blockchain alone doesn't address all IoT security requirements	Azure Sphere's layered security; Samsung ARTIK's multi-level protection	Overreliance on blockchain's inherent security properties
	Design for interoperability with existing systems	Most deployments must integrate with legacy infrastructure	Hyperledger Cactus cross-chain solution; Mulesoft blockchain connectors	Blockchain silos disconnected from enterprise systems

continued on following page

Table 3. Continued

Development Phase	Best Practices	Rationale	Implementation Examples	Potential Pitfalls
Implementation	Leverage established blockchain frameworks	Avoid security risks of custom implementations	Ethereum Enterprise solutions; Hyperledger Fabric deployments	"Not invented here" syndrome leading to security vulnerabilities
	Implement secure key management with hardware protection	Key compromise is a primary attack vector	IoTeX Secure Hardware integration; Ledger hardware security integration	Software-only key storage vulnerable to extraction
	Develop comprehensive identity and access management	Fine-grained control over who can access what	Sovrin's decentralized identity for IoT; uPort integration patterns	Coarse-grained access control leading to excessive permissions
Testing and Validation	Conduct threat modeling specific to blockchain-IoT integration	Unique threat landscape requires specialized analysis	STRIDE analyses of IoBc systems; MITRE ATT&CK framework application	Generic security testing missing blockchain-specific vulnerabilities
	Test data migration and recovery scenarios	Data immutability requires careful planning for changes and failures	ChainLink's fault-tolerant oracle design; Quorum's disaster recovery patterns	Inability to correct data errors or handle major failures
Deployment and Operation	Implement automated monitoring across blockchain and IoT layers	Detect anomalies that cross traditional boundaries	Hyperledger Caliper monitoring; ConsenSys Diligence tools	Siloed monitoring missing cross-layer attack patterns
	Establish secure update mechanisms	Both blockchain and IoT require secure updates	IoTeX secure firmware updates; Ethereum smart contract upgrade patterns	Inability to patch vulnerabilities due to immutability concerns
	Plan for scalability beyond initial deployment	IoT deployments often grow exponentially	Corda Network scalability designs; IOTA Chrysalis scalability enhancements	Performance degradation with increased scale

The analysis of implementation patterns reveals several critical success factors for IoBc deployments:

1. **Clear Value Proposition**: Successful implementations clearly articulate the specific value that blockchain adds to IoT security and privacy beyond what could be achieved with traditional technologies. This clarity helps justify the additional complexity and resource requirements of blockchain integration. For example, Everledger's diamond tracking platform demonstrates clear value in providing immutable provenance verification that prevents fraud in ways traditional databases cannot (Everledger, 2020).
2. **Appropriate Blockchain Selection**: Rather than defaulting to the most popular blockchain platforms, effective implementations select technologies based on specific requirements analysis. This selection considers factors including privacy needs, throughput requirements, governance models, and ecosystem maturity. Walmart's food safety tracking system, for instance, selected Hyperledger Fabric specifically for its private channels and permissioned model appropriate for supply chain partners (Hyperledger Foundation, 2020).
3. **Comprehensive Security Approach**: Successful deployments recognize that blockchain represents only one component of a comprehensive security strategy. They implement multiple security layers addressing device security, communication protection, data privacy, and physical security in addition to blockchain-based verification. The Trusted IoT Alliance reference architecture exemplifies this

layered approach, integrating blockchain with hardware security modules, encrypted communications, and physical tamper protection (TIoTA, 2019).

Common implementation challenges include balancing on-chain versus off-chain processing, managing the complexity of cross-chain interactions, ensuring appropriate key management across diverse devices, addressing the immutability-flexibility tension, and maintaining performance at scale. Successful implementations typically start with clearly defined scope boundaries, implement proofs of concept to validate key assumptions, and expand incrementally based on demonstrated success and lessons learned.

Synthesis of Comparative Findings

The comparative analysis of blockchain platforms, architectural approaches, security models, and implementation patterns reveals important insights about the current state and evolutionary trajectory of Internet of Blockchain. This synthesis identifies convergent trends, divergent approaches, unresolved challenges, and emerging opportunities in this rapidly developing field.

Several key findings emerge from this comparative analysis:

1. **Convergence toward Hybrid Architectures**: Despite their different starting points, IoBc implementations increasingly converge toward hybrid architectures that combine blockchain features with traditional technologies. Pure blockchain approaches struggle with IoT constraints including resource limitations, real-time requirements, and data volume. Similarly, traditional centralized architectures fail to address trust and security challenges in distributed IoT ecosystems. Hybrid approaches leverage the strengths of each technology while mitigating their individual limitations. This convergence suggests that effective IoT security requires both blockchain's decentralized trust and traditional systems' processing efficiency.
2. **Specialization for Vertical Applications**: While early blockchain implementations attempted to serve multiple use cases with generic platforms, newer approaches increasingly specialize for specific vertical applications such as supply chain, energy trading, healthcare, or industrial automation. This specialization enables optimization for domain-specific requirements, threat models, and regulatory frameworks. For example, supply chain implementations emphasize provenance tracking and stakeholder coordination, while energy applications focus on high-throughput transaction processing and complex marketplace operations. This trend indicates that blockchain's security and trust benefits manifest differently across application domains, requiring tailored implementations.
3. **Evolution from Technical to Sociotechnical Focus**: Initial IoBc research emphasized technical challenges such as consensus efficiency, scalability, and device integration. While these technical aspects remain important, recent developments increasingly address sociotechnical dimensions including governance models, economic incentives, regulatory compliance, and stakeholder alignment. This broadened focus acknowledges that successful security implementations depend not only on technological capabilities but also on organizational structures, economic models, and human factors. The evolution toward sociotechnical approaches reflects the complexity of securing distributed systems involving multiple independent actors.
4. **Shift from Centralized to Distributed Data Governance**: Traditional IoT security models rely on centralized data governance where single organizations establish and enforce security policies. Blockchain integration enables more distributed governance models where multiple stakeholders

collectively establish and enforce policies through consensus mechanisms and smart contracts. This shift addresses limitations of centralized governance including single points of failure, trust asymmetries, and potential for unilateral policy changes. However, distributed governance introduces new challenges in decision-making efficiency, responsibility allocation, and conflict resolution. This tension between centralized efficiency and distributed resilience remains incompletely resolved in current implementations.

These findings highlight both significant progress and remaining challenges in IoBc development. The field has matured from conceptual explorations and proof-of-concept implementations to production deployments addressing real-world security and privacy challenges. However, several fundamental tensions and unresolved questions persist:

1. **Scalability-Security Tradeoff**: Despite advances in consensus efficiency and layer-2 scaling solutions, fundamental tensions remain between security guarantees and scalability requirements. High-security consensus mechanisms typically impose transaction throughput limitations incompatible with large-scale IoT deployments. Alternative approaches that prioritize throughput often make security compromises that may be inappropriate for critical applications. Finding the optimal balance between these competing objectives remains challenging, particularly as IoT deployments continue to grow in scale and complexity.
2. **Privacy-Transparency Balance**: Blockchain's inherent transparency conflicts with privacy requirements in many IoT applications, particularly those involving personal data or proprietary information. While privacy-enhancing technologies offer partial solutions, they often increase complexity and resource requirements. The fundamental tension between verifiability through transparency and confidentiality through privacy protection remains incompletely resolved, requiring application-specific tradeoffs rather than universal solutions.
3. **Centralization Tendencies**: Despite blockchain's decentralization ethos, practical IoBc implementations often reintroduce centralization through consortium structures, managed services, or dominant node operators. This pragmatic centralization addresses resource constraints and governance challenges but potentially undermines blockchain's fundamental security proposition. The balance between decentralization ideals and practical implementation realities continues to evolve as the field matures.

These unresolved challenges represent important areas for future research and development in IoBc. Addressing them requires interdisciplinary approaches combining technical innovation with economic models, governance frameworks, and regulatory considerations. The ongoing evolution of both blockchain and IoT technologies will continue to reshape how these challenges are approached and potentially resolved.

IMPLICATIONS AND FUTURE RESEARCH DIRECTIONS

The integration of blockchain with IoT ecosystems has significant implications across multiple domains, from technological development to economic models, regulatory frameworks, and societal impacts. This section examines these implications and identifies critical directions for future research that will shape the evolution of Internet of Blockchain (IoBc) as a paradigm for secure and private IoT ecosystems.

Theoretical and Practical Implications

The development of IoBc extends and challenges existing theoretical frameworks in several domains, including distributed systems theory, cryptoeconomics, information security, and privacy engineering. These theoretical implications guide future research and provide a foundation for understanding the fundamental principles governing secure blockchain integration with IoT.

In distributed systems theory, IoBc implementations challenge traditional assumptions about node homogeneity, reliable connectivity, and resource availability. IoT environments typically feature heterogeneous devices with widely varying capabilities, intermittent connectivity, and significant resource constraints. These characteristics necessitate extensions to distributed consensus theory that accommodate such heterogeneity while maintaining essential security properties. Theoretical work by Sayeed et al. (2019) has begun addressing these challenges through adaptive consensus models that adjust security-performance tradeoffs based on current system conditions and threat levels. Future theoretical development should formalize these adaptive approaches and establish clear bounds on their security guarantees under different conditions.

Information security theory faces significant challenges from the hybrid nature of IoBc systems, which combine blockchain's distributed trust model with traditional security approaches. While substantial theoretical work exists for each domain independently, formal security models that span this boundary are limited. The theoretical framework of "composable security" (Canetti et al., 2002) offers potential approaches for analyzing how security properties compose across these boundaries, but its application to IoBc remains largely unexplored. Future theoretical work should develop formal security models specifically for hybrid systems that combine blockchain and traditional security mechanisms, with particular attention to complex attack vectors that traverse these boundaries.

For industry stakeholders, IoBc has significant practical implications. For device manufacturers, blockchain integration creates both opportunities and challenges. The possibility of creating devices with verifiable security properties provides competitive differentiation in security-sensitive markets (Huh et al., 2017). However, implementing blockchain capabilities increases device complexity, development costs, and potentially bill-of-materials costs due to enhanced security requirements. The industry trend toward security co-processors and trusted execution environments aligns with blockchain integration needs but requires significant investment (Johnson et al., 2019).

For network operators and service providers, blockchain integration influences infrastructure requirements, service offerings, and business models. Traditional centralized IoT platforms must adapt to incorporate blockchain capabilities, either through direct integration or through interoperability with separate blockchain networks. Platform providers such as AWS, Azure, and Google Cloud have developed blockchain integration services that connect their IoT platforms with various blockchain networks (Amazon, 2020; Microsoft, 2021; Google, 2021), but these integrations often remain superficial rather than fundamentally restructuring the platform architecture.

The potential disintermediation effect of blockchain on traditional platform business models represents a significant disruption risk for established IoT service providers. Blockchain-based alternatives enable direct device-to-device interactions and peer-to-peer marketplaces that bypass centralized platforms, potentially reducing their value capture opportunities (Christidis & Devetsikiotis, 2016). Forward-thinking service providers are exploring how to transition from controlling middleware to providing value-added services on decentralized infrastructure, but this business model transition remains challenging.

From a security perspective, blockchain integration introduces both new capabilities and novel challenges. Traditional IoT security models rely heavily on perimeter defenses and trusted central authorities, creating single points of failure vulnerable to targeted attacks. Blockchain's distributed architecture eliminates these single points of failure, enhancing resilience against certain attack vectors such as central server compromise or certificate authority failure (Khan & Salah, 2018). However, this distribution also creates new attack surfaces related to consensus vulnerabilities, smart contract weaknesses, and cross-chain interaction points.

The concept of "security through transparency" represents a significant paradigm shift introduced by blockchain integration. Traditional security often emphasizes "security through obscurity" where system details are hidden to complicate attacks. Blockchain approaches instead rely on full transparency of protocols and operations, with security deriving from cryptographic guarantees rather than information hiding (Antonopoulos, 2014). This transparency enables broader security analysis by diverse stakeholders but requires fundamentally different security design patterns focused on cryptographic verification rather than access limitation.

From a privacy perspective, blockchain integration creates complex implications that vary based on architecture choices and implementation details. The inherent transparency of many blockchain implementations conflicts with privacy principles such as data minimization and the right to be forgotten. However, privacy-enhancing technologies including zero-knowledge proofs, secure multi-party computation, and off-chain storage with cryptographic references can mitigate these concerns when properly implemented (Zyskind et al., 2015).

The regulatory implications of blockchain for IoT privacy remain complex and evolving. Regulations such as the General Data Protection Regulation (GDPR) were not designed with blockchain's immutability in mind, creating compliance challenges particularly regarding the right to erasure ("right to be forgotten"). Various technical approaches have emerged to address these challenges, including storing personal data off-chain with revocable access, using mutable pointers to immutable but encrypted data, and implementing cryptographic deletion through key destruction (Truong et al., 2019). These approaches represent pragmatic compromises between blockchain's immutability and privacy regulation requirements, though their legal sufficiency remains untested in many jurisdictions.

Future Research Directions

Based on the theoretical foundations, comparative analysis, and implications discussed previously, several critical research directions emerge that will shape the future development of Internet of Blockchain. These research priorities address current limitations, unresolved challenges, and emerging opportunities at the intersection of blockchain and IoT.

1. Lightweight Consensus for Heterogeneous IoT Environments

Current consensus mechanisms struggle to accommodate the extreme heterogeneity of IoT environments, where devices range from ultra-constrained sensors to powerful edge servers. Future research should develop adaptive consensus frameworks that dynamically adjust security-performance tradeoffs based on device capabilities, network conditions, and security requirements. This research direction includes:

- **Context-Aware Consensus Selection**: Frameworks that automatically select appropriate consensus mechanisms based on current application requirements, threat levels, and available resources. Initial work by Wang et al. (2019) on dynamic consensus selection provides a foundation, but comprehensive frameworks integrating security metrics with resource optimization remain underdeveloped.
- **Hierarchical Consensus Models**: Advanced theoretical frameworks for multi-tier consensus where different mechanisms operate at different system layers while maintaining cryptographic links between tiers. Current hierarchical approaches such as Plasma (Poon & Buterin, 2017) provide practical implementations, but formal security models defining their guarantees and limitations require further development.

2. Privacy-Preserving Verification for Regulated IoT Domains

The tension between blockchain's transparency and privacy regulations represents a critical challenge for IoBc adoption in sensitive domains such as healthcare, smart homes, and industrial systems. Future research should develop frameworks that enable verification of essential properties without exposing sensitive data. This research direction includes:

- **Zero-Knowledge IoT**: Optimization of zero-knowledge proof systems for resource-constrained IoT environments, enabling verification of properties such as data integrity, policy compliance, and identity without revealing the underlying information. Current zero-knowledge systems such as zk-SNARKs and Bulletproofs (Bünz et al., 2018) require significant computational resources unsuitable for many IoT devices.
- **Privacy-Compliant Blockchain Architectures**: Comprehensive architectural frameworks specifically designed to address regulatory requirements such as GDPR and CCPA while maintaining blockchain's security benefits. Research by Truong et al. (2019) has identified key tensions, but complete architectural solutions balancing compliance with security remain underdeveloped.

3. Cross-Chain IoT Security and Interoperability

As IoT ecosystems increasingly span multiple blockchain networks and traditional systems, security vulnerabilities at interaction boundaries represent critical concerns. Future research should develop frameworks for secure cross-chain operations and heterogeneous system integration. This research direction includes:

- **Formal Security Models for Cross-Chain Interactions**: Theoretical frameworks defining security properties and potential vulnerabilities when IoT systems span multiple blockchain networks. Current approaches such as atomic swaps and hash timelock contracts provide basic cross-chain capabilities, but comprehensive security models addressing IoT-specific requirements remain limited.

- **Standardized Cross-Chain Communication Protocols**: Development and security analysis of standardized protocols for IoT devices to securely interact across different blockchain networks. Initial efforts such as the InterBlockchain Communication Protocol (Cosmos Network, 2021) provide foundation elements, but IoT-specific extensions addressing resource constraints and security requirements need further development.

4. Sustainable Tokenomics for IoT Security

The long-term sustainability of blockchain-based security for IoT depends on economic models that align incentives across diverse stakeholders including device manufacturers, service providers, and users. Future research should develop sustainable tokenomic models specific to IoT security enhancement. This research direction includes:

- **Lifecycle Security Incentives**: Economic models that incentivize security enhancement throughout the complete IoT lifecycle from secure manufacturing through secure decommissioning. Current tokenomic models typically focus on operational incentives without adequately addressing manufacturing security or end-of-life concerns.
- **Resource-Based Tokenomics**: Token models based on actual resource consumption (computation, bandwidth, storage) rather than artificial scarcity, enabling more efficient market operation in resource-constrained IoT environments. Research by Yang et al. (2019) on resource pricing in decentralized IoT networks provides foundation elements, but comprehensive models balancing efficiency with sustainable security incentives require further development.

5. Autonomous Security Governance for IoBc

The distributed nature of IoBc systems creates governance challenges for security management, particularly regarding incident response, vulnerability management, and policy enforcement. Future research should develop frameworks for effective security governance across organizational boundaries. This research direction includes:

- **Decentralized Security Response**: Governance frameworks and technical mechanisms enabling coordinated security incident response across multiple independent stakeholders without requiring centralized authority. Current incident response models typically assume organizational control boundaries that poorly match distributed blockchain-IoT ecosystems.
- **On-Chain Governance for Security Policies**: Mechanisms for collaboratively establishing, updating, and enforcing security policies through on-chain governance processes with appropriate checks and balances. Research by Azouvi et al. (2019) on governance for blockchain security provides foundational concepts, but complete frameworks specifically addressing IoT security governance remain underdeveloped.

These research directions require interdisciplinary collaboration spanning computer science, cryptography, economics, law, and domain-specific expertise across various IoT application areas. Progress in these directions will shape how IoBc evolves from promising technology to mainstream implementation, addressing current limitations while leveraging blockchain's unique capabilities to enhance IoT security and privacy.

CONCLUSION

The integration of blockchain with Internet of Things represents a significant paradigm shift in approaches to securing distributed device ecosystems. This chapter has examined the theoretical foundations, key concepts, implementation approaches, and future directions of Internet of Blockchain (IoBc) as a framework for creating secure and private IoT ecosystems. The comprehensive analysis reveals both the transformative potential of this integration and the significant challenges that must be addressed to realize this potential across diverse application domains.

The theoretical foundations of IoBc span multiple disciplines including distributed systems theory, cryptography, game theory, trust models, systems theory, and information security. These foundations provide essential concepts for understanding how blockchain can enhance IoT security through distributed trust, immutable record-keeping, and cryptographic verification without centralized authorities. The evolution of these theoretical foundations continues as researchers develop new models addressing the unique requirements and constraints of IoT environments, particularly regarding resource efficiency, scalability, and privacy protection.

The key concepts and constructs enabling practical IoBc implementation include lightweight consensus protocols, decentralized identity frameworks, privacy-preserving mechanisms, smart contracts for autonomous operations, tokenization models, and edge computing integration. These building blocks provide the technical foundation for blockchain-enhanced IoT security, though their effective implementation requires careful consideration of application requirements, deployment constraints, and stakeholder objectives. The ongoing development of these concepts continues to enhance their efficiency, security, and applicability across diverse IoT domains.

Comparative analysis of blockchain platforms, architectural approaches, security models, and implementation patterns reveals important insights about effective IoBc deployment. The diversity of approaches reflects the heterogeneity of IoT environments and application requirements, with no single solution optimal for all scenarios. However, patterns of convergence toward hybrid architectures, domain-specific specialization, and integrated privacy-security frameworks indicate emerging best practices in this rapidly evolving field. This comparative analysis provides guidance for selecting appropriate approaches based on specific application requirements and deployment constraints.

The implications of IoBc span theoretical domains, industry practices, security and privacy frameworks, and societal considerations. These implications highlight both the transformative potential of blockchain for enhancing IoT security and the complex challenges involved in its practical implementation. The interdisciplinary nature of these implications necessitates collaboration across traditional boundaries between computer science, business, law, and domain-specific expertise to fully realize the benefits of blockchain-IoT integration.

Future research directions for IoBc focus on addressing current limitations and unresolved challenges while exploring emerging opportunities at the intersection of blockchain and IoT. Priority areas include lightweight consensus for heterogeneous environments, privacy-preserving verification for regulated domains, cross-chain security and interoperability, sustainable tokenomics for security incentives, and autonomous security governance. Progress in these research directions will shape how IoBc evolves from promising concept to mainstream implementation across diverse IoT applications.

As IoT deployments continue to expand in scale, complexity, and criticality, the need for robust security and privacy frameworks becomes increasingly urgent. Traditional centralized security approaches struggle to address the distributed nature, massive scale, and heterogeneous composition of IoT ecosystems. Blockchain offers promising approaches to these challenges through its fundamental properties of decentralization, immutability, and cryptographic verification. However, realizing this potential requires not only technological innovation but also careful consideration of economic incentives, governance structures, and regulatory frameworks.

The ongoing development of Internet of Blockchain represents a critical contribution to securing the digital infrastructure that increasingly permeates physical environments, critical systems, and daily life. By providing transparent, verifiable, and distributed security mechanisms, IoBc has the potential to enhance trust in IoT systems while protecting the privacy and autonomy of individuals and organizations. Achieving this potential requires continued research, thoughtful implementation, and collaborative governance across the complex ecosystem of stakeholders involved in creating, deploying, and using connected devices.

REFERENCES

Ali, M. S., Dolui, K., & Antonelli, F. (2018). IoT data privacy via blockchains and IPFS. In Proceedings of the 7th International Conference on the Internet of Things (pp. 14:1-14:7). ACM.

Allen, C. (2016). The path to self-sovereign identity. Retrieved from https://www.lifewithalacrity.com/2016/04/the-path-to-self-sovereign-identity.html

Anderson, R., & Moore, T. (2006). The economics of information security. *Science, 314*(5799), 610–613. DOI: 10.1126/science.1130992 PMID: 17068253

Androulaki, E., Barger, A., Bortnikov, V., Cachin, C., Christidis, K., De Caro, A., & Muralidharan, S. (2018). Hyperledger Fabric: A distributed operating system for permissioned blockchains. In *Proceedings of the 13th EuroSys Conference* (pp. 1-15). ACM. DOI: 10.1145/3190508.3190538

Antonopoulos, A. M. (2014). *Mastering Bitcoin: Unlocking digital cryptocurrencies.* O'Reilly Media, Inc.

Atzei, N., Bartoletti, M., & Cimoli, T. (2017). A survey of attacks on Ethereum smart contracts (SoK). In *Principles of Security and Trust* (pp. 164–186). Springer. DOI: 10.1007/978-3-662-54455-6_8

Azouvi, S., Maller, M., & Meiklejohn, S. (2019). Egalitarian society or benevolent dictatorship: The state of cryptocurrency governance. In *Financial Cryptography and Data Security* (pp. 127–143). Springer. DOI: 10.1007/978-3-662-58820-8_10

Ben-Sasson, E., Chiesa, A., Garman, C., Green, M., Miers, I., Tromer, E., & Virza, M. (2014). Zerocash: Decentralized anonymous payments from Bitcoin. In *IEEE Symposium on Security and Privacy* (pp. 459-474). IEEE.

Bhargavan, K., Delignat-Lavaud, A., Fournet, C., Gollamudi, A., Gonthier, G., Kobeissi, N., & Zanella-Béguelin, S. (2016). Formal verification of smart contracts: Short paper. In *Proceedings of the 2016 ACM Workshop on Programming Languages and Analysis for Security* (pp. 91-96). ACM. DOI: 10.1145/2993600.2993611

Christidis, K., & Devetsikiotis, M. (2016). Blockchains and smart contracts for the Internet of Things. *IEEE Access : Practical Innovations, Open Solutions, 4*, 2292–2303. DOI: 10.1109/ACCESS.2016.2566339

Diffie, W., & Hellman, M. E. (1976). New directions in cryptography. *IEEE Transactions on Information Theory, 22*(6), 644–654. DOI: 10.1109/TIT.1976.1055638

Dorri, A., Kanhere, S. S., Jurdak, R., & Gauravaram, P. (2017). Blockchain for IoT security and privacy: The case study of a smart home. In *IEEE International Conference on Pervasive Computing and Communications Workshops* (pp. 618-623). IEEE. DOI: 10.1109/PERCOMW.2017.7917634

Fernández-Caramés, T. M., & Fraga-Lamas, P. (2018). A review on the use of blockchain for the Internet of Things. *IEEE Access : Practical Innovations, Open Solutions, 6*, 32979–33001. DOI: 10.1109/ACCESS.2018.2842685

Fischer, M. J., Lynch, N. A., & Paterson, M. S. (1985). Impossibility of distributed consensus with one faulty process. *Journal of the Association for Computing Machinery, 32*(2), 374–382. DOI: 10.1145/3149.214121

Huh, S., Cho, S., & Kim, S. (2017). Managing IoT devices using blockchain platform. In *International Conference on Advanced Communication Technology* (pp. 464-467). IEEE.

Khan, M. A., & Salah, K. (2018). IoT security: Review, blockchain solutions, and open challenges. *Future Generation Computer Systems*, *82*, 395–411. DOI: 10.1016/j.future.2017.11.022

Nakamoto, S. (2008). Bitcoin: A peer-to-peer electronic cash system. Retrieved from https://bitcoin .org/bitcoin.pdf

Novo, O. (2018). Blockchain meets IoT: An architecture for scalable access management in IoT. *IEEE Internet of Things Journal*, *5*(2), 1184–1195. DOI: 10.1109/JIOT.2018.2812239

Poon, J., & Buterin, V. (2017). Plasma: Scalable autonomous smart contracts. White Paper, 1-47.

Popov, S. (2018). The tangle. White Paper, 1-28.

Raj, H., Saroiu, S., Wolman, A., Aigner, R., Cox, J., England, P., & Jaeger, T. (2016). fTPM: A software-only implementation of a TPM chip. In *25th USENIX Security Symposium* (pp. 841-856). USENIX Association.

Reyna, A., Martín, C., Chen, J., Soler, E., & Díaz, M. (2018). On blockchain and its integration with IoT. Challenges and opportunities. *Future Generation Computer Systems*, *88*, 173–190. DOI: 10.1016/j. future.2018.05.046

Truong, N. B., Sun, K., Lee, G. M., & Guo, Y. (2019). GDPR-compliant personal data management: A blockchain-based solution. *IEEE Transactions on Information Forensics and Security*, *15*, 1746–1761. DOI: 10.1109/TIFS.2019.2948287

Wang, X., Zha, X., Ni, W., Liu, R. P., Guo, Y. J., Niu, X., & Zheng, K. (2019). Survey on blockchain for Internet of Things. *Computer Communications*, *136*, 10–29. DOI: 10.1016/j.comcom.2019.01.006

Werbach, K. (2018). *The blockchain and the new architecture of trust*. MIT Press. DOI: 10.7551/mit-press/11449.001.0001

Xiong, Z., Zhang, Y., Niyato, D., Wang, P., & Han, Z. (2018). When mobile blockchain meets edge computing. *IEEE Communications Magazine*, *56*(8), 33–39. DOI: 10.1109/MCOM.2018.1701095

Xu, X., Weber, I., & Staples, M. (2018). *Architecture for blockchain applications*. Springer.

Yang, X., Wang, T., Wen, X., & Cai, W. (2019). Pricing mechanism for data trading in IoT data marketplace based on blockchain. In ICC 2019-2019 IEEE International Conference on Communications (pp. 1-6). IEEE.

Yao, A. C. (1982). Protocols for secure computations. In *23rd Annual Symposium on Foundations of Computer Science* (pp. 160-164). IEEE. DOI: 10.1109/SFCS.1982.38

Zyskind, G., Nathan, O., & Pentland, A. S. (2015). Decentralizing privacy: Using blockchain to protect personal data. In *IEEE Security and Privacy Workshops* (pp. 180–184). IEEE. DOI: 10.1109/SPW.2015.27

Chapter 8
Blockchain-Enabled Grid Resilience and Disaster Response

Amjad Yousef Aldweesh
https://orcid.org/0000-0001-9358-1323
College of Computing and IT, Shaqra University, Saudi Arabia

Saad Alateef
https://orcid.org/0009-0002-3219-0833
Newcastle University, UK

Mohammad Alauthman
https://orcid.org/0000-0003-0319-1968
Petra University, Jordan

Ahmed Al-Qerem
https://orcid.org/0000-0003-2187-6194
Zarqa University, Jordan

ABSTRACT

The reliability of modern power grids is vital for sustaining essential services,especially amid threats like natural disasters and cyberattacks. Blockchain technology offers a decentralized, tamper-resistant infrastructure to enhance grid resilience and disaster response. Resilience relies on real-time monitoring, fault detection, and rapid recovery,all of which blockchain streamlines through secure energy transactions, transparent data sharing, and automated smart contracts. In disasters, blockchain improves coordination with clear audit trails for resource allocation and efficient peer-to-peer energy trading, reducing single points of failure.This paper proposes a blockchain-enabled framework to enhance grid resilience, analyzing its security, scalability, and efficiency through case studies and simulations. Combining blockchain with IoT and AI, the framework creates a decentralized,transparent ecosystem to transform power grid recovery processes, reinforcing global energy security.

DOI: 10.4018/979-8-3373-2439-5.ch008

INTRODUCTION

In recent years, global power grids have encountered an upsurge of extreme weather phenomena—such as hurricanes, fires, and floods—as well as advanced cyber threats, all of which expose critical flaws in existing infrastructure (Deb, 2021)(Chen, Yang, & Liu, 2023). In response, the notion of grid resilience has gained traction among policymakers, electrical utilities, academic researchers, and regional stakeholders (Brayton & O'Neill, 2020). Unlike reliability, resilience emphasizes not only sustained power delivery under normal conditions but also the grid's adaptability when confronted by unexpected stressors, its ability to shorten recovery timelines, and its capacity to evolve in ways that mitigate future disruptions.

Conventional grid architectures rely heavily on centralized monitoring and control—a model that, despite its historical success, can produce central points of failure and limited real-time awareness across distributed assets (Fox, 2021). Large storms or cyber intrusions targeting key communication channels may lead to cascading blackouts and long-term damage (Ventura & King, 2020). Hence, modernizing infrastructure to improve physical robustness and incorporate advanced digital services—IoT-based sensing, distributed intelligence, and robust communications—has become an integral part of grid enhancement strategies (International Energy Agency [IEA], 2021).

Motivation and Relevance

Emerging blockchain or distributed ledger technologies provide a decentralized data platform with consensus-based validation, potentially reducing the vulnerabilities found in single control centers (Hansen & Rodriguez, 2019). Blockchain further supports self-executing smart contracts for automated processes such as fault detection, load shedding, and P2P electricity marketplaces (Michel, Yang, & Clark, 2020). These capabilities promise quicker recovery times, transparent coordination, and tamper-proof data trails, beneficial for everyday grid operation and crisis conditions alike (Elkind, 2018).

Problem Statement

Although considerable technological advancements have been made, many power networks are still inadequately prepared for large-scale disturbances (Prakash, Bose, & Suh, 2022). Traditional disaster recovery frequently involves multi-layered and opaque processes, hampering efficient mobilization of resources, slowing isolation of problematic feeders, and hindering equitable distribution of electricity to essential loads (Fox, 2021). In terms of cybersecurity, centralized schemes are particularly susceptible to large-scale compromise if a malicious actor infiltrates a central control server (Wang, Ching, & Qu, 2021). Blockchain systems inherently reduce such risks by dispersing the ledger among numerous validator nodes, yet practical concerns—including scalability, energy consumption (in some consensus algorithms), privacy, and integration barriers—complicate its use in mission-critical environments (Holmes, Bond, & Kimmel, 2021).

Objectives and Contributions

This paper focuses on conceiving and evaluating a blockchain-based framework for enhanced grid resilience and disaster mitigation:

1. Propose a multi-layered architecture combining physical grid assets, IoT sensing, blockchain nodes, and smart contracts.
2. Analyze security characteristics (e.g., protection from cyberattacks, data integrity) and performance aspects (e.g., throughput, latency).
3. Present simulated scenarios for coordinating resources, detecting faults, and restoring service to critical loads during disasters.
4. Outline policy, socioeconomic, and technical barriers to implementation while suggesting guidelines for real-world deployment.

The main novel points include:

- A broad-based review of decentralized approaches for grid management.
- A blockchain-centric architecture emphasizing resilience and protection.
- Simulation-supported evidence that blockchain designs can outperform classical grids under disruptive conditions.
- Discussion of future research obstacles, from regulatory adaptation to next-generation AI and edge computing.

Paper Organization

The remainder of this paper is structured as follows:

- Section **2** surveys existing techniques for grid resilience and key blockchain applications in energy.
- Section **3** details the research design, simulation approach, and performance metrics.
- Section **4** explores blockchain use-cases that strengthen grid operations.
- Section **5** presents the architecture of our blockchain-based solution for grid resilience.
- Section **6** addresses practical issues such as hardware, communication protocols, and interoperability.
- Section **7** illustrates how the system accelerates disaster recovery and coordinates resources.
- Section **8** reviews cybersecurity measures and methods of scaling to large or real-time grids.
- Section **9** discusses simulation results and case studies, comparing blockchain to standard grid models under disruptions.
- Section **10** delves into socio-economic and policy perspectives: incentives, legal frameworks, and equitable access.
- Section **11** pinpoints limitations that merit further study.
- Section **12** highlights research frontiers, including AI integration and cross-sector expansions.
- Section **13** concludes with the overarching significance of blockchain in crafting resilient energy systems.

BACKGROUND AND RELATED WORK

This section provides an overview of grid resilience tactics, decentralized energy operations, and current blockchain deployments in the power sector. We also identify open research areas requiring additional examination.

Grid Resilience Practices

Strategies for building resilient power systems often include:

- **Infrastructure Hardening:** Reinforcing transmission and distribution assets against severe weather (Brayton & O'Neill, 2020).
- **Backup and Redundancy:** Deploying batteries, diesel gensets, and redundant pathways to ensure continued delivery of electricity to critical consumers (Sun, Li, & Shen, 2019).
- **Centralized Monitoring (SCADA):** Employing real-time supervisory control to locate and isolate electrical faults (U.S. Department of Energy [DOE], 2019).

However, such methods can be undercut by communication or control center failures, motivating additional distributed or decentralized features (Torres, Li, & Brown, 2022).

Microgrids and Decentralized Systems

Microgrids, featuring local generation and storage assets with the ability to disconnect (island) from the main grid, boost community-level resilience (Xu & Alstone, 2019). They minimize transmission losses, offer local autonomy, and empower households to become active prosumers. Yet, coordinating high volumes of DERs in real time can be computationally intricate, with many microgrids still relying on centralized logic that may be vulnerable during disasters (Yamada & Huang, 2020)(Baker & Chen, 2021).

Blockchain Applications in Energy

In principle, blockchain provides secure record-keeping and automation in a decentralized setting. Demonstrations in the energy domain include:

- **P2P Trading:** Neighbors or small businesses exchange surplus solar energy with minimal intermediaries (Lee, Ryu, & Jeong, 2022).
- **Green Certificates:** Tracking renewable energy certificates or carbon credits through an immutable ledger (Gao & Singh, 2022).
- **Event Logging:** Logging grid events to a blockchain for robust auditing and forensic analysis after outages or cyber breaches (Ortiz, Park, & Hauser, 2021).

Despite excitement around these use-cases, most blockchain pilots remain small-scale or at the proof-of-concept stage, with unresolved challenges related to throughput, regulatory ambiguity, and integration into legacy systems (Holmes, Bond, & Kimmel, 2021).

Security in Decentralized Infrastructures

Distributing state information across many nodes inherently reduces single points of failure. However, consensus mechanisms vary in robustness. Proof of Work (PoW) is energy-intensive, while Proof of Stake (PoS) or Byzantine Fault Tolerance (BFT) protocols might sacrifice some level of decentralization or face partial vulnerabilities if a subset of nodes colludes (Chauhan, Li, & Zhou, 2022)(Knapp, Novak, & Rossi, 2021). Additionally, smart contracts must be carefully implemented to avoid code exploits that could compromise critical grid operations (Zhao, Li, & Thompson, 2021).

Research Gap

While blockchain for energy trading and certificate tracking has been studied, its potential for bolstering fast response and restoration in large-scale disruptions remains understudied. Literature often overlooks real-time constraints, microgrid islanding control, or how to synchronize among multiple agents (utilities, government, communities) when calamities strike. This gap underpins our effort to propose and test a blockchain scheme oriented specifically toward resilience and disaster scenarios.

METHODOLOGY

In this section, we outline our research approach, including system design, simulation methods, and the metrics chosen to assess performance in normal and disruptive conditions.

Research Design

We employed an iterative, multi-phased process:

1. **Literature Review:** Surveyed existing resilience methods, blockchain consensus approaches, and relevant standards (see Section **2**).
2. **Framework Development:** Devised an architectural blueprint combining physical assets, IoT sensors, a blockchain platform, and application-level logic.
3. **Prototype and Simulation:** Used MATLAB/Simulink or DIgSILENT for power flow/emulation, plus a permissioned blockchain (e.g., Hyperledger Fabric) for transaction validation.
4. **Scenario Testing:** Subjected the simulated grid to cyberattacks and natural disasters, examining the differences between a blockchain-enabled approach and a more centralized model.
5. **Analysis and Iteration:** Collected data on fault isolation times, restoration durations, transaction throughput, etc. Tweaked blockchain parameters (consensus, block size) and repeated tests.

Simulation Tools

- **Power Grid Emulator:** Evaluated distribution networks featuring solar, wind, battery storage, and controllable loads.
- **Blockchain Platform:** Primarily used Hyperledger Fabric, with alternative runs on an Ethereum-based Proof of Authority (PoA) chain for comparative analysis.

- **Disaster Modules:** Generated random line or node failures and, in some runs, introduced spurious load signals to test robustness.
- **IoT Gateways:** Simulated sensor readings for voltage, current, and fault indicators. Data was packaged into transactions posted to the blockchain.

Evaluation Metrics

- **Critical Load Restoration (hours):** Time needed to resume power at crucial facilities (hospitals, communication nodes).
- **Transaction Throughput (transactions/s):** Volume of settled transactions (e.g., energy trades, fault detection signals).
- **Latency (seconds):** Delay from initiating a transaction to its inclusion in a validated block.
- **Fault Isolation Time (seconds):** Speed of identifying and isolating the faulted segment.
- **Scalability Index:** Relationship between node count, transaction volume, and sustained throughput/latency.
- **Qualitative Observations:** Including user experiences, regulatory compliance, and the clarity of the interface.

Data Analysis

We recorded data logs from both the simulation environment and the blockchain ledger. After each run, we computed statistical means, standard deviations, and confidence intervals. For scenario comparisons, we used standard tests (e.g., t-tests) to measure the significance of observed differences. Python libraries (**pandas**, **matplotlib**) facilitated data processing and visualization.

With this methodology in place, the following section explores the main blockchain functionalities applied to increase security and resilience in power systems.

BLOCKCHAIN APPLICATIONS IN GRID RESILIENCE

Blockchain offers a decentralized, cryptographically secured approach that addresses some of the core challenges in modern grid operations. This section highlights how blockchain-based methods can improve energy transactions, system intelligence, security, and disaster preparedness.

Decentralized Energy Trading

Traditional trading relies on utilities and centralized clearing houses. Blockchain-based platforms enable direct settlements between prosumers and consumers (Lee, Ryu, & Jeong, 2022). Key benefits include:

- Automated pricing through smart contracts.
- Reduced transaction costs, faster settlement cycles.
- Transparent records that prevent double-counting of energy.

Secure and Transparent Data Sharing

Blockchain ledgers can hold operational data—load curves, frequency fluctuations, voltage levels—in a tamper-proof manner (Gao & Singh, 2022). Authorized stakeholders (utilities, regulators) access time-stamped records, allowing:

- Forensic tracking of grid anomalies.
- Proof-of-origin for renewable energy.
- Reduced risk of data manipulation by malicious actors.

Self-Executing Smart Contracts

Smart contracts can automate response actions, such as:

- **Demand Response:** Triggering load adjustments based on real-time grid stress indicators.
- **Fault Isolation:** Immediately opening or closing circuit breakers when sensor data indicate a line fault (Ul Haq, Zhang, & Tohid, 2023).
- **Disaster Protocols:** Prioritizing local microgrids and hospital loads if main feeders fail.

Cybersecurity Enhancement

Distributed consensus ensures no single node can unilaterally change ledger data (Chauhan, Li, & Zhou, 2022). Moreover, cryptographic hashing makes data tampering easily detectable. Additional layers (digital signatures, identity management) reinforce security. Though blockchain technology does not eliminate all risks—especially those related to physical sensors—it significantly reduces reliance on one central hub (Knapp, Novak, & Rossi, 2021).

Disaster-Resilient Operation

During hurricanes or earthquakes, rapid information sharing is paramount (Ahmed, Lee, & Park, 2020). A blockchain ledger accessible to all authorized entities (utility operators, government agencies) can synchronize resource deployment and re-energization strategies. Immutable logs also facilitate transparent post-event audits, revealing bottlenecks and fraud attempts.

Armed with these blockchain functionalities, we now present in Section **5** a blockchain-centered framework that aims to bolster grid resilience end to end.

PROPOSED BLOCKCHAIN-ENABLED FRAMEWORK FOR GRID RESILIENCE

We now introduce a multi-layered system architecture that merges physical power infrastructure, blockchain-based data management, and application-level smart contracts to improve preparedness, real-time response, and post-event recovery.

Conceptual Architecture

Figure **[fig:framework_architecture_extended]** illustrates three layers:

1. **Physical Layer:** Centralized plants, T&D infrastructure, and decentralized DERs (solar, wind, battery).
2. **Blockchain Layer:** A permissioned or consortium network in which multiple participants (utilities, operators, major prosumers) validate transactions and maintain a shared ledger.
3. **Application Layer:** dApps governing P2P energy exchanges, load management, and disaster-specific protocols encoded as smart contracts.

Figure 1. High-level design of the proposed blockchain-enhanced grid architecture

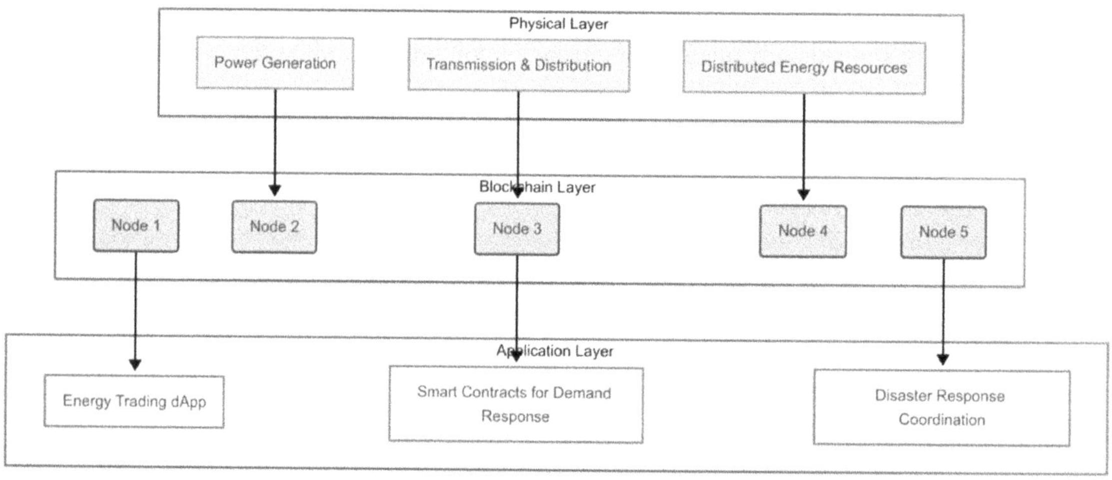

Main Components

DERs and IoT Sensors

DERs feed data on generation/consumption via secure IoT devices. This data is formatted into transactions and broadcast to blockchain nodes for consensus validation.

Blockchain Nodes

Various stakeholders host validator nodes that verify incoming transactions (e.g., energy trades, event logs). Node operators can be public utilities, microgrid managers, or government agencies.

Smart Contracts

Embedded logic handles tasks such as automatically isolating a faulty feeder, implementing demand response events, or executing targeted load rationing in disasters.

Communications and Edge Devices

Resilient communication networks (e.g., mesh radio, cellular) connect sensors and blockchain nodes. Edge gateways can preprocess sensor data to limit bandwidth usage and reduce latency.

Operational Flow

1. **Data Capture:** Sensors record power flows and fault indicators at regular intervals.
2. **Transaction Creation:** These measurements, or commands from operators, form transactions that are proposed to the blockchain network.
3. **Consensus:** Nodes run a consensus mechanism (e.g., pBFT, PoS) to validate each batch of transactions.
4. **Contract Execution:** Confirmed transactions trigger contract operations such as reconfiguring distribution paths or approving local energy trades.
5. **Ledger Update:** Finalized blocks are added to the chain, forming an immutable audit trail.

Resilience Advantages

- **Redundancy:** Multiple nodes maintain the ledger, avoiding reliance on a single controller.
- **Quick Fault Response:** Automated contract logic rapidly isolates or reconfigures sections of the grid.
- **Shared Transparency:** Stakeholders receive uniform, real-time data, improving cooperative response.
- **Immutable Records:** Critical events and decisions are documented for thorough post-incident evaluations.

Next, Section **6** explores the practicalities of hardware, networking, and system compatibility, offering insights on how to realize this framework in real-world power grids.

IMPLEMENTATION CONSIDERATIONS

Moving from conceptual design to an operational system necessitates resolving several technical, regulatory, and organizational issues. Below, we address hardware specifications, communication strategies, consensus choices, and legacy integration.

Hardware and Connectivity

Blockchain Node Requirements

A permissioned approach typically involves server-grade machines (multi-core CPUs, adequate RAM, high-availability storage) for running validator nodes (Knapp, Novak, & Rossi, 2021). For smaller participants or edge sites, lightweight nodes may be sufficient to handle partial ledger data.

IoT Gateways and Edge Computing

IoT sensors in transformers or substations communicate securely with gateway devices. Gateways can preprocess sensor data—filtering, aggregating, encrypting—before sending it to validator nodes (Baker & Chen, 2021).

Redundant Communication Paths

Reliance on a single network (e.g., 4G cellular) could be risky in disaster contexts. Hybrid designs (fiber, satellite, mesh radio) can preserve connectivity under extreme conditions (Fox, 2021).

Consensus Mechanisms

Table 1. Representative consensus algorithms for energy blockchain systems

Algorithm	Advantages	Drawbacks
PoW (Proof of Work)	Well-tested security	High energy use, slow
PoS (Proof of Stake)	Lower energy use	Depends on stake distribution
pBFT (Practical BFT)	Low latency, high throughput	Limited to smaller networks
Raft/IBFT	Simple for private chains	Not as robust for large-scale public

As summarized in Table **1**, PoW is often impractical for energy systems due to excessive computational overhead and latency (Holmes et al., 2021). pBFT or Raft-style protocols may be more appropriate for permissioned blockchains, enabling quicker finality.

Smart Contract Security

Coding Best Practices

Mission-critical contracts require rigorous testing, code audits, and possibly formal proofs against known vulnerabilities (re-entrancy, integer overflow) (Zhao, Li, & Thompson, 2021). Standardized libraries can help.

Fail-Safe Options

Because real-time grid operations demand reliability, fallback strategies—like manual overrides or contract pauses—must be built in for emergency or bug-related scenarios.

SCADA Integration

Interface Layers

Many utilities operate conventional SCADA and EMS solutions. API-based adapters or middleware solutions can bridge SCADA data into blockchain transactions, ensuring phased adoption (Chen, Yang, & Liu, 2023).

Gradual Deployment

A pilot in a microgrid or non-critical feeder might be used to test the blockchain system's reliability and acceptance before broader rollout (Prakash, Bose, & Suh, 2022).

Policy and Regulation

Compliance with Data Privacy

Energy usage data can be sensitive. Permissioned ledgers can employ access control or zero-knowledge approaches to comply with privacy regulations (e.g., GDPR) (Fox, 2021).

Market Design

Blockchain-based P2P exchanges demand new tariff structures and guidelines for compensation. Regulatory bodies may need to address potential conflicts with established utility revenue models (Deb, 2021).

These considerations provide a practical roadmap for deploying blockchain-based solutions in complex energy infrastructures. The subsequent section (Section **7**) focuses on how the system facilitates quick and transparent disaster response.

BLOCKCHAIN FOR DISASTER RESPONSE COORDINATION

In disaster settings, such as large storms or earthquakes, speed and efficiency of response are paramount. Blockchain systems can unify multiple parties—utilities, local governments, and relief organizations—by offering real-time data sharing, automated resource allocation, and immutable records.

Multi-Stakeholder Coordination

Crisis response often involves overlapping jurisdictions and agencies (Ahmed, Lee, & Park, 2020). Using a shared blockchain ledger:

- **Disaster Nodes:** Temporary validator nodes can be spun up by emergency agencies to track event-driven transactions.
- **Prioritized Loads:** Smart contracts implement distribution policies ensuring, for example, hospitals and emergency shelters are the first to receive available power (Ventura & King, 2020).
- **Single Source of Truth:** Transaction logs become a consistent reference point for all stakeholders, reducing confusion and disputes.

Resource Allocation and Tokenization

Limited resources (mobile generators, fuel, repair crews) can be tokenized on-chain, with location and status updated in near real time. Automated matching via contracts can ensure the closest available resource is quickly deployed to where it is most needed (Elkind, 2018). Blockchain's transparency also helps to prevent fraudulent claims or hoarding.

Energy and Microgrid Tracking

Local microgrids can form islands if the main grid is compromised. Once islanded, they rely on internal generation (solar, batteries) or external shipments of resources. Blockchain logs transactions in real time, verifying that each microgrid receives power equitably and can trade surpluses with neighboring regions (Chen, Yang, & Liu, 2023).

Post-Event Analysis

Because every action—switching, load allocation, distribution decisions—is stored on-chain, authorities can conduct transparent post-incident reviews. This forensic data can guide future improvements in contingency plans, ensuring that any identified bottlenecks or misallocations are addressed (Elkind, 2018).

SECURITY AND SCALABILITY ANALYSIS

To operate effectively on large, mission-critical power networks, a blockchain-based approach must exhibit robust defenses against attacks and handle high volumes of real-time data.

Security Aspects

Fewer Single Points of Failure

Distributed consensus diminishes the impact of breaching any single node (Fox, 2021). Attackers have to subvert a large portion of the network to alter the ledger undetected.

Data Immutability

Blocks are cryptographically chained. Any data tampering in older blocks becomes evident unless the attacker re-mines or re-validates all subsequent blocks, which is highly resource-intensive (Chauhan, Li, & Zhou, 2022).

Smart Contract Weaknesses

Vulnerabilities like re-entrancy or integer overflow can be exploited if not caught during development (Zhao, Li, & Thompson, 2021). Formal verification and careful design help minimize these risks.

Scalability Obstacles

Transaction Throughput

Energy IoT networks can generate thousands of data points per second. Conventional blockchains (e.g., Ethereum) may struggle without additional layers or specialized consensus (Knapp, Novak, & Rossi, 2021).

Latency Constraints

Protective relaying often calls for sub-second reaction times. Blockchain-based confirmations might introduce seconds of delay. Proposed solutions include off-chain local decision loops with on-chain final settlement (Michel, Yang, & Clark, 2020).

Scaling Solutions

Permissioned Chains

Restricting validator participation to known, trusted entities allows the network to use faster protocols (pBFT, Raft) with finality in seconds (Holmes, Bond, & Kimmel, 2021).

Layer-Two Networks or Sharding

High-frequency transactions (e.g., sensor updates) can be aggregated in "micro-blockchains" or sidechains, then periodically anchored to the main chain (Knapp, Novak, & Rossi, 2021).

Trade-Offs

Designers must balance security, speed, and decentralization. A purely public chain might be highly trustworthy but too slow. Conversely, a permissioned chain may deliver speed at the expense of broader trust. Risk analyses should guide choices that align with the grid operator's tolerance and regulatory requirements (Fox, 2021).

Section **9** demonstrates how these considerations play out in practice, using case simulations to highlight the advantages and limitations of our proposed framework.

CASE STUDIES OR SIMULATED APPLICATIONS

We present two scenarios—one focusing on cyber intrusion and another on natural disaster response—that compare performance of traditional grids with our blockchain-enhanced model.

Scenario 1: Cyberattack on Centralized vs. Blockchain-Based Grids

Attack Setup: A sophisticated adversary gains entry to a conventional control center, feeding erroneous load commands that mislead operators and lead to widespread outages (Prakash, Bose, & Suh, 2022).

Outcome with Centralized Grid: Operators struggle to isolate false data quickly, resulting in substantial blackouts. Restoration requires full system reboots and manual verifications, lasting over 24 hours (Fox, 2021).

Outcome with Blockchain-Enabled Grid: Fraudulent commands are invalidated by consensus checks, as legitimate sensor nodes outvote the compromised control center. The attack is detected early, and only minor local disruptions occur, leading to restoration in under 6 hours (Elkind, 2018).

Scenario 2: Disaster-Struck Region and Microgrid Coordination

Setting: A coastal region suffers a Category 4 hurricane that disables major transmission lines and one large substation (Deb, 2021).

Blockchain-Managed Microgrids:

- Smart contracts prioritize emergency services—hospitals, water plants—based on pre-agreed rules.
- Surplus energy from unaffected segments is tokenized and redirected to neighborhoods in urgent need.
- Restoration times for essential facilities decrease significantly compared to conventional methods.

Figure **[fig:recovery_time_comparison]** illustrates how blockchain-based coordination can outperform a conventional grid setup. Critical loads under the blockchain model are back online in roughly half the time.

Figure 2. Estimated restoration times after a major hurricane event.

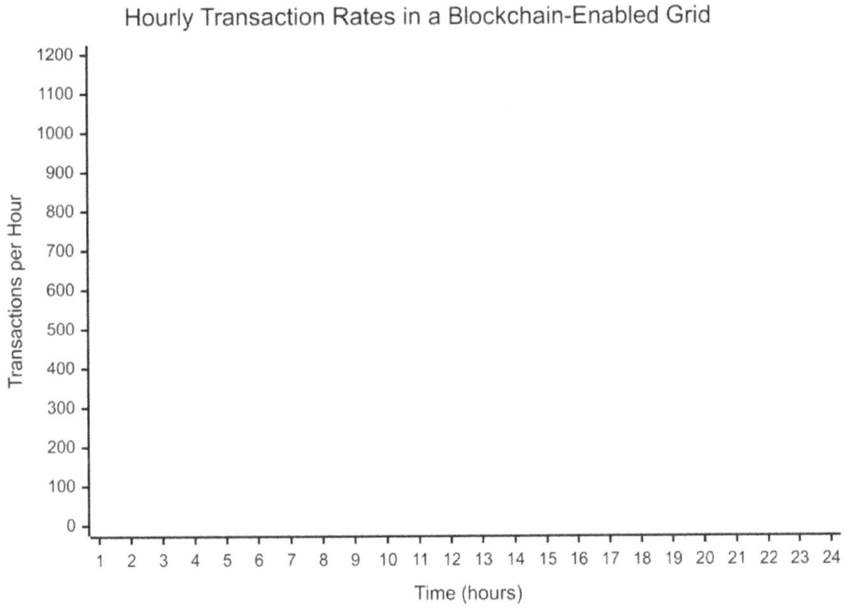

SOCIO-ECONOMIC AND POLICY PERSPECTIVES

Blockchain adoption is influenced not only by technical feasibility, but also by market structures, regulatory contexts, and social acceptance. Below, we consider these broader factors.

Market Design and Utility Models

P2P Trading Frameworks: Blockchain-enabled trading may shift consumer roles, enabling them to become prosumers, which can threaten traditional utility profit models if not regulated properly (Deb, 2021).

Utility Involvement: Some utilities may embrace blockchain, running nodes and providing data analytics or load-balancing services. Others may be slow to adapt, questioning revenue stability (Hansen & Rodriguez, 2019).

Regulatory Considerations

Ledger Legitimacy: Energy exchanges recorded on a blockchain must be recognized legally for billing and dispute resolution (Fox, 2021). Clarity on how digital tokens or smart contracts fit into existing electricity regulations is crucial.

Privacy Protections: Sensitive load data can reveal behavioral patterns. Encryption, zero-knowledge proofs, or permissioned chain frameworks help meet consumer privacy laws (Brayton & O'Neill, 2020).

Community Acceptance and Equity

- **Inclusivity:** Low-income communities might struggle to finance DER installations. Policies or community-led projects can ensure no group is left behind (International Energy Agency [IEA], 2021).
- **Engagement:** Collaborative development with local cooperatives or municipalities can increase trust and readiness to adopt novel blockchain tools (Chen, Yang, & Liu, 2023).

Cost-Benefit Analysis

Start-up expenses (hardware, software, operational training) may be offset by efficiency gains, reduced downtime, and potential new market revenues. Funding mechanisms could involve public-private partnerships, grants, or targeted utility rate adjustments (Ventura & King, 2020). A thorough financial analysis is needed to justify extensive deployment.

LIMITATIONS

While our proposed framework showcases substantial benefits, it also faces several constraints:

1. **Simulation Boundaries:** Real-world complexities like human decision errors or unanticipated hardware breakdowns are difficult to replicate fully in a lab setting.
2. **Regulatory Diversity:** Regions and nations vary in market rules, data privacy standards, and acceptance of blockchain-based commerce.
3. **Scalability and Interoperability:** Handling millions of distributed devices with time-critical processes is nontrivial, even with advanced consensus protocols and edge computing.
4. **Security Assumptions:** The framework assumes robust cryptographic measures and honest majority nodes. Attacks focusing on physical infrastructure or IoT sensor spoofing remain possible.
5. **Economic Viability:** Integrating blockchain into large-scale grid systems could be costly. Comprehensive economic analyses and pilot programs are needed to confirm sustainable return on investment.

FUTURE CHALLENGES AND OPPORTUNITIES

Blockchain in energy systems is evolving rapidly. Key directions include:

Advanced Consensus and Scalability

Hybrid Protocols: Methods that adapt consensus to grid conditions (e.g., partial BFT combined with staking) may provide better performance at scale. **Sharding and Layer-Two Solutions:** Parallelizing transactions or offloading microtransactions can improve throughput.

AI and Predictive Analytics

ML at the Edge: Machine learning models can forecast faults or load spikes, feeding smart contracts that preemptively reconfigure assets. **Privacy-Preserving Analytics:** Federated or encrypted learning can maintain confidentiality while extracting valuable insights.

Cross-Infrastructure Collaboration

Water, transport, and telecom networks often interconnect with power systems. Shared blockchains could coordinate multi-sector resilience, but require consistent data standards and governance.

Regulatory Sandboxes and Policy Evolution

Governments may establish controlled environments to test P2P energy markets or microgrid blockchains, refining policies around digital assets, consumer protection, and system reliability.

Social Equity and Community-Driven Grids

Tokenized incentives or communal ownership of DERs could expand access to clean, resilient energy. Ongoing research on inclusive financing models will be essential.

CONCLUSION

This paper introduced a blockchain-based framework designed to enhance grid resilience and expedite disaster response in modern power networks. Blockchain's decentralized consensus, immutable data, and smart contract automation help mitigate typical vulnerabilities found in centralized systems. Simulation findings indicate faster fault isolation, reduced outage time, and more equitable resource allocation compared to conventional approaches.

Nonetheless, vital hurdles remain. Achieving rapid transaction throughput in real-time scenarios, maintaining strict cybersecurity, integrating with legacy SCADA, and designing equitable market structures are complex tasks. Policymakers and industry stakeholders must navigate regulatory uncertainties—particularly around data privacy and digital energy transactions.

Looking ahead, merging blockchain with cutting-edge IoT devices, AI-driven analytics, and advanced microgrid technologies could produce self-healing and collaborative energy ecosystems robust against both environmental catastrophes and cyber threats. While still in an early adoption phase, blockchain's potential to revolutionize crucial energy infrastructure underscores its importance for future research and innovation.

REFERENCES

Ahmed, N., Lee, D., & Park, S. (2020). Rapid Disaster Response and the Role of Blockchain-based Grids. *IEEE Transactions on Engineering Management*, *67*(3), 725–736.

Baker, J., & Chen, M. (2021). IoT-Enabled Microgrid Coordination: A Review of Methods and Technologies. *Energy Informatics*, *4*(1).

Brayton, K., & O'Neill, M. (2020). Infrastructure Hardening and Resilience: A Comprehensive Review. *Resilient Systems Quarterly*, *2*(4), 55–66.

Chauhan, A., Li, Y., & Zhou, X. (2022). A Survey on Consensus Mechanisms for IoT-Blockchain Integration. *Computer Networks*, *203*, 108623.

Chen, J., Yang, D., & Liu, K. (2023). Emerging Threats and Resilience Measures for Next-Generation Grids. *IEEE Transactions on Smart Grid*, *14*(1), 101–112.

Deb, H. (2021). Cyber Resilience in Modern Power Systems. *Energy Security Journal*, *5*(2), 21–32.

Elkind, P. (2018). Blockchain and Disaster Recovery: A Blueprint for Rapid Response. *International Journal of Disaster Resilience*, *5*(1), 22–33.

Fox, R. (2021). *Critical Challenges in Power Grid Modernization*. EnergyTech Press.

Gao, P., & Singh, A. (2022). Blockchain-Enhanced Traceability for Renewable Energy Certificates. *Applied Energy*, *311*, 118664.

Hansen, T., & Rodriguez, E. "Decentralized Ledgers in Critical Energy Infrastructure: Opportunities and Pitfalls," in *Proc. IEEE Intl. Conf. on Future Power Systems*, 2019, pp. 11–18.

Holmes, L., Bond, S., & Kimmel, D. (2021). Challenges of Blockchain Adoption in Energy Markets. *Computer Standards & Interfaces*, *73*, 103469.

International Energy Agency (IEA). (2021). *World Energy Outlook 2021*. IEA Publications.

Knapp, C., Novak, E., & Rossi, M. (2021). Scaling Blockchains in Energy Systems: Layered Designs and Sidechain Approaches. *IEEE Access : Practical Innovations, Open Solutions*, *9*, 118954–118965.

Lee, D., Ryu, S., & Jeong, D. (2022). Peer-to-Peer Energy Transactions in a Blockchain Environment: A Case Study of Solar Communities. *IEEE Transactions on Smart Grid*, *13*(5), 3671–3681.

Michel, B., Yang, X., & Clark, R. (2020). Smart Contracts for Automated Demand Response in Microgrids. *IEEE Transactions on Industrial Informatics*, *16*(12), 7687–7696.

Ortiz, P., Park, S., & Hauser, G. "Using Blockchain for Power Grid Event Logging and Forensic Analysis," in *Proc. IEEE PES General Meeting*, 2021, pp. 1–5.

Prakash, A., Bose, C., & Suh, B. (2022). Vulnerability Assessments of Electrical Grids Under Extreme Events. *Energy Studies Review*, *34*(3), 77–88.

Sun, D., Li, L., & Shen, C. (2019). Physical Reinforcements for Resilient Distribution Networks. *IEEE Access : Practical Innovations, Open Solutions*, *7*, 99532–99545.

Torres, M., Li, Y., & Brown, T. (2022). Redundancy Strategies for Grid Continuity During Weather Extremes. *Renewable & Sustainable Energy Reviews*, *142*, 110877.

Ul Haq, F., Zhang, J., & Tohid, R. (2023). Smart Contract-Based Fault Isolation in Autonomous Grids. *Electronics (Basel)*, *12*(4), 897.

U.S. Dept. of Energy (DOE). (2019). Roadmap to Achieve Energy Delivery Systems Cybersecurity.

Ventura, D., & King, J. (2020). Cascading Failures in Centralized Grids: Causes, Consequences, and Mitigations. *IEEE Power & Energy Magazine*, *18*(5), 45–57.

Wang, T., Ching, C., & Qu, M. (2021). A Taxonomy of Cyber Threats in Energy Systems. *Journal of Power Security*, *9*(2), 31–47.

Xu, L., & Alstone, P. (2019). Islandable Microgrids for Community Energy Resilience: A Holistic Perspective. *Energy Reports*, *5*, 121–130.

Yamada, T., & Huang, Q. (2020). Distributed Optimization for Microgrid Management. *IEEE Transactions on Power Systems*, *35*(2), 789–800.

Zhao, B., Li, K., & Thompson, J. (2021). Smart Contract Security in Energy Trading Platforms. *Future Generation Computer Systems*, *120*, 117–129.

Chapter 9
Blockchain for Transparent Governance:
Enhancing Accountability and Trust in E-Government

Mohammad Alauthman
https://orcid.org/0000-0003-0319-1968
Department of Information Security, University of Petra, Jordan

Amjad Aldweesh
https://orcid.org/0000-0001-9358-1323
College of Computing and Information Technology, Shaqra University, Saudi Arabia

Ahmad Al-Qerem
https://orcid.org/0000-0003-2187-6194
Computer Science Department, Faculty of Information Technology, Zarqa University, Jordan

Saad Alateef
https://orcid.org/0009-0002-3219-0833
School of Computing Science, Newcastle University, UK

Ammar Almomani
https://orcid.org/0000-0002-8808-6114
Department of Computer Information Science, Higher Colleges of Technology, Sharjah, UAE

ABSTRACT

Blockchain technology has emerged as a transformative tool for enhancing accountability, transparency, and efficiency in e-government systems. This study explores the integration of blockchain-based architectures within startup-driven e-government models, focusing on secure identity management, automated public service workflows, and tamper-resistant record-keeping. By leveraging smart contracts and permissioned distributed ledgers, blockchain can mitigate corruption, streamline public procurement, and improve citizen trust in digital governance. However, challenges such as scalability, interoperability, legal compliance, and data privacy concerns must be addressed to ensure sustainable adoption. Empirical findings from pilot initiatives suggest that blockchain enhances process integrity, transaction

DOI: 10.4018/979-8-3373-2439-5.ch009

traceability, and interdepartmental collaboration, while startup-driven innovation accelerates agile prototyping and deployment.

1. INTRODUCTION

Public sector functions rely on stable processes. Administrations handle documentation, credentials, and services that affect constituencies. E-government initiatives introduce efficiencies that expedite interactions between public agencies and citizens. Growth of information technology has shifted administrative procedures from analog to digital. This shift generates questions about trust, accountability, and data protection. Blockchain offers a ledger-based approach for storing and verifying public transactions. Researchers propose blockchain as a way to strengthen public-sector efficiency (Khan et al., 2022). Blockchain structures can involve distributed ledgers, smart contracts, and consensus mechanisms. These elements enable secure data integrity, traceability, and process automation.

Startup-driven e-government is a focal point of modern innovation. New ventures bring dynamic tools and agile methods that adapt to shifting environments. Startups in e-government contexts experiment with emerging solutions to improve public service offerings (Judijanto, 2023). Blockchain adoption in the public sector has gained attention for improving processes and records management (Sousa, 2023). Governments explore blockchain to protect medical data, voting systems, and procurement channels (Bustamante et al., 2022).

Trust is central to digital government approaches. Citizens assess the reliability of e-government channels by examining data security and accountability measures (Cengiz, 2023). A recognized challenge involves verifying that digital platforms uphold privacy standards. Blockchain structures append data in a manner that is not easily modified. Each transaction block references a cryptographic hash of previous information. This design introduces integrity in records management. Observers note that blockchain's transparent features may limit corruption in procurement (Kademeteme & Bvuma, 2023). This effect may arise from immutable logs that show each stage of a procurement cycle. Startups can introduce efficiency by customizing blockchain-based software modules.

Public agencies benefit from shared services when they collaborate with smaller, agile solution providers. These collaborations can limit bureaucratic overhead. Blockchain-based systems also offer potential to enhance interoperability across government departments (Liu et al., 2023). Distributed ledgers unify data across boundaries. This can facilitate streamlined workflows in licensing, taxation, and welfare systems. However, the introduction of blockchain in the public sector faces technical, legal, and policy obstacles. Questions include: how should agencies fund initial implementations, how do they ensure compliance with existing regulations, and how do they train civil servants to operate distributed ledger platforms (Polcumpally et al., 2024)?

Startup-driven innovation benefits from minimal friction in prototyping. Startups address user needs with quick iteration cycles. Government processes, on the other hand, adhere to strict guidelines. Collaboration requires alignment of regulatory frameworks with agile development (Kassen, 2023). Such alignment can be difficult when compliance procedures involve rigorous approvals.

Implementation of blockchain solutions in e-government can improve accountability through public ledgers. This approach can reduce manipulation of official records. A multi-stakeholder environment often sees distrust between agencies. Blockchain reduces the need for intermediaries that maintain separate copies of documents (Meyers et al., 2023). Government systems can incorporate distributed ledgers to track resource allocations, audits, and program outcomes (Mahula et al., 2025). The immutability of entries can foster trust among citizens, especially when data is accessible through open platforms (Ekanem & Kim, 2023).

Startup-driven e-government revolves around digital innovation. Integration of blockchain services can produce synergy for sustainable ecosystems. These ecosystems combine public and private actors. The economic dimension involves new ventures, while the administrative dimension involves public agencies. Blockchain-based e-governance solutions aim to restructure data flows. This can offer real-time updates for stakeholders. Transparent workflows may mitigate confusion, especially when multiple departments and citizen groups are involved (Kshetri, 2020).

Public sector corruption has been a frequent concern across various regions. Blockchain's tamper-resistant architecture deters malicious manipulation of government records (Trequattrini et al., 2024). This can apply to land registries, identity management systems, and welfare disbursements. Blockchain-based land registries have been examined in pilot initiatives in different countries (Alketbi et al., 2018). These attempts highlight potential improvements in clarity of property ownership. Some governments have tried e-residency systems that rely on cryptographic identities. The usage of such solutions requires stable policy environments and robust infrastructure (Tan, 2023).

Startup-based approaches to e-government proceed with pilot projects, testing code in limited environments. Successful prototypes can scale to broader user bases. Blockchain systems can provide transaction logs that are shared across nodes. This fosters data-driven analysis of e-government processes. For instance, a local authority might track the entire cycle of building permits on a blockchain. Citizens can confirm that steps remain consistent, as each record is publicly viewable. This promotes trust in the issuing office (Casino et al., 2019).

Adoption challenges persist. Many government employees lack expertise to administer blockchain networks. Some administrations exhibit hesitation due to uncertain regulatory guidelines. There is also the question of data privacy. Blockchain's immutability may conflict with the right to erasure in certain jurisdictions (Kwok & Koh, 2019). The existence of sensitive data in an unalterable format raises legal and ethical dilemmas. Startups must design solutions that store only essential, hashed information on-chain, while keeping personal data in secure off-chain repositories.

Coordination among stakeholders becomes necessary for large-scale blockchain adoption. National agencies, local offices, private vendors, nonprofit groups, and academic institutions might collaborate. The governance model for a blockchain network can define roles for each node. The network can use a permissioned approach, restricting who can run full nodes. Governments often require permissioned networks for compliance reasons (Kassem, 2023). This approach allows for partial transparency, where certain transactions remain confidential, but an oversight body can audit them when needed.

Efficient e-government services demand clarity in architecture. Blockchain solutions should integrate with existing digital identity frameworks. In some countries, these frameworks use digital certificates that are accepted by multiple entities. This design can expedite the deployment of blockchain-based records management. Citizens can sign transactions with digital certificates, generating tamper-evident records. Government portals can read the transaction data from the blockchain, verifying that no alteration has occurred (Rodriguez Müller et al., 2024). Startups that deliver these solutions can adjust front-end interfaces to accommodate citizens with diverse skill levels.

Evaluation approaches measure the impact of blockchain-based e-government solutions. Researchers examine metrics such as user satisfaction, transaction throughput, operational costs, and error rates (Piccardo et al., 2024). These metrics can be compared before and after implementation. Transparent governance may correlate with reduced corruption. The measurement of corruption is complicated, but some indices capture changes in perceived corruption levels. Blockchain logs may deter unethical activities by exposing them to public view.

Discussions of scalability persist in academic literature. Startups seeking e-government solutions face performance issues if the blockchain network handles large transaction volumes. Some solutions adopt sidechains or layer-2 protocols for improved throughput (Mustafa et al., 2024). Interoperability challenges arise when multiple blockchains or legacy systems exchange data. Governments may run pilot projects on one platform, then face difficulties merging records with other networks that use different protocols. Startup-driven models emphasize iterative development, so they can adapt protocols to meet cross-platform requirements.

Security is a critical domain in e-government. Blockchain's cryptographic features protect data, but no system is immune to vulnerabilities in smart contract code or private key storage (Atzori, 2017). Attackers might exploit permission configurations or zero-day vulnerabilities. Governments must incorporate robust risk assessments before large-scale deployment. Insurance frameworks might mitigate some risks. However, the objective is to prevent data breaches. Institutions must define roles and responsibilities for node operators, ensuring that staff follow strict security protocols (Tan, 2023).

Policy aspects cannot be overlooked. Blockchain adoption affects privacy laws, data retention mandates, and cross-border data flows. E-government programs often entail user data from multiple jurisdictions. The question is whether blockchain solutions can comply with these rules while maintaining transparency. Some jurisdictions have introduced regulatory sandboxes that allow startups to test blockchain solutions under supervised conditions. This fosters experimentation while mitigating undue risk to citizens (Meyers et al., 2023).

There is a need to explore sustainability. Startup-driven e-government thrives when solutions scale beyond pilot phases. Blockchain solutions can contribute to cost savings if they reduce paperwork and administrative overhead. However, initial investments can be high. The cost of re-training staff and updating legacy systems can exceed budgets. Startups must demonstrate clear return on investment to public agencies. Some governments partner with international organizations to secure funding for innovative digital transformation projects. These partnerships help sustain development.

The synergy between blockchain and e-government extends to advanced functionalities. One possibility is self-sovereign identity, where citizens manage their own credentials on a distributed ledger (Bustamante et al., 2022). This removes reliance on centralized identity providers. Another possibility involves decentralized autonomous organizations, which could inform future governance models. These concepts remain largely theoretical in a public-sector context. Yet, they reflect new directions in digital governance.

Citizen engagement emerges as a pivotal factor. Transparent blockchain-based systems encourage public participation in governance processes. Citizens can track legislation or policy implementations in near real-time. They can verify official statements against blockchain records (Khan et al., 2022). This dynamic approach contrasts with traditional government portals, where citizens passively consume information. Engagement might improve when individuals see that their data is being managed in an accountable way. This fosters a sense of trust.

Despite these opportunities, caution remains. Experts mention that blockchain is not a universal remedy (Crosby et al., 2016). Some e-government functions might not benefit from distributed ledgers. Complexity, performance constraints, or privacy demands may outweigh potential gains. Policymakers must identify use cases that align well with blockchain features. Startups may help in this assessment by conducting feasibility studies.

Interoperability emerges as a recurring issue. Government agencies rely on multiple legacy databases. Data sets are stored in assorted formats with variable quality. Blockchain layers can unify some processes, but bridging them to offline systems requires robust data governance frameworks (Staples et al., 2018). The role of standardization is pertinent. Without standards, local solutions may remain confined to pilot status.

Research notes that transparency can generate open data sets for analysis. These sets might feed analytics pipelines that inform public policy decisions. Blockchain transaction data can reveal patterns in resource distribution, citizenship applications, or benefit claims. Policymakers can use these patterns to optimize allocation. Startups can build analytics dashboards that draw on blockchain APIs. The final outcome is e-government that is data-driven and open to oversight.

The subsequent sections explore these themes in detail. The chapter covers the conceptual foundations of startup-driven e-government, blockchain principles relevant to public administration, transparency and accountability, and the integration between blockchain and startup ecosystems. Later sections address implementation aspects, governance frameworks, empirical findings, sustainability measures, security considerations, evaluation strategies, ethical concerns, and potential future directions. The chapter ends with a conclusion that synthesizes insights from the analysis.

2. CONCEPTUAL FOUNDATIONS OF STARTUP-DRIVEN E-GOVERNMENT

Startup-driven e-government refers to the application of entrepreneurial methods for public-sector digital services. Startups apply iterative development cycles, user-centered design, and venture funding models in e-government contexts (Judijanto, 2023). E-government aims to offer faster, simpler, and more accessible channels for delivering public services. Traditional government departments often rely on established vendors. In contrast, startup-driven models encourage smaller teams to propose novel solutions. These solutions can be more adaptable to emerging technologies, such as blockchain (Meyers et al., 2023).

Startup-driven e-government ecosystems involve collaboration between public agencies, technology firms, and incubators. Incubators may support startups through mentorship, networking, and resources. Public agencies offer problem statements and regulatory contexts. Startups develop prototypes that address identified problems. This dynamic fosters innovation. Stakeholders perform evaluations on solution feasibility. Agile methodologies allow for continuous refinement of product features.

Inter-organizational networks form a core part of e-government transformation (Mahula et al., 2025). Departments share common processes, such as citizen registration, licensing, and tax collection. These processes can be re-engineered for digital workflows. Startups introduce fresh perspectives on data management and user experience. Over time, the best prototypes can expand across multiple jurisdictions or service lines.

Blockchain aligns with the startup-driven approach. Early-stage ventures can develop minimum viable products that showcase blockchain's strengths in verification and data integrity (Kademeteme & Bvuma, 2023). Public administrators may integrate these prototypes into existing portals. If initial pilots indicate operational benefits, the solution can be expanded. This process avoids large-scale procurement risks associated with untested systems.

Venture capital (VC) influences startup behavior. Some VC funds specialize in technology for government. These funds focus on companies offering solutions that address public policy needs. The potential synergy with e-government arises from the possibility of scaling solutions across multiple governments. Blockchain-based e-government solutions could scale globally if they prove robust in one region. This cross-border perspective encourages standardization.

Public trust remains a key issue. Citizens want reliable digital services. Startup-driven approaches rely on a track record of successful prototypes, user engagement, and stable infrastructure. Collaboration with government bodies can help startups navigate legal constraints. For example, if an initiative involves blockchain-based identity verification, the startup must ensure compliance with relevant identity management laws (Cengiz, 2023). Without legal clarity, the project may stall.

Sustainability of startup-driven e-government depends on several factors. One factor is acceptance by stakeholders. Another factor is resource availability for scaling. A further factor is how effectively solutions handle real-world complexities. Government processes involve diverse user groups, legacy data, and unpredictable policy shifts. Blockchain solutions require stable system architectures to handle large transaction volumes. Startups must demonstrate that their technology can accommodate these conditions without high failure rates (Liu et al., 2023).

The next section discusses blockchain in e-government, focusing on core principles like distributed ledgers, smart contracts, and cryptographic security. These principles are essential for understanding how blockchain can enhance accountability and trust.

3. BLOCKCHAIN IN E-GOVERNMENT: CORE PRINCIPLES

Blockchain technology rests on a decentralized network architecture. Nodes store a replica of the ledger. Each transaction is validated through a consensus mechanism, such as proof of work or proof of stake (Khan et al., 2022). In a public sector context, permissioned blockchains are often chosen for privacy and governance reasons (Kassem, 2023).

Smart contracts are a key innovation. They are scripts that run automatically when predefined conditions are met (Alketbi et al., 2018). In e-government, a smart contract might handle the release of welfare payments once eligibility data is confirmed. This eliminates manual steps, reducing administrative bottlenecks. The contract's code is stored on the blockchain, making its logic verifiable. Parties can confirm that the contract does not deviate from the agreed parameters (Crosby et al., 2016).

Cryptographic elements provide data integrity. Hashing ensures that any tampering in a block is detected. Each block references the hash of the previous block, creating a chain structure. Attempting to alter a past block invalidates subsequent hashes. This design feature assists in guaranteeing that official records remain consistent over time. If an attacker wishes to falsify a transaction, they must rewrite the entire chain.

Consensus mechanisms ensure that only valid transactions enter the ledger. In a public sector setting, participants may be authorized agencies or nodes with specific roles (Meyers et al., 2023). The consensus protocol can be set to achieve finality within seconds or minutes, depending on the system. Public ledgers might allow open participation, but e-government contexts typically restrict node operation to verified entities. This approach reduces the risk of malicious or unknown participants.

Blockchain data can be open or partially restricted. Public blockchains allow anyone to view transaction data. Private or permissioned blockchains limit read and write access to designated nodes (Sousa, 2023). Governments may prefer permissioned ledgers to comply with data confidentiality requirements. Citizens, however, may still access a subset of data, such as final records of policy decisions or official bulletins.

The immutability feature of blockchain is relevant for auditing. Regulators can analyze transaction flows to detect irregularities. Departments can rely on the blockchain record as a single source of truth. This mitigates duplication or discrepancies in separate databases. If a dispute arises about a transaction, the blockchain ledger can serve as evidence. The cryptographic timestamps and digital signatures authenticate the transaction's origin and time.

Adoption in e-government contexts has potential but faces challenges. Integration with existing databases is a technical hurdle. The alignment of blockchain logic with administrative rules is another obstacle. Some e-government processes involve complex workflows, involving multiple sign-offs and forms. These processes may require carefully designed smart contracts, each referencing policy conditions (Staples et al., 2018).

Resource consumption is a noted issue in certain blockchain systems. Proof-of-work mechanisms demand significant computational power. Many e-government applications prefer proof-of-authority or other consensus algorithms with lower energy consumption. This ensures that solutions align with sustainability goals.

Next, transparency and accountability in public administration are discussed, focusing on how blockchain addresses known shortcomings in existing systems.

4. TRANSPARENCY AND ACCOUNTABILITY IN PUBLIC ADMINISTRATION

Transparency is a principle that encourages openness of government actions. Accountability refers to the obligation of public officials to justify decisions and accept responsibility for outcomes. E-government attempts to make information more accessible to the public. However, digital platforms still require trust in the managing authority. Blockchain solutions distribute trust across nodes, reducing reliance on a single central entity (Kademeteme & Bvuma, 2023).

Public procurement is one area with transparency concerns. Traditional processes can involve concealed negotiations, bribery, or favoritism. Blockchain-based procurement logs each bid, contract award, and delivery milestone. Once recorded, these entries cannot be secretly changed. This structure discourages corrupt practices because any alteration attempt is visible in the ledger. The presence of multiple authorized nodes further ensures that no single entity can rewrite history (Cengiz, 2023).

Identity management is another domain. Government identities provide access to services. A blockchain-based identity system can allow citizens to control credential sharing. This can reduce bureaucratic overhead and enable more efficient data verification. By design, the citizen's identity record is stored in a tamper-resistant ledger (Polcumpally et al., 2024). If a government official tries to retroactively alter personal data, it leaves a trace.

Financial transparency emerges in budgeting and expenditures. Governments handle large sums of public funds. Blockchain-based budgeting platforms could detail how funds move from central agencies to local offices, then onward to contractors or service providers. Citizens and oversight bodies can track the flow of funds and detect discrepancies. This fosters accountability, as officials cannot easily manipulate digital transactions (Queiroz et al., 2024).

Legislative processes also benefit from transparency. Certain parliamentary or council decisions can be documented on a blockchain. Votes and amendments are appended to the ledger. Stakeholders can review the final records without concerns of hidden modifications. This approach can encourage civic engagement, since constituents have verifiable information about legislative actions (Ekanem & Kim, 2023).

However, achieving transparency through blockchain also raises privacy challenges. Government data may include sensitive personal or policy details. Deciding what to store on-chain and what to keep off-chain requires deliberate design. Overexposure can threaten individual privacy or national security. A possible solution involves storing only cryptographic proofs on-chain, with off-chain data accessible under controlled conditions (Kwok & Koh, 2019).

Accountability improves when officials cannot deny past decisions. In certain contexts, historical records might vanish or remain incomplete. Blockchain-based archives limit such tampering. Researchers note that immutability promotes robust audit trails (Hajian Berenjestanaki et al., 2023). As a result, wrongdoing is easier to trace, and those responsible face greater likelihood of detection.

This section shows how blockchain can address transparency and accountability concerns in multiple domains of public administration. The next section explains how blockchain intersects with startup-driven e-government models, focusing on system design, resources, and user-centric strategies.

5. INTEGRATING BLOCKCHAIN WITH STARTUP-DRIVEN E-GOVERNMENT MODELS

Blockchain integration in startup-driven e-government models depends on system architecture, stakeholder objectives, and policy constraints. Startup-driven e-government initiatives rely on agile development to prototype, test, and refine solutions. Blockchain offers verifiable data storage and automated processes. The synergy emerges when startups leverage distributed ledgers for streamlined workflows.

Table 1, which outlines typical integration points in startup-driven e-government solutions.

Table 1. Integration points

	Description
Identity Verification	Using blockchain for digital identities and credential management
Payment and Disbursement	Automating fund transfers and subsidies through smart contracts
Document Authentication	Storing and verifying official records in tamper-resistant ledgers
Procurement Processes	Handling tender invitations, bids, contract awards in a shared ledger
Voting and Civic Participation	Recording votes, petitions, or referenda in a transparent blockchain setup

In startup-driven models, identity verification is often a critical first step (Polcumpally et al., 2024). Citizens must log in securely to access e-services. A startup may build a blockchain-based identity management service that local municipalities adopt. Smart contracts can automate the verification procedure. In this way, the municipality can confirm eligibility for benefits or licenses with minimal overhead.

Payment and disbursement processes can also benefit from blockchain. Welfare payments or tax rebates can be transferred automatically once conditions are met (Alketbi et al., 2018). The ledger records each transaction. Auditors can examine these entries. This approach can prevent duplication of benefit claims or unauthorized payments.

Document authentication is vital for notarized records, certificates, permits, and licenses (Casino et al., 2019). Startups can create a module that anchors digital documents to a blockchain. Citizens can verify authenticity by comparing a document's hash with the blockchain record. This method reduces reliance on manual checks and eliminates possibilities of forging official documents.

Procurement processes become more transparent when startup-driven portals incorporate blockchain. Bidders submit proposals that are time-stamped on the ledger. The awarding process is visible. Any attempt to modify evaluation records triggers an auditable discrepancy. This fosters fair competition and trust among participating parties (Kademeteme & Bvuma, 2023).

Finally, voting and civic participation can use a blockchain foundation. A startup might propose a secure e-voting platform. Citizens receive unique credentials that allow them to cast ballots. The ledger ensures that votes remain unaltered, reinforcing trust in the outcome (Hajian Berenjestanaki et al., 2023). Civic initiatives, such as petitions, can follow similar logic. This fosters continuous engagement beyond election periods.

Integrating blockchain with startup-driven models requires attention to user interfaces, as government digital services must cater to all demographics. Startups tend to excel in user experience design. They can deliver streamlined portals and mobile apps. Through iterative testing, they refine workflows to ensure minimal friction.

Resources are a concern. Blockchain infrastructure requires server nodes, cryptographic libraries, and integration layers. Government systems may require specialized hardware security modules for key storage. Funding constraints limit widespread deployment. Pilot projects often rely on external grants or partnerships with larger technology firms.

Policy alignment is essential. When a startup proposes a blockchain-based identity system, it must align with data protection regulations (Tan, 2023). Startups must coordinate with legal advisors to confirm compliance with local rules. This alignment may slow the speed of innovation, but it prevents later disputes or shutdowns.

In summary, integration focuses on identity, payments, documentation, procurement, and voting. Each area benefits from the ledger's immutability and the startup's capacity for iterative product development. The chapter moves next to implementation factors and typical blockchain architectures in e-government scenarios.

6. IMPLEMENTATION FACTORS AND ARCHITECTURE

Implementing blockchain in e-government depends on certain technical and administrative elements. Hardware, software, governance policies, and training programs are central components (Meyers et al., 2023). A typical architecture involves a permissioned blockchain network managed by authorized entities. Government departments, private contractors, and relevant oversight bodies might run nodes.

Below is **Figure 1** illustrating a simplified blockchain architecture in an e-government context:

Figure 1. Blockchain architecture for transparent e-government

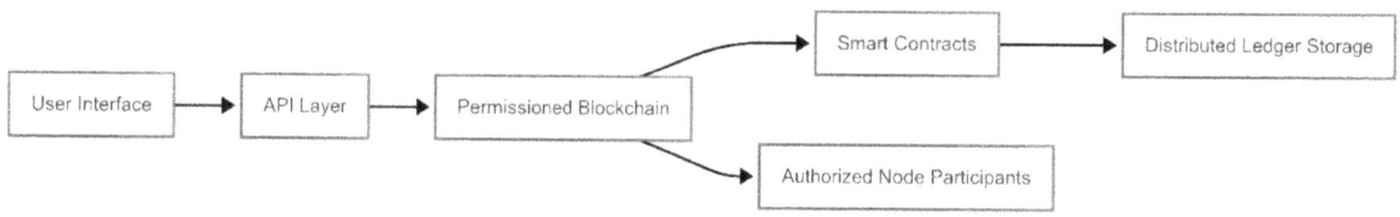

This diagram depicts the flow from user input to the blockchain network. The permissioned blockchain enforces access controls. Smart contracts automate predefined tasks. The distributed ledger stores transaction data. Authorized node participants validate transactions.

Startup-driven solutions often rely on cloud infrastructures to host front-end services and API layers. The blockchain node layer is typically distributed among stakeholders. Communication uses secure protocols. Node operators must be trained to manage cryptographic keys. Implementation factors include:

1. **Node Setup and Maintenance**: Entities must install and maintain node software.
2. **Smart Contract Development**: Code must reflect administrative rules accurately.
3. **Scalability**: The network must handle the expected transaction load.
4. **Interoperability**: Legacy systems must integrate with blockchain APIs.
5. **Governance**: Formal agreements define how stakeholders approve updates and protocol changes.

Smart contracts in e-government solutions encode processes such as permit approvals or financial disbursements. Each contract references identity data, which must be validated. If the contract logic references external data sources, oracles are necessary. Oracles supply real-world information, for instance, a certificate from a licensing authority. The contract logic triggers subsequent steps if the input data matches required conditions (Staples et al., 2018).

Deployment typically involves test environments. Startups develop proof-of-concept solutions in a sandbox. Government partners review compliance with relevant regulations (Kassem, 2023). During pilot deployment, the system may handle a limited volume of transactions, or focus on a single type of service. The pilot's success can determine whether further expansion happens.

Training programs are a significant component. Government employees must learn the basics of blockchain, including ledger review procedures and key management. Mismanagement of cryptographic keys can undermine security (Polcumpally et al., 2024). Administrative procedures may need to be adjusted to align with automated logic. For example, a department that previously required multiple paper signatures might shift to a multi-signature smart contract.

Financial considerations appear at multiple stages. Setup costs cover hardware, licensing, and training. Startups can offer flexible payment models, including subscription services for ongoing maintenance. This structure can reduce initial capital outlay for government agencies. However, it may introduce long-term dependencies on the startup for support and upgrades (Mahula et al., 2025).

Once implementation is underway, governance and policy dimensions become significant. The next section explores how policy frameworks interact with blockchain adoption in a public sector context.

7. GOVERNANCE AND POLICY DIMENSIONS

Blockchain governance extends beyond technical consensus. It includes stakeholder management, regulatory alignment, and conflict resolution (Cengiz, 2023). In e-government contexts, official mandates might define how nodes are distributed and who has authorization to validate transactions. Policy instruments regulate data classification, privacy obligations, and compliance standards.

Some governments adopt pilot programs backed by legislation or executive orders. These mandates legitimize blockchain trials and secure funding. Policymakers outline goals, such as reducing transaction times or improving accountability in specific agencies (Kassem, 2023). The public can hold officials responsible if targets are unmet. This approach fosters performance-based governance.

Regulatory sandboxes allow startups to experiment with blockchain solutions under monitored conditions. Authorities waive or relax certain rules temporarily. This environment encourages risk-taking, but includes reporting obligations. If a pilot demonstrates viability, policymakers may formalize the solution through updated regulations (Sousa, 2023).

International standards also influence blockchain governance. Cross-border interactions rely on interoperability in identity management and data formats. Supranational bodies can promote frameworks that governments can adopt. These frameworks define best practices for security and compliance (Rodriguez Müller et al., 2024). Startups can then develop solutions that align with those standards, increasing acceptance across multiple jurisdictions.

Governance extends to network upgrades. Over time, new features or patches may be needed. The relevant stakeholders must agree on changes to the protocol or smart contracts (Staples et al., 2018). In a typical environment, a committee representing government departments and private partners reviews upgrade proposals. Approval may require consensus or a supermajority vote. The final outcome is integrated into the ledger governance rules.

Policy must also address liability. If a smart contract malfunctions and causes financial losses, clarity is needed. Startups may carry liability for coding errors, or government agencies might bear responsibility for the business logic. An established legal framework can reduce disputes and maintain trust.

The next section moves into a review of existing initiatives to show real-world outcomes.

8. EMPIRICAL FINDINGS FROM EXISTING INITIATIVES

Multiple pilot studies illustrate blockchain adoption in e-government. Governments in different regions have tested prototypes in land registration, licensing, and welfare systems (Alketbi et al., 2018). Pilot results reveal improvements in data validation and record retrieval. However, they also demonstrate challenges in user training and system maintenance.

Some empirical data indicates that blockchain can shorten processing times for permit applications. This conclusion is based on small-scale deployments in local municipalities (Judijanto, 2023). Another key finding is that start-up involvement accelerated the testing phase. Innovative methods reduced bureaucratic red tape, speeding up solution deployment.

Corruption reduction is a central theme. Transparent procurement processes deter unlawful behavior. Observers noted improved trust among bidders in a pilot environment (Kademeteme & Bvuma, 2023). Officials reported fewer complaints about awarding decisions. Still, some attempts faced resistance from staff used to older systems. Skilled change management eased the transition.

Studies of e-voting experiments found that blockchain-based systems can record ballots securely, but user experience was mixed (Hajian Berenjestanaki et al., 2023). Some voters were unfamiliar with digital interfaces. Rural areas encountered internet connectivity issues. These logistical concerns highlight that technology alone does not guarantee success.

Land registration pilots revealed that uploading records to a blockchain is straightforward if the data is organized. However, legacy data often contains errors or missing pieces. Migrating inaccurate records onto an immutable ledger can embed mistakes permanently. Thus, data cleaning is a preliminary step (Casino et al., 2019).

Analyses of cost structures show that adopting blockchain may lead to short-term expense increases. Over the long term, reduced paperwork and automation may lower operational costs (Sousa, 2023). The return on investment depends on solution scope and adoption rates. If only a small fraction of citizens use the service, cost savings remain limited. A broad deployment may justify the initial expenditure.

Overall, empirical findings reveal incremental benefits in transparency and efficiency. However, success often depends on supportive leadership, user adoption, policy alignment, and adequate digital infrastructure. The next section discusses frameworks for building sustainable ecosystems.

9. FRAMEWORK FOR SUSTAINABLE ECOSYSTEMS

A sustainable ecosystem in e-government involves stable funding, institutional backing, technical resilience, and stakeholder engagement (Mahula et al., 2025). Blockchain solutions must operate long-term and remain adaptable. Rapid technological shifts can leave systems obsolete if not maintained. **Table 2**, illustrating key elements of a sustainable e-government ecosystem that relies on blockchain.

Table 2. Ecosystem Elements

	Description
Policy Support	Clear regulations and supportive legal frameworks
Technological Infrastructure	Reliable networks, servers, and node management
Funding and Investment	Public budgets, venture capital, or mixed financing
Skilled Workforce	Trained developers, officials, and end users
Collaborative Governance	Multi-stakeholder committees for decision-making
User Engagement	Citizen awareness, accessibility, and feedback loops

Policy support is foundational. A stable legal environment encourages adoption and risk-taking. Technology infrastructure requires robust network connections, cybersecurity measures, and well-maintained hardware for nodes. Funding may originate from government budgets, multilateral agencies, or private investments (Rodriguez Müller et al., 2024). Skilled personnel are needed to handle blockchain deployment. Collaborative governance ensures that changes occur with agreement among relevant parties.

User engagement includes citizen outreach. Training programs can familiarize the public with blockchain-based services. Feedback loops gather input from users. Startups can incorporate user suggestions into subsequent software updates, aligning with agile principles. If adoption rates increase, the solution's overall viability improves.

Ecosystem sustainability emerges when each element functions harmoniously. A deficiency in any area can undermine progress. The next section investigates security, trust, and how they affect adoption and system operations.

10. SECURITY, TRUST, AND CITIZEN ENGAGEMENT

Security is paramount in blockchain-based e-government. The ledger holds official records, identity credentials, and transaction data. A breach can compromise citizen privacy or financial stability. Trust in e-government solutions emerges when systems withstand threats. Citizens must believe that their data is safe (Khan et al., 2022).

Blockchain's cryptographic underpinnings reinforce security. Each transaction is signed by a private key, reducing the risk of impersonation. However, if private keys are stolen, attackers can manipulate accounts. Governments and startups can mitigate such risks by using secure hardware, multi-factor authentication, and regular audits (Staples et al., 2018).

Network-level attacks might aim to disrupt consensus by controlling a majority of nodes. Permissioned networks reduce this likelihood because participants undergo vetting. Denial-of-service attacks can still occur, affecting node operation. Startups and government agencies must prepare incident response plans. Backup procedures and redundancy are also important.

Trust influences adoption. If citizens suspect that blockchain records can be manipulated or sold to third parties, they will avoid the platform. Clear communication strategies can detail how blockchain stores data, who can access it, and how immutability protects against tampering (Queiroz et al., 2024). Focused outreach through community forums or social media can clarify misconceptions.

Citizen engagement grows when services demonstrate tangible benefits, such as faster processing and transparent documentation. E-government portals can include dashboards that show real-time updates of transactions or service status (Judijanto, 2023). This visibility prompts users to trust the underlying blockchain. As usage increases, the system's legitimacy strengthens.

The next section examines scalability and interoperability constraints, focusing on technical barriers and solutions.

11. SCALABILITY AND INTEROPERABILITY CHALLENGES

Scalability refers to the capacity of a blockchain network to process growing transaction volumes without significant performance degradation. Interoperability is the ability of different systems to exchange and use information seamlessly (Mustafa et al., 2024). Both concepts are crucial for broad e-government adoption.

Public blockchains face throughput bottlenecks. Some networks process limited transactions per second. Enterprise or permissioned blockchains can implement more efficient consensus methods, but must maintain security. E-government projects might expect large spikes in transactions during election periods or tax seasons (Tan, 2023). Startups must design solutions that handle these peaks without service disruptions.

Sidechains or layer-2 solutions can offload transactions from the main chain. This approach has been proposed for e-government use cases that require high throughput. Another solution is to partition data across multiple chains. However, fragmentation can complicate data retrieval.

Interoperability involves linking blockchain solutions with existing government databases. Some data resides in legacy mainframes. Others may follow older standards. Adapters or middleware can facilitate data exchange. Standardized data schemas can streamline these processes (Staples et al., 2018). If the e-government ecosystem includes several blockchains or cross-border services, bridging solutions might be needed.

Below is **Figure 2**, showing an interoperability scenario:

Figure 2. Blockchain interoperability in e-government

In this figure, the adapter API bridges legacy databases with one blockchain network. An interoperability layer links multiple blockchain networks. Distributed applications built on top can access data from both networks.

Scalability solutions must align with privacy and compliance rules. Sharding or sidechains might store partial data sets, limiting exposure. However, concurrency issues can arise when multiple shards attempt to update a single record. Test environments can explore best practices for e-government contexts.

The next section considers how to evaluate blockchain-based e-government solutions.

12. EVALUATION APPROACHES

Evaluation of blockchain-based e-government solutions examines performance, security, user acceptance, cost-effectiveness, and compliance (Piccardo et al., 2024). Researchers may adopt a mixed-method approach, combining quantitative metrics with qualitative feedback.

1. **Transaction Metrics**: Throughput, latency, finality time, and block size can show efficiency.
2. **Security Audits**: Code reviews of smart contracts, penetration testing, and node security checks.
3. **User Acceptance Testing**: Surveys or interviews gauge citizen satisfaction and ease of use (Ekanem & Kim, 2023).
4. **Cost-Benefit Analysis**: Compares development and operational expenses with observed savings in manual labor.
5. **Compliance Assessment**: Verifies adherence to data protection, financial regulations, and sector-specific rules.

Table 3, summarizing possible metrics and methods:

Table 3. Evaluation dimensions

	Indicators	Methods
Technical Performance	TPS, latency, node reliability	Benchmark tests, load simulations
Security and Privacy	Vulnerabilities, encryption strength	Security audits, threat modeling
User Adoption and Satisfaction	Number of registered users, usage rate	Surveys, focus groups, usage logs
Economic Efficiency	Return on investment, overhead reduction	Financial analysis, cost tracking
Legal and Regulatory Compliance	Data protection adherence, liability rules	Audit reports, legal review

Stakeholders can use these metrics to inform decisions on whether to continue, expand, or revise blockchain-based solutions. Ongoing evaluations capture the dynamic nature of e-government. Periodic reviews can track changes in policy, user behavior, and technological capabilities.

The next section discusses ethical and legal considerations that arise when blockchain intersects with public-sector administration.

13. ETHICAL AND LEGAL ASPECTS

Blockchain's immutability raises ethical and legal considerations. Permanent data storage conflicts with rights that allow individuals to request data deletion. Some jurisdictions have "right to be forgotten" clauses (Kwok & Koh, 2019). E-government solutions must handle personal data with care. A design approach may store hashed references on-chain, while the actual personal data resides off-chain and can be deleted if required.

Consent is another ethical issue. Citizens must understand how their data is processed in a blockchain system. If they cannot withdraw consent, or if they lose control of their data, trust erodes. Clear user agreements can outline data-handling practices. Public bodies might create oversight committees to ensure compliance with data ethics standards (Staples et al., 2018).

Accountability for blockchain-driven decisions is also complex. Smart contracts execute automatically, but who is liable if an error occurs? Government bodies may claim that the code is the source of the error, while startups might argue that it was an implementation detail. Clear contractual agreements and legislative frameworks can reduce ambiguity (Cengiz, 2023).

Political contexts may also affect blockchain adoption. Decentralization can challenge traditional power structures. Some officials may be reluctant to relinquish control. Policymakers can address such concerns by highlighting the public benefits of transparency.

The following section outlines avenues for future research in blockchain-based e-government.

14. FUTURE RESEARCH DIRECTIONS

Future research may investigate novel consensus mechanisms tailored to e-government, focusing on low-energy consumption and fast finality. Studies could explore how to embed policy logic in smart contracts for real-time compliance checks (Mustafa et al., 2024). Researchers may also examine advanced privacy solutions, such as zero-knowledge proofs, that reveal minimal data while verifying transaction validity.

There is scope for interdisciplinary work involving political science, law, and technology to refine governance models for permissioned networks (Meyers et al., 2023). The role of stakeholder committees in protocol updates warrants closer scrutiny. Another area is the integration of artificial intelligence with blockchain-based e-government. AI could automate large-scale analytics of on-chain data to detect patterns in public service usage or corruption.

Comparative studies could evaluate different national approaches. Some countries prioritize transparency, while others emphasize security or privacy. Cross-case analyses may highlight how cultural norms and legal systems shape blockchain adoption. The cost-benefit profiles of various e-government functions can guide policymakers seeking to allocate resources effectively (Kassem, 2023).

Pilot expansions could focus on advanced identity frameworks. Digital identities anchored in blockchain can incorporate multiple credentials: health records, tax information, and educational certificates. Researchers can examine citizen acceptance, the potential for data misuse, and the viability of decentralized identity management.

Scaling solutions beyond single municipalities to entire regions or cross-border contexts is another area for investigation (Rodriguez Müller et al., 2024). Large-scale networks must handle concurrency, standardization, and multi-jurisdictional governance. More robust prototypes can validate whether blockchain truly delivers sustainable benefits at scale.

Finally, user-centric design remains an area requiring research. Studies can evaluate how interface design affects citizen trust in blockchain services. Iterative improvements, supported by user feedback, can refine the e-government experience, enhancing inclusive access for populations with varying technical skills.

The closing section synthesizes these ideas and reflects on the significance of blockchain for transparent governance and trust in e-government.

15. CONCLUSION

Blockchain technology offers potential to enhance accountability and trust in e-government by providing immutable records, transparent workflows, and automated processes. Startup-driven approaches bolster this potential through agile development, rapid prototyping, and user-centered design. The synergy of blockchain and startup-driven e-government can address corruption, reduce inefficiencies, and improve service delivery. Evidence from pilot initiatives shows incremental improvements in speed, transparency, and citizen trust (Judijanto, 2023; Kademeteme & Bvuma, 2023).

Achieving broad adoption necessitates alignment of policy, technical architectures, and stakeholder incentives. Permissioned blockchains managed by authorized nodes appear suitable for public-sector contexts. Smart contracts can encode regulatory requirements and streamline routine processes (Polcumpally et al., 2024). Implementation challenges involve legacy integration, regulatory compliance, financial planning, and workforce training (Liu et al., 2023).

Scalability and interoperability remain obstacles. Large transaction volumes require efficient consensus protocols, sidechains, or layered solutions. Interfacing with legacy databases and multiple blockchain platforms demands standardized frameworks and robust middleware (Staples et al., 2018). Governance models must define how updates occur, who is liable for errors, and how privacy rights are upheld (Cengiz, 2023).

Sustainable ecosystems rely on policy support, technological infrastructure, secure funding, skilled personnel, collaborative governance, and user engagement. Startups thrive when these elements align, enabling them to refine services, attract user bases, and deliver tangible public benefits (Mahula et al., 2025). Blockchain solutions can promote inclusive, transparent, and accountable governance, but only if they address real-world complexities in data management, privacy, and institutional reform.

Future research points to specialized consensus mechanisms, advanced privacy features, cross-jurisdictional cooperation, and AI-driven analytics of blockchain data. These developments could elevate e-government from basic digital channels to dynamic, trusted platforms for civic participation and public administration. Collaborations across government agencies, startups, academic institutions, and policy bodies can guide effective blockchain integration in e-government.

The interplay of blockchain and startup-driven e-government represents a promising path toward digital innovation for sustainable ecosystems. Although hurdles remain, continued exploration, experimentation, and policy development can refine approaches that deliver value to citizens and contribute to robust governance practices.

REFERENCES

Alketbi, A., Nasir, Q., & Talib, M. A. (2018). Blockchain for government services: Benefits and challenges. *IEEE Access : Practical Innovations, Open Solutions*, *7*, 43426–43433.

Bustamante, P., Cai, M., Zambrano, R., & Weiss, M. (2022). Government by code? Blockchain applications to public sector governance. *Frontiers in Blockchain, 5*, Article 869665.

Casino, F., Dasaklis, T. K., & Patsakis, C. (2019). A systematic literature review of blockchain-based applications: Current status, classification and open issues. *Telecommunication Systems*, *71*(1), 53–74.

Cengiz, F. (2023). Blockchain governance and governance via blockchain: Decentralized utopia or centralized dystopia? *Policy Design and Practice*, *6*(4), 446–464. DOI: 10.1080/25741292.2023.2247203

Crosby, M., Pattanayak, P., Verma, S., & Kalyanaraman, V. (2016). Blockchain technology: Beyond bitcoin. *Applied Innovation*, *2*, 6–10.

Ekanem, U., & Kim, Y. S. (2023). Empirical study of factors influencing adoption of blockchain technology converged with e-government in Nigerian public sector. *International Journal of Contents*, *19*(1), 67–81. DOI: 10.5392/IJoC.2023.19.1.067

Hajian Berenjestanaki, M., Barzegar, H. R., El Ioini, N., & Pahl, C. (2023). Blockchain-based e-voting systems: A technology review. *Electronics (Basel)*, *13*(1), 17. DOI: 10.3390/electronics13010017

Judijanto, L. (2023). Implementation of blockchain technology in improving transparency of public services: A case study on government service delivery in Indonesia. *West Science Information System and Technology*, *1*(2), 63–71. DOI: 10.58812/wsist.v1i02.477

Kademeteme, E., & Bvuma, S. (2023). Using blockchain technology to improve the integrity and transparency of procurement processes between SMMEs and government: A systematic literature review. *Journal of The British Blockchain Association*, *6*(1), 1–13. DOI: 10.31585/jbba-7-1-(1)2024

Kassem, M. (2023). Blockchain and digital governance: Decentralization of decision making policy. *The Review of Policy Research*, *42*(1), 45–67.

Khan, S., Shael, M., Majdalawieh, M., Nizamuddin, N., & Nicho, M. (2022). Blockchain for governments: The case of the Dubai Government. *Sustainability (Basel)*, *14*(11), 6576. DOI: 10.3390/su14116576

Kwok, J. S. H., & Koh, L. H. (2019). Is blockchain technology a watershed for e-government? The case of Seoul's blockchain 2.0 initiatives. *Asia Pacific Journal of Public Administration*, *41*(1), 37–50.

Liu, Y., Lu, Q., Zhu, L., Paik, H. Y., & Staples, M. (2023). A systematic literature review on blockchain governance. *Journal of Systems and Software*, *197*, 111576. DOI: 10.1016/j.jss.2022.111576

Mahula, S., Tan, E., Crompvoets, J., & Timmers, P. (2025). What motivates public sector organisations to use blockchain? [in press]. *International Journal of Public Sector Management*, *38*(1), 120–136. DOI: 10.1108/IJPSM-12-2023-0361

Meyers, G., van Oirsouw, C. C. K., Keymolen, E. L. O., & Goossens, J. (2023). After the announcement: An interdisciplinary analysis of blockchain development in governments. *Policy Design and Practice*, *6*(4), 433–447. DOI: 10.1080/25741292.2023.2272332

Mustafa, G., Rafiq, W., Jhamat, N., Arshad, Z., & Rana, F. A. (2024). Blockchain-based governance models in e-government: A comprehensive framework for legal, technical, ethical and security considerations. *International Journal of Law and Management*. (Advance online publication)

Piccardo, G., Conti, L., & Martino, A. (2024). Blockchain technology and its potential to benefit public services provision: A short survey. *Future Internet*, *16*(8), 290. DOI: 10.3390/fi16080290

Polcumpally, A. T., Pandey, K. K., Kumar, A., & Samadhiya, A. (2024). Blockchain governance and trust: A multi-sector thematic systematic review and exploration of future research directions. *Heliyon*, *10*(12), e32975. DOI: 10.1016/j.heliyon.2024.e32975 PMID: 38984295

Queiroz, M. M., Fosso Wamba, S., Wamba-Taguimdje, S. L., & Lu, Q. (2024). How emerging technologies can solve critical issues in organizational operations: An analysis of blockchain-driven projects in the public sector. *Government Information Quarterly*, *41*(1), 101912. DOI: 10.1016/j.giq.2024.101912

Rodriguez Müller, A. P., Martin Bosch, J., & Tangi, L. (2024). An overview of the expected public values arising from blockchain adoption in the European public sector.* [forthcoming]. *International Journal of Public Sector Management*, *37*(2), 240–257.

Sousa, M. J. (2023). Blockchain as a driver for transformations in the public sector. *Policy Design and Practice*, *6*(4), 415–432. DOI: 10.1080/25741292.2023.2267864

Staples, M., Chen, S., Falamaki, S., Ponomarev, A., Rimba, P., Tran, A. B., & Zhu, J. (2018). Risks and opportunities of blockchain technologies in the public sector. In *Proceedings of the 51st Hawaii International Conference on System Sciences (HICSS)* (pp. 5470–5479). IEEE.

Tan, E. (2023). The missing piece: The link between blockchain and public policy design. *Policy Design and Practice*, *6*(4), 488–504. DOI: 10.1080/25741292.2023.2233160

Tanquattrini, R., Palmaccio, M., Turco, M., & Manzari, A. (2024). The contribution of blockchain technologies to anti-corruption practices: A systematic literature review. *Business Strategy and the Environment*, *33*(1), 4–18. DOI: 10.1002/bse.3327

Trequattrini, R., Palmaccio, M., Turco, M., & Manzari, A. (2024). The contribution of blockchain technologies to anti-corruption practices: A systematic literature review. *Business Strategy and the Environment*, *33*(1), 4–18. DOI: 10.1002/bse.3327

Chapter 10
AI and Blockchain Integration for Advanced Metering Infrastructure (AMI)

Mohammad Alauthman
https://orcid.org/0000-0003-0319-1968
Department of Information Security, University of Petra, Jordan

Ammar Almomani
https://orcid.org/0000-0002-8808-6114
Higher Colleges of Technology, Sharjah, UAE

Ahmad Al-Qerem
https://orcid.org/0000-0003-2187-6194
Computer Science Department, Faculty of Information Technology, Zarqa University, Jordan

Saad Alateef
https://orcid.org/0009-0002-3219-0833
School of Computing Science, Newcastle University, UK

Amjad Aldweesh
https://orcid.org/0000-0001-9358-1323
College of Computing and Information Technology, Shaqra University, Saudi Arabia

Mouhammd Alkasassbeh
https://orcid.org/0000-0001-8396-7441
Princess Sumaya University for Technology, Jordan

ABSTRACT

This chapter examines how Artificial Intelligence (AI) and blockchain converge in the context of Advanced Metering Infrastructure (AMI) within smart grids. It presents theoretical concepts of AI for data analytics and automation, and explains blockchain's potential to create transparent, decentralized, and secure data environments. The discussion explores the synergy between AI-driven intelligence and blockchain-based trust frameworks for managing metering data, demand-side response, energy trading, and anomaly detection. Several use cases demonstrate how AI-led approaches elevate predictive and optimization tasks, while blockchain safeguards meter data integrity and coordinates distributed market transactions. The chapter presents operational, security, and interoperability challenges and emphasizes the scope for future research on standardizing AI–blockchain modules and addressing scalability constraints. Findings highlight the strategic importance of pairing blockchain-based trust with AI-based intelligence for achieving automated, reliable, and consumer-focused AMI.

DOI: 10.4018/979-8-3373-2439-5.ch010

1. INTRODUCTION

Advanced Metering Infrastructure (AMI) in smart grids involves networks of intelligent meters and communication channels that collect and disseminate energy data to utilities, prosumers, and other stakeholders. AMI data include near-real-time usage measurements, power quality statistics, and historical consumption profiles. AMI supports many functions, including billing, outage management, and demand-side response. It is a foundational component of the modern smart grid because it establishes a bidirectional communication system between utilities and end consumers (Pop, Cioara, Antal, Anghel, Salomie, & Bertoncini, 2018).

Researchers have investigated AMI since the early 2000s, but recent technological advances—particularly in Artificial Intelligence (AI) and blockchain—promise to raise the level of data accuracy, security, and automation for AMI-based operations (Andoni et al., 2019). Conventional AMI systems optimize grid operations through centralized, utility-controlled data acquisition and processing. However, shifting consumer expectations, distributed energy resources, and dynamic pricing schemes require more flexible, secure, and intelligent approaches to data management. This transforms AMI from a mere automated billing system into a multi-functional layer of data-driven insights and real-time decision support.

AI refers to the suite of computational techniques that replicate or enhance human-like intelligence in tasks such as prediction, learning, or optimization. Machine learning models process meter data to detect anomalies, forecast loads, guide pricing, and coordinate distributed energy resources. AI can improve operational efficiency by automating load balancing and scheduling supply to match consumption patterns (Khan et al., 2023). Additionally, AI techniques can identify abnormal usage or meter data manipulation by analyzing consumption patterns at the household or microgrid level. Anomalies often disrupt AMI-based services because they signal equipment failure, unauthorized data tampering, or malicious attacks (Xue, Yu, Cen, & Li, 2025).

Blockchain is a distributed ledger technology recognized for decentralized transaction validation, immutability of stored records, and transparent consensus. Blockchain provides trust in multi-party settings without relying on a central coordinator. It achieves this through a peer-to-peer network of nodes that validate new blocks of data through consensus algorithms (Al Shareef, Seçkiner, Eid, & Abumeteir, 2024). Smart contracts can programmatically automate business logic, enabling self-enforcing contracts for energy trading or demand response. Integrating blockchain into AMI can deliver tamper-resistant data storage, auditable transactions, and secure machine-to-machine exchanges (Alladi, Chamola, Rodrigues, & Kozlov, 2019).

The convergence of AI and blockchain in AMI can address data-driven decision-making needs and trust deficits. AI depends on consistent data streams for training and inference, while blockchain secures these streams and ensures that data remain authentic. This synergy benefits a range of use cases: (1) anomaly detection that uses blockchain-logged meter data to identify suspicious usage patterns with AI-based analytics; (2) decentralized energy trading where prosumers rely on secure ledgers for peer-to-peer transactions and AI-driven market mechanisms for price discovery; and (3) demand-side management that automates load shifting or aggregator-driven energy curtailment through blockchain-based smart contracts triggered by AI forecasts (Rahman, Chowdhury, Shorfuzzaman, Hossain, & Hammoudeh, 2023).

However, implementing AI–blockchain solutions in AMI faces technical and organizational challenges. High volumes of meter data increase computational demands on blockchain networks, and consensus protocols can become bottlenecks. AI training processes can consume substantial computational resources, and blockchain-based distributed architectures might slow down real-time grid controls if not carefully designed (Singh & Singha, 2024). Furthermore, privacy issues arise when user consumption data reside on public or partially public blockchain networks, necessitating cryptographic mechanisms or permissioned blockchains for compliance with data protection regulations (Jithish, Alangot, Mahalingam, & Yeo, 2023). Despite these challenges, the potential for improved transparency, security, and data-driven efficiency motivates ongoing research and pilot projects worldwide.

This chapter explores how AI and blockchain can reinforce each other in AMI, offering a theoretical foundation and a detailed review of practical applications and case studies. Researchers continue to test pilot systems that combine advanced machine learning models for load forecasting or anomaly detection with secure ledger technologies that coordinate distributed transactions and data sharing. Such efforts shape the future of decentralized grid management, where each node—whether it is a household, electric vehicle (EV), or microgrid—actively participates in the overall system's stability and efficiency.

1.1 Rationale for AI and Blockchain Integration in AMI

AMI is a cornerstone for digitalizing the energy sector. It enables near-real-time communication for monitoring and controlling distributed resources. However, standard AMI setups often rely on centralized, proprietary data platforms prone to outages, security breaches, and single points of failure (Hasan et al., 2022). Traditional data systems may also lack robust approaches for reconciling the large volume of data from thousands or millions of meters. These barriers motivate the integration of robust technologies like AI and blockchain, which can process and secure large-scale data in distributed environments.

AI applications in AMI range from advanced predictive maintenance to real-time load balancing. They rely on consistent, high-fidelity data from meters and sensors. AI can autonomously initiate actions, such as adjusting inverter settings for distributed generation or signaling consumers to reduce consumption during peak periods (Kuzior, Sira, & Brożek, 2022). However, malicious actors can tamper with meter data, or external factors can corrupt data streams. Blockchain integration addresses data integrity vulnerabilities by logging meter data on a tamper-resistant ledger (Andoni et al., 2019). Each data entry becomes immutable once validated through consensus, deterring attempts to alter usage records.

Blockchain-based smart contracts also enable new business logic. For instance, an AI algorithm can forecast next-day solar generation, and a smart contract can automatically execute peer-to-peer energy trades when certain conditions arise (Han, Meng, & Luo, 2024). Such an approach eliminates intermediaries, reduces transaction costs, and promotes transparency. This synergy supports an advanced AMI ecosystem that is more responsive, secure, and consumer-oriented than legacy systems.

1.2 Scope and Objectives

This chapter aims to:

1. Present core AI and blockchain concepts in the context of AMI.
2. Examine how AI-driven analytics can improve metering, billing, anomaly detection, and resource management.
3. Illustrate blockchain's role in decentralizing data management, energy trading, and security.
4. Identify the synergy between AI-based decision tools and blockchain-based trust mechanisms.
5. Highlight real-world applications and research breakthroughs in anomaly detection, demand response, and decentralized trading.

6. Outline security, privacy, and performance challenges.
7. Suggest areas for future exploration that can strengthen AI–blockchain integration in AMI.

1.3 Chapter Structure

The chapter continues with a review of AI concepts relevant to AMI, followed by an examination of blockchain fundamentals. It then delves into how these technologies merge to address critical AMI functions such as billing automation, data integrity, load management, and security monitoring. The text integrates tables that summarize key AI techniques, blockchain components, synergy points, and use cases. Figures coded in Mermaid illustrate high-level architectures and process flows. Security and privacy receive special attention, due to the sensitivity of meter data. This comprehensive review concludes with implementation challenges, lessons from existing pilot projects, and future directions that encourage standardization, scalability, and open research.

2. OVERVIEW OF AI CONCEPTS IN AMI

AI describes a suite of methods that enable systems to learn patterns, make predictions, and perform tasks that would otherwise require human-level cognition. AI applications in AMI include forecasting, anomaly detection, automated control, and intelligent optimization of resources. AI approaches vary in complexity, from simple linear models to deep reinforcement learning. The wide availability of AMI data motivates the use of advanced algorithms that process continuous streams of meter readings, environmental variables, and historical usage profiles (Hua et al., 2022).

2.1 Machine Learning for AMI

Machine learning refers to algorithms that learn relationships within data and make data-driven decisions with minimal human intervention. AMI data often arrive in large volumes from smart meters at different intervals (e.g., 15-minute or hourly). Machine learning models can handle these volumes by training on historical consumption, identifying consumption patterns, or detecting novel usage behaviors (Li, Herdem, Wen, & Kuzior, 2023).

2.1.1 Supervised Learning

Supervised learning algorithms use labeled data to learn input-output mappings. For instance, a model may learn to predict daily consumption based on weather data, occupant behavior, and historical usage. In a typical scenario, a set of meter readings is paired with known consumption outputs, and the model adjusts its internal parameters to minimize error. Regression models (linear, polynomial, or more advanced forms like random forests) are common for load forecasting. Classification approaches can label data as normal or anomalous. **Table 1** provides an overview of common machine learning tasks and their relevance to AMI.

Table 1. Summarized common machine learning tasks in AMI

Machine Learning Task	Description	Application in AMI
Load Forecasting	Predicts future energy consumption based on historical usage and external factors.	Enables scheduling, dynamic pricing, and demand response optimization.
Anomaly Detection	Identifies deviations from normal consumption patterns.	Detects fraud, meter tampering, or faulty sensor behavior.
Clustering	Groups consumers based on similar energy usage profiles.	Supports segmentation for targeted demand-side programs.
Classification	Assigns labels (e.g., normal or abnormal) based on input features.	Used in identifying outage events or customer behavior types.
Regression	Estimates continuous output variables such as energy load or cost.	Supports consumption prediction and pricing models.
Reinforcement Learning	Learns optimal actions through interaction with the environment.	Guides automated control of distributed energy resources.
Feature Selection	Identifies key variables influencing model outcomes.	Improves model performance and interpretability in AMI analytics.

In Figure 1, the flowchart shows a typical machine learning pipeline, from data collection through deployment, with feedback to continuous monitoring.

Figure 1. Generic machine learning process flow in AMI context

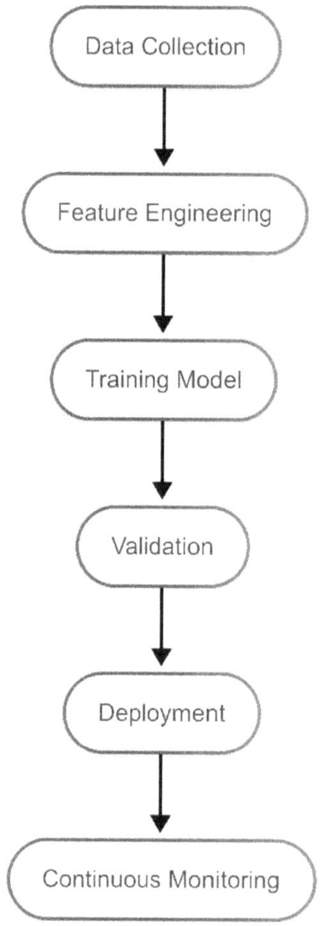

Supervised methods can process large amounts of training data if the data are accurately labeled. In AMI, acquiring reliable labels for anomalies can be challenging, but if well-designed, these methods are effective at capturing consumption patterns (Wang, 2025).

2.1.2 Unsupervised Learning

Unsupervised learning deals with unlabeled data, seeking hidden structures or clusters. In AMI, consumption data can be high-dimensional. Unsupervised methods, such as clustering or dimensionality reduction, can group consumers by consumption similarity. These groupings can inform targeted demand response or personalized pricing. Anomaly detection is also prevalent, where the model flags patterns that deviate from the norm (Xue et al., 2025).

Unsupervised learning can detect subtle abnormalities. A meter's consumption might slightly deviate from peer meters with similar usage habits. Through clustering or density-based anomaly detection, the AI system can highlight suspicious activity for further investigation (Jithish et al., 2023).

2.1.3 Semi-supervised Learning

Semi-supervised learning combines labeled and unlabeled data. This approach is valuable when extensive unlabeled data exist but labeled instances are scarce. In an AMI anomaly detection scenario, labeling every meter reading as normal or abnormal can be cumbersome. Semi-supervised methods leverage limited labeled anomalies to guide learning across many unlabeled samples.

2.2 Deep Learning

Deep learning refers to neural networks with multiple layers that learn hierarchical feature representations. Deep learning can capture complex, non-linear consumption patterns in time-series AMI data (Hasan et al., 2022). Recurrent neural networks (RNNs) and their variants, such as LSTM (Long Short-Term Memory) networks, are well-suited for sequential data. Convolutional neural networks (CNNs) are also utilized, transforming time-series data into images or employing sliding windows.

Deep learning models achieve strong performance in load forecasting and anomaly detection. They extract complex relationships among variables, including temperature, occupancy, economic activity, and consumer habits (Rahman et al., 2023). However, they demand large training datasets and significant computational resources. This requirement can strain local computing in AMI systems, prompting the adoption of distributed or cloud computing frameworks.

2.3 Reinforcement Learning

Reinforcement learning (RL) focuses on agents that learn optimal actions through trial and error within an environment. In AMI, RL can manage distributed resources, schedule demand response, or decide on real-time pricing signals (Han et al., 2024). An RL agent receives states (meter readings, grid conditions) and executes actions (load control, price adjustments). A reward function measures how well the action meets objectives, such as cost minimization or peak load reduction.

RL can adapt to dynamic grid environments, where consumer demand, generation from renewables, and market prices change frequently. Multi-agent RL extends this idea to large-scale systems with many interacting agents (Singh & Singha, 2024). Agents representing households or microgrids coordinate actions to maximize collective objectives.

2.4 Federated Learning

Federated learning (FL) distributes training across multiple devices or edge nodes without centralizing raw data. This technique addresses privacy concerns, since individual meters or microgrids only share model updates, not private consumption data (Jithish et al., 2023). Blockchain can manage the aggregation of local models or updates securely, enabling FL processes in decentralized AMI contexts.

2.5 AI Limitations and Considerations in AMI

AI models rely on high-quality, consistent data. AMI data can contain missing values, sensor errors, or maliciously manipulated entries. Preprocessing and validation are critical to prevent model degradation or inaccurate decisions (Kumar et al., 2020). Model interpretability is another concern, especially when black-box methods like deep learning or ensemble models are used. Stakeholders require explanations for critical decisions, especially in regulated energy sectors.

Scalability and latency matter because AMI can involve thousands or millions of devices. Real-time or near-real-time control imposes strict latency constraints. AI pipelines must efficiently handle continuous streams of data, potentially requiring edge computing solutions. Additionally, there are privacy concerns because consumption profiles can reveal personal behaviors. AI developers must apply privacy-preserving methods, potentially in conjunction with permissioned blockchain or secure multiparty computation.

2.6 Relevance of AI to Modern AMI Challenges

Modern AMI challenges include high data velocity and volume, cybersecurity threats, and complex integration with distributed energy resources. AI methods address these areas by automating data analysis, detecting suspicious patterns, and coordinating decisions across the grid. Researchers emphasize AI's potential to enhance situational awareness, reduce manual oversight, and enable dynamic, consumer-centric services.

AMI operators can benefit from advanced machine learning or RL methods to optimize operations, while cost reductions in sensor technology and edge devices facilitate scaled AI deployments. However, data privacy regulations and the need for robust security solutions drive interest in blockchain for protecting AI data pipelines. The next section discusses blockchain fundamentals, emphasizing how blockchain complements AI in securing data exchanges and forming trustless transaction networks.

3. BLOCKCHAIN FUNDAMENTALS FOR AMI

Blockchain is a decentralized ledger that allows a network of participants to record transactions in an immutable and transparent manner. Once a transaction is validated, the record cannot be easily altered. This architecture is relevant to AMI because it ensures meter data integrity and facilitates decentralized energy trading or other multi-party collaborations (Alladi et al., 2019).

3.1 Core Blockchain Concepts

Blockchain's core idea is that no central authority controls the ledger. Instead, nodes in a peer-to-peer network validate transactions using a consensus mechanism. Each block in the chain references its predecessor, creating a chronological record. Modifying any block invalidates subsequent blocks unless the network reconsiders them under the consensus algorithm (Andoni et al., 2019). **Table 2** highlights the main components of a blockchain system, focusing on how they apply to AMI deployments.

Table 2. Core blockchain components

Components	Application in AMI
Blocks and Hash Links	Store meter data or transactions. Hash links prevent tampering of past records.
Consensus Mechanisms	Decide valid blocks among participants. Examples: Proof of Work (PoW), Proof of Stake (PoS), or Practical Byzantine Fault Tolerance (PBFT).
Smart Contracts	Enable automated execution of business logic for demand response, billing, or energy trading.
P2P Network	Connect AMI nodes (utilities, prosumers, microgrids) in a decentralized manner.
Immutable Ledger	Guarantees a transparent audit trail for meter readings, trade transactions, or aggregator instructions.
Cryptographic Tools	Maintain confidentiality of energy usage data and manage access permissions.

Blockchain can operate as a public or permissioned network. Public blockchains allow any node to join. Permissioned blockchains constrain membership to known entities. Public networks risk exposure of sensitive meter data, but permissioned networks with robust access control can safeguard consumer privacy (Li et al., 2023).

3.2 Smart Contracts and Decentralized Applications

Smart contracts are self-executing programs stored on the blockchain. They permit developers to encode business logic (e.g., energy price triggers, or penalty conditions) that run automatically when triggered (Binyamin & Ben Slama, 2025). In an AMI scenario, a smart contract might settle payments among prosumers or adjust energy consumption limits during peak events.

Decentralized applications (dApps) build on smart contracts for broader functionalities. In an AMI environment, a dApp can track meter data, coordinate distributed generation, or manage user access to aggregated consumption analytics. These applications rely on the underlying blockchain's immutability and consensus for trust, reducing reliance on a single authority.

3.3 Blockchain for Data Integrity and Security

AMI data must be accurate and tamper-free. A blockchain system can log meter readings in chronological blocks, preventing retroactive modifications (Pop et al., 2018). This tamper-resistant ledger can serve as a credible source during billing disputes or regulatory audits. Additionally, the distributed structure mitigates single points of failure.

Security services provided by blockchain include:

- **Integrity**: Hash-based links protect data from unauthorized edits.
- **Non-repudiation**: Digital signatures confirm that transactions originate from specific parties.
- **Availability**: Replication across nodes ensures data accessibility, even if some nodes fail.
- **Auditability**: A historical record of every transaction or data entry remains visible for authorized reviewers.

3.4 Consensus Mechanisms for AMI Scalability

Selecting a consensus mechanism is critical for balancing security and throughput in AMI. Traditional blockchains use Proof of Work, which is energy-intensive and slow. Alternative mechanisms like Proof of Stake or PBFT reduce resource consumption and allow higher transaction throughput, fitting real-time or near-real-time AMI operations (Han et al., 2024).

Permissioned blockchains often use PBFT or Raft to finalize blocks with minimal latency. Utilities, meter operators, and other stakeholders form a consortium and validate blocks through a known set of validators. This arrangement can handle frequent meter data entries without overwhelming the network (Singh, Rathee, & Kumar, 2020).

3.5 Integrating Blockchain with Existing AMI Architecture

AMI typically includes a meter data management system (MDMS), which stores and processes consumption data from multiple meters. Integrating blockchain requires hooking the MDMS or smart meters into the blockchain network, possibly through edge devices or APIs (Kumar et al., 2020). Access control must be configured, and cryptographic key management must ensure that only authorized entities write data.

Scalability remains a challenge because standard blockchain solutions cannot support massive data throughput without specialized optimizations (Andoni et al., 2019). Sidechains, off-chain computation, or layer-two solutions can offload data to reduce on-chain load. Data compression and interval-based batching can further optimize performance.

3.6 Relevance of Blockchain to Modern AMI Challenges

Blockchain aligns with current AMI needs for security, decentralization, and multi-party collaboration. AMI is not only about capturing data but also exchanging data among prosumers, aggregators, and utilities. Blockchain-based frameworks address trust issues by guaranteeing data integrity and enabling automated, traceable transactions. Researchers also explore how AI models benefit from the robust data environment that blockchain creates, since the data's authenticity and completeness are less suspect (Khan et al., 2023).

When combined with smart contracts, blockchain automates tasks such as issuing incentives for demand response or reconciling energy trades among peer consumers (Rahman et al., 2023). This approach reduces administrative overhead, lowers transaction costs, and facilitates transparent rule enforcement. The next section examines how AI and blockchain can converge in AMI to deliver data-driven intelligence and trust-enforced autonomy.

4. CONVERGENCE OF AI AND BLOCKCHAIN IN AMI

AI thrives on high-quality data streams, while blockchain ensures data's trustworthiness. Together, they allow more robust and efficient AMI processes. Blockchain secures the input data, making AI's outputs more reliable for decision-making. Meanwhile, AI can manage blockchain networks, adjusting parameters such as smart contract rules or identifying faulty nodes.

4.1 Architectural Overview

A combined AI–blockchain AMI solution might include:

1. **Data Collection Layer**: Smart meters measure consumption and send data to local gateways.
2. **Blockchain Layer**: Validates, stores, and secures data. Smart contracts enforce logic for billing or incentives.
3. **AI Layer**: Models analyze data for forecasting, anomaly detection, or dynamic pricing. Results can trigger on-chain actions through smart contracts.
4. **Application Layer**: User-facing portals, aggregator dashboards, or utility controls present insights, manage exceptions, or authorize trades.

Figure 2 illustrates a high-level synergy where AI modules connect to a blockchain ledger for data reading and transaction writing.

Figure 2. High-level AI–blockchain synergy in AMI

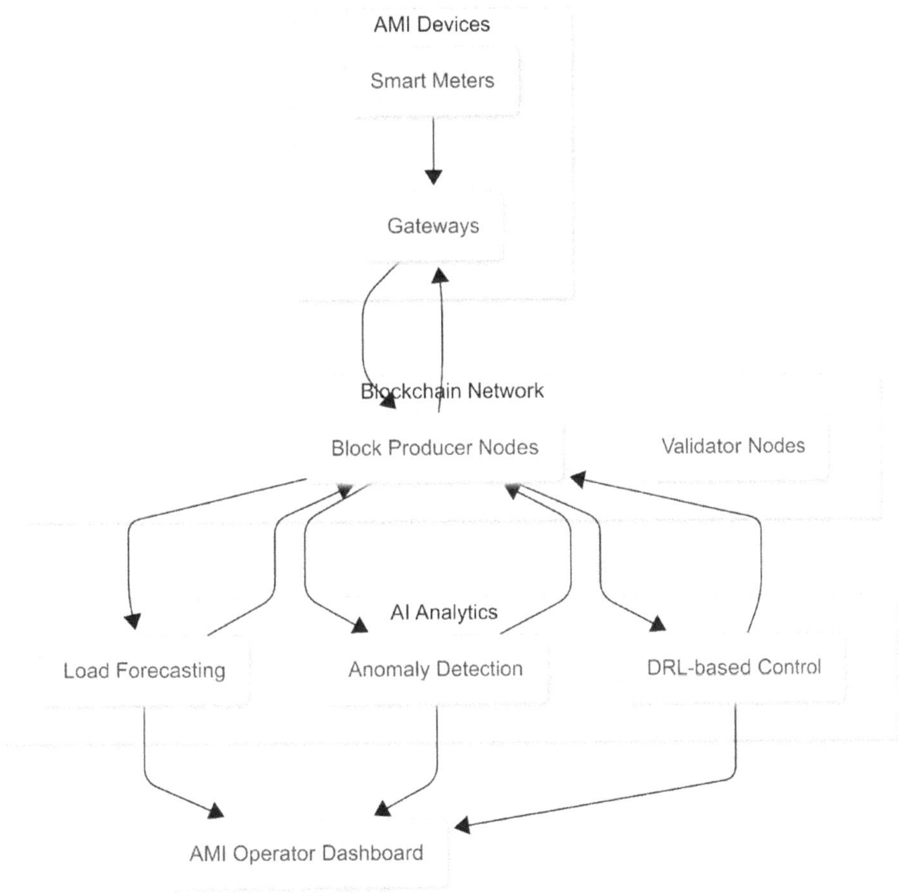

This figure shows how AI analytics read data from the blockchain and send instructions or detection results back to the blockchain.

4.2 AI-Enhanced Blockchain Operations

Although blockchain provides a secure ledger, misconfiguration or malicious nodes can still threaten the network. AI can detect anomalies in block production or node behavior (Kuzior et al., 2022). A reinforcement learning agent can optimize consensus parameters (e.g., block size, scheduling, or node validation load). AI-based traffic analysis can also deter distributed denial-of-service attacks.

4.3 Blockchain-Secured AI Data Pipelines

AI depends on consistent and trustworthy data. By hosting meter data on blockchain or on an interoperable layer, the system guarantees immutability (Xue et al., 2025). Smart contracts can record metadata about the data source, timestamps, and any transformations performed. This approach helps auditors confirm that predictions or automated decisions rest on authentic data.

Blockchain's consensus ensures multiple parties agree on data correctness. This mitigates data poisoning, where false entries could degrade model performance or produce inaccurate forecasts (Han et al., 2024). In a blockchain-backed AI pipeline, any suspicious data point triggers an automated rollback or an alarm for manual inspection.

4.4 Automated Execution with Smart Contracts

Smart contracts automate tasks that rely on AI outputs. For example, an AI model may forecast high demand for the next day. The forecast, once verified, activates a smart contract that adjusts time-of-use rates or dispatches load curtailment signals. This mechanism is transparent and does not rely on manual oversight (Rahman et al., 2023).

Table 3. summarizes use cases for AI–blockchain convergence in AMI.

Table 3. AI–blockchain use cases in AMI; AI role; blockchain role

Energy Trading	**Price prediction, transaction matchmaking**	**Smart contracts for trade settlement, decentralized ledger**
Demand Response	Load forecasting, automated control signals	Incentive disbursement, event logging, tamper-proof records
Anomaly Detection	Identifying suspicious consumption or device behavior	Immutable data storage, consensus on suspicious events
Billing and Settlement	Pattern recognition, cost optimization	Transparent metering records, automatic payment enforcement
Distributed Resource Coordination	Reinforcement learning for distributed scheduling	Secure communication channels, contract-based resource sharing
Data Marketplace	Quality evaluation of data providers, AI-based aggregator	Tokenized access rights, immutability, distributed transactions

4.5 Privacy and Federated Learning

Privacy is a top priority when dealing with consumption data. Blockchain can reveal aggregated data in the ledger, but personal details must remain private. Off-chain methods or encrypted channels within permissioned blockchains can address privacy. Federated learning further reduces data leakage by confining raw data to local devices. Blockchain coordinates model updates among devices, preserving privacy without centralizing data (Jithish et al., 2023).

4.6 Benefits of AI–Blockchain Integration

- **Data Integrity**: AI processes reliable data, enhancing the accuracy of forecasts or decisions.
- **Automation**: Smart contracts combined with AI predictions lower administrative overhead.
- **Enhanced Security**: Blockchain's consensus mechanisms protect data from unauthorized changes, and AI can further protect the blockchain from malicious actors.
- **Transparency**: Users can audit blockchain transactions, building trust in AI-driven decisions.

4.7 Limitations and Trade-Offs

- **Computational Overhead**: Blockchain consensus adds latency, and AI model training is resource-intensive. Achieving real-time or near-real-time responsiveness can be challenging.
- **Data Privacy**: Detailed consumption data on a public ledger risk privacy violations. Permissioned blockchains or cryptographic techniques can mitigate this concern, but add complexity.
- **Scalability**: High data volumes from millions of meters require architectural optimizations, such as off-chain processing.
- **Interoperability**: AI–blockchain solutions must interface with existing grid protocols and standards to avoid silos.

5. KEY APPLICATIONS IN DEMAND RESPONSE, ENERGY TRADING, AND ANOMALY DETECTION

This section delves into specific applications of AI–blockchain technologies that improve AMI capabilities. The focus areas include demand response, decentralized energy trading, and anomaly detection.

5.1 Demand Response

Demand response (DR) programs encourage consumers to shift or reduce consumption during peak demand. AMI data inform DR operators about real-time or near-real-time load conditions (Pop et al., 2018). AI models forecast demand spikes or capacity constraints and issue control signals that reduce consumption or shift loads. Blockchain can automate incentive payouts or penalize noncompliance without manual interventions.

5.1.1 AI-Driven Load Forecasting

Load forecasting is central to demand response. Machine learning models, such as RNNs, can predict hourly or daily consumption. Accuracy increases with high-quality historical data, which is validated on-chain for integrity (Kuzior et al., 2022). AI models then generate price signals or scheduling instructions.

5.1.2 Smart Contract-Based DR Programs

Smart contracts on a permissioned blockchain can register participants and track their compliance with load reduction commitments. When an AI forecast indicates a shortage, the contract initiates a DR event. Meters confirm consumption reduction, and the contract disburses predefined incentives (Han et al., 2024). This reduces overhead associated with manual verification of consumption.

Example Flow of Events

1. AI Engine forecasts a grid stress event.
2. Smart contract triggers a DR event and notifies participants.
3. Participants reduce consumption if feasible.
4. Meters record consumption changes on the blockchain.
5. Consensus validates changes.
6. Smart contract calculates incentives or penalties.
7. Settlement is executed automatically.

5.1.3 DR Effectiveness and Transparency

This approach boosts transparency because all consumption data relevant to the DR event are immutable and traceable (Singh & Singha, 2024). Auditors can confirm the DR timeline and measure compliance. AI further refines DR strategies by analyzing the historical performance of participants and adjusting future incentives to ensure reliability.

5.2 Decentralized Energy Trading

Distributed energy resources (DERs), such as rooftop solar and battery storage, transform consumers into "prosumers." They can sell surplus energy to neighbors. Traditional power markets rely on centralized entities for trading and settlements, which can be slow and opaque. Blockchain-based trading platforms enable direct transactions between prosumers, while AI optimizes pricing and matching (Rahman et al., 2023).

5.2.1 Role of Blockchain in Peer-to-Peer (P2P) Trading

In P2P trading, prosumers list available capacity on a blockchain marketplace. Buyers submit bids, and a matching algorithm clears trades. The ledger records transactions with trust enforced through consensus. Users do not rely on a central broker (Alladi et al., 2019).

Smart contracts manage settlement. If a prosumer delivers the contracted energy, payment is executed automatically. If the prosumer fails, penalties apply. This transparency encourages more participation (Andoni et al., 2019).

5.2.2 AI for Price Forecasting and Market Clearing

AI models estimate dynamic prices based on supply-demand factors, solar generation, and consumption patterns. By learning from historical trades, the model converges on fair prices, maximizing social welfare. Some projects use multi-agent reinforcement learning, where each prosumer is an agent aiming to maximize profit. Equilibrium emerges through repeated interactions (Singh & Singha, 2024).

5.2.3 Scalability and Practical Considerations

Large-scale P2P trading demands quick settlement and high throughput. Permissioned blockchains with optimized consensus can process trades at scale (Singh, Rathee, & Kumar, 2020). Off-chain order matching can reduce on-chain load, with only final trades recorded. AI-based forecasting must also adapt to changes in weather, local regulations, and participant behaviors.

5.3 Anomaly Detection

AMI data anomalies can originate from cyberattacks, meter faults, or fraud. Early detection is crucial for grid stability and revenue protection. Anomaly detection methods attempt to recognize unusual consumption patterns or suspicious meter signals.

5.3.1 Machine Learning Models for Anomaly Detection

Unsupervised or semi-supervised techniques detect outliers. These techniques group normal usage patterns, then flag deviations. Labels for fraudulent or tampered data are often scarce, making such approaches viable (Xue et al., 2025).

Neural network-based methods such as autoencoders reconstruct normal consumption profiles and measure reconstruction errors. A high error indicates potential anomaly. Feature engineering can incorporate time-of-day usage, seasonal variations, and cross-customer comparisons (Wang, 2025).

5.3.2 Blockchain Logging of Meters

If meter data appear on a distributed ledger, unauthorized modifications are harder to execute. Attackers cannot retroactively alter consumption records. This discourages tampering attempts. In a scenario where a malicious entity tries to modify data in transit, consensus rules can reject the unverified data.

5.3.3 Automated Alerts and Smart Contract Escalation

When an AI model identifies an anomaly, a smart contract can place the suspect meter or account under review. The contract freezes certain transactions until an operator verifies authenticity. This automated escalation reduces the time to respond and prevents further damage (Hasan et al., 2022).

5.3.4 Real-World Effectiveness

Researchers report high detection rates with AI-based anomaly detection on real AMI datasets (Rahman et al., 2023). Integrating blockchain adds an extra layer of integrity. The combination can significantly reduce theft, meter tampering, and advanced persistent threats that exploit centralized vulnerabilities.

5.4 Case Illustrations

Studies from pilot projects show how the synergy of AI and blockchain addresses AMI challenges. Although many are in proof-of-concept stages, the results indicate potential improvements in data trust, operational efficiency, and user empowerment.

6. SECURITY AND PRIVACY CONSIDERATIONS

Security is vital in AMI because compromised meter data can lead to incorrect billing, supply-demand imbalances, or even catastrophic grid failures. AI and blockchain each enhance security, but they also introduce new vulnerabilities.

6.1 Cyber Threats in AMI

- **Data Tampering**: Attackers can manipulate meter readings to reduce bills or cause supply-demand misalignment.
- **Denial of Service**: Flooding communication channels can disrupt data collection or system coordination.
- **Man-in-the-Middle**: Interception and modification of consumption data during transmission.
- **Replay Attacks**: Re-sending old meter data to mask real consumption.

6.2 Blockchain Security Properties

Blockchain's immutability and distributed consensus reduce many forms of data tampering. Attackers must alter multiple nodes to rewrite the chain. For permissioned blockchains, known validators increase accountability. Private or consortium blockchains can also enforce stricter identity checks (Andoni et al., 2019).

6.3 AI for Intrusion Detection

Machine learning models can monitor network traffic patterns for suspicious spikes or abnormal transaction activity. A blockchain-based ledger can store network flow logs. AI then correlates logs from multiple segments of the grid, detecting complex multi-stage attacks (Khan et al., 2023).

6.4 Privacy Challenges

The ledger's transparency conflicts with consumer privacy. Fine-grained consumption reveals occupancy, lifestyle habits, or household composition. Public blockchains expose these data to all participants. Solutions involve encryption, zero-knowledge proofs, or storing only summarized or hashed data on-chain (Pop et al., 2018).

6.5 Regulatory Compliance

Energy regulators and consumer protection agencies often require strict data protection. Permissioned blockchains that store only necessary data reduce exposure. Smart contracts must implement compliance logic for data retention or erasure. AI developers must also respect privacy by design, masking or aggregating sensitive meter data (Hua et al., 2022).

6.6 Coordinated Security Strategies

Security in AI–blockchain systems involves multi-layered defenses. AI detects anomalies or intrusions, while blockchain ensures transaction integrity and data immutability. Layering cryptographic techniques, secure key management, and robust consensus protocols creates a resilient architecture (Hasan et al., 2022).

7. IMPLEMENTATION CONSIDERATIONS AND CHALLENGES

Realizing an AI–blockchain AMI platform involves addressing technical, operational, and economic factors. This section outlines major considerations.

7.1 Scalability

Smart grids contain millions of meters, each generating data. On-chain storage is costly. High transaction throughput demands efficient consensus. Solutions include layering and sidechain approaches, where the main chain records critical checkpoints, and most data remain off-chain (Singh, Rathee, & Kumar, 2020).

7.2 Latency

Demand response or real-time control decisions cannot endure long latencies. Blockchain consensus can slow transaction finality. AI-based solutions often require timely data for accurate decisions. Hybrid architectures buffer or preprocess data at the edge, then record aggregated outcomes on-chain (Kumar et al., 2020).

7.3 Data Quality and Integration

AMI data vary in format, frequency, and reliability. Implementation requires standardizing data ingestion and cleaning procedures (Pop et al., 2018). AI models also need robust data pre-processing pipelines that handle missing or corrupted values.

7.4 Hardware and Infrastructure

AI training demands computational power, often provided by GPUs or cloud clusters. Blockchain nodes require stable connectivity, secure key storage, and sufficient processing resources for consensus. Edge or fog computing solutions balance resources between local devices and the cloud.

7.5 Regulatory and Organizational Hurdles

Utilities may resist changing legacy billing or operations processes. Regulators may restrict how consumer data are used or how energy trades are settled. Organizations must establish governance rules for permissioned blockchain networks and identify validators or consensus participants (Al Shareef et al., 2024).

7.6 Costs

Integration of AI and blockchain can increase capital and operational expenses. AI experts, blockchain developers, and specialized hardware raise costs. However, these expenses may be offset by reduced fraud, improved efficiency, and new revenue streams (Han et al., 2024).

7.7 Interoperability and Standards

The energy sector uses many protocols and data models. AI–blockchain solutions must align with standards such as IEC 61850 or CIM. Efforts to develop common frameworks or APIs for exchanging data between legacy systems and decentralized platforms continue (Singh & Singha, 2024).

8. FUTURE RESEARCH DIRECTIONS

AI–blockchain integration in AMI is a growing field. The following directions provide opportunities for innovative research.

8.1 Federated and Privacy-Preserving AI

Future work can refine federated learning approaches that use blockchain for coordination among distributed meters or microgrids. Privacy-enhancing techniques (homomorphic encryption, secure enclaves) can protect sensitive data while enabling training (Jithish et al., 2023).

8.2 Lightweight Consensus for High-Frequency Meter Data

Research can focus on consensus protocols optimized for AMI's data frequency. Hybrid systems that log critical events on-chain while using an off-chain data store for less critical events could support real-time applications.

8.3 Multi-Agent Reinforcement Learning for Grid-Wide Coordination

Researchers can explore multi-agent reinforcement learning where each grid node runs an AI agent. The blockchain mediates interactions and ensures honest reward calculation. This approach may enhance distributed scheduling or microgrid cooperation (Han et al., 2024).

8.4 Standardization and Interoperability

Energy stakeholders need robust standards for data formats, identity management, and smart contract logic. Collaborative efforts between standards bodies and industry can accelerate adoption.

8.5 Integration with Other Emerging Technologies

Future efforts can combine AI–blockchain frameworks with Internet of Things (IoT) sensors or edge computing solutions to further automate distribution. EV charging networks and sector coupling (electricity-heat-gas) also present opportunities (Binyamin & Ben Slama, 2025).

9. CONCLUSION

AI and blockchain offer complementary strengths in addressing AMI's challenges. AMI data drive AI-based analysis for forecasting, anomaly detection, and control. Blockchain secures data flows and transaction processes, removing many trust barriers. Integrating these technologies supports advanced functionalities such as decentralized energy trading, automated demand response, and robust cybersecurity defenses.

Although technical, organizational, and regulatory hurdles persist, proof-of-concept demonstrations confirm the promise of this combined approach. The synergy lies in AI's capacity for real-time learning from large datasets, supported by blockchain's immutable and decentralized nature. Operators, regulators, and end-users can benefit from reduced fraud, transparent incentives, and more reliable grid operations.

Continued research should tackle concerns about scalability, privacy, and interoperability to move beyond pilots and into widespread deployment. AI–blockchain integration can transform passive electricity consumers into active grid participants, accelerating the evolution toward decentralized, intelligent, and resilient energy networks. AMI stands at the forefront of this transition, and the combined potential of AI and blockchain suggests a new paradigm for data-centric energy systems that uphold security, efficiency, and user empowerment.

REFERENCES

Al Shareef, A. M., Seçkiner, S., Eid, B., & Abumeteir, H. (2024). Integration of blockchain with artificial intelligence technologies in the energy sector: A systematic review. *Frontiers in Energy Research, 12*, 1377950. DOI: 10.3389/fenrg.2024.1377950

Alladi, T., Chamola, V., Rodrigues, J. J. P. C., & Kozlov, S. (2019). Blockchain in smart grids: A review on different use cases. *Sensors (Basel), 19*(22), 4862. DOI: 10.3390/s19224862 PMID: 31717262

Andoni, M., Robu, V., Flynn, D., Abram, S., Geach, D., Jenkins, D., McCallum, P., & Peacock, A. (2019). Blockchain technology in the energy sector: A systematic review of challenges and opportunities. *Renewable & Sustainable Energy Reviews, 100*, 143–174. DOI: 10.1016/j.rser.2018.10.014

Binyamin, S., & Ben Slama, S. (2025). IntelliGrid AI: A blockchain and deep-learning framework for optimized home energy management with V2H and H2V integration. *AI, 6*(2), 34. DOI: 10.3390/ai6020034

Han, Y., Meng, J., & Luo, Z. (2024). Multi-agent deep reinforcement learning for blockchain-based energy trading in decentralized electric vehicle charger-sharing networks. *Electronics (Basel), 13*(21), 4235. DOI: 10.3390/electronics13214235

Hasan, M. K., Alkhalifah, A., Islam, S., Ali, R., & Beg, M. O. (2022). Blockchain technology on smart grid, energy trading, and big data: Security issues, challenges, and recommendations. *Wireless Communications and Mobile Computing, 2022*, 9065768. DOI: 10.1155/2022/9065768

Hua, W., Chen, Y., Qadrdan, M., Jiang, J., Sun, H., & Wu, J. (2022). Applications of blockchain and artificial intelligence technologies for enabling prosumers in smart grids: A review. *Renewable & Sustainable Energy Reviews, 161*, 112308. DOI: 10.1016/j.rser.2022.112308

Jithish, J., Alangot, B., Mahalingam, N., & Yeo, K. S. (2023). Distributed anomaly detection in smart grids: A federated learning-based approach. *IEEE Access : Practical Innovations, Open Solutions, 11*, 7157–7179. DOI: 10.1109/ACCESS.2023.3237554

Khan, A. A., Laghari, A. A., Rashid, M., Li, H., Javed, A. R., & Gadekallu, T. R. (2023). Artificial intelligence and blockchain technology for secure smart grid and power distribution automation: A state-of-the-art review. *Sustainable Energy Technologies and Assessments, 57*, 103282. DOI: 10.1016/j.seta.2023.103282

Kumar, N. M., Chand, A. A., Malvoni, M., Prasad, K. A., Mamun, K. A., Islam, F. R., & Chopra, S. S. (2020). Distributed energy resources and the application of AI, IoT, and blockchain in smart grids. *Energies, 13*(21), 5739. DOI: 10.3390/en13215739

Kuzior, A., Sira, M., & Brożek, P. (2022). Using Blockchain and Artificial Intelligence in Energy Management as a Tool to Achieve Energy Efficiency. *Virtual Economics, 5*(3), 69–90. DOI: 10.34021/ve.2022.05.03(4)

Li, J., Herdem, M. S., Wen, S., & Kuzior, A. (2023). Methods and applications for Artificial Intelligence, Big Data, Internet of Things, and Blockchain in smart energy management. *Energy and AI, 11*, 100208. DOI: 10.1016/j.egyai.2022.100208

Pop, C., Cioara, T., Antal, M., Anghel, I., Salomie, I., & Bertoncini, M. (2018). Blockchain-based decentralized management of demand response programs in smart energy grids. *Sensors (Basel), 18*(1), 162. DOI: 10.3390/s18010162 PMID: 29315250

Rahman, M., Chowdhury, S., Shorfuzzaman, M., Hossain, M. K., & Hammoudeh, M. (2023). Peer-to-peer power energy trading in blockchain using efficient machine learning model. *Sustainability (Basel), 15*(18), 13640. DOI: 10.3390/su151813640

Singh, K., & Singha, N. (2024). Credit-based energy trading system using blockchain and machine learning. *The Journal of Supercomputing, 80*(11), 1–22. DOI: 10.1007/s11227-024-06073-1

Singh, M., Rathee, G., & Kumar, P. (2020). Blockchain and AI amalgamation for energy cloud management. *Journal of Systems Architecture, 110*, 101746.

Wang, J. (2025). Multi-agent system based smart grid anomaly detection using blockchain machine learning model in mobile edge computing network. *Computers & Electrical Engineering, 121*, 109825. DOI: 10.1016/j.compeleceng.2024.109825

Xue, Q., Yu, Y., Cen, S., & Li, Z. (2025). Secure and transparent energy management using blockchain and machine learning-based anomaly detection: A case study of the Ausgrid dataset. *Computers & Industrial Engineering, 171*, 111040.

Chapter 11
AI-Powered Smart Contracts for Energy Grid Optimization:
A Hybrid Intelligence Approach

Abdelraouf Ishtaiwi
https://orcid.org/0000-0001-8013-7151
Data Science and Artificial Intelligence, University of Petra, Jordan

Amjad Aldweesh
https://orcid.org/0000-0001-9358-1323
College of Computing and Information Technology, Shaqra University, Saudi Arabia

Ahmad Al-Qerem
https://orcid.org/0000-0003-2187-6194
Computer Science Department, Faculty of Information Technology, Zarqa University, Jordan

Mouhammd Alkasassbeh
https://orcid.org/0000-0001-8396-7441
Princess Sumaya University for Technology, Jordan

ABSTRACT

Societies worldwide are reshaping their energy infrastructures to address evolving sustainability targets, economic requirements, and regulatory goals. In this chapter, I examine how artificial intelligence (AI) and blockchain-enabled smart contracts can unify in hybrid intelligence systems to optimize energy grids. This exploration highlights key concepts of distributed ledger technology, AI-driven algorithms, and decision-making processes that combine human expertise with machine-driven insights. The emphasis is on theoretical foundations, yet I also integrate practical viewpoints and selected real-world scenarios to connect these innovations with tangible energy challenges. By bridging smart contracts and AI, the proposed model aims to advance cost-efficient management of grids, enhance trust among participants, and improve operational reliability. I propose that hybrid intelligence facilitates balanced data-driven strategies and human input, creating resilient solutions to energy distribution, consumption, and market interaction.

DOI: 10.4018/979-8-3373-2439-5.ch011

1. INTRODUCTION

Energy grids have grown more interconnected, reflecting pressing concerns around efficient resource management, shifting demand profiles, and an urgent need to accommodate renewable sources (Bouckaert et al., 2021). Historically, centralized control structures maintained stable supplies, set market prices, and scheduled conventional power generation. But the diffusion of distributed energy resources (DERs)—including photovoltaics, microturbines, and battery storage—has added complexity to grid oversight. As these resources proliferate, new dynamics challenge traditional operations: power flows can reverse, local supply can exceed local demand, and market mechanisms must adapt rapidly to evolving conditions (Ali et al., 2021).

Several techniques have emerged to cope with these transformations. Operators are installing monitoring devices at various nodes, implementing advanced communication standards, and studying dynamic pricing. Growing interest in blockchain further suggests an appetite for decentralized market coordination and heightened security (Andoni et al., 2019). Yet these systems, despite offering strong cryptographic capabilities, can still benefit from the intelligence AI provides—especially for predictive modeling and real-time optimization.

This setting paves the way for AI-powered smart contracts, where decentralized agreements on a blockchain can self-execute based on advanced analytic triggers, drawing on real-time data from sensors, meters, and grid control systems (Kirli et al., 2022). The synergy of AI with blockchain-based contracts fosters robust data-driven decision-making—one that can react nimbly to load variations, identify cost-effective resource allocations, and handle complex negotiations among stakeholders.

Still, technology alone rarely suffices to address the complexity of an entire energy sector. Hybrid intelligence, which merges human expertise with algorithmic insights, promises more balanced solutions (Dellermann et al., 2019). This chapter unfolds by exploring the theoretical foundations of this integrated approach. I delve into the elements that make AI-driven smart contracts attractive for grid optimization, potential pitfalls and regulatory challenges, and the broader conceptual framework that ties each piece together.

The discussion proceeds through an organic structure. After considering the broader energy landscape, I explore how hybrid intelligence emerged as a collaborative model. I then focus on the building blocks of AI-powered smart contracts, emphasizing relevant blockchain mechanics and advanced data-analysis techniques. I discuss how domain knowledge from human experts interacts with machine learning and blockchain to shape decisions. I illustrate potential grid applications and note how decentralized energy transactions, demand-response programs, and market clearing can be improved through AI-augmented contracts. Finally, I address possible obstacles, from regulatory to cybersecurity concerns, and speculate about the future of this dynamic domain.

Across this chapter, I regularly reference published literature to ground the discussion. Methodologically, the chapter draws on systematic reviews in blockchain-energy (Al Shareef et al., 2024; Chen et al., 2022) and AI-based grid management (Ahmad et al., 2022; Al-Fatlawi et al., 2024). Integrating these insights, I aim for a coherent, theory-focused narrative that examines the feasibility and conceptual underpinnings of hybrid intelligence for the energy sector.

By merging AI techniques with blockchain-based smart contracts, researchers and practitioners stand poised to reshape conventional energy system management. This approach could lead to resource optimization that is transparent, cost-effective, and flexible enough to accommodate the evolving energy mix. And by extending these innovations with human knowledge and oversight, I contend that energy solutions can become more inclusive and reliable over time.

2. THE EVOLVING ENERGY LANDSCAPE

The modern energy sector has reached a turning point. Traditional large-scale power plants and one-way energy flows are giving way to decentralized grids that incorporate renewables, battery systems, microgrids, and intelligent control devices (Bouckaert et al., 2021). This shift is not only technological but also socio-economic, as more stakeholders become prosumers who both consume and generate energy.

The push to reduce greenhouse gas emissions is accelerating these developments, with governments issuing regulations that prioritize renewables and incentivize energy efficiency (U.S. Department of Energy, 2024). As new types of generation come online, the grid must dynamically balance volatile supply from resources like solar and wind with demand that is no longer merely passive. Households can adjust consumption in response to real-time price signals, or they might feed locally produced electricity back into the grid, creating unforeseen fluctuations.

Conventional supervisory control and data acquisition (SCADA) systems have historically handled grid monitoring and control, but they can struggle with these new complexities. Data volumes are swelling, and local decision-making becomes crucial to accommodate diverse energy producers and consumers (Ali et al., 2021). This broader context underpins the demand for novel solutions: operators need real-time analytics, predictive maintenance, and autonomous processes that can respond to grid conditions swiftly.

At the same time, blockchain has emerged as a secure, tamper-resistant technology that can decentralize trust (Andoni et al., 2019). For energy, decentralized ledger frameworks promise transparent peer-to-peer trades, efficient settlement, and data integrity. And yet, blockchain alone is not enough to handle the predictive or adaptive aspects of grid operations. The synergy between AI-based optimization and blockchain-based security is taking shape (Ferrag & Maglaras, 2020). This synergy, in certain contexts, is operationalized through the vehicle of smart contracts: self-executing agreements whose conditions are algorithmically enforced on a distributed ledger (Belotti et al., 2019).

2.1 Decentralization Trends

One defining aspect of the evolving energy sector is decentralization. Households, small businesses, and larger industrial sites often install solar arrays, wind turbines, or combined heat-and-power units. Municipalities develop microgrids to secure local resilience, particularly in remote or disaster-prone areas (Ali et al., 2021). As these resources multiply, the central operator may not even see each micro-scale generator. Some city neighborhoods might rely more on local transactions than on large external power plants.

Decentralization also changes who makes decisions about energy supply and demand. Aggregators bundle thousands of small-scale DERs to bid in energy markets, while decentralized finance (DeFi) platforms show that users can pool resources for a variety of applications. If we apply these ideas to energy, a homeowner's battery could be traded or collectively managed in real time through a platform that ensures price fairness and reliability. Indeed, the pursuit of such fine-grained solutions has generated an appetite for distributed data analytics and blockchain-based transaction settlement (Al-Abri et al., 2022).

2.2 The Demand for Real-Time Insights

Legacy control architectures often rely on scheduling practices that are updated once or twice a day. With the infusion of renewables, these time horizons become inadequate. Weather patterns shift quickly, changing solar production. Electric vehicles (EVs) can add or remove large loads with minimal notice. Real-time or near-real-time analytics can address these fluctuations through dynamic price signals and automated load adjustments (Al-Fatlawi et al., 2024). AI-driven algorithms can forecast short-term load and generation, enabling better planning.

But forecasting alone does not solve the coordination problem. Actors in this ecosystem need binding mechanisms to enforce commitments, track performance, and manage exceptions if something goes wrong. This is where smart contracts can play a role. An EV might sell grid services, for instance, committing to feed power back when the grid is stressed. The contract that handles these transactions can incorporate forecasts from an AI model to determine if, for example, the EV has enough charge to fulfill that commitment. If it does, the contract triggers a payment. If not, the contract enforces relevant penalties (Al-Saif et al., 2021).

2.3 Emerging Role of Hybrid Intelligence

While AI can automate many tasks, energy systems remain heavily regulated and complex. Grid operators and policymakers bring domain knowledge that pure data-driven models might miss (Dellermann et al., 2019). A hybrid intelligence framework ensures that humans oversee critical decisions, interpret ambiguous outcomes, and override algorithms when needed. This approach leverages the consistency and speed of AI while mitigating the risk of blind spots.

In the remainder of the chapter, I bring these ideas together: the demand for decentralization, real-time coordination, and a balanced intelligence structure that melds human expertise with advanced AI. Blockchain-based smart contracts will emerge as a central theme, forming a foundation on which many of these capabilities can be built.

3. HYBRID INTELLIGENCE: CONCEPTUAL UNDERPINNINGS

A strict reliance on machine learning or algorithmic decision-making often overlooks real-world nuances. Rules are shaped by regulatory guidelines, historical infrastructure constraints, and public interest considerations. Hybrid intelligence merges human cognitive strengths with computational assets, creating solutions that incorporate contextual judgment and data-driven optimization (Dellermann et al., 2019).

3.1 Defining Hybrid Intelligence

Dellermann et al. (2019) describe hybrid intelligence as a "socio-technological arrangement" in which humans and AI collaborate, expanding each other's capabilities. Humans bring domain knowledge, intuition, ethical awareness, and interpretive skills. Machines contribute speed, pattern recognition, and predictive power.

In an energy context, grid operators often have decades of experience with local weather patterns, reliability standards, and political constraints that shape energy policy. Meanwhile, AI can analyze enormous datasets and produce real-time insights. Combining these strengths yields better operational decisions.

3.2 Role of Experts in AI-Driven Systems

Designers of AI-based solutions sometimes forget the crucial role of human oversight. Even advanced systems can produce erroneous forecasts or fail to accommodate unforeseen events (Altin & Eyimaya, 2023). Expert knowledge is vital in refining AI outputs, checking reasonableness, and shaping decision boundaries.

For instance, consider an AI model that predicts future electricity prices. If the model misses a regulatory shift that changes how carbon credits are priced, the output could lead to misguided strategies. A grid operator monitoring the system might spot that mismatch. That operator's feedback can feed back into the AI model, adjusting it to incorporate new policies. This continuous human-in-the-loop approach is fundamental to hybrid intelligence (Dellermann et al., 2019).

3.3 Balancing Data-Driven and Policy-Driven Decisions

Energy markets rest on more than just supply and demand. They reflect broader societal goals around decarbonization, system reliability, and equitable access. AI can discover patterns in consumption data, but may not grasp that certain policies favor local renewable integration over short-term cost minimization.

Hybrid intelligence can weave such considerations into algorithms by implementing rule-based constraints or weighting factors that reflect policy priorities (Bishaw, 2024). A municipality might embed higher social value to community solar projects or prioritize resilience for critical public facilities. By combining computational efficiency with domain insights, hybrid intelligence ensures that AI-driven smart contracts do not solely chase cost metrics.

3.4 Communication Channels in Hybrid Systems

Hybrid intelligence needs effective communication channels between human stakeholders and AI algorithms. In many energy settings, specialized dashboards display real-time forecasts and recommended actions, and operators can override or refine those recommendations (Al Shareef et al., 2024). This feedback loop fosters trust and fine-tunes the learning process.

In blockchain-based markets, such communication extends to on-chain governance. Smart contracts can incorporate procedures that let authorized operators pause or modify contract execution if unanticipated conditions arise (Belotti et al., 2019). This creates a system that is partially autonomous but still subject to human intervention.

Table 1 presents an overview of how hybrid intelligence can address various challenges in the energy domain. It outlines typical problems, the AI techniques commonly employed, and the specific contributions of human expertise.

Table 1. Overview of hybrid intelligence contributions to energy challenges

Energy Challenge	AI Technique	Human Contribution
Demand Forecasting	Machine learning (time-series)	Provide local knowledge of consumer behavior and trends
Distributed Generation Integration	Optimization (linear, metaheuristics)	Validate feasibility of resource dispatch schedules
Dynamic Pricing	Reinforcement learning	Incorporate policy constraints, social equity concerns
Real-Time Grid Balancing	Predictive analytics	Oversee failsafe measures, manage contingencies
Renewables Scheduling	Stochastic modeling	Input for unique weather anomalies, site-specific info
Transactive Energy Management	Blockchain-based auctions	Adjust contract terms to ensure regulatory compliance

This table highlights the interplay between computational tools and human insight. Hybrid intelligence thrives by integrating big data analytics with the contextual awareness of experts, forging comprehensive answers to energy problems.

4. THE BUILDING BLOCKS OF AI-POWERED SMART CONTRACTS

Smart contracts are programmable scripts deployed on a blockchain network. When pre-defined conditions are met, these scripts automatically execute agreed-upon transactions (Belotti et al., 2019). In the energy domain, they might settle payments, update supply schedules, or reconfigure grid settings.

AI-powered smart contracts link these self-executing scripts to external data streams and machine learning models, turning them into dynamic agents. Through an oracle or data feed, the contract can incorporate real-time grid data, weather forecasts, or market indices. AI logic can then trigger actions—such as adjusting prices, toggling loads, or activating energy storage—without direct human intervention.

4.1 Core Principles of Blockchain

Blockchains, whether permissioned or permissionless, store a sequence of records (blocks) that are cryptographically linked (Andoni et al., 2019). Each block references the hash of the previous one, forming an immutable chain. The consensus mechanism—be it Proof of Work, Proof of Stake, or another protocol—ensures agreement across nodes on the validity of new blocks.

- **Decentralization**: Instead of a single database, nodes maintain identical copies of the ledger.
- **Transparency**: Transactions are visible to participants, though encryption can safeguard sensitive details.
- **Immutability**: Altering a block would require redoing the consensus for subsequent blocks, deterring tampering.
- **Smart Contracts**: Scripts stored on the chain that auto-execute upon receiving specific inputs.

Figure 1 captures the fundamental architecture of a blockchain network, with nodes, blocks, and the consensus process. We can see how each component builds on the previous, ensuring data integrity.

Figure 1. Simplified representation of blockchain data flow.

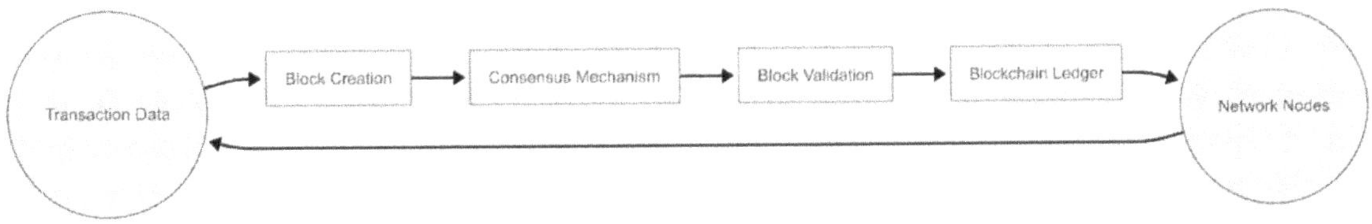

4.1.1 Permissioned vs. Permissionless Blockchains

Energy applications may choose permissioned blockchains, in which only approved participants can validate transactions (Chen et al., 2022). This approach can handle regulatory demands for auditing, and it often provides faster transaction throughput. By contrast, permissionless chains, like Ethereum, allow anyone to join. They may offer broader participation, but also face scalability and privacy challenges.

4.2 Smart Contract Mechanics in Energy

Smart contracts hold rules about energy trade conditions, price triggers, penalty clauses, or other operational details. For example, if a neighborhood microgrid's production surpasses local demand, a contract can automatically sell excess power at a pre-negotiated rate to a connected substation or aggregator (Al-Saif et al., 2021).

- **Execution**: Once conditions are verified (e.g., "house battery has 5 kWh available"), the contract issues the transaction.
- **Settlement**: Payment tokens or digital assets are automatically debited from the buyer's wallet and credited to the seller.
- **Auditing**: The transaction is recorded on-chain, creating a permanent audit trail.

4.3 AI Integration and Oracles

To function effectively, smart contracts need reliable external data or "oracles." In energy contexts, oracles deliver real-time meter readings, price signals, and weather data (Vionis & Kotsilieris, 2024). AI models can run off-chain or on specialized layers, interpreting these inputs to produce predictions or recommended actions. The contract then uses these results to decide if certain conditions are met.

For instance, a forecasting model might determine that next hour's load is projected to exceed local capacity. This triggers a condition in the contract, prompting the system to import power from a neighboring microgrid at a pre-arranged rate. AI-based analytics can refine these decisions by factoring in historical consumption patterns, user preferences, and dynamic pricing forecasts (Ferrag & Maglaras, 2020).

5. AI IN ENERGY GRID OPTIMIZATION

AI offers powerful tools for handling the growing complexity of modern grids. From forecasting to optimization, these methods promise more agile and automated system control.

5.1 Machine Learning for Forecasting

Machine learning approaches, like neural networks or gradient-boosted trees, can predict short-term load, generation, or market prices (Ahmad et al., 2021). While linear regression models once dominated, advanced techniques can process non-linearities and high-dimensional data, leading to improved accuracy.

- **Load Forecasting**: Patterns in historical consumption, combined with real-time weather data and occupancy information, let machine learning models anticipate short-term demand fluctuations.
- **Renewable Generation Forecasting**: Neural networks trained on solar irradiance, cloud cover, wind speed, and historical power output can approximate future output for solar or wind plants.
- **Price Forecasting**: Reinforcement learning or supervised algorithms can evaluate market data, supply/demand imbalances, and macro-level factors to predict dynamic energy prices.

5.2 Optimization and Control

Beyond forecasting, AI-based optimization strategies can fine-tune dispatch schedules for generators, coordinate battery storage, and suggest demand-response measures (Al-Fatlawi et al., 2024). Heuristic methods like genetic algorithms or particle swarm optimization handle discrete and non-linear constraints typical of energy systems. These algorithms can run repeatedly, incorporating real-time data to update recommendations.

An example scenario might involve optimizing the charge/discharge cycle of household batteries. The AI algorithm receives electricity price forecasts, user consumption patterns, and constraints on battery performance. It then recommends when to store or release energy to minimize costs or maximize grid support. If integrated with a smart contract, these recommendations can trigger automated transactions for energy exchange with peers.

5.3 AI-Driven Demand Response

Demand-response programs have traditionally relied on time-of-use tariffs and manual adjustments by consumers. AI can automate this by detecting imminent demand peaks and signaling connected devices—like thermostats, EV chargers, or industrial machinery—to reduce loads (ElHusseini et al., 2020). The advantage is speed and precision: if an AI model predicts a peak in 15 minutes, the contract can dispatch curtailment signals and reward participants promptly.

5.4 Multi-Agent Systems and Decentralized Control

Energy grids are distributed, with many actors making local decisions. Multi-agent system models treat each node—such as a microgrid, industrial plant, or aggregator—as an agent pursuing specific goals (Ahmad et al., 2022). AI guides these agents as they negotiate power exchanges or respond to grid conditions.

Smart contracts facilitate secure interactions among agents. If an agent commits to deliver 100 kW in the next time slot, that agreement can be enforced through the contract logic. The AI-based agent observes whether the resource is indeed available and triggers either a delivery or a penalty. This structure reduces the transaction overhead typically seen in centralized clearinghouses.

5.5 Data Governance and Security

More data can improve forecasts, but it also raises privacy and security issues. AI models need extensive historical records of consumption, meteorological data, and device status. Blockchain-based networks can secure this data by cryptographically hashing it and restricting access to authorized parties (Al-Saif et al., 2021).

Yet, there is a trade-off: blockchains are transparent, so sensitive information might be exposed if not properly masked or stored off-chain. Zero-knowledge proofs, ring signatures, and other privacy-preserving methods can reduce these risks (Belotti et al., 2019).

6. BLOCKCHAIN MECHANISMS FOR ENERGY TRANSACTIONS

Blockchain-enabled energy platforms aim for more direct, peer-to-peer transactions. Rather than relying on a central utility, prosumers can sell surplus energy to neighbors or purchase from local microgrids. This fosters competition, potentially lowering prices and encouraging green energy adoption (Al-Abri et al., 2022).

6.1 Tokenization of Energy

Tokenization converts energy units into digital tokens that can be traded on a blockchain (Andoni et al., 2019). Each token might represent a kilowatt-hour of electricity or a renewable energy certificate. Smart contracts track token ownership and automatically execute exchanges when certain triggers occur.

Some initiatives propose linking physical energy flows to tokens. A meter reading from a rooftop solar array can mint new tokens, reflecting the produced kilowatt-hours. Those tokens can then be sold to interested buyers, ensuring an auditable record of energy production (Chen et al., 2022).

6.2 Consensus Protocol Considerations

Energy transactions require rapid validation to avoid bottlenecks. High-throughput consensus protocols, like Practical Byzantine Fault Tolerance (PBFT) or delegated Proof of Stake, can expedite block creation (Dhillon et al., 2017). The trade-off is decentralization: permissioned or consortium blockchains might be more suitable, as they allow only trusted validators.

For real-time grid applications, some designs experiment with off-chain channels or sidechains that process transactions quickly, then anchor final states onto the main blockchain (Chen et al., 2022). The result is lower latency for near-instant settlement.

6.3 Smart Contracts for Settlement and Penalties

In an energy trading scenario, settlement is automated. A prosumer's smart meter records energy injected into the grid. The blockchain node receives this data and checks any relevant contract. If the agreed price is 0.10 tokens per kWh, the seller's account is credited, and the buyer's account is debited.

If performance deviates from the contract—for example, if the prosumer promises 5 kWh but only delivers 3 kWh—a penalty clause might trigger, docking tokens from the seller's collateral. This automation is especially relevant in demand-response and ancillary service markets, where reliability is paramount (Al-Saif et al., 2021).

6.4 Security and Resilience

Blockchain networks rely on robust cryptography, but energy systems face additional security threats, such as spoofed meter readings or denial-of-service attacks on communication channels (Badidi, 2022). Layered architectures can mitigate these issues. For instance, data from meters could be signed by trusted hardware devices before it even hits the blockchain. AI-based anomaly detection can flag suspicious patterns, such as abrupt spikes in consumption or duplication of meter IDs.

Proper design also accounts for resilience. Energy grid disruptions might sever communication, but the decentralized nature of blockchain can keep local nodes operational. Once connectivity is restored, transactions can be reconciled (Belotti et al., 2019).

7. SYNERGIZING AI AND SMART CONTRACTS

So far, we have examined AI's capacity for optimization and forecasting, and the role of blockchain in facilitating transparent, self-executing trades. The convergence of these technologies yields AI-powered smart contracts, an arrangement in which the contract's logic extends beyond static if–then statements (Kirli et al., 2022).

7.1 Intelligent Contract Logic

In many blockchains, smart contracts are deterministic: their outputs follow predefined logic. AI, on the other hand, often relies on models with probabilistic or approximate reasoning (Bishaw, 2024). Merging these approaches requires bridging deterministic on-chain rules with off-chain model execution.

In typical architectures, AI runs in an off-chain environment or "oracle layer," generating decisions or triggers. These triggers feed the on-chain contract. One might call it "semi-autonomous," given that the actual contract execution is strictly deterministic, but the condition checks rely on AI predictions. For instance, if an off-chain system predicts 2 MW of surplus in the next 15 minutes, the contract triggers a specific event.

7.2 Adaptive Pricing

Energy markets can adapt prices to supply and demand in near-real-time. An AI model identifies the expected load curve, generation patterns, and user profiles. It then suggests a dynamic pricing schedule that aims to balance the local grid. A smart contract enforces these tariffs automatically, adjusting them at each time interval and settling trades accordingly (ElHusseini et al., 2020).

This approach can support the integration of renewables by incentivizing load shifting when solar production peaks, or by rewarding consumers who reduce demand during peak load times.

7.3 Automated Demand-Response Contracts

In demand-response, participants agree to reduce consumption in exchange for financial benefits. AI can detect system stress or forecast that demand will exceed certain thresholds (Al-Fatlawi et al., 2024). Once the model signals a need for curtailment, the contract instructs connected devices—like smart thermostats or EV chargers—to reduce load. If participants comply, they receive tokenized rewards.

Figure 2 depicts a simplified flow: an AI-based module monitors grid conditions; it passes instructions to a smart contract, which triggers device-level actions and issues incentives.

Figure 2. Automated demand-response concept using AI triggers and smart contracts.

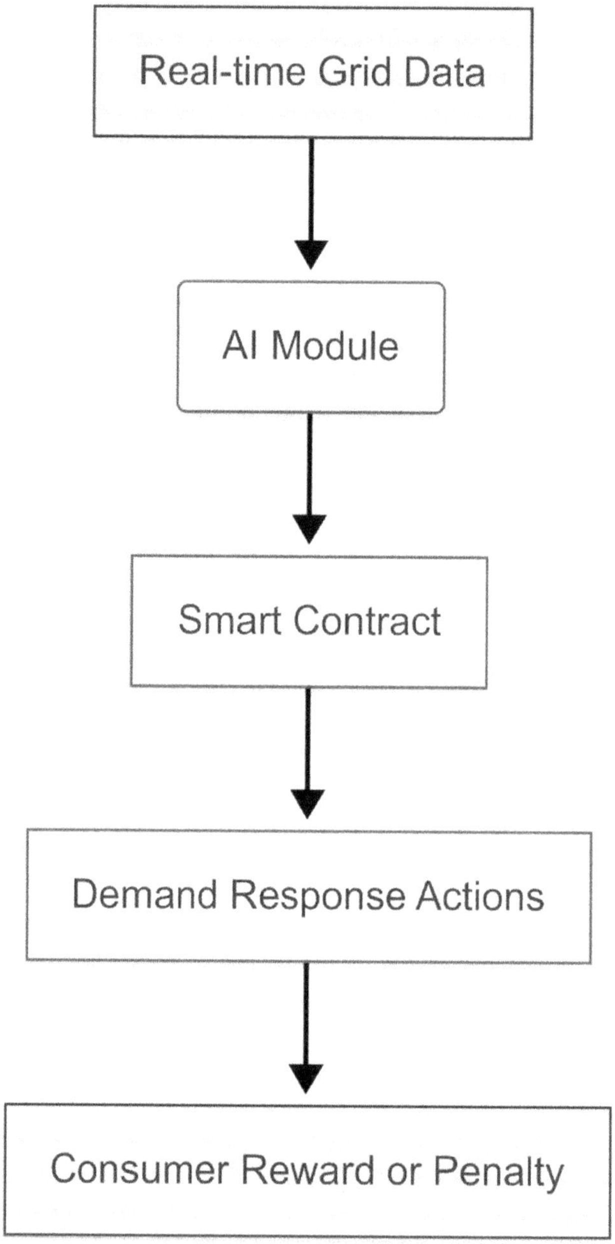

7.4 Market Integration and Auctions

Energy systems often rely on bidding and auction mechanisms for capacity or ancillary services. AI-driven agents can formulate bids based on predicted conditions. The contract records these bids in a decentralized manner, awarding capacity to the lowest-cost or best-fit providers. Payment settlement occurs automatically (Kirli et al., 2022).

Because everything is on-chain, the process is transparent, reducing the possibility of collusion. Nonetheless, smaller participants may still face barriers if the bidding structure is too complex. Hybrid intelligence can involve human reviewers who design simpler user interfaces or moderate auctions in early adoption phases (Dellermann et al., 2019).

7.5 Compliance and Auditing

One noteworthy advantage is that each contract's operations are visible on the blockchain, creating immutable audit trails (Chen et al., 2022). Regulators can track compliance with energy quotas, environmental goals, or demand-response obligations. AI can continuously analyze transaction data to spot anomalies or potential fraud.

In regulated environments, partial anonymity or permissioned access might be required to protect sensitive data (Al-Abri et al., 2022). But the underlying principle remains: combining AI's capacity for large-scale data analysis with blockchain's integrity ensures that trust is encoded in the system.

8. HYBRID INTELLIGENCE FOR DECISION-MAKING

While automated smart contracts and AI predictions are essential, final judgments about critical grid operations or policy decisions often rest with human agents. Hybrid intelligence frameworks give them the means to interpret AI outputs, provide domain expertise, and refine rules.

8.1 Human-in-the-Loop Models

Human-in-the-loop design ensures that operators remain engaged. For instance, an AI might propose reconfiguration of a substation to accommodate sudden changes in voltage from solar infeed. Rather than executing instantly, the model's recommendation appears on an operator's interface. The operator can accept, modify, or reject it, with the final input passing to the smart contract for on-chain execution.

In urgent cases, the system might shift to an autonomous fail-safe mode if human input is not available. This approach handles emergencies such as severe weather events or grid instability (Dellermann et al., 2019).

8.2 Expert Oversight in Smart Contracts

Some energy regulators have mandated that certain transactions or market changes require explicit human authorization (Andoni et al., 2019). Smart contracts can reflect these rules, requiring multiple signatures before large-scale resource dispatch occurs. AI can recommend the best course of action, but a domain expert must sign off.

8.3 Collective Governance

An alternative is collective governance via decentralized autonomous organizations (DAOs). In theory, stakeholders—like prosumers, grid operators, and local authorities—vote on proposals. AI helps structure the proposals or evaluate their outcomes, while smart contracts tally votes and enforce results (Belotti et al., 2019). This approach can democratize decision-making, though it also demands robust user education and participation.

8.4 Ethical and Social Considerations

It is possible that AI-driven pricing might disadvantage certain socioeconomic groups or that fully automated contracts might override consumer preferences. Hybrid intelligence ensures that ethical considerations remain part of the decision process (Dellermann et al., 2019). Human experts can set fairness constraints or intervene if vulnerabilities arise.

Table 2 provides examples of how hybrid intelligence can enhance decision-making across typical energy scenarios.

Table 2. Hybrid intelligence in energy decision-making scenarios

Scenario	AI Contribution	Human Role
Real-Time Outage Management	Rapid localization, data fusion to identify fault zones	Approve final switchgear operations if safety protocols demand
Demand-Side Bidding in Markets	Automated forecasting and bidding algorithms	Validate fairness, ensure compliance with market rules
Dynamic Network Reconfiguration	Multi-objective optimization of distribution feeders	Integrate local knowledge about line constraints, environmental factors
Renewable Integration Planning	Predictive analytics on resource availability	Balance community interests, verify compliance with environmental rules
Microgrid Islanding Decisions	Anomaly detection, risk assessment of external conditions	Confirm thresholds for islanding, coordinate local facility readiness

This underscores the synergy between computational intelligence and human expertise in shaping robust, context-aware decisions.

9. TECHNOLOGICAL INFRASTRUCTURE AND ARCHITECTURE

Implementing AI-powered smart contracts requires a technical stack that integrates sensors, communication networks, data processing layers, machine learning frameworks, and blockchain systems. Each layer must cooperate securely and efficiently.

9.1 Edge and Cloud Computing

Energy grids distribute across extensive geographical areas. Data from sensors might be aggregated at local substations (edge computing), where AI models run to address time-critical operations (Badidi, 2022). Aggregated data or summarized insights can then flow to cloud platforms for broader analytics or policy-based decisions.

9.2 Communication Standards

Reliability in data exchange is central. Standards like IEC 61850 define how substation devices communicate. Interoperability among hardware vendors ensures that AI-based solutions can access consistent data. Cryptographic protocols safeguard these exchanges (Al Shareef et al., 2024).

9.3 Off-Chain AI Execution

Smart contracts on many blockchain platforms are limited in computational scope to keep transaction costs manageable. Complex AI computations usually happen off-chain, with only final results or key parameters passing on-chain (Belotti et al., 2019). Off-chain data feeds, oracles, or "Layer 2" solutions connect advanced AI modules to the blockchain.

9.4 Interoperability

Energy markets involve multiple actors, sometimes on different blockchain networks or legacy systems. Interoperability frameworks allow transactions to flow across networks. A distribution grid's permissioned chain might link to a regional aggregator's chain, which interacts with a national-level wholesale market. Smart contracts or bridging protocols coordinate these cross-chain operations (Chen et al., 2022).

9.5 Monitoring and Diagnostics

Continuous monitoring is crucial to spot anomalies in hardware or data feeds. AI-based diagnostics can detect early signs of sensor drift or malicious attacks. The system might flag suspicious variations in device data, triggering a human operator or halting contract execution until the anomaly is resolved (Badidi, 2022). **Figure 3** illustrates a conceptual architecture for AI-powered energy applications. It shows data capture from field devices, edge analytics, and a blockchain layer for settlement and contract execution.

Figure 3. Conceptual architecture illustrating data flow, AI components, and blockchain settlement.

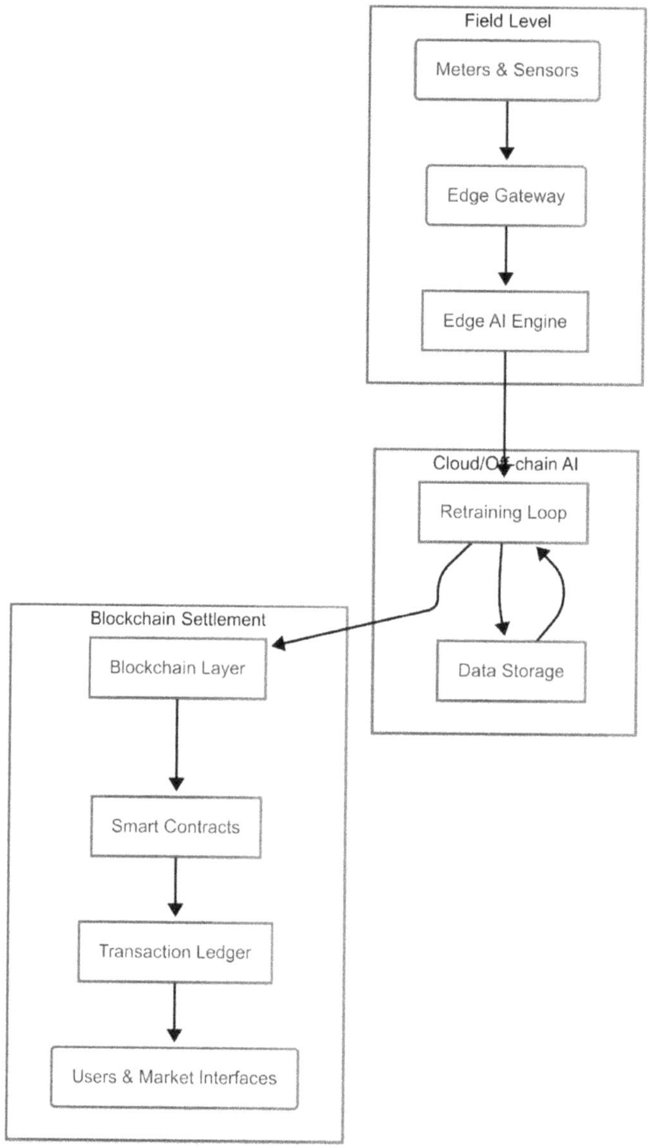

10. REAL-WORLD IMPLEMENTATIONS AND CASE STUDIES

Though much attention remains theoretical, pilot projects and emerging platforms offer insights into the feasibility of AI-powered smart contracts for energy systems. I focus on selected examples that illustrate the interplay of technology, business models, and regulatory frameworks.

10.1 Microgrid Peer-to-Peer Trading Pilot

A municipal microgrid in Europe implemented a peer-to-peer energy trading pilot. Households with rooftop solar installed smart meters that recorded real-time production. The pilot used a permissioned blockchain to track all energy exchanges. AI predicted hourly generation and consumption, adjusting local prices automatically (Al-Saif et al., 2021).

Residents could set personal preferences, such as a maximum or minimum selling price. A smart contract matched offers and demands, clearing trades every 15 minutes. The pilot found that the AI-based approach reduced reliance on external grid imports by 20%.

10.2 Demand-Response in Industrial Parks

An industrial park in Asia tested an automated demand-response platform, combining AI forecasting with blockchain-based settlements (ElHusseini et al., 2020). Factories consented to let the system reduce certain loads during peak events. AI scanned real-time data on the park's total power draw, anticipating peak times. A smart contract enforced curtailment obligations and distributed financial rewards proportionally.

Operators highlighted that the system improved predictability for both the park and the regional utility. However, they noted challenges around the reliability of old metering infrastructure, emphasizing the importance of robust device calibration.

10.3 Integrating Electric Vehicle Fleets

A technology firm collaborated with a local power utility to manage a growing fleet of electric vehicles (Al-Saif et al., 2021). The utility used AI to forecast short-term load spikes from EV charging. A blockchain-based contract brokered price incentives for EV owners who delayed charging or fed battery power back into the grid.

Participants reported financial savings, while the utility found that peak loads were mitigated. This arrangement highlights how AI and blockchain synergy can accommodate flexible resources like EVs, turning them into dynamic assets rather than passive loads.

10.4 Renewable Energy Certificates and Green Tokenization

Some energy companies tokenize green energy certificates to prove that certain kilowatt-hours come from renewable sources (Chen et al., 2022). AI ensures that generation meets required thresholds and verifies the location/time data. Smart contracts issue tokens in proportion to validated renewable output. Purchasers of these tokens can claim environmental benefits, supporting compliance with corporate sustainability goals.

11. IMPLEMENTATION CHALLENGES AND LIMITATIONS

While promising, AI-powered smart contracts face constraints in scalability, reliability, regulation, and user acceptance. Recognizing these hurdles is vital for designing viable systems.

11.1 Scalability Concerns

High transaction volumes strain blockchain networks. Energy systems generate thousands of meter readings per second. Many blockchain platforms cannot handle this throughput, leading to slow confirmations and high fees (Andoni et al., 2019). Off-chain solutions or sidechains alleviate these bottlenecks, but add architectural complexity.

AI-based operations also depend on large datasets. Retraining models or running real-time analytics can be computationally heavy. Edge computing partially solves this, but large-scale integration across multiple sites remains a challenge (Badidi, 2022).

11.2 Data Quality and Integrity

Smart contracts depend on external oracles. If these oracles provide inaccurate data—whether by malfunction or malicious tampering—the contract will produce flawed results (Al-Abri et al., 2022). Energy meters in older buildings can be less reliable or vulnerable to tampering. Consistent data governance, including auditing and hardware security modules, is essential.

11.3 Regulatory and Legal Constraints

Energy markets are heavily regulated. National laws may restrict peer-to-peer trading, dictate electricity tariffs, or require licensed intermediaries. Some jurisdictions mandate that the central utility remains the sole buyer of electricity, limiting direct blockchain transactions (Al Shareef et al., 2024).

Smart contracts also raise legal questions around liability. If an automated contract incorrectly charges a user, who is responsible? If an AI-based system triggers a grid reconfiguration that leads to equipment damage, can the developer be held liable? Clear legal frameworks are still emerging (Andoni et al., 2019).

11.4 Security Threats

Decentralization can deter some attacks by removing single points of failure. Yet, vulnerabilities persist. A majority attack on a blockchain network could rewrite transactions. AI modules could be subverted to produce spurious signals (Badidi, 2022). Real-world systems need layered security, including robust identity management and intrusion detection.

11.5 User Adoption and Trust

Complex interfaces and unpredictable price swings can deter new users. Many households or small business owners are unfamiliar with blockchain wallets or dynamic pricing. Projects must prioritize user-friendly designs, transparent incentives, and educational outreach (Ali et al., 2021).

Table 3 summarizes typical challenges encountered in AI-powered blockchain energy applications, with potential mitigation strategies.

Table 3. Key challenges in AI-powered smart contracts and mitigation approaches

Challenge	Description	Mitigation
Scalability	High transaction and data throughput requirements	Use sidechains, adopt efficient consensus, batch transactions
Data Integrity	Risk of erroneous or tampered data feeding smart contracts	Deploy secure hardware modules, robust oracles, redundant checks
Regulatory Limitations	Legal restrictions on direct energy trading, unclear liability frameworks	Work with authorities, run sandbox pilots, refine policy
Security Threats	Blockchain or AI modules can face hacking, majority attacks, or data poisoning	Enforce strong cryptography, multi-layer verification
Usability Barriers	Complex technology can deter participation	Create intuitive UIs, educate users, provide clear incentives

Addressing these challenges requires collaboration among regulators, industry stakeholders, researchers, and local communities.

12. REGULATORY AND POLICY DIMENSIONS

Energy systems do not exist in a vacuum. They operate within policy frameworks that set market rules, environmental goals, and consumer protection standards. The same is true for blockchain and AI applications, which face emerging data governance and digital rights legislation in various regions (Al Shareef et al., 2024).

12.1 Licensing and Market Design

Many jurisdictions stipulate that energy retailers must hold licenses, meet reliability standards, and adhere to specific pricing regulations. Blockchain-based peer-to-peer models can collide with these requirements, especially if the platform functions like an unregulated marketplace (Andoni et al., 2019). Regulators may need to adapt licensing regimes to accommodate decentralized trading while preserving consumer protection.

12.2 Data Protection and Privacy

AI relies on large-scale data collection, but privacy laws (e.g., GDPR in Europe) restrict how personal data can be processed. Even anonymized consumption patterns could reveal lifestyle details (Badidi, 2022). Hybrid intelligence approaches must ensure that any data stored on-chain or passed to AI models does not violate privacy obligations. Zero-knowledge proofs or aggregated data usage are potential solutions.

12.3 Incentive Structures

Policy can incentivize or disincentivize the uptake of AI and blockchain solutions. Feed-in tariffs for renewables, net metering policies, or carbon pricing schemes directly influence the viability of peer-to-peer transactions. Some governments may offer grants or tax breaks for adopting AI-based grid management tools (U.S. Department of Energy, 2024).

12.4 Legal Status of Smart Contracts

Smart contracts operate autonomously once deployed. If a contract inadvertently charges excessive fees or incorrectly issues tokens, reversing these actions may be cumbersome. In some regions, legislators have begun drafting regulations that define the legal enforceability of smart contracts (Chen et al., 2022). Achieving clarity on dispute resolution and liability is paramount for mainstream acceptance.

12.5 International Coordination

Energy systems, especially in interconnected regions, cross national borders. Harmonizing blockchain-based transaction frameworks among different regulatory regimes is complex (Al-Abri et al., 2022). International organizations could provide guidance or standards that unify technical and legal aspects.

13. FUTURE DIRECTIONS AND INNOVATIONS

AI-powered smart contracts remain at an early stage of maturity in the energy sector. Several trends are poised to expand their influence in the coming years.

13.1 Next-Generation AI Models

Deep reinforcement learning, transfer learning, and federated learning hold promise for energy optimization. Federated learning allows multiple agents to train a shared model without exchanging raw data, preserving privacy (Altin & Eyimaya, 2023). This approach could gain traction in multi-utility scenarios where data must remain local.

13.2 Advanced Consensus Mechanisms

As computational costs and carbon footprints become more significant, eco-friendly consensus methods—like Proof of Authority or Proof of Stake—will likely evolve (Dhillon et al., 2017). For energy applications, carbon neutrality might become an important design goal for the underlying blockchain, especially if the platform aims to promote environmental sustainability.

13.3 Integration with the Internet of Things (IoT)

Many devices—from smart thermostats to EV charging stations—are part of the IoT ecosystem. Smart contracts that interface with these devices via AI can orchestrate loads in real time, offering granular control. Standardized IoT protocols and robust security solutions will be critical for large-scale adoption (Shahinzadeh et al., 2022).

13.4 Domain-Specific Smart Contract Languages

Mainstream languages like Solidity or Vyper might not fully meet the specialized needs of energy trading and regulatory compliance. We may see domain-specific languages that incorporate built-in safety checks or grid constraints, making it easier to design and verify complex transactions (Kirli et al., 2022).

13.5 Hybrid Intelligence Governance Models

Beyond technology, hybrid intelligence can expand the scope of how communities govern local energy systems. DAOs, or similar community-driven frameworks, could involve direct democratic control of local microgrids, with AI providing real-time data and optimization suggestions (Belotti et al., 2019). Over time, these models may mature, offering localized solutions to global energy challenges.

14. TOWARD A COMPREHENSIVE ANALYTICAL MODEL

Having surveyed technical, economic, and policy aspects, I propose a conceptual model that integrates AI and smart contracts within a hybrid intelligence framework. This model aims to align with real-world constraints while taking advantage of emerging innovations.

14.1 Model Components

1. **Data Collection Layer**: Meters, sensors, SCADA systems, and IoT devices gather continuous data on generation, load, and market indicators.
2. **Edge AI Processing**: Local nodes use lightweight AI models for fast, real-time decisions, especially in demand-response or microgrid islanding.
3. **Cloud AI and Modeling**: Larger forecasting and optimization tasks run in centralized or distributed cloud environments, updated regularly with new data.
4. **Blockchain Settlement Layer**: Smart contracts hold energy trading logic, enforce compliance, and record transactions.
5. **Hybrid Governance**: Human experts, regulators, or community stakeholders oversee critical decisions. They maintain the right to override or refine AI-driven recommendations.

14.2 Operational Flow

- **Step 1**: AI model forecasts near-term load and generation.

- **Step 2**: Smart contract checks forecast data from the oracle. If supply shortfall is likely, it triggers a demand-response event or requests additional capacity from a local aggregator.
- **Step 3**: If human oversight is required, a domain expert reviews the AI's recommendation before it is finalized.
- **Step 4**: The blockchain processes the transaction, updating ledger balances and applying relevant penalties or rewards.
- **Step 5**: The system logs performance metrics, feeding them back into the AI models for retraining and improved accuracy.

14.3 Performance Metrics

Users might evaluate the model on metrics like cost savings, peak load reduction, greenhouse gas emissions, or resilience to outages (Ali et al., 2021). Social acceptance could be measured by user participation rates in peer-to-peer markets or satisfaction surveys.

Figure 4 outlines the feedback loops between AI, smart contracts, and human supervision.

Figure 4. Feedback loops in an AI-powered, blockchain-settled energy ecosystem with human oversight.

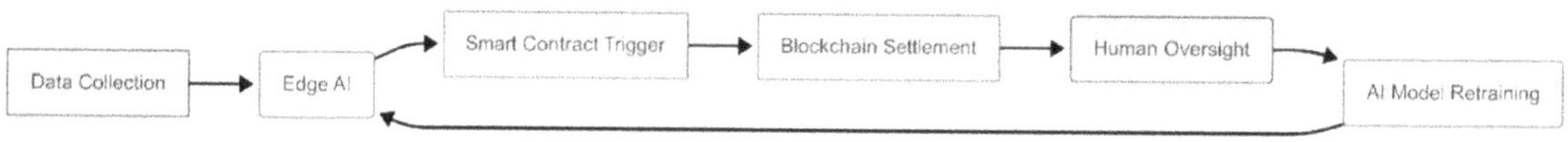

15. PRACTICAL DESIGN GUIDELINES

The path from concept to reality requires careful planning. Here, I present guidelines for developers and policymakers aiming to deploy hybrid intelligence solutions for energy grids.

15.1 Start Small

Pilot projects or limited-scale microgrids allow experimentation without risking widespread grid instability (ElHusseini et al., 2020). Gathering real operational data is invaluable for refining AI models and identifying technical pitfalls early.

15.2 Ensure Interoperability

In practice, existing legacy systems will remain for years. Solutions must interface with standard protocols, such as Modbus or IEC 61850, and be prepared to integrate data from multiple vendors (Al Shareef et al., 2024).

15.3 Implement Clear Off-Chain Governance

While a smart contract might handle basic transactions, complex or high-stakes scenarios demand human input. A governance policy specifying how and when operators intervene prevents confusion or legal disputes (Kirli et al., 2022).

15.4 Prioritize Security

Energy infrastructure is critical. Implement robust cybersecurity measures, from secure hardware for meter readings to multi-signature approvals for large transactions. Consider external audits of smart contract code to reduce vulnerabilities (Badidi, 2022).

15.5 Communicate Benefits to Stakeholders

Households or firms might view AI and blockchain technology with suspicion. Demonstrate tangible benefits—lower bills, potential revenue from excess energy, improved reliability—through clear, transparent communication (Al-Saif et al., 2021).

Table 4 offers a concise checklist for those planning an AI-powered, blockchain-enabled energy project, summarizing key design considerations.

Table 4. Implementation checklist for AI-powered smart contracts in energy

Item	Consideration	Example Action
Data Infrastructure	Secure data acquisition from meters and sensors	Calibrate and seal meters, use encryption
AI Model Lifecycle	Ongoing training and validation on updated data	Set up MLOps pipeline, define update frequency
Smart Contract Design	Incorporate fallback mechanisms and error handling	Use multi-sig for critical functions
Blockchain Scalability	Evaluate throughput requirements, consider sidechains	Pilot on testnet, measure transaction speed
Governance & Oversight	Define roles for domain experts, define override powers	Draft governance charter, hold periodic reviews
User Engagement	Provide simple interfaces, user support, and guidance	Develop a mobile app or web portal for monitoring
Security Audits	Perform code reviews, vulnerability scans	Hire third-party auditors, set bug bounties

16. CASE ILLUSTRATION: A HYPOTHETICAL REGIONAL GRID DEPLOYMENT

To demonstrate how these guidelines might apply in practice, imagine a regional grid in which a utility manages both rural and urban distribution, and numerous prosumers own solar installations.

16.1 Setting the Stage

- **Location**: A region with moderate solar potential and existing demand-response incentives.

- **Actors**: The utility company, local aggregator, prosumers with rooftop solar, commercial sites with battery storage, and city officials.
- **Regulatory Framework**: Partial liberalization allows peer-to-peer energy trading, though the utility remains central for balancing and reliability.

16.2 Phased Deployment

1. **Pilot Microgrid**: The utility selects a suburban neighborhood of 200 homes for a pilot. Smart meters connect to a permissioned blockchain. AI forecasts next-day solar production, adjusting real-time prices. Prosumers trade among themselves, with any surplus sold back to the main grid.
2. **Scaling to Commercial Loads**: Commercial buildings with large rooftop arrays join the network. AI-based optimization schedules battery discharges, capturing peak price differentials.
3. **Demand-Response Integration**: The aggregator extends AI-driven demand-response to participants with HVAC systems or EV chargers. Smart contracts reward load reductions during peak hours.
4. **Broader Rollout**: As confidence grows and data accumulates, the program expands region-wide, adding multiple neighborhoods and small industrial facilities.

16.3 Hybrid Intelligence Example

When AI forecasts an imminent peak, a demand-response event is triggered, offering a certain reward token rate. The contract notifies participating devices. However, an operator spots a potential conflict with local substation constraints. She adjusts the parameters to account for newly installed equipment that the AI had not yet factored in. This change leads to revised instructions on the next contract update, ensuring the system remains stable.

16.4 Performance Outcome

Metrics indicate a reduction of net peak demand by 10%, with prosumers reporting modest but meaningful revenue gains. The utility uses data from the pilot to refine forecasts and implement additional strategies. City officials promote the system as part of a green urban development initiative, driving further user adoption.

This hypothetical scenario illustrates how the convergence of AI, smart contracts, and human expertise can transform grid operations, delivering economic and social benefits if well-orchestrated.

17. DISCUSSION OF BROADER IMPLICATIONS

Implementing AI-powered smart contracts in real-world energy grids sparks a range of ramifications—some beneficial, others less certain. These revolve around market dynamics, social equity, and environmental sustainability.

17.1 Market Efficiency and Innovation

By lowering transaction costs and automating settlements, decentralized energy markets could become more vibrant. Suppliers and consumers can respond rapidly to price signals, potentially leading to more efficient resource allocation (Ferrag & Maglaras, 2020). At the same time, smaller participants may struggle with technical complexities, leading to a digital divide if not addressed (Ali et al., 2021).

17.2 Social Equity

Dynamic pricing, while efficient, can disproportionately impact vulnerable consumers who cannot shift demand easily. Hybrid intelligence must consider these realities when designing AI algorithms. A balanced approach might incorporate targeted subsidies or social tariffs that ensure low-income households are not penalized (Dellermann et al., 2019).

17.3 Environmental Impact

Automation and real-time trading can facilitate higher penetration of renewables, as the grid can adapt more flexibly to intermittent solar or wind. Blockchain-based carbon credit systems might spur green investments, further accelerating decarbonization (Chen et al., 2022). Conversely, energy consumption from certain consensus mechanisms or large-scale AI computations could offset some sustainability gains if not carefully managed (Dhillon et al., 2017).

17.4 Regulatory Evolution

Over time, policymakers may adopt new regulations that codify how AI-based transactions are monitored, how data privacy is ensured, and how liabilities are assigned. The interplay between decentralized technology and existing top-down regulations is complex, but it has the potential to yield more adaptive and future-proof policies (Andoni et al., 2019).

17.5 Ethical Dimensions

Providing AI with extensive control over critical infrastructure raises moral questions about accountability. Humans must remain able to interpret system logic and intervene when algorithms fail (Dellermann et al., 2019). Additionally, large-scale data collection can infringe on personal privacy if not safeguarded by strict protocols (Badidi, 2022).

18. CONCLUSION

This chapter explored the theoretical constructs and emerging possibilities of AI-powered smart contracts for optimizing energy grids. It also examined how hybrid intelligence—an integrated model of human expertise and machine-driven insights—can address the complexities and uncertainties characteristic of the modern power sector.

Energy systems worldwide are shifting toward decentralized architectures that integrate renewables, batteries, and demand-responsive loads. AI methods provide predictive prowess and optimization capabilities, yet they require a transparent and automated way to enforce market rules and deliver settlements. Blockchain-based smart contracts fill this gap, offering trustless, immutable platforms for transactions and contract execution.

At the same time, purely autonomous systems often neglect contextual factors. By weaving in human oversight, domain knowledge, and community input, hybrid intelligence ensures that technological innovation remains aligned with social and regulatory realities. This synergy aims to make the grid more agile, sustainable, and inclusive.

However, implementing these solutions in practice faces obstacles like scalability limitations, data quality issues, and unclear regulatory frameworks. Addressing these requires collaboration among utilities, technology developers, policymakers, and end users. While pilot programs demonstrate promising results, widespread adoption will hinge on designing robust technical architectures, refining policy, and building public trust.

Looking to the future, advanced consensus mechanisms, federated AI approaches, and domain-specific smart contract languages could further enhance system resilience. Tools such as decentralized autonomous organizations (DAOs) may eventually democratize local energy governance, blending computational optimization with communal decision-making. The ultimate outcome depends on ongoing dialogue between technical innovation and societal values.

In sum, AI-powered smart contracts represent a compelling pathway for grid modernization. They combine the predictive intelligence of advanced algorithms with the trustless execution of blockchains, underpinned by human expertise that safeguards adaptability and accountability. As the energy sector continues to evolve, these hybrid solutions have the capacity to reshape market dynamics, reinforce environmental commitments, and deliver tangible benefits to consumers and communities.

REFERENCES

Ahmad, T., Zhang, D., Huang, C., Zhang, H., Dai, N., Song, Y., & Chen, H. (2021). Artificial intelligence in sustainable energy industry: Status quo, challenges and opportunities. *Journal of Cleaner Production, 289*, 125834. DOI: 10.1016/j.jclepro.2021.125834

Ahmad, T., Zhu, H., Zhang, D., Tariq, R., Bassam, A., Ullah, F., AlGhamdi, A. S., & Alshamrani, S. S. (2022). Energetic systems and artificial intelligence: Applications of Industry 4.0. *Energy Reports, 8*, 334–361. DOI: 10.1016/j.egyr.2021.11.256

Al-Abri, T., Onen, A., Al-Abri, R., Hossen, A., Al-Hinai, A., Jung, J., & Ustun, T. S. (2022). Review on energy application using blockchain technology with an introduction to the pricing infrastructure. *IEEE Access : Practical Innovations, Open Solutions, 10*, 80119–80137. DOI: 10.1109/ACCESS.2022.3194161

Al-Fatlawi, H. H., Khayoon, Q. M., Albhadly, A. J., & Ibrahim, A. K. (2024). AI-enhanced demand response strategies in smart grids: Toward a sustainable energy future. *Journal of Electrical Systems, 20*(7s), 1054–1064. DOI: 10.52783/jes.3535

Al-Saif, N., Ahmad, R., Salah, K., Yaqoob, I., Jayaraman, R., & Omar, M. (2021). Blockchain for electric vehicles energy trading: Requirements, opportunities, and challenges. *IEEE Access : Practical Innovations, Open Solutions, 9*, 156947–156961. DOI: 10.1109/ACCESS.2021.3130095

Al Shareef, A. M., Seçkiner, S., Eid, B., & Abumeteir, H. (2024). Integration of blockchain with artificial intelligence technologies in the energy sector: A systematic review. *Frontiers in Energy Research, 12*, 1377950. Advance online publication. DOI: 10.3389/fenrg.2024.1377950

Ali, M., Prakash, K., Hossain, M. A., & Pota, H. R. (2021). Intelligent energy management: Evolving developments, current challenges, and research directions for a sustainable future. *Journal of Cleaner Production, 314*, 127904. DOI: 10.1016/j.jclepro.2021.127904

Altin, N., & Eyimaya, S. E. (2023). Artificial intelligence applications for energy management in microgrid. In *Proceedings of the 2023 11th International Conference on Smart Grid (icSmartGrid)* (pp. 1–6). IEEE. DOI: 10.1109/icSmartGrid58556.2023.10170860

Andoni, M., Robu, V., Flynn, D., Abram, S., Geach, D., Jenkins, D., McCallum, P., & Peacock, A. (2019). Blockchain technology in the energy sector: A systematic review of challenges and opportunities. *Renewable & Sustainable Energy Reviews, 100*, 143–174. DOI: 10.1016/j.rser.2018.10.014

Badidi, E. (2022). Edge AI and blockchain for smart sustainable cities: Promise and potential. *Sustainability (Basel), 14*(13), 7609. DOI: 10.3390/su14137609

Belotti, M., Božić, N., Pujolle, G., & Secci, S. (2019). A vademecum on blockchain technologies: When, which, and how. *IEEE Communications Surveys and Tutorials, 21*(4), 3796–3838. DOI: 10.1109/COMST.2019.2928178

Bishaw, F.Faisal Ghazi Bishaw. (2024). Artificial intelligence applications in renewable energy systems integration: A review. *Journal of Electrical Systems, 20*(3), 566–582. DOI: 10.52783/jes.2983

Bouckaert, S., Pales, A., Mekhilef, S., McGlade, C., Remme, U., & Wanner, B.. (2021). *Net Zero by 2050: A roadmap for the global energy sector*. International Energy Agency.

Chen, S., Ping, J., Yan, Z., Li, J., & Huang, Z. (2022). Blockchain in energy systems: Values, opportunities, and limitations. *Frontiers in Energy*, *16*(1), 9–18. DOI: 10.1007/s11708-022-0818-8

Dellermann, D., Ebel, P., Söllner, M., & Leimeister, J. M. (2019). Hybrid intelligence. *Business & Information Systems Engineering*, *61*(5), 637–643. DOI: 10.1007/s12599-019-00595-2

Dhillon, V., Metcalf, D., & Hooper, M. (2017). *Blockchain-enabled applications: Understand the blockchain ecosystem and how to make it work for you*. Apress. DOI: 10.1007/978-1-4842-3081-7

ElHusseini, H., Assi, C., Moussa, B., Attallah, R., & Ghrayeb, A. (2020). Blockchain, AI and smart grids: The three musketeers to a decentralized EV charging infrastructure. *IEEE Internet of Things Magazine*, *3*(2), 24–29. DOI: 10.1109/IOTM.0001.1900081

Ferrag, M. A., & Maglaras, L. (2020). DeepCoin: A novel deep learning and blockchain-based energy exchange framework for smart grids. *IEEE Transactions on Engineering Management*, *67*(4), 1285–1297. DOI: 10.1109/TEM.2019.2922936

Kirli, D., Couraud, B., Robu, V., Salgado-Bravo, M., Norbu, S., Andoni, M., Antonopoulos, I., Negrete-Pincetic, M., Flynn, D., & Kiprakis, A. (2022). Smart contracts in energy systems: A systematic review of fundamental approaches and implementations. *Renewable & Sustainable Energy Reviews*, *158*, 112013. DOI: 10.1016/j.rser.2021.112013

Li, Y., Zhao, Y., Wu, L., & Zeng, Z. (2024). *Artificial intelligence enabled computational methods for smart grid forecast and dispatch*. Springer.

Shahinzadeh, H., Zanjani, S., & Moradi, J. (2022). The transition toward merging big data analytics, IoT, and artificial intelligence with blockchain in transactive energy markets. In *Proceedings of the 2022 IEEE Global Energy Conference (GEC)*. IEEE. DOI: 10.1109/GEC55014.2022.9986604

U.S. Department of Energy. (2024). *AI for Energy: Opportunities for a modern grid and clean energy economy*. U.S. DOE Office of Critical and Emerging Technologies.

Vionis, P., & Kotsilieris, T. (2024). The potential of blockchain technology and smart contracts in the energy sector: A review. *Applied Sciences (Basel, Switzerland)*, *14*(1), 253. DOI: 10.3390/app14010253

ADDITIONAL READING

Boumaiza, A., & Sanfilippo, A. (2022). AI for energy: A blockchain-based trading market. In *Proceedings of the 2022 IEEE Industrial Electronics Society Annual Conference (IECON)* (pp. 1–6). IEEE.

Dargan, J., Gupta, N., & Singh, L. (2021). Blockchain based energy management system: A proposed model. In *Proceedings of the 2021 International Conference on Technological Advancements and Innovations (ICTAI)* (pp. 510–514). IEEE. DOI: 10.1109/ICTAI53825.2021.9673233

International Renewable Energy Agency (IRENA). (2019). *Innovation landscape brief: Blockchain*. IRENA.

Tiwari, S., & Tyagi, A. K. (Eds.). (2025). *AI and blockchain in smart grids: Fundamentals, methods, and applications*. CRC Press. DOI: 10.1201/9781003500391

Chapter 12
Decentralized Energy Markets Transforming Energy Distribution and Trading With Blockchain Technology

Amjad Yousef Aldweesh
https://orcid.org/0000-0001-9358-1323
College of Computing and IT, Shaqra University, Saudi Arabia

Mohammad Alauthman
https://orcid.org/0000-0003-0319-1968
Petra University, Jordan

Saad Alateef
https://orcid.org/0009-0002-3219-0833
Newcastle University, UK

Ahmed Al-Qerem
https://orcid.org/0000-0003-2187-6194
Computer Science Department, Faculty of Information Technology, Zarqa University, Jordan

ABSTRACT

Decentralized energy markets are transforming energy trading by leveraging blockchain to enhance transparency, efficiency, and resilience. These markets allow producers and consumers to interact in flexible, decentralized systems, disrupting traditional energy supply chains. However, this innovation highlights the critical need for robust cybersecurity in startup-driven digital governance ecosystems. Startups face challenges like limited resources, rapid scaling, and regulatory complexities, leaving them vulnerable to breaches, disruptions, and financial losses. This chapter explores how such threats can undermine decentralized energy platforms and governance frameworks. Addressing these risks requires a multi-layered approach, including industry-specific regulations, collaboration between stakeholders, and technological innovation. Partnerships strengthen defenses, while advanced technologies improve threat detection and response.

DOI: 10.4018/979-8-3373-2439-5.ch012

INTRODUCTION

Decentralized energy markets are reshaping how individuals and businesses produce, distribute, and consume energy. At the heart of this transition lies blockchain technology, a distributed ledger framework designed to record and verify transactions in a transparent, immutable, and secure manner (Sikorski, Haughton, & Kraft, 2017; Zhang et al., 2018). While traditional energy markets depend on centralized authorities for transactions and settlement, decentralized energy markets allow prosumers—entities that both produce and consume energy—to directly trade excess energy with peers. This reconfiguration of market power disrupts the conventional, top-down approach to energy distribution, creating new possibilities for cost reduction, efficiency gains, and improved sustainability (Wang & Su, 2020; Aitzhan & Svetinovic, 2018).

Nevertheless, the emergence of decentralized energy trading platforms is not without complications, particularly in the realm of cybersecurity. The rapid ascent of startups offering blockchain-based energy solutions has introduced novel vulnerabilities, from smart meter tampering to sophisticated hacking attempts aimed at manipulating blockchain-based contracts or data. Startups, often operating with limited technical and financial resources, may inadvertently expose energy systems to malicious actors. In such an environment, cybersecurity becomes paramount. A compromised decentralized energy platform can disrupt not only local energy trades but also broader grid stability, leading to severe economic and societal ramifications (Li et al., 2018).

Digital governance refers to the frameworks, regulations, and practices guiding the use of digital technologies for public and private administration. In the context of decentralized energy markets, digital governance involves ensuring that blockchain-based energy trading platforms comply with standards of security, reliability, and transparency. Effective digital governance must address the inherent risks associated with emerging technologies, balancing innovation with the need for robust security protocols (Park, Lee, & Chang, 2018). Ecosystems, in this regard, extend beyond mere technology infrastructure; they encapsulate the stakeholders, policies, and cultural norms that shape the development and application of these technologies. Startups, policymakers, large energy utilities, and consumers all play pivotal roles in co-creating an environment where technology innovation can flourish without compromising on cybersecurity and system integrity.

Cybersecurity in startup-driven digital governance ecosystems is thus crucial for multiple reasons. First, startups are often the harbingers of innovative solutions that challenge established market norms (Morstyn et al., 2018). Their agility and willingness to experiment can lead to breakthroughs that established corporations might overlook. However, the same agility can result in insufficient security measures if cybersecurity is viewed as secondary to product development and market entry (Sousa, Silvestre, & Vardasca, 2019). Second, as decentralized energy markets become more widespread, the interconnectedness of various systems (e.g., smart grids, IoT devices, and blockchain networks) increases the overall attack surface. A single compromised node or application in the startup ecosystem can cascade into a large-scale disruption. Third, ensuring public trust in decentralized energy trading platforms is critical for broader adoption. High-profile security incidents can erode confidence and hamper the progress of these nascent markets (Munsing, Mather, & Moura, 2017; Zhang & Wen, 2017; Pop et al., 2018).

In examining the importance of cybersecurity within these startup-driven ecosystems, it is essential to go beyond purely technical considerations. While blockchain technology provides intrinsic security benefits, such as immutability and consensus mechanisms, these features do not guarantee complete protection. Human factors, regulatory frameworks, and organizational policies also determine the security posture of decentralized energy markets. Stakeholder collaboration across the public and private sectors is instrumental in designing and implementing policies that hold startups accountable for security while still allowing room for innovation. Additionally, consumer education plays a crucial role in fostering secure practices, from the management of private keys to understanding the risks and benefits of decentralized energy trading.

Research Objectives

The overarching goal of this chapter is to explore how cybersecurity challenges and solutions manifest within startup-driven digital governance ecosystems that support decentralized energy markets. The specific objectives include:

1. **Identify key cybersecurity challenges** that startups face in developing and deploying blockchain-based decentralized energy solutions.
2. **Examine the role of digital governance** and policy frameworks in shaping these markets, highlighting their relevance to cybersecurity.
3. **Present strategies and best practices** for enhancing cybersecurity, including both technological and non-technological measures.
4. **Discuss real-world insights and future directions** by analyzing selected case studies and exploring pathways for sustained security and innovation.

Chapter Overview

To address these objectives, this chapter is organized as follows:

- **Section 1** provides a background on decentralized energy markets and the underpinning blockchain technology, setting the stage for the cybersecurity discourse.
- **Section 2** delves into cybersecurity challenges specific to startup ecosystems, focusing on the unique vulnerabilities and constraints startups face.
- **Section 3** explores how digital governance frameworks intersect with cybersecurity needs in the context of decentralized energy markets.
- **Section 4** outlines strategies for enhancing cybersecurity, discussing practical solutions, partnerships, and emerging frameworks.
- **Section 5** highlights technological and non-technological solutions, ranging from advanced tools like AI to training initiatives and compliance measures.
- **Section 6** features case studies illustrating how startups have navigated cybersecurity hurdles in decentralized energy projects.
- **Section 7** examines future research trajectories and evolving policy considerations.
- **Conclusion** recaps the chapter's key contributions, emphasizing the role of robust cybersecurity in ensuring the viability of decentralized energy markets.

Through this structure, the chapter aims to offer a comprehensive understanding of the challenges and solutions at the intersection of decentralization, blockchain technology, and cybersecurity within startup-driven digital governance ecosystems.

BACKGROUND ON DECENTRALIZED ENERGY MARKETS AND BLOCKCHAIN TECHNOLOGY

Decentralized energy markets represent a fundamental shift in how power generation and distribution are conceptualized. Traditionally, energy systems have been centralized, with large utility companies owning significant infrastructure and maintaining control over production, distribution, and pricing (Khaqqi et al., 2018). In such systems, end-users have limited or no active role beyond consumption. By contrast, decentralized energy markets enable more localized production and direct trading, often utilizing renewable energy sources such as solar and wind. These markets rely on peer-to-peer (P2P) trading mechanisms where producers and consumers negotiate transactions in near real time (Wörner & von Bomhard, 2018; LO3 Energy, 2017).

The Emergence of Decentralized Energy Systems

The rising interest in decentralized energy systems is driven by multiple factors. The rapid decline in the cost of renewable energy technologies has made local generation increasingly attractive . Additionally, concerns about climate change, energy security, and grid resiliency have fueled policy initiatives that encourage distributed generation (Di Silvestre et al., 2018). From a social perspective, decentralized systems empower communities to manage their energy resources and potentially gain financial benefits from surplus energy sales (Power Ledger, 2017; Aziz et al., 2019; Kuzlu et al., 2014; Liang et al., 2018).

In parallel, the evolution of information and communication technologies (ICT) has facilitated more sophisticated energy management systems. Smart meters, for instance, allow granular data collection on consumption and production, enabling more dynamic pricing and load balancing . This connectivity also serves as the foundation for new market mechanisms, wherein individual entities can trade energy on digital platforms. However, the expansion of such interconnected systems has inevitably broadened the cybersecurity threat landscape (Gai et al., 2019; Zhou et al., 2019; Kang et al., 2017).

The Role of Blockchain

Blockchain technology has emerged as a key enabler for decentralized energy markets . Initially popularized by cryptocurrencies, blockchain operates as a distributed ledger maintained by a network of nodes. These nodes validate and record transactions using a consensus mechanism, ensuring data integrity and making it nearly impossible for any single entity to unilaterally alter the ledger (Tapscott & Tapscott, 2016).

In the context of energy trading, blockchain's tamper-proof nature provides a reliable record of transactions between prosumers. Smart contracts, self-executing scripts that run on the blockchain, automate the execution of energy trades based on predefined conditions, eliminating the need for intermediaries . Additionally, blockchain can facilitate micro-payments for small quantities of energy, which might be economically unviable in traditional systems due to transaction costs (Kshetri, 2017).

Despite these advantages, blockchain-based platforms also have potential vulnerabilities. If nodes in the network are compromised, attackers can disrupt the consensus mechanism or manipulate smart contracts. Furthermore, blockchain's transparency, while beneficial for auditability, can expose sensitive transaction data if not properly secured . Hence, as startups build these platforms, they must incorporate robust cybersecurity measures to protect both the trading platform and the underlying grid infrastructure (Rathor & Saxena, 2020).

Figure 1 illustrates a simplified blockchain-enabled energy trading model. Producers and consumers transact directly with each other, while the blockchain network records every transaction in a secure, distributed manner. Although the architecture appears straightforward, complexities arise in practice, such as grid balancing, regulatory compliance, and the need for robust cybersecurity protocols.

Figure 1. Basic blockchain-enabled energy trading model

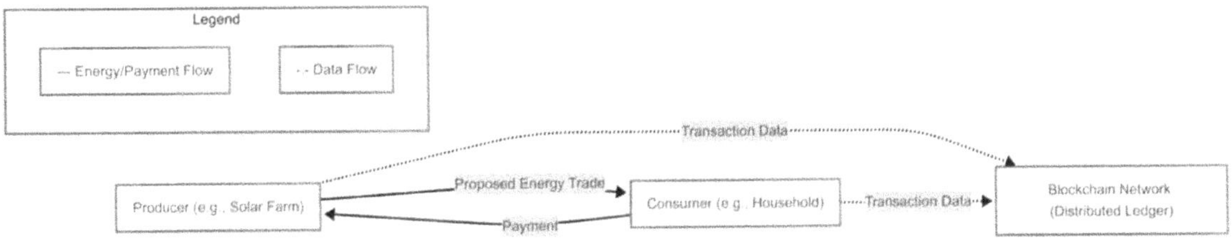

CYBERSECURITY CHALLENGES IN STARTUP ECOSYSTEMS

The push toward decentralized energy markets, often spearheaded by startups, raises unique cybersecurity concerns . Unlike established corporations with sizable budgets and specialized security teams, startups typically operate under resource constraints, making them more vulnerable to cyberattacks. Moreover, their rapid development cycles and the pressure to achieve market traction can lead to security being deprioritized.

Resource Constraints and Rapid Scaling

Startups face the dual challenge of securing their platforms while also needing to scale quickly. Engineering teams may focus on feature delivery and user acquisition, leaving security as an afterthought. This approach creates gaps that attackers can exploit . Furthermore, hiring experienced cybersecurity professionals can be financially prohibitive, pushing many startups to rely on outsourcing or generic security solutions that may not be fully adapted to their specific contexts.

Lack of Expertise in Blockchain Security

Blockchain technology, while increasingly popular, remains a specialized field. Many developers transitioning from traditional application development to blockchain platforms grapple with concepts like consensus mechanisms, smart contract vulnerabilities, and cryptographic key management . This knowledge gap can introduce architectural flaws and coding errors that weaken system security. In the energy sector, where blockchain solutions must interface with physical infrastructure, risks multiply, as compromise can lead to physical damages or large-scale energy disruptions.

Regulatory and Compliance Complexities

Decentralized energy markets do not neatly fit into conventional regulatory frameworks . Startups often confront a patchwork of energy regulations, data privacy laws, and financial compliance requirements. Navigating these regulations is complicated by varying standards across jurisdictions. The complexity multiplies when technology components cross international borders, as data sovereignty regulations can limit the manner in which data is stored or transferred . Startups that fail to address regulatory and compliance demands expose themselves to legal liabilities and reputational risks, which in turn can magnify cybersecurity vulnerabilities.

Insider Threats and Social Engineering

Beyond external cyberattacks, startups are also susceptible to insider threats and social engineering . As teams grow rapidly, employees, contractors, or partners with varying levels of access to critical systems may unintentionally or maliciously compromise security. Phishing attacks aimed at staff who lack sufficient cybersecurity training can lead to unauthorized access, revealing private keys or other sensitive data used in blockchain-based energy transactions .

Case Studies of Startup Vulnerabilities

Table 1. Selected case studies of startup cybersecurity challenges

Startup/Platform	Incident Type	Key Lessons Learned
EnergyXchange	Smart Contract Exploit	Inadequate code audits can lead to vulnerabilities in the contract layer, resulting in financial losses for participants.
GreenPowerNet	Phishing and Social Engineering	Lack of employee training and two-factor authentication allowed attackers to compromise admin credentials.
EcoGrid	Insider Access	An ex-employee with retained system privileges manipulated energy credit balances, highlighting gaps in offboarding procedures.

In Table 1, we see real-world examples of how vulnerabilities in blockchain-based energy startups can manifest. These incidents underscore the importance of thorough code reviews, employee training, and robust administrative protocols. For instance, EnergyXchange experienced a high-profile exploit of its smart contracts due to insufficient code audits . As a result, users lost trust in the platform. Similarly, GreenPowerNet suffered from a targeted phishing attack, resulting in compromised administrative credentials . These examples reveal how attackers often exploit the weakest link in the security chain—whether that is code integrity or human factors.

Figure 2 provides a visual summary of the key cybersecurity challenges facing startups in decentralized energy markets. Each challenge is interlinked, illustrating how rapid scaling under resource constraints can lead to inadequate training or system checks, which in turn heightens the risk posed by social engineering and insider threats.

Figure 2. Key cybersecurity challenges in startup ecosystems

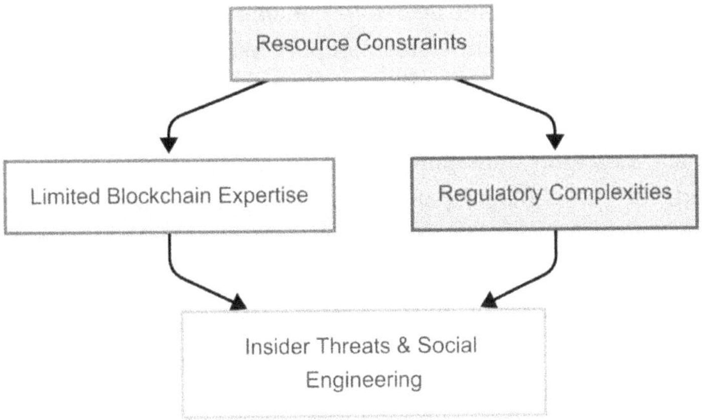

DIGITAL GOVERNANCE AND ITS SECURITY IMPLICATIONS

Digital governance frameworks are crucial for orchestrating how new technologies, including blockchain-based decentralized energy solutions, are integrated into society. They encompass regulations, policies, standards, and oversight mechanisms . For decentralized energy markets, effective governance ensures that the economic and environmental benefits of blockchain technology can be realized without compromising security or consumer rights.

Policy Frameworks and Blockchain Adoption

Government agencies and international bodies have begun to recognize the transformative potential of blockchain in energy markets . Some jurisdictions provide regulatory sandboxes to encourage experimentation, while others enact legislation that addresses data privacy, consumer protection, and transactional transparency. However, mismatches between technological development speed and regulatory processes can create gaps that malicious actors exploit. For instance, in some regions, the legal status of blockchain-based energy transactions remains ambiguous, complicating efforts to prosecute cybercriminals or set uniform security standards.

Standardization Efforts

Standards organizations like the Institute of Electrical and Electronics Engineers (IEEE) and the International Organization for Standardization (ISO) are developing frameworks specific to blockchain and decentralized systems . These include guidelines for secure coding, key management, and consensus protocols. The existence of industry-wide standards can streamline cybersecurity measures, but alignment among startups remains challenging. Compliance can be resource-intensive, and the rapid evolution of blockchain technology risks rendering standards outdated .

Accountability and Transparency

Decentralized energy markets distribute decision-making and transaction validation among multiple nodes, theoretically reducing the risk of unilateral manipulation . However, diffusion of responsibility can complicate accountability. Who is liable if a smart contract malfunctions, leading to widespread disruptions or financial losses? Policymakers are grappling with how to assign liability in decentralized systems . Transparent governance structures, including well-defined roles and responsibilities, are essential to ensure that startups and other stakeholders maintain secure practices throughout the platform's lifecycle.

Figure 3 outlines the key stakeholder interactions that shape digital governance in decentralized energy markets. Governments set the regulatory tone, while standards bodies provide technical guidance. Startups operationalize these frameworks, and consumers ultimately validate the system's efficacy and trustworthiness. Security lapses at any point in this chain can undermine the entire ecosystem .

Figure 3. Stakeholder interactions in digital governance of decentralized energy

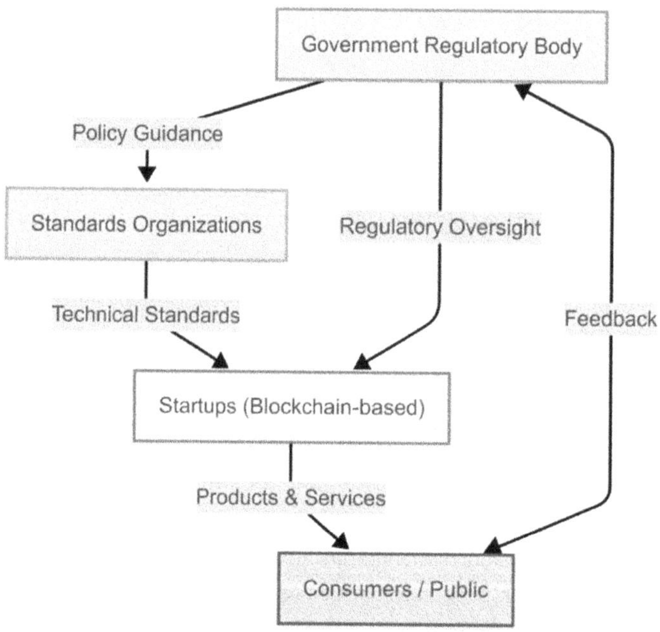

Role of Policy-Making in Mitigating Risks

Policy-making can mitigate cybersecurity risks through multiple channels. One approach is mandating security-by-design principles, ensuring that startups incorporate security features from the earliest stages of development . Another approach involves public funding for cybersecurity research and training programs, addressing the shortage of skilled professionals. Policymakers can also incentivize threat intelligence sharing among startups, utilities, and government agencies, creating a collaborative environment conducive to rapid threat detection and mitigation .

Table 2. Policy instruments for enhancing cybersecurity in decentralized energy markets

Policy Instrument	Level	Expected Outcome
Security-by-Design Mandates	National	Encourages startups to integrate security features early in the software development lifecycle.
Regulatory Sandboxes	Regional	Allows controlled experimentation under supervised environments, reducing risks of large-scale failures.
Grant Funding for Cyber R&D	National/International	Addresses the talent gap and fosters development of cutting-edge security solutions specific to decentralized systems.
Threat Intelligence Sharing	International	Facilitates real-time exchange of threat data among stakeholders, accelerating incident response.

Table 2 summarizes several policy instruments used to enhance cybersecurity in decentralized energy markets. These instruments span national to international levels, reflecting the global reach of energy and blockchain systems. The effectiveness of these measures relies on the willingness of startups and other stakeholders to adhere to, and meaningfully engage with, regulatory and technical guidelines.

STRATEGIES FOR ENHANCING CYBERSECURITY

Given the multifaceted threats described, cybersecurity strategies in startup-driven ecosystems must be both comprehensive and adaptable. Ranging from rigorous technological implementations to broader governance and collaborative efforts, these strategies aim to secure the entire lifecycle of blockchain-based energy platforms.

Security Architecture and Layered Defenses

A robust security architecture entails deploying multiple layers of defenses, including network segmentation, intrusion detection systems, and zero-trust models . Startups in decentralized energy markets benefit from layered security because it helps isolate critical components (e.g., smart contract modules, user data repositories) from less secure parts of the system . For instance, segmenting the blockchain infrastructure from public-facing interfaces reduces the potential attack surface, forcing attackers to navigate multiple security checkpoints .

Secure Coding and Smart Contract Audits

Smart contracts are the backbone of blockchain-based energy trading, automating transactions under predefined conditions. However, coding vulnerabilities can lead to exploits such as reentrancy attacks and overflow errors . Conducting regular audits of smart contracts is therefore essential. Third-party audit firms or open-source communities can scrutinize the contract code, identifying weaknesses and recommending fixes . Automated tools like Slither and Mythril further streamline the process of detecting common vulnerabilities .

Identity and Access Management (IAM)

In a decentralized energy ecosystem, multiple stakeholders, including energy producers, consumers, utilities, and regulators, require different levels of system access . Implementing robust IAM ensures that only authorized parties can modify smart contracts, read sensitive data, or control specific network nodes. Startups should adopt multi-factor authentication (MFA) and periodic key rotation to mitigate risks associated with stolen credentials .

Incident Response and Business Continuity Planning

Cyber incidents can and do happen, so having a comprehensive incident response plan is paramount . Startups should define incident response teams, roles, and escalation pathways. Regular drills and simulations help prepare staff to handle real-world crises. Business continuity planning is equally important, ensuring that critical functions, such as energy trading and settlement processes, can be restored quickly to mitigate financial and social impacts .

Figure 4 presents a lifecycle approach to cybersecurity. By treating security as an ongoing process—encompassing prevention, detection, response, and continuity—startups can remain agile in addressing emerging threats. The cyclical nature of this approach underscores the reality that cybersecurity is never a one-time endeavor but a continual progression of improvement and adaptation .

Figure 4. Lifecycle approach to cybersecurity in decentralized energy startups

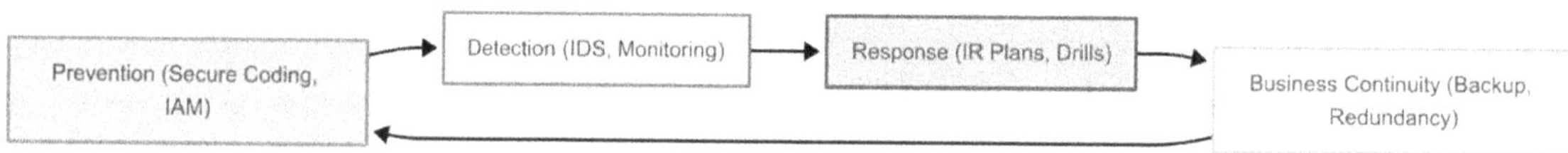

Public–Private Partnerships and Collaborative Models

Startup-driven ecosystems benefit significantly from public–private partnerships (PPPs). Government agencies often possess threat intelligence that may not be readily available to private entities, while startups bring innovation and market responsiveness . Collaboration between these sectors can yield shared cybersecurity infrastructure, standardized audit frameworks, and more effective regulation enforcement. In many regions, energy regulators also serve as liaisons to the public, helping to build consumer trust in new technologies .

Table 3. Collaborative strategies for cybersecurity enhancement

Strategy	Stakeholders Involved	Key Benefits
Joint Threat Intelligence Platforms	Startups, Governments, Utilities	Accelerated detection of emerging threats; pooled resources for analysis.
Co-Investment in Security R&D	Public Agencies, Private Companies	Financial risk-sharing for advanced security technologies. Spurs innovation in niche areas like secure smart metering.
Regulatory Guidance Workshops	Regulatory Bodies, Startups	Clarifies compliance requirements, fosters open dialogue, and reduces misunderstandings.

As summarized in Table 3, collaborative efforts can take multiple forms, including threat intelligence sharing and co-investment in security research. Such synergies can lead to more resilient decentralized energy markets, as potential risks are tackled collectively rather than in isolation .

TECHNOLOGICAL AND NON-TECHNOLOGICAL SOLUTIONS

Effective cybersecurity strategies in decentralized energy markets require a blend of advanced technological tools and non-technological measures. While blockchain technology itself offers certain security advantages, complementary systems and practices are necessary to secure the entire ecosystem.

Advanced Tools: AI, Blockchain, and Intrusion Detection

Artificial Intelligence (AI) can augment cybersecurity by automating threat detection and response . Machine learning models can analyze large volumes of network traffic data to identify anomalies, such as unusual login patterns or data flows. By integrating AI into intrusion detection systems (IDS), startups can proactively mitigate threats, reducing the time between detecting an intrusion and resolving it .

Blockchain technology remains a critical element of decentralized energy markets, but the concept of multi-layer or hybrid blockchains is gaining traction for additional security benefits . In such architectures, a private blockchain layer can handle sensitive transactions or user data, while a public layer manages broader market transactions. This approach can address privacy and scalability concerns .

Non-Technological Measures: Employee Training and Compliance

Even the most sophisticated technology stack is susceptible to human error . Therefore, continuous employee training on cybersecurity best practices, such as recognizing phishing attempts and handling cryptographic keys, is imperative. Clear policies on password management, endpoint security, and device usage reduce the likelihood of inadvertent breaches .

Compliance with data protection laws (e.g., the General Data Protection Regulation) and industry-specific standards provides an additional layer of governance . Regular compliance checks and audits reinforce best practices and offer a safety net against legal complications. These audits can also reveal oversights in technology deployments, ensuring that recommended security controls are effectively implemented .

Table 4. Integration of technological and non-technological solutions

Solution Type	Example Tools/Methods	Primary Impact
Technological	Intrusion Detection Systems, AI-based Analysis	Rapid threat detection, minimal human intervention for routine tasks.
Non-Technological	Cybersecurity Training, Policy Enforcement	Mitigation of human error, legal compliance, establishment of security culture.

As shown in Table 4, a holistic approach combines both technological and non-technological measures. Startups stand to benefit from adopting this approach early, building a cybersecurity-conscious culture alongside robust technical defenses .

Figure 5 illustrates how technological and non-technological measures feed into an integrated cybersecurity framework. This synergy is particularly relevant for decentralized energy startups, where the complexity of the underlying technology stack calls for a multifaceted security strategy .

Figure 5. Integration of technological and non-technological measure

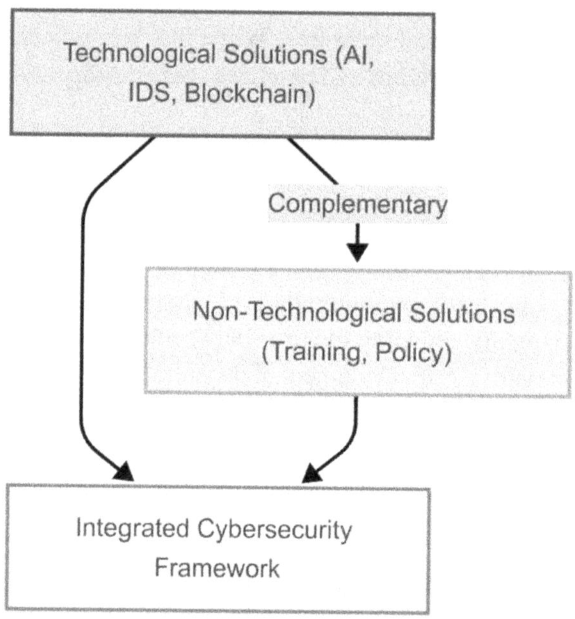

CASE STUDIES: STARTUP APPROACHES TO CYBERSECURITY IN DECENTRALIZED ENERGY

To understand the practical applications of cybersecurity strategies discussed so far, this section examines illustrative case studies of startups actively engaged in decentralized energy markets. These examples highlight common pitfalls, successful interventions, and emerging best practices.

Case Study 1: SolarTrade

Background. SolarTrade is a startup offering a blockchain-based platform that enables prosumers to sell excess solar energy to neighbors. Operating primarily in an urban area with high rooftop solar adoption, the company gained traction for its user-friendly mobile application .

Cybersecurity Challenge. Early in its launch, SolarTrade experienced multiple phishing attempts targeting its customer support portal. Attackers sought to gain admin-level access, potentially to alter energy credit balances. Investigations revealed that staff were inadequately trained in recognizing suspicious emails .

Interventions. SolarTrade implemented mandatory cybersecurity training for all employees, introducing guidelines for email scanning and multi-factor authentication on administrative accounts . Additionally, the startup collaborated with a third-party security firm for penetration testing and improved its incident response plan.

Outcomes. No successful breaches occurred after these measures were introduced. SolarTrade also received positive feedback from local energy regulators for adopting a proactive security stance, aiding the platform's reputation and user trust .

Case Study 2: WindChain

Background. WindChain is a newcomer focusing on peer-to-peer wind energy trading. The platform leverages a hybrid blockchain model, allowing private transactions for large wind farm operators and a public ledger for smaller, community-driven projects .

Cybersecurity Challenge. WindChain faced a smart contract exploit that allowed bad actors to double-spend energy credits. The vulnerability originated in an integer overflow bug within a critical payment function .

Interventions. WindChain promptly froze suspicious transactions and performed a full audit of its smart contracts. Collaboration with blockchain experts helped identify similar latent bugs, leading the company to establish a dedicated audit workflow. WindChain also utilized bug bounty programs, incentivizing external researchers to disclose vulnerabilities responsibly .

Outcomes. By taking swift action, WindChain minimized financial losses and restored platform integrity. The hybrid blockchain model was further refined to segregate high-risk functions into a more secure, private chain. This approach significantly reduced the potential attack surface .

Case Study 3: HydroConnect

Background. HydroConnect aims to connect hydropower micro-grids with local communities through a decentralized marketplace, focusing on rural regions with limited grid infrastructure .

Cybersecurity Challenge. The startup's main issue centered on regulatory compliance. Operating across multiple jurisdictions, HydroConnect was obligated to adhere to diverse data protection and energy trading regulations. Compliance costs were significant, straining the company's limited resources .

Interventions. HydroConnect established a legal advisory board composed of energy regulators, privacy experts, and local community representatives. This board helped streamline processes to meet regulatory requirements without compromising the platform's accessibility. Simultaneously, the company introduced a role-based access control system to ensure that data handling aligned with each jurisdiction's guidelines .

Outcomes. The advisory board's insights enabled HydroConnect to maintain compliance and avoid legal penalties. Their robust governance structure also reassured investors, who contributed additional funding for security enhancements and market expansion .

Table 5. Comparative summary of case studies

Startup	Key Challenge	Main Solution	Results
SolarTrade	Phishing Attacks	Mandatory Staff Training	Zero breaches after training, improved reputation.
WindChain	Smart Contract Exploit	Comprehensive Code Audits & Bug Bounties	Minimized losses, enhanced contract security, refined hybrid blockchain.
HydroConnect	Regulatory Compliance	Legal Advisory Board & Role-based Access Control	Maintained compliance, secured investor confidence.

Table 5 offers a high-level comparison of the three case studies, illustrating varied approaches to cybersecurity challenges. While each startup confronted different issues, a common thread emerges: proactive, well-structured strategies that combine technology solutions with policy or training measures were most effective in mitigating risks.

These case studies demonstrate that successful cybersecurity strategies in decentralized energy startups are multifaceted, context-specific, and adaptive. Whether the challenge is smart contract integrity, staff awareness, or compliance, addressing it early and comprehensively fosters trust, investment, and long-term sustainability in the emerging decentralized energy market sector.

FUTURE DIRECTIONS

Decentralized energy markets are still in a nascent stage, with many startups experimenting with business models, regulatory approaches, and technological solutions. As the ecosystem matures, new trends will likely reshape cybersecurity practices.

Convergence with Other Emerging Technologies

The intersection of decentralized energy markets and technologies like the Internet of Things (IoT), edge computing, and 5G networks will introduce new security dimensions . IoT devices such as smart meters and sensors, already integral to decentralized systems, will become more sophisticated, potentially increasing vulnerabilities if not properly secured.

Regulatory Evolution

Energy and data privacy regulations will continue to evolve to address the complexities of blockchain-based platforms . Future frameworks may require real-time monitoring or reporting of transactions to regulatory bodies, further emphasizing the need for robust security. The global push for sustainability could also lead to incentives that reward secure, environmentally friendly energy solutions.

Cyber Insurance and Risk Management Services

As financial stakes grow, cyber insurance specifically tailored to decentralized energy platforms may gain traction . Startups could benefit from specialized policies that cover liabilities related to smart contract exploits, data breaches, or system downtimes. Complementing insurance, risk management consultancies may emerge, focusing on unique vulnerabilities in blockchain-enabled energy ecosystems.

Continued Research and Collaboration

Academic and industry researchers will play a pivotal role in advancing blockchain security protocols, especially as they relate to large-scale energy trading . Collaborative research initiatives, possibly funded by government grants or international organizations, can accelerate the adoption of best practices.

Overall, the future landscape of decentralized energy markets will be shaped by technological innovation, regulatory adaptation, and collaborative problem-solving. Startups will remain key drivers of this transformation, necessitating a proactive, well-rounded approach to cybersecurity.

CONCLUSION

Decentralized energy markets, enabled by blockchain technology, are set to reshape the ways in which energy is produced, distributed, and traded. By granting consumers greater agency and promoting sustainable practices, these markets hold the potential to disrupt traditional energy structures. However, the innovations that make decentralized energy trading so appealing also introduce substantial cybersecurity challenges.

Startups at the forefront of this revolution operate under intense pressure to scale quickly and outmaneuver competitors. This environment often leads to a deprioritization of security measures. Yet, as demonstrated throughout this chapter, robust cybersecurity protocols are foundational to sustaining trust, ensuring regulatory compliance, and preserving the integrity of decentralized energy trading networks. From resource constraints to complex compliance landscapes, startups grapple with a wide spectrum of threats ranging from external hacking attempts to insider vulnerabilities.

Key insights from the chapter underscore the need for multilayered defense strategies, encompassing both technological tools and non-technological solutions like employee training and regulatory adherence. Partnerships between public agencies, standards organizations, and private entities can enhance threat intelligence sharing and foster a unified approach to security. Moreover, emerging trends suggest that convergence with IoT, the evolution of energy regulations, and the rise of cyber insurance will further shape the cybersecurity priorities in decentralized energy markets.

In summary, while decentralized energy markets present promising opportunities for efficiency and sustainability, they also necessitate a stringent focus on cybersecurity. By integrating secure coding practices, layered defenses, policy compliance, and stakeholder collaboration, startups can better safeguard their platforms. Future research could expand on these strategies, exploring deeper technical solutions, broader international regulations, and evolving business models. Above all, the digital governance of decentralized energy must continue to adapt, ensuring that innovation proceeds hand in hand with secure, reliable market operations.

REFERENCES

Aitzhan, N. Z., & Svetinovic, D. (2018). Security and privacy in decentralized energy trading through multi-signatures, blockchain and anonymous messaging streams. *IEEE Transactions on Dependable and Secure Computing*, *15*(5), 840–852. DOI: 10.1109/TDSC.2016.2616861

Aziz, A. F. A., Khalid, S. N., Mustafa, M. W., & Aliyu, N. (2019). Blockchain technology for renewable energy: Principles, applications and prospects. *IOP Conference Series. Earth and Environmental Science*, *268*, 012113.

Di Silvestre, M. L., Gallo, P., Guerrero, J. M., Musca, R., Sanseverino, E. R., Sciumè, G., & Zizzo, G. (2018). Blockchain for power systems: Current trends and future applications. *Renewable & Sustainable Energy Reviews*, *113*, 109268.

Gai, K., Wu, Y., Zhu, L., Qiu, M., & Choo, K. K. R. (2019). Privacy-preserving energy trading using consortium blockchain in smart grid. *IEEE Transactions on Industrial Informatics*, *15*(6), 3548–3558. DOI: 10.1109/TII.2019.2893433

Kang, J., Yu, R., Huang, X., Maharjan, S., Zhang, Y., & Hossain, E. (2017). Enabling localized peer-to-peer electricity trading among plug-in hybrid electric vehicles using consortium blockchains. *IEEE Transactions on Industrial Informatics*, *13*(6), 3154–3164. DOI: 10.1109/TII.2017.2709784

Khaqqi, K. N., Sikorski, J. J., Hadinoto, K., & Kraft, M. (2018). Incorporating seller/buyer reputation-based system in blockchain-enabled emission trading application. *Applied Energy*, *209*, 8–19. DOI: 10.1016/j.apenergy.2017.10.070

Kshetri, N. (2017). Blockchain's roles in strengthening cybersecurity and protecting privacy. *Telecommunications Policy*, *41*(10), 1027–1038. DOI: 10.1016/j.telpol.2017.09.003

Kuzlu, M., Pipattanasomporn, M., & Rahman, S. (2014). Communication network requirements for major smart grid applications in HAN, NAN and WAN. *Computer Networks*, *67*, 74–88. DOI: 10.1016/j.comnet.2014.03.029

LO3 Energy. (2017). *Brooklyn Microgrid Project*. Retrieved from https://lo3energy.com/

Li, Z., Kang, J., Yu, R., Ye, D., Deng, Q., & Zhang, Y. (2018). Consortium blockchain for secure energy trading in industrial Internet of Things. *IEEE Transactions on Industrial Informatics*, *14*(8), 36903700.

Liang, G., Weller, S. R., Luo, F., Zhao, J., & Dong, Z. Y. (2018). Distributed blockchain-based data protection framework for modern power systems against cyber attacks. *IEEE Transactions on Smart Grid*, *10*(3), 3162–3173. DOI: 10.1109/TSG.2018.2819663

Morstyn, T., Farrell, N., Darby, S. J., & McCulloch, M. D. (2018). Using peer-to-peer energy-trading platforms to incentivize prosumers to form federated power plants. *Nature Energy*, *3*(2), 94–101. DOI: 10.1038/s41560-017-0075-y

Munsing, E., Mather, J., & Moura, S. (2017). Blockchain for a transactive energy system: An experimental case study. *Proceedings of the 1st IEEE International Conference on Energy Internet (ICEI)*, 85-90.

Park, L. W., Lee, S., & Chang, H. (2018). A sustainable home energy prosumer-chain methodology with energy tags over the blockchain. *Sustainability (Basel), 10*(3), 658. DOI: 10.3390/su10030658

Pop, C., Cioara, T., Antal, M., Anghel, I., Salomie, I., & Bertoncini, M. (2018). Blockchain based decentralized management of demand response programs in smart energy grids. *Sensors (Basel), 18*(1), 162. DOI: 10.3390/s18010162 PMID: 29315250

Power Ledger. (2017). *White paper: An introduction to the Power Ledger platform.*

Rathor, S. K., & Saxena, D. (2020). Energy management system for smart grid: An overview and key issues. *International Journal of Energy Research, 44*(6), 4067–4109. DOI: 10.1002/er.4883

Sikorski, J. J., Haughton, J., & Kraft, M. (2017). Blockchain technology in the chemical industry: Machine-to-machine electricity market. *Applied Energy, 195*, 234–246. DOI: 10.1016/j.apenergy.2017.03.039

Sousa, S., Silvestre, J., & Vardasca, R. (2019). Blockchain-based peer-to-peer energy trading and billing platform for microgrids. *International Journal of Electrical Power & Energy Systems, 112*, 232–241.

Tapscott, D., & Tapscott, A. (2016). *Blockchain revolution: How the technology behind Bitcoin is changing money, business, and the world.* Penguin.

Wang, Q., & Su, Z. (2020). Integrating blockchain technology into the energy sector—From theory of blockchain to research and application of energy blockchain. *Computer Science Review, 37*, 100275. DOI: 10.1016/j.cosrev.2020.100275

Wörner, D., & von Bomhard, T. (2018). When your sensor earns money: Exchanging data for cash with Bitcoin and blockchain technologies. *IEEE Sensors Applications Symposium (SAS)*, 1-6.

Zhang, C., Wu, J., Zhou, Y., Cheng, M., & Long, C. (2018). Peer-to-peer energy trading in a microgrid. *Applied Energy, 220*, 1–12. DOI: 10.1016/j.apenergy.2018.03.010

Zhang, Y., & Wen, J. (2017). The IoT electric business model: Using blockchain technology for the internet of things. *Peer-to-Peer Networking and Applications, 10*(4), 983–994. DOI: 10.1007/s12083-016-0456-1

Zhou, Z., Wang, B., Guo, Y., & Zhang, Y. (2019). Blockchain and computational intelligence inspired incentive-compatible demand response in internet of electric vehicles. *IEEE Transactions on Emerging Topics in Computational Intelligence, 3*(3), 205–216. DOI: 10.1109/TETCI.2018.2880693

ADDITIONAL READING

Andoni, M., Robu, V., Flynn, D., Abram, S., Geach, D., Jenkins, D., McCallum, P., & Peacock, A. (2019). Blockchain technology in the energy sector: A systematic review of challenges and opportunities. *Renewable & Sustainable Energy Reviews, 100*, 143–174. DOI: 10.1016/j.rser.2018.10.014

Burger, C., Kuhlmann, A., Richard, P., & Weinmann, J. (2016). *Blockchain in the energy transition: A survey among decision-makers in Germany*. ESMT Berlin.

Dorri, A., Kanhere, S. S., & Jurdak, R. (2017). Towards an optimized blockchain for IoT. *IEEE/ACM Second International Conference on Internet-of-Things Design and Implementation (IoTDI)*, 173-178.

Hassan, M. U., Rehmani, M. H., & Chen, J. (2019). Privacy preservation in blockchain-based IoT systems: Integration issues, prospects, and challenges. *Future Generation Computer Systems*, *97*, 512–529. DOI: 10.1016/j.future.2019.02.060

Mengelkamp, E., Gärttner, J., Rock, K., Kessler, S., Orsini, L., & Weinhardt, C. (2018). Designing microgrid energy markets: A case study: The Brooklyn Microgrid. *Applied Energy*, *210*, 870–880. DOI: 10.1016/j.apenergy.2017.06.054

Nakamoto, S. (2008). *Bitcoin: A peer-to-peer electronic cash system*. Bitcoin Whitepaper.

Noor, S., Yang, W., Guo, M., van Dam, K. H., & Wang, X. (2018). Energy demand side management within micro-grid networks enhanced by blockchain. *Applied Energy*, *228*, 1385–1398. DOI: 10.1016/j.apenergy.2018.07.012

Tushar, W., Yuen, C., Mohsenian-Rad, H., Saha, T., Poor, H. V., & Wood, K. L. (2020). Transforming energy systems via peer-to-peer energy trading: Potential of game-theoretic approaches. *IEEE Signal Processing Magazine*, *37*(4), 90–111. DOI: 10.1109/MSP.2018.2818327

Chapter 13
Smart Building Energy Management:
Blockchain Solutions for Urban Efficiency

Ali Al Maqousi
https://orcid.org/0000-0002-3242-804X
Department of Information Security, University of Petra, Jordan

Kashinath Basu
https://orcid.org/0000-0002-0831-3752
Department of Computing and Communication Technologies, Oxford Brookes University, UK

Amjad Aldweesh
https://orcid.org/0000-0001-9358-1323
College of Computing and Information Technology, Shaqra University, Saudi Arabia

ABSTRACT

Abstract This chapter examines how blockchain technology can advance energy management in smart buildings and urban areas. It analyzes the theoretical foundations of blockchain solutions and their integration into energy monitoring, control, and optimization strategies. It describes how blockchain can manage peer-to-peer transactions, coordinate distributed energy resources, and facilitate trust-minimized data sharing for building operators, occupants, and regulators. It then explores digital twin models, cybersecurity considerations, interoperability approaches, and energy market mechanisms that rely on immutable ledgers and smart contracts. The chapter reviews challenges and outlines directions for future research and deployment. The discussion underscores that blockchain presents potential for transparent and automated energy management. Realizing these benefits depends on supportive policies, reliable technology, cross-sector engagement, and continued system refinement.

1. INTRODUCTION

Smart buildings use sensing, control, and data analysis to enhance operational performance and occupant welfare. They regulate heating, ventilation, air conditioning, lighting, security, and other essential services. This optimization requires continuous data input from interconnected devices, integrated an-

DOI: 10.4018/979-8-3373-2439-5.ch013

alytics, and real-time decision-making (Sun, 2024). In many cases, these systems must coordinate with external stakeholders such as utilities, regulatory agencies, and energy suppliers. Blockchain technology has emerged as a technique to manage data exchange in a secure, decentralized manner. Blockchain offers a way to track and verify each transaction, including electrical energy transfers, occupant billing, equipment maintenance logs, and system alarms (Khatoon et al., 2019).

Blockchain is a distributed ledger that stores data in blocks linked with cryptographic hashes. Users append transactions to the ledger after network nodes verify them via consensus algorithms (Andoni et al., 2019). The irreversibility of recorded information establishes traceability and transparency. These properties can help resolve trust issues related to verifying energy data, automating settlements, and securing contracts between unknown or partially trusted parties. Smart contracts allow coded instructions to be stored on-chain and triggered by data inputs, so they execute automatically when certain conditions are met. These features are relevant for building energy management, where distributed energy systems, occupant behaviors, and operational demands shape the flow of resources.

In existing building management solutions, centralized servers process sensor and meter data. Third parties handle billing, authentication, and analytics. When these solutions scale to large building clusters and microgrids, they may become slow or vulnerable to single points of failure (Zhao et al., 2022). Blockchain-based architectures avoid centralized gatekeepers and can provide near real-time transaction validation. Blockchain nodes can be distributed across buildings, city authorities, and local energy companies. By distributing validation authority, the entire system acquires resilience if one node fails or becomes compromised.

Blockchain also broadens possibilities for peer-to-peer energy trading among building prosumers. Prosumers are users who both consume and generate electricity, often through rooftop solar photovoltaics or other on-site renewable technologies (Fu et al., 2023). In conventional grid structures, prosumers who produce excess electricity sell it back to the utility. Peer-to-peer trading enables direct transactions with neighboring buildings or households, often at flexible rates determined by supply-demand conditions. Blockchain ledgers record each trade, guaranteeing that counterparties have accurate records of volumes and prices. Such local trading supports distributed generation and may reduce system congestion (Zhang et al., 2018).

This chapter reviews the theoretical frameworks behind blockchain adoption in building energy management. It shows how system configurations can integrate with existing or evolving energy infrastructure. It also discusses the emerging role of digital twin modeling, which creates real-time, data-driven representations of physical systems. Integrating blockchain with digital twins can improve reliability in simulated energy flows. The chapter addresses cybersecurity, as the large network of IoT devices in buildings needs robust solutions for data integrity and privacy. Finally, it presents potential regulatory and economic implications, acknowledging that blockchain-based systems must operate within local legal structures and occupant acceptance levels (Savur & Emekci, 2024). The discussion underscores that blockchain alone is not a panacea but can serve as an enabling technology for transparent and automated energy management.

The chapter has several sections. Section 2 defines key aspects of smart buildings, exploring how they collect and process operational data. Section 3 presents the fundamentals of blockchain technology in the context of decentralized transactions. Section 4 addresses theoretical frameworks that justify blockchain's role in building energy management. Section 5 explains how blockchain applications can facilitate urban energy efficiency, including peer-to-peer trading and interoperability with broader grids. Section 6 discusses security and privacy considerations. Section 7 shows the potential of digital twin integration with blockchain. Section 8 examines regulatory and policy dimensions. Section 9 explores economic and social impacts. Section 10 outlines best practices and strategies for implementation. Section 11 identifies challenges and limitations. Section 12 provides future outlook. Section 13 summarizes the findings. The references appear at the end, drawn only from the list supplied in the verified reference set.

2. SMART BUILDING ENERGY MANAGEMENT: FOUNDATIONS AND KEY PRINCIPLES

Smart building energy management involves the coordinated operation of devices, subsystems, and building occupants. Traditional building management systems (BMS) focused on improving energy efficiency by automating lighting and temperature controls. Modern approaches go further by integrating occupant schedules, external weather data, and microgrid interactions. These systems adopt sensing and communication technologies to gather data on occupancy, temperature, humidity, and device performance (Sun, 2024). They then rely on analytics to optimize energy demand based on occupant comfort or cost goals.

Building sensors transmit data to a central management platform, which often uses machine learning algorithms to detect anomalies, schedule maintenance, and adjust setpoints. These management platforms also have user interfaces that facility managers or occupants can use to monitor energy consumption. In large complexes, the BMS must handle significant data flows, coordinate multiple floors or zones, and integrate energy generation or storage devices. The reliance on a single management server can introduce vulnerabilities. If that server fails, building controls may degrade or stop. Cyberattacks can also target that central node, risking occupant safety or property damage (Meng & Zhu, 2024).

Integration with distributed energy resources adds complexity. Rooftop solar systems or shared battery storage facilities need to communicate with the BMS. The system must decide when to use stored energy, when to draw from the grid, and when to export excess power. Real-time decisions can reduce electricity costs by capitalizing on time-of-use tariffs or demand response programs (Andoni et al., 2019). Demand response involves adjusting building loads during peak grid demand to receive financial compensation. Large commercial buildings can save on electricity bills by turning off non-essential devices for short periods, but such coordination requires accurate metering, verification, and swift settlement.

The physical layout of buildings poses additional challenges. Different floors or sections may have distinct temperature requirements or occupant densities. Equipment such as heating, ventilation, air conditioning (HVAC) systems, lighting arrays, and security cameras can each produce separate data streams. Harmonizing these subsystems under one platform is key to real efficiency gains (Khalifa & Marzouk, 2024). In practice, buildings might adopt open communication standards or vendor-specific systems. Each approach requires cybersecurity measures that protect sensor data from unauthorized manipulation.

Thus, modern building energy management has become a complex cyber-physical system. It merges hardware sensors, software analytics, occupant preferences, and external grid signals. This environment needs secure data handling, verifiable transactions, and the ability to process events in near real time. Blockchain solutions, with features such as distributed consensus and transparent transaction records, appear compatible with these requirements (Sun, 2024). The next sections show how blockchain aligns with or complements the fundamental processes of smart building energy management.

3. BLOCKCHAIN FUNDAMENTALS IN ENERGY MANAGEMENT CONTEXT

Blockchain technology is a distributed ledger maintained by a network of nodes. Each node stores the entire ledger, which is composed of cryptographically linked blocks. A block groups multiple validated transactions, each referencing data such as the sender, recipient, and transaction amount. Nodes reach consensus on the validity of new transactions before creating a new block (Andoni et al., 2019). This process differs from centralized databases, which store data on a server with an authority that can add or modify records at will. With blockchain, each participant may propose transactions, but no single entity can alter confirmed blocks.

Public blockchains like Bitcoin or Ethereum rely on consensus algorithms such as proof-of-work or proof-of-stake. These methods secure the network by ensuring that malicious parties cannot easily rewrite the ledger. However, public chains can be slow or resource-intensive, making them less suitable for many building energy applications (Khatoon et al., 2019). Instead, permissioned or consortium blockchains often gain traction. In these designs, only authorized nodes can validate transactions. Permissioned blockchains tend to handle higher throughput with lower energy consumption.

Smart contracts enable programmable transactions on a blockchain. They are pieces of code deployed on the network that self-execute under specified conditions. For instance, a contract could state that if a building's solar generation meets a threshold, any surplus is automatically offered to local buyers. Once the data feed confirms that the threshold is reached, the contract executes, and the ledger logs the result (Hahn et al., 2017). This approach lowers administrative overhead, reduces settlement time, and can enhance trust. It also introduces new security considerations, as vulnerabilities in smart contract code may be exploited.

Within energy management, the ledger can record meter readings, occupant data, or transactions among distributed energy resources. Participants share a single version of the truth, removing the need for multi-party reconciliation or trusted third parties (Zhang et al., 2018). Buildings can host blockchain nodes, so they each have local copies of relevant transactions. Regulators, utilities, or service providers can run additional nodes. This distributed structure mitigates some single-point-of-failure risks, although the system's overall reliability depends on the number and location of nodes. Proper node management is essential; too few nodes could result in centralization, while too many might slow the network.

Blockchain's immutability is another relevant property. Once data is appended, it becomes infeasible to modify it without the consensus of the network (Andoni et al., 2019). This attribute suits compliance-related tasks such as verifying carbon offset trades or ensuring that building performance data meets local regulations. The ledger can serve as an audit trail, giving regulators direct access to verified data. Project developers can also use blockchain to assure investors or lenders about the building's operational metrics. When integrated with advanced monitoring hardware, these ledgers can deliver near real-time transparency.

Nonetheless, blockchains are not free from limits. Storing large datasets on-chain can be expensive and inefficient. Many systems store small transaction records on the blockchain while large data is kept off-chain, with cryptographic hashes linking it to the on-chain record. Confidentiality can also pose challenges because transaction details on public ledgers are typically transparent to all. Private or permissioned architectures can restrict visibility to authorized parties (Bhushan et al., 2020). Zero-knowledge proofs or other privacy-enabling technologies can further protect sensitive data about occupant behavior or building-level financial transactions.

In short, blockchain is not a simple data store but a decentralized method for data validation and stateful logic execution (Khatoon et al., 2019). It enhances trust among parties without a need for a central arbiter, which is valuable for complex energy networks where many stakeholders handle generation, consumption, and trading. The next section delves into the theoretical reasoning behind blockchain's adoption in building energy management, offering insights from economics, technology acceptance, game theory, and systems thinking.

4. THEORETICAL FOUNDATIONS FOR BLOCKCHAIN ADOPTION IN SMART BUILDINGS

Blockchain solutions in building energy management can be examined through multiple theoretical frameworks. Each provides a rationale for how and why these systems may be adopted or resisted.

4.1 Technology Acceptance Perspectives

Many technology adoption models, including the technology acceptance model, highlight perceived usefulness and perceived ease of use as factors determining whether new technologies are integrated into daily operations (Savur & Emekci, 2024). In building energy management, blockchain's perceived usefulness arises from secure record-keeping, automated settlements, and transparent energy flows. Perceived ease of use depends on user interfaces, training programs, and how well the blockchain integrates with existing BMS platforms. If building managers or occupants view blockchain as overly complex, adoption may stall.

4.2 Transaction Cost Economics

Transaction cost economics posits that organizations adopt mechanisms that reduce the costs of transactions, including monitoring, enforcement, and bargaining (Khatoon et al., 2019). In building energy systems, verifying meter data, billing occupants, and managing distributed renewable generation can involve high administrative overhead. Blockchain can reduce these transaction costs by automating settlements through smart contracts and providing trustworthy records, thus lowering enforcement costs. Participants no longer rely on third-party verification, which might save time and mitigate disputes.

4.3 Game Theory

Game theory examines how individuals and organizations act in strategic situations. In blockchain-based building energy systems, participants seek to maximize benefits, such as cost savings, while cooperating on resource sharing. Blockchain establishes transparent rules for how trades occur and how prices respond to supply and demand (Zhang et al., 2018). This clarity can reduce uncertainty and discourage cheating. Participants trust the immutable ledger, so they have fewer incentives to misreport or delay data updates. Over time, stable equilibrium strategies can emerge, in which prosumers cooperate to balance local grids.

4.4 Systems Theory

Systems theory treats buildings as nodes in a larger socio-technical environment. Building energy management does not operate in isolation but interacts with municipal utilities, occupant needs, and city planning initiatives (Bhushan et al., 2020). Blockchain can serve as a shared substrate that integrates these interactions, providing a layer of data coordination that crosses organizational boundaries. When each stakeholder manages a node, the system might self-organize more effectively. Blockchain also supports modular growth, so a few buildings can pilot the system, then more buildings can join, forming an expanding energy network.

Each framework highlights different aspects of blockchain adoption. Together, they explain why building managers may view blockchain as a strategic asset that reduces costs, automates complex exchanges, and builds trust among stakeholders (Savur & Emekci, 2024). This theoretical underpinning underlies practical applications in peer-to-peer trading, digital twins, cybersecurity, and beyond.

5. BLOCKCHAIN-ENABLED APPROACHES TO URBAN ENERGY EFFICIENCY

Blockchain-based smart building solutions shape how energy is generated, distributed, and consumed in urban areas (Esfandi et al., 2024). These solutions contribute to efficiency targets by creating transparent markets, incentivizing resource optimization, and coordinating distributed sources. Urban environments contain varied building stocks, including commercial complexes, residential towers, and mixed-use developments. Many of these buildings use IoT sensors or advanced metering infrastructure, offering abundant data for energy optimization. Blockchain can unify this data under a decentralized ledger, improving accountability.

5.1 Peer-to-Peer Energy Trading

One high-profile use of blockchain in energy management is peer-to-peer (P2P) trading among building occupants. In conventional models, utilities serve as intermediaries for selling surplus power. In P2P arrangements, building owners with solar panels or small wind turbines sell excess electricity directly to neighbors (Zhang et al., 2018). A blockchain ledger records each kWh traded and transfers payment automatically via smart contracts. This local trading reduces transmission losses and can yield better prices for prosumers. Occupants with high power needs at certain times can benefit from real-time local supply.

These trades often rely on microgrids that operate in parallel with the main grid. When local generation is insufficient, the grid provides backup. When generation exceeds local demand, surplus flows back to the grid or local buyers (Mengelkamp et al., 2018). Blockchain solutions streamline these transactions by establishing consensus on meter readings, verifying each participant's identity, and automating settlement. A P2P exchange might align with dynamic pricing that reflects local supply-demand conditions. Such real-time signals encourage load shifting, which can flatten peaks and reduce stress on central power stations.

5.2 Demand Response and Aggregation

Demand response programs compensate building owners for adjusting energy use during peak times. Large buildings can shut down non-critical loads to decrease strain on the grid. Blockchain-based systems can automate the verification of load reductions via smart meters. When the building meets its reduction target, the ledger records the result, and a smart contract dispenses the agreed-upon incentive (Hahn et al., 2017). Demand aggregators can coordinate multiple buildings, bundling their load reductions. The aggregator's credibility improves if each building's contribution is tamper-proof.

At the same time, smaller buildings or households can join these programs by pooling resources. In a blockchain-based approach, participants do not have to rely solely on aggregator reports, since each building's performance is recorded on-chain (Andoni et al., 2019). This transparency reduces disputes about whether participants delivered the promised load shift. Coordinating these actions via smart contracts also cuts administrative delays, letting the program respond in real time to sudden peak loads.

5.3 Citywide Energy Coordination

Urban governments often seek carbon neutrality or reduced emissions. They encourage building owners to adopt on-site renewables, energy storage, or energy-efficient devices (Bhushan et al., 2020). Blockchain solutions can help them verify compliance with sustainability targets. A city might deploy a permissioned blockchain that each building connects to. This blockchain tracks relevant metrics such as monthly consumption, renewable generation, or greenhouse gas (GHG) emissions. City officials have a node on the network, accessing real-time updates. This structure simplifies carbon accounting by having all data on a unified ledger.

Cities can also use blockchain to manage local energy markets at scale. Buildings with surplus power receive tokens or credits from those that need extra energy. These transactions become part of a broader market that includes electric vehicle charging stations or local energy storage sites (Vakil et al., 2018). The city might integrate dynamic tariffs that encourage energy conservation or investment in clean technologies. Because the system is decentralized, building managers have more direct control over their trades. The city benefits from aggregated data that helps it plan for grid upgrades, load balancing, and resource allocation.

Figure 1. Peer-to-peer energy trading in a local microgrid.

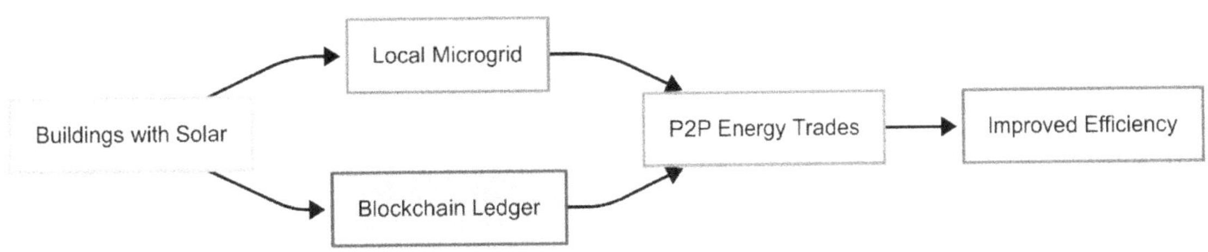

Figure 1. A simplified diagram of peer-to-peer energy trading in a local microgrid. Buildings with solar panels feed data into a blockchain ledger. The local microgrid infrastructure coordinates trades, and the blockchain manages settlements. The result is improved local efficiency and occupant engagement.

6. CYBERSECURITY AND DATA PRIVACY CHALLENGES

Smart building energy management relies on IoT devices, network connections, and sensor platforms that send large volumes of data to management systems or blockchains. While blockchain can reduce some central points of failure, it cannot eliminate all risks. Attackers may target IoT devices or communication channels to alter meter readings (Meng & Zhu, 2024). If enough devices are compromised, false data could enter the ledger and undermine trust in the system. Proper device authentication and data encryption are crucial in this context.

Nodes on a blockchain must also protect private keys that control transaction signing. If attackers gain control of these keys, they can manipulate transactions or disrupt services. Permissioned blockchains restrict node membership, which can lower the chance of malicious nodes. However, permissioned systems must enforce robust rules for node admission and revocation (Khatoon et al., 2019). They also need governance structures to resolve disputes if a node is suspected of wrongdoing. The practicalities of node management become more complex if many stakeholders and buildings join the network.

Data privacy is critical. Building managers want to ensure occupant data remains confidential, especially if it includes occupant schedules or personal behavioral patterns (Saeed et al., 2022). A transparent ledger can reveal more information than desired. Privacy-preserving techniques such as zero-knowledge proofs can prove that a transaction is valid without disclosing the underlying data. Alternatively, systems can store only hashed or aggregated data on-chain, with the full dataset kept off-chain in secure repositories. Access controls ensure that only authorized parties can retrieve occupant or building-level details.

Regulatory environments also shape security and privacy requirements. If data protection laws mandate the right to erasure, implementing this on an immutable blockchain can be difficult (EU Blockchain Observatory and Forum, 2022). One solution is to store personal data off-chain and reference it on-chain via unique identifiers. That design allows partial compliance with erasure requests by deleting off-chain records, although references remain on-chain. Despite these strategies, designing a robust privacy-respecting architecture often involves trade-offs in system complexity, performance, and cost.

Thus, blockchain for building energy management must rest on a layered security model. This includes device authentication, encrypted data transmission, secure node management, and privacy-preserving transaction protocols (Meng & Zhu, 2024). Adoption depends on addressing occupant and building owner concerns that the system will protect sensitive data and maintain reliable operations.

7. DIGITAL TWIN INTEGRATION AND REAL-TIME ANALYTICS

Digital twin models create virtual representations of physical assets or processes. They update in real time by receiving data from sensors and other sources. In building energy management, a digital twin might replicate HVAC systems, lighting, occupant flows, and structural elements. The twin then uses simulation algorithms to predict how adjustments in setpoints or occupant behavior will affect power consumption or comfort (Khalifa & Marzouk, 2024).

Blockchain strengthens digital twin environments by providing a trustworthy record of sensor readings, operational logs, and transaction data. The twin's analytics depend on accurate inputs. If these inputs are manipulated or lost, the twin's predictions might be invalid. A blockchain ledger can store hashed sensor data or system states, making them tamper-evident. This arrangement can prove helpful if building owners or regulators need an audit trail of how the twin's predictions were derived (Khatoon et al., 2019).

Blockchain-based smart contracts can also automate feedback loops. If the digital twin detects that an HVAC subsystem is drifting from optimal performance, it can trigger a blockchain transaction that arranges maintenance or initiates an alert. In a multi-tenant building, the twin might track occupant-level consumption to see if loads exceed agreed-upon baselines (Khalifa & Marzouk, 2024). When conditions are met, a smart contract can adjust occupant billing rates or dispatch incentives for energy-saving actions. This integration of predictive analytics and automated execution can streamline building operations.

Digital twins can extend beyond individual buildings. City planners can model entire districts, including microgrids, renewable resources, and transport networks (Esfandi et al., 2024). Multiple buildings feed blockchain-validated data into a district-level digital twin. This twin examines scenarios for demand response, energy storage allocation, or new policy interventions. Stakeholders trust that the twin's data inputs are secure and verified on-chain, which reduces doubts about data integrity. The system can optimize strategies, test them in a simulated environment, and then propose real-world changes that can be encoded as new rules in the blockchain-based energy market.

Figure 2. Digital twin integration with blockchain.

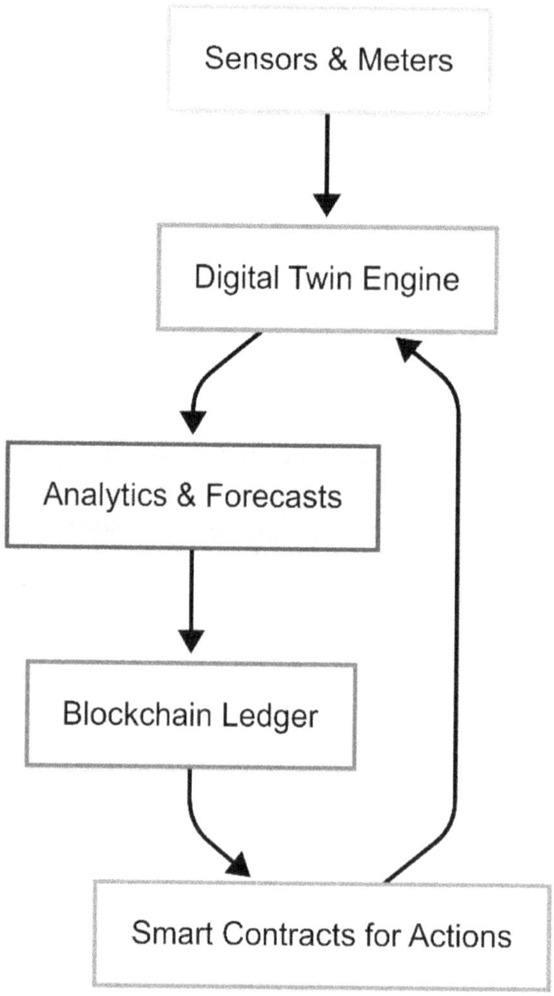

Figure 2. Digital twin integration with blockchain. Sensor data informs the digital twin engine, which generates forecasts and analytics. The blockchain ledger records data and triggers smart contracts that implement actions. A feedback loop refines analytics in near real time.

8. REGULATORY AND POLICY PERSPECTIVES

Blockchain-based building energy management must align with existing regulations related to energy markets, consumer protection, and privacy. Many countries require that only licensed entities buy or sell electricity. Peer-to-peer trading platforms may face legal barriers unless they operate under specific regulatory exemptions or pilot programs (Andoni et al., 2019). Policymakers in some regions have created sandboxes for blockchain energy applications. These sandboxes give innovators a controlled environment to test new models, gather data, and refine regulatory frameworks. If results are positive, authorities may expand permissible use or formalize new rules.

Data privacy laws also influence how occupant information can be processed on a blockchain (EU Blockchain Observatory and Forum, 2022). Regulations such as the General Data Protection Regulation (GDPR) in the European Union grant individuals rights over their personal data. Managing these rights in an immutable ledger requires creative solutions, such as storing personal data off-chain. Compliance with these rules can introduce extra steps that raise system overhead. Building owners must understand how these privacy constraints affect data flows, occupant consent, and third-party access to consumption or behavioral information.

Local building codes can also shape how blockchain systems integrate into building management processes. In some jurisdictions, building codes require advanced metering or specify the data retention period for energy records. Blockchain solutions that comply with these codes could streamline audits and certifications (Batool & Channi, 2024). For example, the system might store building energy performance data for several years on-chain to ensure traceability. Regulators can then query the chain to confirm that the building meets efficiency standards. If local or national governments encourage net-zero construction or green retrofitting, blockchain could support those programs by automating the distribution of renewable energy credits and verifying compliance (Fu et al., 2023).

Policy also intersects with cost considerations. Decision-makers could provide incentives or subsidies for buildings that join blockchain-based peer-to-peer trading platforms. They could adopt time-of-use tariffs that motivate energy trades or load shifting. On the other hand, policies might limit the usage of blockchain in certain contexts if they consider it too novel or unproven. Effective implementation strategies often involve stakeholder engagement, pilot demonstrations, and alignment with broader energy transition goals (World Economic Forum, 2018). As public awareness of blockchain grows, policies may evolve, balancing innovation with consumer protection.

Table 1. Regulatory factors influencing blockchain adoption in building energy management

Regulatory Factor	**Influence on Blockchain Deployment**
Electricity Market Rules	May allow or restrict peer-to-peer trading and aggregator-led demand response.
Data Protection Laws	Require privacy safeguards, data minimization, and careful record handling.
Building Codes	Establish metering standards and record retention requirements.
Pilot/Sandbox Programs	Offer controlled frameworks for experimentation and collecting performance data.
Environmental Regulations	Can encourage integration of renewables and energy efficiency tracking.

9. ECONOMIC AND SOCIAL IMPLICATIONS

Blockchain-based energy management systems can reconfigure economic relationships in the building sector. They enable direct energy exchanges among occupants, building owners, and external energy markets, reducing dependence on centralized utilities. This change can lower operating costs and encourage prosumers to invest in on-site renewable energy (Zhang et al., 2018). Efficiency improvements may reduce building utility bills by optimizing loads and capturing time-of-use rate advantages. Aggregation of smaller resources via a blockchain platform can also open up revenue streams from ancillary services, such as frequency regulation or voltage support (Andoni et al., 2019).

Social implications emerge from how these systems distribute costs and benefits. Building occupants who adopt efficient devices or generate surplus energy can see direct financial gains. Occupants who cannot invest in renewables might worry about disparities in cost savings. A robust design can include incentives that distribute benefits fairly. For instance, community-based token systems might reward overall energy conservation, not just individual generation (Vakil et al., 2018). By formalizing these arrangements through smart contracts, building managers can promote collaborative energy strategies. Transparent transactions can also reduce disputes over billing or energy usage (Hahn et al., 2017).

The sense of engagement can increase among occupants. Seeing real-time data on energy consumption may prompt them to modify behaviors, such as using less electricity during peak hours (Bhushan et al., 2020). In multi-tenant buildings, blockchain-based metering can refine how costs are shared, moving from approximate allocations to occupant-specific consumption. Although transparency can foster accountability, it may raise privacy concerns if occupant usage patterns become too visible. The design of user interfaces and privacy protocols thus influences occupant acceptance (Saeed et al., 2022). Strong occupant support, combined with local policy incentives, can lead to broader acceptance of blockchain solutions.

Table 2. Economic and social outcomes of blockchain-based building energy management

Outcome	Description
Reduced Operational Costs	Automated settlements, minimal third-party fees, and dynamic pricing can lower expenses.
New Revenue Streams	Surplus energy sales, demand response programs, or ancillary grid services can generate income.
Occupant Engagement	Real-time data and incentives encourage active participation in energy efficiency.
Social Equity Concerns	Differing access to technologies might create disparities in cost savings or service benefits.
Transparency and Trust	Immutable records reduce billing disputes and increase trust among occupants, managers, and regulators.

10. IMPLEMENTATION STRATEGIES AND BEST PRACTICES

Implementing blockchain in building energy management requires aligning technical capabilities with stakeholder needs. Organizations can approach the process by defining clear objectives, selecting suitable blockchain frameworks, and integrating with existing BMS platforms (Khatoon et al., 2019). The following strategies capture the main phases involved:

10.1 Goal Definition and Stakeholder Engagement

Managers must clarify whether their primary goal is cost reduction, carbon footprint reduction, occupant empowerment, or grid reliability. Each goal influences how transactions are structured and how incentives are defined. It is also important to engage occupants, regulators, utilities, and technology providers early. This fosters realistic expectations about data privacy, technical feasibility, and economic returns. Pilot projects at a small scale, such as one building or campus, can demonstrate tangible benefits before wider deployment (Hahn et al., 2017).

10.2 Selecting the Blockchain Framework

Many building projects adopt permissioned blockchains for higher throughput, easier governance, and privacy controls. Some implementations rely on recognized enterprise platforms that integrate well with IoT devices. Decisions about consensus algorithms should balance energy consumption, latency, and resilience. Proof-of-stake or practical Byzantine fault tolerance (PBFT) often suit enterprise contexts better than proof-of-work (Andoni et al., 2019). Testing multiple configurations in a pilot environment helps refine the final choice.

10.3 Smart Contracts and Data Integration

Effective use of blockchain depends on well-designed smart contracts that handle billing, trading, and maintenance tasks. These contracts must receive data from reliable oracles, such as calibrated meters or sensors. Developers should code carefully and test for vulnerabilities. Some building owners use multi-signature protocols requiring approval from multiple parties before transactions finalize, which helps avoid unilateral contract execution errors (Teng et al., 2021). Data storage must also be planned, with large datasets kept off-chain and essential reference data on-chain.

10.4 Cybersecurity Protocols and Privacy

Operators must adopt secure key management practices. Each node or device needs unique credentials stored in tamper-resistant hardware if possible (Saeed et al., 2022). Network communication protocols should encrypt data from sensor to blockchain node. Permissioned ledgers often have fewer nodes, simplifying monitoring but creating potential single points of infiltration if an attacker compromises the node admission process. Security audits and intrusion detection systems can provide additional protection. Data privacy is another priority. Designers can use privacy-preserving techniques or enforce granular access control on a need-to-know basis (Meng & Zhu, 2024).

10.5 Iterative Evaluation and Expansion

Once the system operates at a small scale, managers can assess performance with metrics such as cost savings, occupant satisfaction, and system uptime (Batool & Channi, 2024). Adjustments might include refining contract logic, adding new devices, or improving the user interface. Scaling up requires careful planning of node infrastructure, network bandwidth, and data throughput. Partnerships with utilities or city authorities may be needed for advanced use cases like microgrid integration or large-scale demand response (Andoni et al., 2019).

Figure 3. Implementation cycle for blockchain-based energy management.

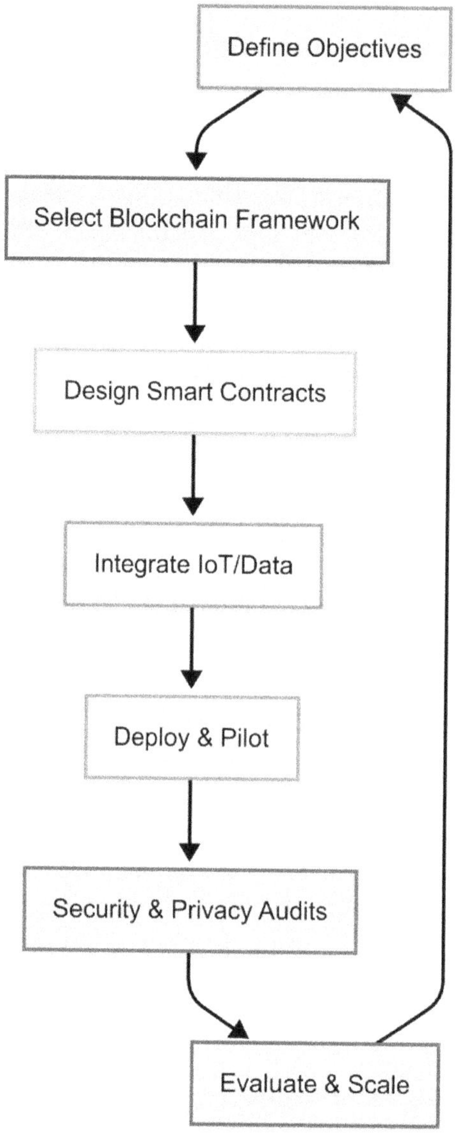

Figure 3. A general implementation cycle for blockchain-based energy management in buildings. Objectives guide framework selection, smart contract design, and data integration. The pilot phase leads to audits and evaluations, which feed back to refine objectives in an iterative manner.

11. CHALLENGES AND LIMITATIONS

Blockchain holds promise for building energy management, but it faces constraints. Scalability remains a major concern. If numerous transactions flow into the ledger every few seconds from sensors and occupants, the blockchain network could struggle to keep up (Zhang et al., 2018). Off-chain aggregation techniques can help by batching multiple sensor readings into a single on-chain update, reducing overhead. Yet real-time applications that require immediate consensus might be difficult to implement with large participant numbers.

Regulatory and legal uncertainties also create obstacles. Many markets lack clear rules on blockchain-based P2P trading or aggregator-run demand response (EU Blockchain Observatory and Forum, 2022). Projects may risk legal conflicts if local legislation treats them as unauthorized energy sales. Some building owners may be unwilling to invest in uncertain territory. Establishing pilot programs with regulatory approval can mitigate this risk. Ongoing policy discussions in different regions will help clarify the permissible scope for blockchain-based energy solutions.

Interoperability is another limitation. Buildings have diverse hardware, software, and communication standards. Converging them in a single blockchain environment can require major integration efforts (Bhushan et al., 2020). The market remains fragmented, with various vendors offering BMS platforms that do not share common APIs. Standardizing how devices report data or how nodes communicate on the blockchain is key. Over time, standards organizations and industry coalitions may define compatible protocols, easing integration.

Cost considerations also matter. Deploying blockchain-based solutions can involve hardware upgrades, software licenses, and training. While there may be long-term savings from streamlined operations, the up-front expenses can be a barrier, especially for small building owners (Andoni et al., 2019). The complexity of building the necessary governance structures for node management and dispute resolution adds to these challenges. Occupant acceptance remains vital: If users do not trust or understand the system, adoption is likely to lag (Savur & Emekci, 2024).

Table 3. Common challenges for blockchain in building energy management

Challenge	Description
Scalability	High transaction throughput is difficult on certain blockchain infrastructures.
Regulatory Uncertainty	Energy trading and data handling face incomplete or unclear rules.
Interoperability	Integrating diverse building systems and blockchains requires standard protocols.
Up-Front Costs	Hardware, software, and training expenditures may deter some building owners.
Occupant Acceptance	Adoption depends on user trust, data privacy, and intuitive interfaces.

These challenges highlight the barriers to widespread adoption of blockchain-based solutions in the building sector (Andoni et al., 2019; Bhushan et al., 2020).

12. FUTURE OUTLOOK

Blockchain in building energy management is likely to evolve in tandem with advancements in IoT, artificial intelligence, and 5G communication. AI-driven analytics can refine how blockchain-based contracts price electricity or schedule demand response events in real time. Data collected from occupant wearables or advanced sensors can feed into AI models that predict usage patterns, enabling more accurate load balancing (Kolahan et al., 2021). Blockchain can then store these forecasts and outcomes, creating accountability around system performance.

Quantum computing poses another dimension. If quantum computers become powerful enough, they might break existing cryptographic schemes. Blockchain developers are investigating quantum-safe algorithms to maintain security. Although significant breakthroughs in quantum computing may still be years away, planning for post-quantum security is part of forward-looking strategies (Meng & Zhu, 2024).

Integration with digital twin frameworks will likely deepen. The next wave of digital twins may incorporate city-scale data, capturing transport networks, water supply, and social infrastructure in addition to building energy systems (Esfandi et al., 2024). Blockchain can anchor the reliability of these massive datasets, ensuring that stakeholders can trust the model's integrity. Such integrated twins could drive policymaking in real time, testing scenarios for greenhouse gas reductions or emergency responses. The blockchain ledger would also log the chain of events leading to policy decisions, improving accountability.

From a market perspective, the emergence of "energy as a service" could reshape how buildings procure or sell electricity (Zhang et al., 2018). Under such models, occupants pay based on actual usage at the moment of consumption. Blockchain-based microtransactions handle these payments, removing the need for monthly billing cycles. This approach might accelerate the shift to renewables and flexible demand profiles. Future developments may also see large-scale aggregator platforms that coordinate thousands of buildings across a city. These platforms could rely on blockchain for distributed governance, allowing participants to vote on policy changes or rate the reliability of each aggregator node (World Economic Forum, 2018).

Collaboration among stakeholders will be essential. Technology vendors, utilities, building owners, and regulators each have distinct roles. Standards bodies may emerge to define how building-level blockchains connect with broader grid management blockchains (EU Blockchain Observatory and Forum, 2022). The success of these systems hinges on bridging technical innovation with institutional acceptance. As climate goals drive the energy transition, blockchain-based solutions may prove valuable in orchestrating distributed resources, verifying carbon credits, and engaging participants in more flexible and resilient energy networks.

Figure 4. Projected evolution of building energy management.

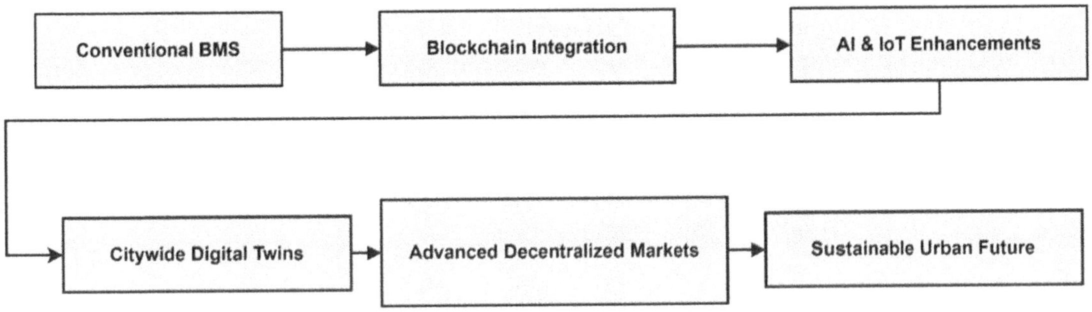

Figure 4. Projected path of building energy management evolution. Conventional BMS moves toward blockchain integration, then incorporates AI and IoT enhancements. Eventually, citywide digital twins and advanced decentralized markets pave the way for sustainable urban futures.

13. CONCLUSION

Blockchain-based energy management in smart buildings can create transparent, automated, and trustworthy systems. By recording sensor data and financial transactions on a decentralized ledger, building managers, occupants, and external partners gain real-time visibility into energy flows. Peer-to-peer trading, demand response, and aggregator services can operate under smart contracts that remove manual settlement processes. Digital twin models, which simulate building performance, can rely on validated on-chain data for accurate predictions.

Despite these potentials, challenges persist. Blockchain scalability, privacy, regulatory uncertainty, and integration with existing building systems require careful attention. Some building operators may resist new technologies, particularly if the up-front costs and complexities appear high. Stakeholder engagement and pilot implementations can address these hesitations by offering tangible evidence of efficiency and cost benefits. Ongoing technological advances, including AI integration, quantum-safe cryptography, and improved IoT device security, will likely shape the trajectory of blockchain in this domain.

In many urban contexts, decarbonization and resilience have become leading priorities. Smart buildings can support these goals by optimizing local resources and enabling occupant participation in energy markets. Blockchain technology, when embedded into these systems with robust governance, has the capacity to coordinate distributed energy resources, reduce administrative overhead, and secure occupant data. Future research can explore more expansive city-scale deployments and refined governance models that handle thousands of participants while protecting privacy and ensuring economic fairness. With thoughtful implementation, blockchain can foster an environment where buildings play active roles in sustainable urban energy ecosystems.

REFERENCES

Andoni, M., Robu, V., Flynn, D., Abram, S., Geach, D., Jenkins, D., McCallum, P., & Peacock, A. (2019). Blockchain technology in the energy sector: A systematic review of challenges and opportunities. *Renewable & Sustainable Energy Reviews*, *100*, 143–174. DOI: 10.1016/j.rser.2018.10.014

Batool, S., & Channi, H. K. (2024). Role of blockchain IoT in smart building for distributed renewable power. In *Advances in Computational Intelligence and Robotics (Chapter 4, pp. 63–80)*. IGI Global. https://doi.org/DOI: 10.4018/979-8-3693-1586-6.ch004

Bhushan, B., Khamparia, A., Sagayam, K. M., Gupta, D., Chowdhary, C. L., & Joshi, G. P. (2020). Blockchain for smart cities: A review of architectures, integration trends, and future research directions. *Sustainable Cities and Society*, *61*, 102360. DOI: 10.1016/j.scs.2020.102360

Esfandi, S., Tayebi, S., Byrne, J., Taminiau, J., Giyahchi, G., & Alavi, S. A. (2024). Smart cities and urban energy planning: An advanced review of promises and challenges. *Smart Cities*, *7*(1), 414–444. DOI: 10.3390/smartcities7010016

EU Blockchain Observatory and Forum. (2022). *Blockchain applications in the energy sector*. European Commission report.

Fu, S., Tan, Y., & Xu, Z. (2023). Blockchain-based renewable energy certificate trade for a low-carbon community of active energy agents. *Sustainability (Basel)*, *15*(23), 16300. DOI: 10.3390/su152316300

Hahn, A., Singh, R., Liu, C.-C., & Chen, S. (2017). Smart contract-based campus demonstration of decentralized transactive energy auctions. In *2017 IEEE Power & Energy Society Innovative Smart Grid Technologies Conference (ISGT)* (pp. 1–5). IEEE. https://doi.org/DOI: 10.1109/ISGT.2017.8086092

Khalifa, F., & Marzouk, M. (2024). Integrated blockchain and digital twin framework for sustainable building energy management. *Journal of Industrial Information Integration*. Advance online publication. DOI: 10.1016/j.jii.2024.100747

Khatoon, A., Verma, P., Southernwood, J., Massey, B., & Corcoran, P. (2019). Blockchain in energy efficiency: Potential applications and benefits. *Energies*, *12*(17), 3317. DOI: 10.3390/en12173317

Kolahan, A., Maadi, S. R., Teymouri, Z., & Schenone, C. (2021). Blockchain-based solution for energy demand-side management of residential buildings. *Sustainable Cities and Society*, *75*, 103316. DOI: 10.1016/j.scs.2021.103316

Meng, X., & Zhu, L. (2024). Augmenting cybersecurity in smart urban energy systems through IoT and blockchain technology within the digital twin framework. *Sustainable Cities and Society*, *106*, 105336. DOI: 10.1016/j.scs.2024.105336

Mengelkamp, E., Gärttner, J., Rock, K., Kessler, S., Orsini, L., & Weinhardt, C. (2018). Designing microgrid energy markets: A case study with the Brooklyn Microgrid. *Applied Energy*, *210*, 870–880. DOI: 10.1016/j.apenergy.2017.06.054

Nai Fovino, I., Andreadou, N., Geneiatakis, D., Giuliani, R., Kounelis, I., Lucas, A., & Soreli, G. (2021). *Blockchain in the energy sector: Use cases identification and analysis (EUR 30782 EN)*. Publications Office of the European Union.

Saeed, M., Amin, R., Aftab, M., & Ahmed, N. (2022). Trust management technique using blockchain in smart building. In *Proceedings of the 2021 International Symposium on Networks, Computers and Communications (Engineering Proceedings Vol. 20, No. 1, p. 24)*. MDPI. https://doi.org/DOI: 10.3390/engproc2022020024

Savur, D., & Emekci, Ş. (2024). A review on smart building and blockchain technology. *Politeknik Dergisi*, *27*(1), 1–13. DOI: 10.2339/politeknik.1403561

Sun, L. (2024). Exploration of blockchain-based building energy management. *Smart Systems and Green Energy*, *6*(1), 60–67. DOI: 10.23977/ssge.2024.060108

Teng, F., Zhang, Q., Wang, G., Liu, J., & Li, H. (2021). A comprehensive review of energy blockchain: Application scenarios and development trends. *International Journal of Energy Research*, *45*(12), 17546–17578. DOI: 10.1002/er.7109

Vakil, S., Petro, T., & Goyne, S. (2018). *Blockchain: A true disruptor for the energy industry – Use cases and strategic questions*. Deloitte Insights White Paper. Deloitte.

World Economic Forum. (2018). *Building block(chain)s for a better planet*. WEF/PwC White Paper.

Zhang, C., Wu, J., Zhou, Y., Cheng, M., & Long, C. (2018). Peer-to-peer energy trading in a microgrid: A blockchain-based market mechanism. *Applied Energy*, *220*, 1–12. DOI: 10.1016/j.apenergy.2018.03.010

Zhao, S., Zhu, S., Wu, Z., & Jiang, B. (2022). Cooperative energy dispatch of smart building clusters based on smart contracts. *International Journal of Electrical Power & Energy Systems*, *138*, 107896. DOI: 10.1016/j.ijepes.2021.107896

ADDITIONAL READING

Egunjobi, O. O., Gomes, A., Egwim, C. N., & Morais, H. (2024). A systematic review of blockchain for energy applications. e-Prime – Advances in Electrical Engineering. *Electronics and Energy*, *9*, 100751. DOI: 10.1016/j.prime.2024.100751

Krishnan, S., Balas, V. E., Golden, J., Robinson, Y., & Kumar, R. (Eds.). (2021). *Blockchain for Smart Cities. Academic Press*. Elsevier.

United Nations Environment Programme. (2022). Blockchain for sustainable energy and climate in the Global South: Use cases and opportunities. UNEP Report.

Chapter 14
Exploring Blockchain Technology:
Concepts, Features, and Industry Transformations

B. Yamini
https://orcid.org/0000-0003-3531-108X
SRM Institute of Science and Technology, India

M. G. Dinesh
https://orcid.org/0009-0009-7643-5338
Easa College of Engineering and Technology, India

L. A. Anto Gracious
https://orcid.org/0000-0003-2386-9182
R.M.K. College of Engineering and Technology, India

R. Siva Subramanian
https://orcid.org/0000-0002-7509-9223
R.M.K. College of Engineering and Technology, India

ABSTRACT

Blockchain which is derived from virtual currency such as Bitcoin is now considered revolutionary. This paper aims at discussing what blockchain is, its attributes, and its application in finance, healthcare, supply chain and governance. It is a decentralized system that provides transparency, security, and a low level of alteration, which makes it reliable and eliminates the need for middlemen. This also comprises blockchain structures, consensus mechanisms, smart contracts as well as the comparison of the cryptocurrency with traditional systems based on decentralization, cryptographic security and scalability. Blockchain is beneficial for strengthening the digital transactions, securing the healthcare data and the traceability of the supply chain. Thus, the issues like scalability, legal aspects and adoption remain to be a concern. It examine present and future studies and development which may help to eliminate these issues and strengthen the position of blockchain in the sphere of digital and decentralized markets.

DOI: 10.4018/979-8-3373-2439-5.ch014

1. INTRODUCTION

1.1 Background and Motivation

In the current world where everything is being transformed by technology, blockchain technology has proven to be one of the most revolutionary innovations that has changed the face of data sharing and storage. Blockchain in its essence is an unalterable, distributed and decentralized database that can act as the key component in enhancing various applications(Zheng et al 2018)(Gadekallu et al 2022). Blockchain was initially designed for the implementation of cryptocurrencies and its applicability expanded to numerous issues in various sectors including finance, supply chain, and health sector and government sectors. It is a decentralized system and as such; it lacks a central control hence avoiding the common problems associated with central control such as fraud, manipulation and inefficiency. The feature of the technology in offering a forgery proof record of the transactions has brought about a radical change in the way trust is built in the systems. The first application of blockchain technology was the bitcoin which aimed at providing a decentralized way of transacting through the use of the internet. With time, people saw that blockchain is not only useful in the financial industry but also in other fields where issues to do with data integrity, transparency, and decentralization are of great importance. With the help of smart contracts, the application of blockchain went beyond the financial transaction and became a basis for apps that could execute various tasks, verify information, and manage ownership. The constant development of the blockchain is attributed to its timely demand and the rising need for secure and efficient solutions in various sectors(Khang et al 2022).

1.2 Objective of the Survey

The purpose of this questionnaire is to gather information on the subject of blockchain technology and its essential principles, characteristics, and uses in various industries. Thus, answering these questions in detail, the survey will help the readers to have a profound understanding of the principles of blockchain, the measures that provide its safety and openness, and the ways it is being utilized to revolutionize different industries. This work will not only explain the technology of blockchain but also focus on the use of the technology and the barriers that have to be crossed to make it widespread. One of the purposes of this paper is to accentuate the possibility of change brought by blockchain technology. It has emerged from the realm of an experimental technology and is now being seen as a valuable tool by both businesses and governments. The survey will explore the topic of how blockchain can solve many of the problems existing in business today, such as the lack of trust, transparency, security of data, and ineffective work in the financial, supply chain, healthcare, and other industries. In addition, the paper will describe the possibilities that blockchain can offer to organizations, for example, DeFi, tokenization and digital identity.

1.3 Structure of the Paper

The structure of this paper is to provide the reader with a step by step approach of understanding blockchain technology and its application in various fields. The first section gives a general idea of what blockchain is and how it evolved from a mere platform for the cryptocurrency to a platform for decentralized applications. This section also lays a basis for the explanation of why blockchain is such an essential concept in the modern digital environment and the driving force behind its popularity. In the second section, the basic ideas of blockchain concept like decentralized ledger system, cryptographic technique and consensus algorithm will be explained. This section will help the readers to understand the concept of blockchain, its basic components and how consensus mechanisms like PoW and PoS enhance the security of the blockchain. Next, the third section will present the features of the blockchain, including decentralisation, openness, non-tampering, and security. It will also show how these features make the use of blockchain secure and efficient in cases where data security and process transparency is paramount. Moreover, this section will discuss the issues like scalability and combination of blockchain with other advanced technologies like Artificial Intelligence (AI) and Internet of Things (IoT)(Aswini et al 2024). Section four will have questions that will seek to determine the extent of applicability of blockchain in various industries. This paper will dissect the impacts of blockchain in fragments like finance where it is being applied through digital currency and decentralized finance, supply chain management via transformative application of transparency and track record, health where blockchain is used to improve record keeping, government especially in identity and elections, and education particularly in certification and accreditation. The section will describe practical cases of the blockchain application that will illustrate the potential of this technology. The fifth section of the paper will discuss the weakness and prospect of blockchain technology. These will be the areas of scalability and the problems associated with it, security, regulatory concerns, and the limitations to the further spread of this technology in the industry. With the help of these challenges, the survey will identify the current state of the blockchain and define further research directions. In the end, the conclusion will restate the major arguments of the paper and provide a prognosis of what the future holds for blockchain technology. It will also provide ideas on the prospects and threats of the future in the development of blockchain. Thus, the structure of the paper will allow for the comprehensive understanding of the subject matter of the paper, which is blockchain technology, its uses and potential for future development.

2. FUNDAMENTAL CONCEPTS OF BLOCKCHAIN TECHNOLOGY

2.1 Definition and Overview

Blockchain can be described as a decentralised and distributed electronic record keeping system that is used in the secure, transparent, and tamper proof recording of transactions over the internet. A blockchain system is different from conventional database systems where a single entity owns and manages the data base, With blockchain, no single entity has full control over the system(Kumar et al 2023). However, there are many nodes that hold a copy of the blockchain, this makes the data to be distributed across the network, thus making it highly secure from manipulation. This means that there will be no third-party involvement in the process of validating the transactions such as banks or governments, thus, decreasing the chances of fraud. In its essence, it allows for peer-to-peer transactions by leveraging the "blocks" of information. The block has a list of validated transactions, and once a block is full, it is linked to the previous block, that is why it is called a chain – the block chain. This makes it a chronological sequence of blocks containing all the information related to the transactions ensuring that there is record of all the transactions done(Tariq, M. U. 2024).

2.2 Core Components

Blockchain technology has several basic constituents such as, block, transaction, node, and consensus.

Blocks: The blocks consist of several transactions and have their unique values known as hash. A block normally conveys the hash from the preceding block, the amount of time when the block was created, the data to be recorded, and a number, the nonce, which is used to check the validity of the block.

Transactions: Transactions are the records that are made on the blockchain like transfer of coins or tokens or the execution of smart contract. These are broadcasted to the network and then validated by the participants or nodes in the network.

Nodes: Nodes are the individual computers or any other entities that have a copy of the blockchain database. They engage in confirming and spreading transactions and ensuring that data is compiled and made available.

Consensus Mechanisms: Consensus mechanisms are the process which is followed in the network to come to a consensus about the state of the blockchain. These mechanisms make sure that all the transactions are legal and all the members are in consensus with the records of the ledger.

Figure 1. Blockchain technology: key components and relationships

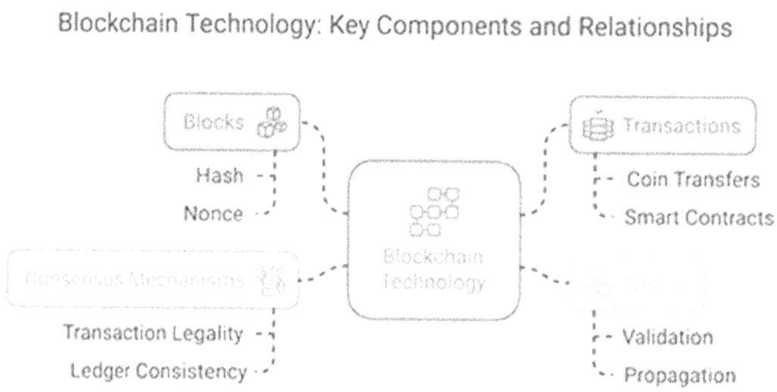

2.3 Types of Blockchain

There are several classifications of the blockchain network and each type has its own characteristics and is designed for certain purposes:

Figure 2. Blockchain types and their characteristics

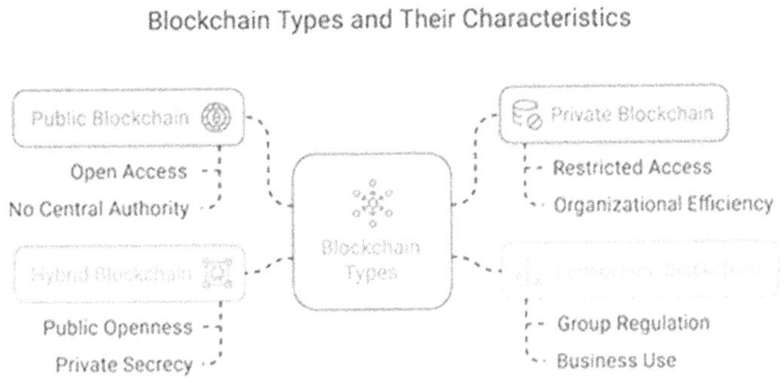

Public Blockchain: A public blockchain is also called an open blockchain since anyone can be a node in the network. It has no central authority and one does not need an invitation to join the network or to gain access to the data. Bitcoin and Ethereum are typical of the so-called public Distributed Ledger Technology because anyone can validate transactions and participate in validating consensus.

Private Blockchain: Private blockchains are those in which the nodes are limited and have access to the network. The participants are usually from the same organization and the flow of information is restricted by the control point or a set of control points. Private blockchains are much more efficient and flexible than public blockchains as they do not need consensus algorithms.

Consensus Blockchain: A consortium blockchain is a form of a private blockchain but it has some features of a public blockchain. It is not controlled by a single company but is under the regulation of a group of organizations. It is restricted for only the selected participants to view and approve the transactions; it is ideal for business entities that need a permissioned distributed ledger for their operations, such as supply chain.

Hybrid Blockchain: Hybrid block chain is a mixture of both public and private block chain. It permits some information to be open to the public but others which are sensitive cannot be accessed by the public. This kind of blockchain is useful when there is a need for equal openness and secrecy of transactions.

2.4 Consensus Mechanisms

Consensus algorithms refer to the procedures followed by nodes in a blockchain network to arrive at a unanimity on the correctness of certain transaction in addition to whether and where to add the same to the chain. Different consensus mechanisms commonly in use are as follows:

Proof of Work (PoW): It is a consensus algorithm used by Bitcoin in which participants known as miners need to solve a given puzzle so as to validate the transactions and add blocks to the chain. This mechanism is costly and requires a lot of power, but it is regarded as very safe.

Proof of Stake (PoS): It is a consensus algorithm that is different from PoW; participants (validators) who create new blocks are selected depending on their quantity of the cryptocurrency tokens they own and are ready to risk for staking. It is less energy-intensive than PoW and is applied to such blockchains as Ethereum 2.0.

Byzantine Fault Tolerance (BFT): BFT is one of the ways to guarantee the blockchain will still be able to perform correctly even when a certain amount of the nodes are bad actors or are faulty. It is most beneficial in private blockchains and permissioned networks since the participants are more trustworthy.

2.5 Smart Contracts and Their Role

Smart contract is an open source computerized agreement that has the terms of the agreement written in the source code. Smart contracts are executed on the blockchain and when predetermined conditions are fulfilled, the contract performs the obligations of the agreement. Smart contracts also remove the need for lawyers or notaries in the transactions thereby lowering the cost and time of the transactions.

The main purpose of smart contracts is to execute operations and follow the rules set without the need to involve a third party. This may include straightforward uses, for example when a payment is made, the program can transfer cryptocurrencies; it can also be used in more complex business scenarios or to initiate actions in supply chains. Smart contracts can be applied in many sectors in the present society such as in the financial sector for decentralized financial products, or in real estate for property agreements, among others due to their efficiency, security, and transparency. Therefore, the general understanding of the blockchain is based on the decentralized and distributed blocks which guarantee openness, security and non- alterability. This paper seeks to explain how the concepts of blocks, transactions, nodes, and consensus mechanism help in the proper functioning of the blockchain system. There are different classifications of blockchains and different methods such as PoW, PoS, DPoS, and BFT that can be used for transaction validation. Smart contracts bring even more benefits of blockchain by automating processes and allowing trustless transactions in various fields.

3. KEY FEATURES OF BLOCKCHAIN

3.1 Decentralization

Another important characteristic of the blockchain technology is decentralization, which means that there is no central point that regulates or approves the transactions. In conventional systems, there is a central authority such as a bank or government who oversees and authenticates the transactions which causes some delay, fraud or weakness. This is unlike blockchain, which is a decentralized system where no one participant has an authority over the entire system. This implies that, every node in the network has a copy of the blockchain and is involved in the validation of transactions through consensus. This makes it difficult for an attacker to gain full control of the data or for data to be lost in the event of a failure. For instance, if a node is suddenly shut down, the blockchain system will still be viable since other nodes are available. Also, decentralization is useful in improving the level of trust among the participants since decisions are made collectively rather than by a single authority. This feature is very important especially for application like the cryptocurrency, where the trust is decentralized and does not rest with any central authority.

Figure 3. Key attributes and challenges of blockchain technology

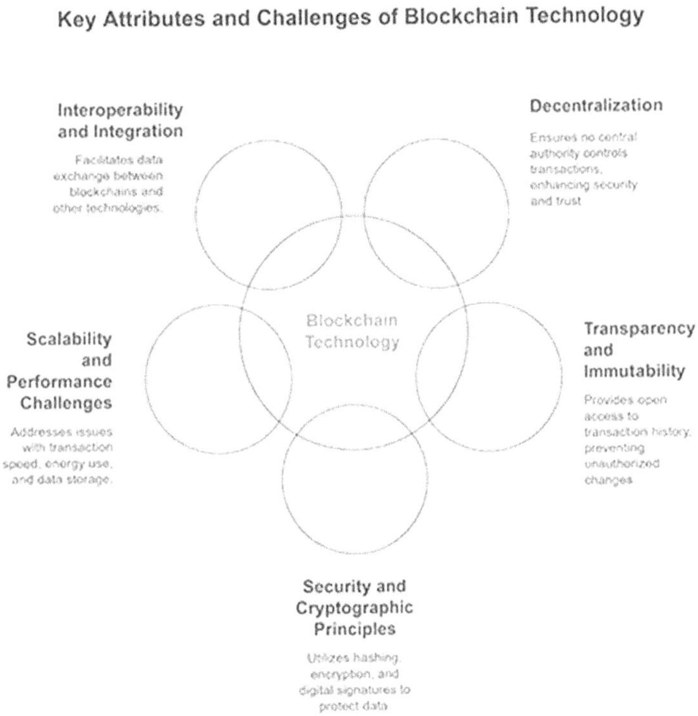

3.2 Transparency and Immutability

Blockchain is based on the foundation of openness and non- alterability, which ensures that the system is credible. Transparency implies that any record that is entered into the block chain is open for the public to view (in the case of public block chains). The transaction history is fully transparent, so everyone can get the information about all the processes which happen in the network. It promotes accountability and minimizes any form of illegitimate activity since all happenings are transparent in the network. Immutability refers to the tamper-proof nature of blockchain data. When a transaction is made and put on the block chain then it is almost impossible to change as it has to go through the consensus of the network. The blocks are connected to the previous block through a hash, and this creates a chain of blocks. If someone tries to modify any transaction in a block, they would have to change the block's hash, and in turn, the hash of the block following that and so on. This is computationally very expensive in large Decentralized network, thus the data remains the same over the time. Therefore, it is clear that blockchain offers a high degree of trust and data credibility which makes it suitable for applications such as financial processes, sharing of physical asset information, and documentation.

3.3 Security and Cryptographic Principles

Blockchain technology uses encryption to ensure that the data is protected and kept away from unauthorized access. These three forms of cryptographic techniques are hashing, encryption, and digital signatures.

- Hashing: It is a unique combination of letters and numbers that is located at the bottom of each block of data in the block chain. Hashing is the process of converting the input data into a string of a fixed number of characters, and modifying the input data will result in a new hash. This makes it possible to notice any alteration of the data recorded in the block chain.
- Secrecy: Like any other secure system, blockchain uses encryption to ensure that the privacy of the users and the transactions is well maintained. The process of public-key cryptography is such that only the intended recipients can read the transactions. Every participant has the pair of keys, the public one is used for encryption and the private one is used for decryption of data so only those who have the right to access information can do so.
- Digital Signatures: It is a type of electronic signature that is used in the validation of a transaction. Such a transaction is signed by a participant using his private key. This serves as proof of ownership and also checks whether the transaction is authentic since only the owner of the private key can produce the right signature.

These principles of cryptography make blockchain safe, unalterable, and immune to fraud and forgery, which explains why it is used to store valuable information and conclude transactions.

3.4 Scalability and Performance Challenges

Despite the benefits that it provides, blockchain has some issues which are related to scalability and performance problems. Various decentralized computerized systems, especially the ones that are based on the Bitcoin and Ethereum platforms, are slow in terms of transactions and are energy-intensive.

Transaction Speed: As the number of users and the number of transactions in the network increases, it may slow down the blockchain. For instance, public blockchains may have longer time to confirm transactions due to the consensus mechanisms such as Proof of Work.

Energy Consumption: Proof of Work that is used by Bitcoin is one of the most energy-consuming consensus algorithms. The mining participants known as 'miners' use their computational power to solve these puzzles in order to verify transactions. This process consumes large quantities of electricity which is a major issue concerning the environment.

The Storage: Rather frequently as the size of the blockchain increases the amount of information that is stored per node is also affected. As the number of blocks increases, the data needed to contain a complete copy of the ledger also increases. This can be disadvantageous to nodes with low computational power while it has the possibilities of weakening the decentralization of the network.

These are the issues which are currently being studied and there are solutions on the way in the form of layer 2 solutions such as the Lightning Network for the Bitcoin or shift to better consensus algorithms such as Proof of Stake.

3.5 Interoperability and Integration with Other Technologies

Interoperability can be defined as the integration of different blockchain systems so that they can exchange data. Since the number of platforms is increasing steadily, the issue of compatibility has emerged as critical. Inter-chain communication is required to enable interoperability between different blockchains so that value and information can be transported between them. For instance, Polkadot and Cosmos are developing the means of connecting different blockchains so that they can communicate and exchange data seamlessly and securely. Blockchain also has the capability of being incorporated with other modern technologies such as AI, IoT and cloud computing. For instance, IoT devices can use blockchain to encrypt data that cannot be altered in any way, for example, data collected from sensors or any device. Likewise, the applications of AI could also benefit from blockchain by using it for storage and decision-making of data. Blockchain technology has the potential to enhance the current cloud-based services because cloud data storage can be more secure and transparent through the use of blockchain. Blockchain's compatibility with other technologies paves way for a world of innovation with possibilities in several fields such as financial, healthcare and more beyond. To summarize the discussion, de-centralization, transparency, non-tampering, protection, and extensibility in the blockchain significance are precious qualities in data organization and digital business transactions. Despite the issues of performance, energy concern, and compatibility, the adoption of blockchain with other technologies would be significant in the development of decentralized systems.

4. POTENTIAL IMPACT OF BLOCKCHAIN ON VARIOUS INDUSTRIES

4.1 Financial Sector and Cryptocurrencies

Cryptocurrencies such as Bitcoin and ether those are created on blockchain technology shifted the financial sector in a very effective way. Cryptocurrencies are decentralized digital means of payment that eliminates the middleman, which is the banks and allows for cross border transactions in a single platform. These transactions are thus safe, secure, and unalterable since blockchain enhances their transparency and traceability.

Apart from cryptocurrencies, Decentralized Finance or DeFi is an even more significant application of blockchain. DeFi is a term that describes the utilization of blockchain and cryptocurrencies aiming to reinvent the existing financial services like lending, borrowing, and trading. Smart contracts enable users to perform financial operations with each other, without requiring in-between special services of banks or brokers. This makes it possible to receive and provide financial services without limited control by central authorities. Another new concept is Central Bank Digital Currencies (CBDCs), in which governments consider the implementation of digital currencies that will be backed by their respective central bank. CBDCs use blockchain as their basis, which leads to increased transparency, reduced expenses, and increased effectiveness in the use of digital money and monetary policy.

4.2 Supply Chain and Logistics

It is thus evident that blockchain is revolutionizing supply chain and logistics by enhancing the aspects of transparency. Businesses rely on supply chain intermediaries or middlemen, and there is always a problem of authenticity of products with the supply chain information. In the case of blockchain, every process in the supply chain right from the manufacturing process to the delivery process can be put on the block chain meaning that there is a clear record of all the processes in the supply chain (Lakumarapu et al 2024). This helps businesses and consumers to confirm the genuineness of the products and are thus able to avoid receiving substandard goods that are counterfeit. Also, blockchain can stop fraud because it will not allow any alteration of the information that has been stored. For instance, it can be used in the food industry to trace the history of a product from the farm to the consumer's table to eliminate instances where the food product is contaminated. Likewise, in the pharmaceutical sector, blockchain can be applied in the tracking of drugs in the market to eliminate the circulation of fake medicine.

4.3 Healthcare Industry

Blockchain has a great potential in the healthcare sector, for instance in the management of patient records and drugs. Another major issue of the current healthcare system is data silos where patient data are scattered in different systems and providers and can be accessed only in an integrated format at the point of care(Subha, S et al 2024). Blockchain has the potential to create a safe environment for storing patients' records and at the same time give the patients the control over their records and grant other healthcare practitioners access to the data that is relevant to their patient. Blockchain also increases the traceability level of drugs to ensure that the circulated drugs are original and not fake. Recording all the activities of a drug, from production to distribution using blockchain ensure that there is a single and accurate record of a drug so that one can easily tell if the drug is real or fake.

4.4 Government and Public Sector

In the government and public domain specifically, this game changing technology has the application in digital identity, voting, records, etc. The current methods of identity are not very secure and there is always a risk of fraud and hacking of the identity database. In this case, people would be able to own their data and provide the government or banks or service providers with verified documents without the need for multiple documents. It also is useful in improving the voting system since it can provide a record of votes that cannot be altered. This is especially the case in countries that harbour fears of electoral fraud. Voters can vote without any problem and the results of the voting can be checked on the blockchain in an automatic manner. In addition, blockchain will facilitate efficient record keeping since it will provide a secure and unalterable record keeping system for government records such as property deeds, birth certificates and business licenses. This minimizes cases of record alteration and makes records authentic and retrievable when required.

4.5 Education and Certification

In the education and certification domain, blockchain can offer a more secure and efficient system for perks for academic records, and handling of decentralized learning. The conventional methods of confirming qualifications take time and may sometimes involve falsified documents or other related issues. Blockchain also offers the possibility for the educational institutions to provide secure diplomas and certificates, which can be easily verified by employers or other institutions. Also, blockchain can be used for the development of the decentralized learning platform where content and certificates can be shared and verified (Ambhika et al 2024). This creates opportunities for such non-formal education as online classes or obtaining the certifications through the non-traditional courses, which makes process of the checking the qualifications more free and clear.

4.6 Energy and Sustainability

Blockchain is quite popular now when it comes to the application in the energy sector, mainly with the help of carbon credits and peer-to-peer electricity trading. Blockchain can provide a reliable mode of ensuring that carbon credits, which are used to balance out emissions of greenhouse gases, are well accounted for. This way, the carbon credit transactions cannot be double-spent and the reduction of emissions is recorded accurately. However, blockchain enables the decentralized trading of electricity from one consumer to another or even from prosumers in a community, thus eliminating middlemen. This makes it possible to increase the use of renewable energy sources such as solar and wind and facilitate the shift to the sustainable energy systems.

4.7 Real Estate and Legal Industry

Blockchain in both, the real estate and legal industries is allowing for improved property and legal document sales and management through smart contracts. Real estate deals take a long time, require substantial documentation, and the use of several middlemen, and possess high odds of fraud. Through the use of blockchain, records of property ownership can be well recorded and any agreements that are facilitated by the use of smart contracts can be well implemented when certain conditions are met. This makes it easier and eliminates the need to involve third parties like notary or escrow services. In the legal field, smart contracts can also improve the effectiveness and security of such things as contract execution and protection of intellectual property. This minimizes the chances of a conflict and makes it possible to implement the provisions of a contract as agreed. Therefore, the effective application of blockchain may bring about revolutionary changes in numerous fields, including finance, supply chain, healthcare, government servicc, cducation, energy and real estate. This way, blockchain fits many problems of traditional systems by providing more reliable, safe, and less centralized approaches.

5. CHALLENGES AND FUTURE DIRECTIONS

5.1 Scalability and Performance Bottlenecks

Currently, one of the most important issues that relate to blockchain technology is scalability. This simply means the trend of the blockchain solution's performance as the number of transactions or users is increased or extended. The majority of the existing blockchain solutions like Bitcoin and Ethereum have been criticized for slow and high-latency transactions during the peak traffic. The main issues that contribute to the performance issues are the consensus mechanisms inherent in blockchains (Abdollahi, A. et al 2023). For example, there is Proof of Work adopted by bitcoin where one has to use a lot of computational power to solve problems which are a mathematical puzzle. This may result to delay in the confirmation of the transaction and high charges every time there are many users transacting. To solve these problems, a number of solutions are being worked on at the moment, including Layer 2 solutions such as the Lightning Network for Bitcoin and different consensus models such as Proof or Stake (PoS) which is adopted by the Ethereum 2. These solutions are designed to enhance the scalability by either handling the transactions outside the blockchain or by decreasing the computational work needed for the validation of a transaction.

5.2 Regulatory and Legal Challenges

With the increasing adoption of blockchain technology, it is advisable to have proper guidelines that can provide some measures of control on its utilization while at the same time preventing the vice versa. The decentralized and anonymous character of blockchain hinders the regulators to monitor and control the transactions that take place under the blockchain network which raises questions about the illicit activities including money laundering, frauds, and tax evasion. In addition, the development of the applications of blockchain technology is much faster than the establishment of laws and regulations to govern the technology. The appearance of Cryptocurrencies has led to the development of many countries' legislation to regulate such matters as market manipulation, protection of investors, and taxation. Nevertheless, it is still unclear whether cryptocurrencies are acceptable and legal in many legal systems of the world, and this legal uncertainty is a drawback in terms of innovation and introduction. In addition, smart contracts, which is another real-world use of blockchain, are legally ambiguous as to contract enforceability, ownership of property rights, and governing legal authority. Government must establish legal structures that can regulate blockchain since the technology is decentralized, and the records are unalterable, while at the same time protecting the privacy of the users.

5.3 Security Concerns and Privacy Issues

Blockchain is seen as a secure and resistant to alterations because of its use of encryption but it is not fully safe from security threats. Although altering the transaction data incorporated in blockchain is quite challenging due to the consensus mechanisms, namely, the Proof of Work, blockchain networks are vulnerable at different levels. 51% attacks, for instance, happen when a certain group try to gain more hashing power with the aim of changing transactions. While such an attack has not been witnessed on larger blockchains, it has happened in other blockchains with a smaller community (Vanmathi, C. et al 2024). However, it is the applications that are constructed on the basis of blockchain, such as exchanges and wallets, that are attacked by hackers. Another concern is privacy. Although the blockchain ensures that the transactions made are transparent and cannot be altered, this transparency is a problem when it comes to the issue of privacy. For instance, in the case of the public blockchains such as that of Bitcoin, details of a transaction are open to everyone. Still, the executed transactions are associated with certain individuals, which may cause certain privacy concerns. The solutions that are considered are zero-knowledge proofs and privacy coins such as Monero to enable private transactions without compromising the blockchains.

5.4 Adoption Barriers and Industry Readiness

Thus, the use of blockchain technology is still limited in many fields even though it has numerous advantages. The first major challenge that has been identified in the integration of blockchain is the low awareness and knowledge of the technology. Most organizations are yet to fully understand the right approach to adopt blockchain technology that will benefit the organization. However, the process of incorporating blockchain with the existing systems can be a challenging task besides being expensive. And so, for a company to invest in new infrastructure, the technology has to be fairly mature, and this is why it is often hard for them to invest in such technologies. Furthermore, there may be some sectors that are reluctant to adopt the new technology due to self-interest or simply because they are used to the current structure such as financial institutions and governments. In addition, blockchain technology's advancement is also hampered by a shortage of experienced individuals who are knowledgeable in matters concerning blockchain technology construction and management as well as its implementation in different organizations. The major challenge that organizations face today is the scarcity of blockchain developers.

5.5 Future Research Trends and Opportunities

The future of the blockchain technology is promising and there are a number of research prospects and trends concerning the existing and potential disadvantages of the blockchain technology. Among the most promising topics for further studies one can identify:

1.Scalability Solutions: In the constant search for a new consensus, such as PoS, PoA and DAGs there is a design approach for improved blockchains of higher throughput.
2.Interoperability: Different blockchains will come up and, therefore, integration of different blockchains is going to be very crucial (Chen, G. et al 2018). Further studies could also be carried out on the integration of blockchain platforms where different blockchains can be connected to enable the sharing of data and transaction between them in a more fluent manner.
3.Blockchain and IoT Integration: The integration of blockchain with IoT is highly possible and would allow secure interaction between various IoT devices. Blockchain can also help to protect IoT because it has no central control point, and the data stored in the blockchain cannot be changed.
4.Applicability of Blockchain for Sustainability: There is growing interest on how blockchain can provide solutions to some of the existing sustainability issues such as carbon credits, renewable energy, and sustainability ratings. Further research may be conducted on how to improve the use of carbon credits and the energy consumption of the blockchain.

Figure 4. Blockchain technology research trends

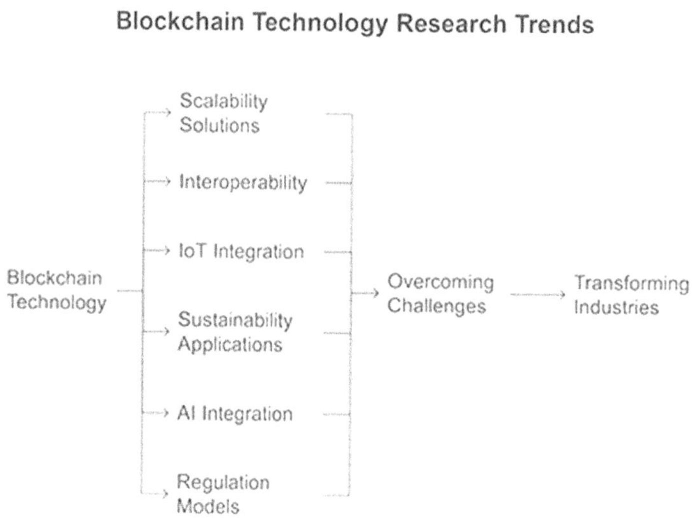

5.AI & Blockchain: There are massive gains from integrating the technology since both cryptocurrencies and AI benefit from decentralization, and combining will lead to better decision-making, improve the accountability of AI models, as well as permitting audits of machine learning algorithms for fairness(Prabha, D et al 2024)(Sreelakshmi et al 2024). Studies in this area may result in the creation of self-contained, self-governed blockchain systems that require few inputs from people.
6.Regulation of blockchain ecosystems: It is important to outline the regulation and governance models of the blockchain. Possible research could be towards legal architectures that will be capable of being implemented within decentralized networks, and which conform to the current legal systems.

On the same note, the vital strengths involved make blockchain technology holds great potential in the future, but it comes with some challenges such as scalability, security, regulation and the final factor the adoption barrier. However, current research and development in scalability solutions, standards, integration with other technologies, and governance models provide a good opportunity to overcome these challenges (Laroiya, C. et al 2020). In the future, the use of blockchain technology can change the industries of the entire world making the processes more effective, transparent and secure.

6. CONCLUSION

In this survey, the basic ideas, important attributes, opportunities, and issues of blockchain technology were discussed. Blockchain technology is essentially a distributed ledger that provides attributes of openness, security, and which cannot be altered, and has paved way for revolutionary changes in various fields. Discussing future work, the major aspects that should be highlighted include decentralization, transparency, security employing cryptography, challenges towards scalability, as well as symbiosis of blockchain with the tendencies of the new generation such as AI, IoT, and cloud. Indeed, it is noteworthy that many industries like financial, medical, and governmental, as well as supply chains also show the possibilities of this technology that is aimed at making work more effective, transparent, and secure. The changes that blockchain brings are long-term and numerous. Blockchain in its full development could revolutionalise interaction in digital space, removing the middleman, increasing reliability and using decentralisation in various fields. It has the possibility to revolutionize industries by presenting them with better secured, more transparent, and more efficient systems which could lead to cutting costs, minimizing frauds, and increasing accountability. In finance, the use of decentralized finance and developing cryptocurrencies may bring financial inclusion and financial services for all, blockchain in supply chain and in healthcare can improve the traceability and the security for the end users as well as for business owners. However, challenges remain. But before forming an overall positive outlook of blockchain technology, four factors have to be considered: scalability, security threats, regulation, and adoption barriers. This brings a lot of opportunities for innovation and research despite the challenges mentioned above. The advancement in scalability and the integration of blockchain with other sectors will go a long way in solving these problems to bring the real change that blockchain is capable of bringing across the world. Thus, blockchain's prospects seem quite favorable, as the technology is constantly being developed and improved. Because of these challenges, blockchain is expected to be central in enhancing decentralisation, effectiveness, and credibility of digital environments in the world.

REFERENCES

Abdollahi, A., Sadeghvaziri, F., & Rejeb, A. (2023). Exploring the role of blockchain technology in value creation: A multiple case study approach. *Quality & Quantity*, *57*(1), 427–451. DOI: 10.1007/s11135-022-01348-2

Ambhika, C., Gayathri, S., Anish, T. P., & Sheena, B. G. (2024, August). Enhancing Predictive Modeling in High Dimensional Data Using Hybrid Feature Selection. In *2024 5th International Conference on Electronics and Sustainable Communication Systems (ICESC)* (pp. 873-879). IEEE. DOI: 10.1109/ICESC60852.2024.10690153

Aswini, J., Sudha, K., Ganesh, G., Subramanian, S., & Ghinea, G. (2024). AI-powered parallel computing architecture and its applications. In *The convergence of self-sustaining systems with AI and IoT* (pp. 23–39). IGI Global. DOI: 10.4018/979-8-3693-1702-0.ch002

Chen, G., Xu, B., Lu, M., & Chen, N. S. (2018). Exploring blockchain technology and its potential applications for education. *Smart Learning Environments*, *5*(1), 1–10. DOI: 10.1186/s40561-017-0050-x

Gadekallu, T. R., Huynh-The, T., Wang, W., Yenduri, G., Ranaweera, P., Pham, Q. V., . . . Liyanage, M. (2022). Blockchain for the metaverse: A review. *arXiv preprint arXiv:2203.09738.*

Khang, A., Chowdhury, S., & Sharma, S. (Eds.). (2022). *The data-driven blockchain ecosystem: fundamentals, applications, and emerging technologies.* DOI: 10.1201/9781003269281

Kumar, P. R., Meenakshi, S., Shalini, S., Devi, S. R., & Boopathi, S. (2023). Soil Quality Prediction in Context Learning Approaches Using Deep Learning and Blockchain for Smart Agriculture. In *Effective AI, Blockchain, and E-Governance Applications for Knowledge Discovery and Management* (pp. 1-26). IGI Global. DOI: 10.4018/978-1-6684-9151-5.ch001

Lakumarapu, S., Nithyanandhan, R., Bhargavi, V. S., Anish, T. P., & Nalini, M. (2024, August). Machine Learning Approaches for Liver Disease Prediction: A Comparative Analysis. In *2024 5th International Conference on Electronics and Sustainable Communication Systems (ICESC)* (pp. 159-164). IEEE. DOI: 10.1109/ICESC60852.2024.10689974

Laroiya, C., Saxena, D., & Komalavalli, C. (2020). Applications of blockchain technology. In *Handbook of research on blockchain technology* (pp. 213–243). Academic press. DOI: 10.1016/B978-0-12-819816-2.00009-5

Prabha, D., Subramanian, R. S., Dinesh, M. G., & Girija, P. (2024). Sustainable Farming Through AI-Enabled Precision Agriculture. In *Artificial Intelligence for Precision Agriculture* (pp. 159–182). Auerbach Publications. DOI: 10.1201/9781003504900-8

Sreelakshmi, S., Babu, A. A., Lakshmipriya, C., Gracious, L. A., Nalini, M., & Subramanian, R. S. (2024, October). Enhancing Intrusion Detection Systems with Machine Learning. In *2024 2nd International Conference on Self Sustainable Artificial Intelligence Systems (ICSSAS)* (pp. 557-564). IEEE. DOI: 10.1109/ICSSAS64001.2024.10760341

Subha, S., Balakrishnan, C., & Anish, T. P. (2024, August). Optimizing Machine Learning Algorithms for Heart Disease Prediction in Healthcare: A Comparative Study. In 2024 5th International Conference on Electronics and Sustainable Communication Systems (ICESC) (pp. 866-872). IEEE. DOI: 10.1109/ICESC60852.2024.10689877

Tariq, M. U. (2024). Application of blockchain and Internet of Things (IoT) in modern business. In *Future of customer engagement through marketing intelligence* (pp. 66–94). IGI Global. DOI: 10.4018/979-8-3693-2367-0.ch004

Vanmathi, C., Farouk, A., Alhammad, S. M., Bhattacharya, S., & Kasyapa, M. S. (2024). The Role of Blockchain in Transforming Industries Beyond Finance. *IEEE Access : Practical Innovations, Open Solutions*, *12*, 148845–148867. DOI: 10.1109/ACCESS.2024.3468611

Zheng, Z., Xie, S., Dai, H. N., Chen, X., & Wang, H. (2018). Blockchain challenges and opportunities: A survey. *International Journal of Web and Grid Services*, *14*(4), 352–375. DOI: 10.1504/IJWGS.2018.095647

Chapter 15
Securing Blockchain Solutions in the Energy Sector:
Mitigating Advanced Persistent Threats (APTs)

Mohammad Alauthman
https://orcid.org/0000-0003-0319-1968
Department of Information Security, University of Petra, Jordan

Amjad Aldweesh
https://orcid.org/0000-0001-9358-1323
College of Computing and Information Technology, Shaqra University, Saudi Arabia

Ala Hamarsheh
https://orcid.org/0000-0002-4736-7331
Computer Science, Faculty of Information Technology, Arab American University, Palestine

Saad Alateef
https://orcid.org/0009-0002-3219-0833
School of Computing Science, Newcastle University, UK

Ammar Almomani
https://orcid.org/0000-0002-8808-6114
Department of Computer Information Science, Higher Colleges of Technology, Sharjah, UAE

Mouhammd Alkasassbeh
https://orcid.org/0000-0001-8396-7441
Princess Sumaya University for Technology, Jordan

ABSTRACT

The energy sector faces expanding risks associated with Advanced Persistent Threats (APTs). These incidents pose challenges for securing power generation, distribution, and other operations that rely on Industrial Control Systems (ICS) and Supervisory Control and Data Acquisition (SCADA) technologies. Entities have employed diversified defensive measures, but attackers continue to refine and perpetuate long-duration intrusions. Blockchain has attracted attention as a potential solution to preserve data integrity, decentralize trust, and ensure tamper resistance, thus helping organizations detect and mitigate advanced cyberattacks. This chapter explores the role of blockchain in protecting energy infrastructures against APTs, drawing on relevant literature, regulatory reports, and technical studies. It reviews blockchain fundamentals, elaborates on how this technology can disrupt the APT intrusion lifecycle, and outlines a conceptual framework for integrating blockchain into ICS architectures. Considerations for scalability, governance, interoperability, and policy are discussed.

DOI: 10.4018/979-8-3373-2439-5.ch015

1. INTRODUCTION

The rapid digital transformation of the energy sector has propelled efficiencies and capabilities in areas such as power generation, transmission, distribution, and renewable integration. Industrial Control Systems (ICS), remote sensors, and advanced data analytics have bolstered operational accuracy and facilitated the emergence of decentralized energy markets. At the same time, this growing reliance on digitally connected infrastructures has broadened the attack surface for malicious actors seeking to compromise or manipulate critical energy operations (Krause, Ernst, Klaer, Hacker, & Henze, 2021).

Among the various cyber threats, Advanced Persistent Threats (APTs) stand out because they blend stealth, persistence, and well-funded capabilities to infiltrate targeted networks and maintain clandestine access. Once these groups or individuals gain unauthorized control or obtain footholds in sensitive systems, they can exfiltrate data, sabotage production, or even inflict physical damage (Department of Energy [DOE], Cybersecurity and Infrastructure Security Agency [CISA], National Security Agency [NSA], & Federal Bureau of Investigation [FBI], 2022). The Stuxnet and Industroyer incidents exemplified how a malware construct could damage or interrupt actual physical processes within ICS environments (Chen & Abu-Nimeh, 2011; Cherepanov & Lipovsky, 2017).

Blockchain technology initially gained recognition through cryptocurrency applications. However, its underlying decentralization, cryptographic hashing, and tamper-evident properties transcend financial use cases. Researchers and industry experts have begun investigating the feasibility of blockchain to secure ICS networks and supply chains, including energy grids (Andoni et al., 2019). By distributing trust among multiple participants, blockchain can diminish the potential for single points of failure. It can also establish immutable logs that make it more difficult for attackers to conceal their intrusion activities. Furthermore, smart contracts can help automate real-time validation of ICS configurations and transactions, limiting the window for sabotage (Choo, Dehghantanha, & Parizi, 2020).

This chapter examines the synergy between blockchain and ICS in mitigating APTs. It first addresses the nature of APTs in the energy sector, the vulnerabilities inherent in ICS architectures, and the specific difficulties in preventing or detecting highly refined threats. Next, it presents blockchain fundamentals, focusing on decentralization, consensus algorithms, and how these properties can curtail advanced intrusions. Subsequent sections detail design considerations for blockchain implementations, including discussions about scalability, governance frameworks, regulatory imperatives, and cross-industry collaboration (Freier, 2024). The chapter then proposes a structured model that illustrates how blockchain can function as a security and monitoring layer integrated with ICS, aiming to dissuade, detect, and impede malicious activities.

Throughout, the discussion draws on case studies that illustrate partial or pilot deployments of blockchain for secure energy trading or sensor data validation. It also assesses challenges, such as integrating with legacy devices and meeting regulatory mandates. By providing a broad view that encompasses technical underpinnings, operational realities, and policy dimensions, this chapter serves as a reference for stakeholders who seek to harness blockchain-based defenses against some of the most persistent and damaging types of cyberattacks afflicting modern energy systems.

2. ADVANCED PERSISTENT THREATS IN ENERGY INFRASTRUCTURES

2.1 Defining APTs in the Context of Energy Operations

Advanced Persistent Threats constitute a category of cyber-incidents characterized by stealth, sophistication, and persistence (Allied Market Research, 2022). Attackers have extensive resources that allow them to probe systems meticulously, exploit zero-day vulnerabilities, and use social engineering to gain an initial foothold. Once inside, APT operators escalate privileges and expand laterally across critical assets, often exfiltrating data or preparing to disrupt operations at a time of their choosing.

In the energy sector, such intrusions can compromise ICS platforms that control turbines, substations, or distribution networks. Energy companies also manage corporate and administrative networks storing intellectual property, supply chain data, and pricing information (Saeed & Gull, 2024). APTs in this domain aim to collect sensitive data, degrade operations, or manipulate energy market conditions. Some intrusions reflect strategic or geopolitical motives, where state-backed entities target foreign energy infrastructures to gather intelligence or threaten system reliability (Jaikaran, 2025).

2.2 Motivations and Typical Targets

Energy enterprises remain vital to the functioning of modern society, making them attractive for attackers seeking leverage. Persistent intrusions into power plants or distribution grids can yield multiple advantages to hostile entities. Some do it for economic benefits, attempting to disrupt markets or extort ransom. Others, including state-sponsored groups, may explore ways to cripple a target nation's critical infrastructure should geopolitical tensions escalate (Prasada & Prasad, 2024). Attackers also target new technologies in renewable integration, energy storage, and carbon credit management, since these areas involve rapidly evolving solutions that may have vulnerabilities.

ICS and SCADA systems are central to delivering power. These systems historically operated as air-gapped or isolated networks, but modernization has increased connectivity to facilitate remote operation, data analytics, and advanced maintenance (Alsayaydeh & Gaid, 2025). This interconnection can ease daily workflows but also augments the cyber-attack surface. Once adversaries compromise a single device with insufficient security patches or guessable credentials, they can move deeper into SCADA modules, distributed control systems, and enterprise resource planning tools.

2.3 Historical Illustrations of APT Incidents

One seminal example is the Stuxnet malware, discovered around 2010, which targeted Iranian nuclear centrifuges. Although not specifically targeting the broader energy grid, Stuxnet brought attention to ICS vulnerabilities. It showed how specialized malware could surreptitiously alter process parameters in ways that degrade or destroy physical machinery (Chen & Abu-Nimeh, 2011). More recent incidents such as the 2015 Ukraine power grid hack underscored the practical possibility of orchestrating widespread blackouts (Lee, Assante, & Conway, 2016). Subsequent threats, including Industroyer and TRITON, displayed advanced automation designed for ICS sabotage (Cherepanov & Lipovsky, 2017).

In many of these episodes, attackers infiltrated central control platforms or engineering workstations and manipulated SCADA commands. This manipulation triggered abrupt changes in physical process setpoints, resulting in partial or extended outages. The compromised utilities struggled to detect malicious traffic. Once the infiltration was realized, it demanded extensive forensic investigations and remediation protocols.

2.4 ICS-Specific Vulnerabilities and Exposure

Industrial Control Systems in the energy sector exhibit vulnerabilities due to outdated hardware and software, inadequate patching, and incomplete security monitoring (Kirkman, Fulton, Hemmes, Garcia, & Wilson, 2024). Many ICS components were developed before cybersecurity became paramount. These devices sometimes have weak authentication or rely on plain text communication protocols. In certain cases, the integration of vendor-specific solutions introduces a complicated patch management process, which delays updates. Some organizations also rely on contractors or third parties for routine maintenance or monitoring. These connections generate additional remote access points that attackers may exploit (Avertium, 2022).

In ICS environments, the priority is operational continuity. Downtimes affect businesses and society, so ICS environments tend to be risk-averse in applying updates or reconfiguring networks frequently. Many utilities do not patch ICS software unless an acute threat emerges, because patch installation might require a complete system restart. This operational constraint offers malicious actors opportunities to exploit known vulnerabilities that remain unpatched for months or years (Chibi, Ait Oualhaj, Fassi Fihri, & El Ghazi, 2024).

2.5 Regulatory Pressures and Shortcomings

Various regions have established requirements for critical infrastructure security, but these often lag behind modern threats (Saeed & Gull, 2024). For instance, the North American Electric Reliability Corporation (NERC) defines Critical Infrastructure Protection (CIP) standards, mandating baseline protections for bulk power systems. Nonetheless, APTs regularly bypass compliance-based controls by exploiting social engineering or zero-day exploits. Standards also vary globally, preventing consistent best practices from being uniformly applied. Combined with the economic and technical complexities inherent in upgrading ICS, the regulatory environment struggles to keep pace with determined adversaries (DOE et al., 2022).

The following table provides an overview of some common APT attack vectors in energy-focused ICS environments and how blockchain-based approaches can help mitigate these tactics.

Table 1. ICS attack vectors and blockchain-based mitigations

Attack Vector	Description	Blockchain-Based Mitigation
Phishing and Credential Theft	Attackers trick employees into revealing credentials or clicking malicious links.	Blockchain-driven identity management with revocation capabilities; distributed auditing of logins.
Exploitation of Unpatched ICS Devices	Attackers capitalize on outdated software in PLCs, RTUs, or SCADA systems.	Secure, immutable update records stored on-chain; proof-of-patch compliance visible to all consortium members.
Lateral Movement via Compromised Nodes	Gaining access to less critical devices, then pivoting to high-value ICS nodes.	Distributed logging that triggers alerts when unusual credential or device usage patterns occur.

continued on following page

Table 1. Continued

Attack Vector	Description	Blockchain-Based Mitigation
Log Tampering and Data Alteration	Attackers delete or falsify event logs to hide their tracks.	Immutable, consensus-validated transaction records that cannot be secretly modified by a single entity.
Malicious Commands to Physical Devices	Sending unauthorized setpoint changes or shutdown signals to turbines/substations.	Smart contract–based policy checks that validate each command against authorized thresholds.

3. FUNDAMENTALS OF BLOCKCHAIN TECHNOLOGY

3.1 Decentralized Architecture and Core Mechanisms

Blockchain technology relies on a distributed ledger that maintains consistent records across a network of nodes (Choo et al., 2020). Each node holds a copy of the ledger, and newly verified transactions are appended to create an unbroken chain of blocks. Every block includes a cryptographic hash referencing the previous block, generating an immutable record of events.

Unlike traditional databases, blockchain does not require a central server. Instead, nodes use consensus algorithms to agree on block validity. This decentralized setup protects the ledger from unilateral tampering or data manipulation. Attackers seeking to alter transaction history must compromise a majority of nodes or computing power, an onerous requirement in well-structured networks.

3.2 Types of Blockchain Networks

Blockchain networks are categorized by their permissioning model. Public blockchains allow open participation without restrictions. Examples include well-known cryptocurrency systems. In contrast, private or permissioned blockchains restrict participation to authorized entities (Borkovcová, Černá, & Sokolová, 2022).

In a consortium blockchain, multiple organizations share administrative roles. Entities collectively manage membership, consensus rules, and transaction validation. Such systems often fit energy use cases where several utilities or regulators collaborate while wanting to maintain control and data confidentiality. The decision regarding whether to employ a fully public or permissioned blockchain depends on regulatory, operational, and privacy considerations (Freier, 2024).

3.3 Consensus Algorithms and Their Implications

The blockchain's security and performance hinge on the chosen consensus protocol. Proof of Work (PoW) involves miners competing to solve computational puzzles, which is effective but energy-intensive. Proof of Stake (PoS) assigns block production based on token ownership or an equivalent stake, thereby reducing computational overhead but introducing other complexities (Choo et al., 2020). Private blockchains often adopt Byzantine Fault Tolerance (BFT) or variants (for example, Practical Byzantine Fault Tolerance, PBFT) to handle node failures more efficiently. Such algorithms allow nodes to validate transactions by reaching an agreement without heavy computation, which can be suitable for ICS contexts that require lower latency (Borkovcová et al., 2022).

3.4 Smart Contracts and Programmable Logic

A smart contract is software code that runs on the blockchain. It automatically executes predefined conditions when triggered by a transaction, removing the need for intermediaries (Meng & Zhu, 2024). These contracts verify data inputs, check compliance, and apply operational rules. In energy systems, a smart contract might confirm that an energy trade aligns with regulation or that a control command for a turbine remains within safe operational bounds.

The immutability of blockchain code can enhance security but also demands rigorous development and testing. Bugs in smart contracts can cause disruptions. In an ICS environment, an accidental bug in a contract controlling an energy resource could produce unexpected downtime or hamper rapid adjustments to supply-demand changes (Bhavana et al., 2024).

3.5 Relevance to Energy Sector Cybersecurity

The energy sector grapples with challenges including data authenticity, trust among multiple stakeholders, and real-time operational needs. Blockchain promises benefits in these areas. If ICS logs are stored on a decentralized ledger, tampering becomes difficult since altering logs on one node does not affect the synchronized copies on other nodes (Chibi et al., 2024). The technology also fosters transparency by providing each stakeholder with a single source of truth for transactions and device states (Krause et al., 2021).

The potential for secure, peer-to-peer energy trading represents another interesting angle. Many energy systems are evolving away from centralized production to distributed generation, where smaller prosumers feed power into local microgrids. Blockchain-based platforms can handle identity, transaction verification, and settlement in near-real time without relying solely on a central broker (Andoni et al., 2019). This approach addresses not only efficiency but also the risk of fraudulent reporting or incorrect billing.

The following table offers a concise comparison of commonly used blockchain consensus protocols, highlighting their suitability for ICS environments where throughput and low latency can be critical.

Table 2. Comparative analysis of blockchain consensus protocols for ICS

Consensus Protocol	Key Advantages	Potential Drawbacks	Use Case in ICS
Proof of Work (PoW)	High security and decentralization	Very energy-intensive; slower transaction throughput	Not ideal for real-time ICS due to latency and cost
Proof of Stake (PoS)	Lower energy consumption; faster finality	Wealth/stake centralization concerns; requires robust penalty mechanisms	Suitable for semi-distributed ICS data logging
PBFT / BFT Variants	Low latency; tolerates some malicious nodes	Requires trust among nodes; limited scalability as network grows	Well-suited for consortium energy networks with fewer participants
Proof of Authority (PoA)	Extremely fast; straightforward governance	Centralized trust in authorities; not fully permissionless	Good for private ICS blockchains with defined node operators

4. APT INTRUSION METHODS IN ICS ENVIRONMENTS

4.1 ICS Architectures and Their Security Gaps

Industrial Control Systems support tasks such as monitoring and controlling physical processes across power plants, refineries, transmission networks, and distribution grids. They include SCADA platforms, Distributed Control Systems (DCS), and Programmable Logic Controllers (PLC), often connected through specialized communication protocols such as Modbus, DNP3, or IEC 61850. These systems have grown increasingly connected to enterprise IT networks, third-party contractors, and cloud-based analytics (Alsayaydeh & Gaid, 2025).

Attackers exploit these interconnections. A single exposed Remote Desktop Protocol (RDP) port or an employee's compromised VPN credentials can be enough to pivot into the ICS network. Once inside, adversaries can modify control logic, alter sensor readings, or issue commands that disrupt physical processes. The historically closed nature of ICS can lull some operators into complacency, but the continuing digitization of energy systems means these networks often remain accessible in ways not fully understood by defenders (Lee et al., 2016).

4.2 Common Tactics, Techniques, and Procedures

APTs generally follow a pattern. They conduct extensive reconnaissance, studying network architecture, employee roles, and vulnerabilities. They craft phishing emails, exploit unpatched software, or employ compromised vendor systems to infiltrate the target. Once a foothold is gained, they often escalate privileges and move laterally, searching for crucial ICS nodes. They may then establish persistence by installing backdoors, forging certificates, or embedding themselves in system processes (Chen & Abu-Nimeh, 2011).

This infiltration and lateral movement can remain undetected for extended periods. Attackers frequently disable or modify logging systems to conceal activities. They also seek administrative credentials to facilitate stealth. Finally, at a moment of strategic choice, they may exfiltrate data or disrupt operations by sending malicious commands to control devices.

4.3 Consequences of Unmitigated APTs

The outcomes of unmitigated APT campaigns in energy infrastructures can be severe. In the case of sabotage, attacks on ICS might cause outages, damage to equipment, or physical harm. Power supply disruptions may result in blackouts, industrial shutdowns, and ripple effects on hospitals, financial institutions, and other critical services (Dragos, Inc., 2025). Even non-disruptive intrusions, such as espionage or theft of intellectual property, can undermine an energy company's competitive edge by exposing proprietary research or trade secrets.

Prolonged intrusions erode trust in the reliability of energy supply. If the public perceives that malicious groups can cause outages, it may weaken confidence in modernizing smart grids or expanding digital solutions. Remediation costs can be high, as compromised equipment sometimes needs replacement, and the entire infrastructure may require reconfiguration (Saeed & Gull, 2024). Insurance premiums can rise, and regulators may levy penalties for failing to safeguard critical infrastructure.

5. LEVERAGING BLOCKCHAIN TO COUNTER APTS

5.1 Mapping Blockchain Security Attributes to APT Risks

Blockchain technology provides distributed consensus and an immutable ledger. These characteristics can interfere with critical elements of an APT's success, particularly in the areas of stealth and log manipulation. Attackers who aim to hide intrusion footprints often delete or alter system logs. By writing events to a decentralized blockchain, defenders can preserve these logs in a tamper-evident manner (Chibi et al., 2024). Even if intruders compromise one node and modify local data, the unchanged ledger copies on other nodes will highlight discrepancies.

Decentralization also eliminates single points of failure. Some ICS rely on central databases or data historians that store details about process states, configurations, and transactions. A well-orchestrated breach of such central repositories can allow attackers to manipulate records and remain undetected. Blockchain-based approaches make such unilateral manipulation far more complex, because an entire group of nodes must be undermined simultaneously (Ulghar, Jahankhani, & Kendzierskyj, 2021).

5.2 Enhancing Device and User Identity Management

In many ICS, identity management remains lax, with shared accounts or default credentials still in place. Blockchain-based identity frameworks can ensure that each device, operator, or service account has a unique cryptographic credential recorded on a ledger that is visible to authorized parties (Ulghar et al., 2021). Whenever a user logs into a control system or a device issues a command, the blockchain can record the transaction along with the identity credential. This arrangement elevates accountability and deters malicious insiders, since forging credentials requires significant effort to tamper with distributed records.

Furthermore, a blockchain-based identity system can distribute certificates or tokens that are updated in near-real time. In typical environments, revoking credentials might be delayed by bureaucracy. With blockchain, once administrators detect suspicious activity, they can revoke or update privileges promptly. The ledger's consensus ensures that revocation is recognized across the network. Attackers who rely on stolen or compromised credentials may thus lose access or find it more difficult to impersonate legitimate accounts.

5.3 Secure Logging and Forensic Advantages

An APT's success often depends on modifying system logs to remove traces of infiltration. Conventional logging solutions store data in centralized servers or clusters that an advanced attacker can eventually compromise. Blockchain-based logging, by contrast, appends every log entry to a chain distributed across multiple nodes. Any attempt to alter or remove a transaction after confirmation is easily detected because the chain's cryptographic linkage breaks if records are manipulated (Prasada & Prasad, 2024).

From a forensic perspective, retaining data in a secure ledger helps investigators reconstruct the sequence of actions leading up to an incident. Since each record references the previous record, the entire ledger's integrity can be verified, producing a robust audit trail for ICS events. Forensic teams can more reliably ascertain the timeline and specific manipulations that occurred, which can speed up remediation and help organizations pinpoint root causes.

5.4 Decentralized Validation in Energy Transactions

Contemporary energy markets are adopting peer-to-peer trading, microgrids, and distributed generation. These arrangements can benefit from blockchain-based transaction validation. In conventional configurations, a central authority or broker verifies trades and sets prices. An attacker compromising that authority could falsify transactions or manipulate pricing data (Bhavana et al., 2024). By distributing these responsibilities among multiple validating nodes, blockchain diminishes the influence of any single compromised point.

In ICS security, data validation extends beyond financial transactions. Commands that instruct a turbine to vary output or adjust voltage levels can also be validated by nodes running consensus protocols (Chibi et al., 2024). If a command does not comply with preconfigured thresholds or user permissions, it is flagged or rejected. This approach can prevent ICS sabotage, assuming response times align with operational necessities.

5.5 Smart Contract-Driven Security Policies

Smart contracts allow defenders to codify security policies that execute automatically. For instance, when certain ICS devices surpass established thresholds, these contracts can trigger alarms or isolate parts of the network (Meng & Zhu, 2024). By embedding rule sets that define acceptable operating boundaries, energy operators reduce dependence on manual checks. If an attacker attempts to push a device beyond safe limits, the smart contract can stop the command before it is processed.

However, care must be taken when implementing such measures. Smart contract code must be accurate, thoroughly tested, and responsive to real-time changes in operational contexts. For example, legitimate operational demands might require short-term changes that exceed normal thresholds, so the system must allow overrides with proper authorization. If these mechanisms become overly rigid, legitimate commands might be blocked, leading to inefficiencies or even physical damage if emergency overrides fail. Nonetheless, properly calibrated smart contracts can add a valuable layer of automated protection against malicious manipulations.

6. ARCHITECTING BLOCKCHAIN SOLUTIONS FOR ICS RESILIENCE

6.1 Choosing Between Public and Consortium Blockchains

Energy systems often involve regulated markets, partnerships among several utilities, and interactions with government oversight agencies. Designing an entirely public blockchain for these sensitive operations may cause data confidentiality concerns and regulatory pushback (Borkovcová et al., 2022). Consortium blockchains, managed by selected stakeholders, offer a middle ground. This arrangement provides decentralization within a trusted circle of participants, ensuring data privacy and control over membership while retaining the security advantages of distributed ledger technology (Freier, 2024).

In a consortium, each entity runs at least one validating node. The group collectively decides how to add new members, handle protocol updates, or address misbehaving nodes. This structure can align with existing governance models in the energy sector, where multiple operators already coordinate operations, manage shared infrastructure, and comply with region-specific rules.

6.2 Implementing Scalable Consensus Mechanisms

In ICS networks, sensor data and control commands can be frequent. If each piece of data requires a blockchain transaction, the system may quickly grow unwieldy. Traditional Proof of Work blockchains can only handle a limited number of transactions per second. Therefore, many energy-specific blockchains adopt consensus protocols like PBFT, which can handle higher throughput, or hybrid solutions that bundle transactions into blocks off-chain and periodically anchor them to the main chain (Andoni et al., 2019).

Scalability is not solely a technical issue. It also encompasses the design of data retention policies. Some ICS logs are very large. Storing all log data on the main blockchain could become unsustainably large and hamper query efficiency. Energy operators might store only cryptographic hashes or summaries on-chain while archiving the full logs in distributed databases (Meng & Zhu, 2024). This hybrid model strikes a balance between immutability and data manageability.

6.3 Governance Structures and Multi-Stakeholder Trust

In the energy sector, governance for blockchain initiatives involves representatives from utilities, regulatory agencies, equipment manufacturers, and sometimes consumer groups. This arrangement must clearly define node ownership, decision-making authority, and the mechanism for protocol evolution (Freier, 2024). If the system needs urgent updates due to a new cyber threat, governance processes must allow timely action without devolving into unilateral control by one party.

Defining how to remove a malicious or compromised node is critical. Since a node might store confidential transaction records, expelling it from the network demands a transparent and secure process. Operators also require contingency plans for the scenario where a node attempts to fork the ledger or refuses to update software. Formal governance frameworks, which can be embedded in smart contracts or anchored in legal agreements, help maintain trust among the consortium (Choo et al., 2020).

Below is a conceptual diagram illustrating how multiple stakeholders—utilities, regulators, and third-party vendors—could share and manage a consortium blockchain. Each node operator takes part in consensus, and a predefined governance policy decides how to add or remove participants.

Figure 1. Multi-stakeholder governance model for ICS blockchain

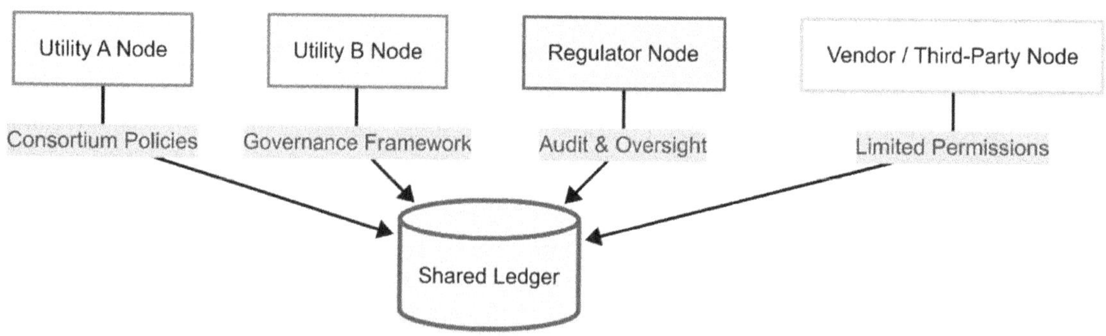

(The figure shows each stakeholder node connecting to a shared blockchain layer. Consortium policies define membership, consensus rules, and updates. Regulators maintain an oversight role, and vendors may hold restricted roles or have partial read/write privileges.)

6.4 Interfacing with Legacy ICS Components

Many ICS components remain entrenched in legacy environments that do not natively support blockchain interactions. Therefore, integration often relies on gateway devices or middleware layers that communicate with controllers and sensors using standard industrial protocols, then translate relevant data to blockchain transactions (Krause et al., 2021). These gateway architectures must guarantee real-time performance so that operational commands and safety instructions are not delayed by consensus processes.

A gradual or phased implementation might be preferable to avoid disruptions. Selected ICS sub-systems, such as sensor data streams or security event logs, could be the first to be anchored on the blockchain. Over time, the scope can expand to more critical components once operators confirm reliability and refine the architecture. As new ICS equipment is installed or older devices are updated, manufacturers can incorporate blockchain compatibility more directly (Kirkman et al., 2024).

6.5 Addressing Privacy and Compliance

Energy data can include market-sensitive information, personal consumption details, or cross-border trade records. Storing these details on a public ledger may violate privacy regulations such as the General Data Protection Regulation (GDPR) in the European Union (Freier, 2024). Private and consortium blockchains mitigate some of these issues by restricting read and write permissions to authorized nodes. Yet even in these setups, designing transaction content must ensure compliance.

Certain solutions employ zero-knowledge proofs or partial encryption to hide transaction details while still allowing verification. Others store only hashed references to off-chain data. The appropriate approach depends on local laws, enterprise preferences, and the sensitivity of the data. Stakeholders should design policies for data retention periods, the right to deletion under GDPR, and procedures for responding to legal inquiries.

7. PROPOSED FRAMEWORK FOR APT MITIGATION WITH BLOCKCHAIN

7.1 Conceptual Model Overview

A layered framework integrates blockchain into the ICS architecture, focusing on synergy between decentralized logging, identity management, and automated threat response. The proposed model sits alongside existing ICS layers rather than replacing them. It uses blockchain nodes that run on dedicated servers within each utility or authorized partner. Data from sensors, controllers, and user activities flow into this blockchain layer through secure gateways. Once on-chain, transactions are validated and become tamper-evident (Prasada & Prasad, 2024).

Smart contracts embedded in the system scrutinize incoming data, device commands, and user credentials. If an anomaly arises, an automated response can isolate suspicious commands, prompt an alert, or revoke credentials. Security information and event management (SIEM) tools then parse the blockchain data to identify patterns indicative of lateral movement or data exfiltration attempts (Chibi et al., 2024).

Figure 2. Conceptual blockchain integration for ICS security

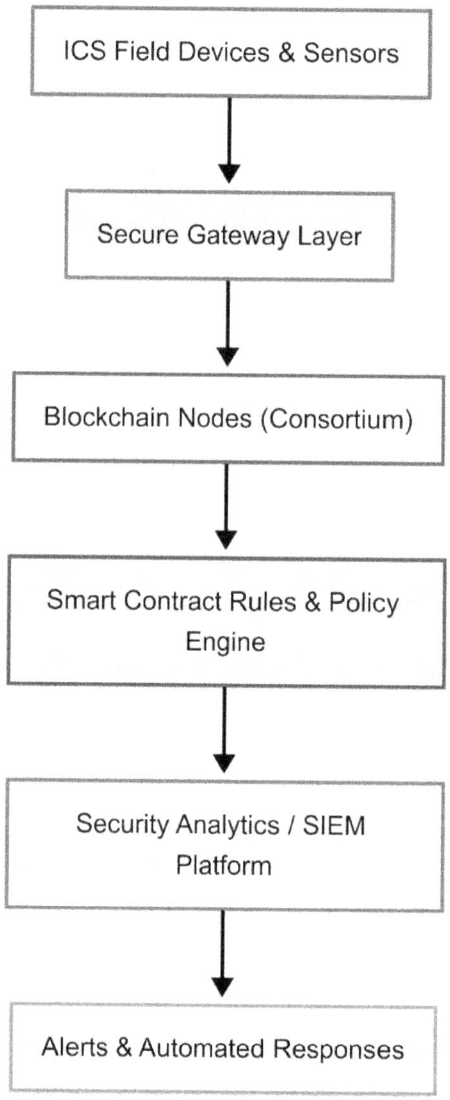

(The figure depicts how ICS data funnels through secure gateways to the blockchain, where consortium nodes verify transactions under established policies. Smart contracts manage the logic for anomaly detection and potential blocking. Security analytics tools monitor the ledger, generating alerts and orchestrating responses when suspicious activity is detected.)

7.2 Identity-Based Access Control

In this framework, each ICS device and each human operator has a unique blockchain-based identity. The credentials map to a public-private key pair, with relevant metadata stored on-chain. Operators in control rooms or remote offices must validate their identities through blockchain transactions before issuing ICS commands. Devices that attempt to register new states or transmit sensor data sign these transactions digitally, verifying authenticity.

If a malicious individual steals an operator's credentials, rapid revocation can occur on the chain. Validators at each node will reject future commands from the revoked identity, limiting the potential impact of compromised credentials. This approach requires robust key management and provisioning processes to replace or rotate keys safely.

7.3 Decentralized Logging of Events and Commands

The framework envisions that ICS events, such as sensor measurements, operator commands, and error logs, are hashed or fully recorded on the blockchain. Full recording might not be feasible for high-frequency data streams, so the system can store summaries or critical event markers, anchoring them to the chain at regular intervals (Meng & Zhu, 2024).

When an event is logged, the transaction includes a timestamp, device ID, and a reference to the data. This log is distributed across all consortium nodes, creating a tamper-resistant historical record. Even if an attacker alters logs on one gateway or ICS server, the distributed ledger ensures a consistent record remains among the other nodes.

7.4 Smart Contract-Driven Anomaly Detection

While specialized intrusion detection systems can analyze raw network data, the proposed blockchain layer can use smart contracts to enforce high-level rules. For instance, a contract could stipulate that any command to alter generator output by more than 20% within a 10-minute window triggers a secondary verification. This verification might require multi-party approval or a second signature from a designated safety officer. If these conditions are not met, the contract declines the command before it propagates to the ICS (Ulghar et al., 2021).

Multiple smart contracts can run in parallel, each addressing distinct operational areas. One contract might govern trading transactions in a peer-to-peer market, while another oversees critical device set-points. The entire set of contracts can be updated through consortium governance procedures, ensuring the rules remain current with evolving operational needs.

7.5 Automated Response and Collaborative Forensics

When the system detects suspicious activity, such as a sudden escalation of privileges, it can automatically enact predefined response strategies. The framework might isolate the compromised node, block suspicious accounts, or freeze trades in a localized sub-network. Security teams then examine the on-chain logs to reconstruct events. Because these logs are hardened by the consensus process, it is less likely that the attacker could have manipulated them to hide intrusion tracks (Prasada & Prasad, 2024).

Over time, forensics data can be pooled across multiple energy entities in a consortium, further enriching threat intelligence. Organizations can analyze patterns of known threat actors or repeated infiltration attempts. This collective knowledge, secured on-chain, helps each participant adapt defenses. For instance, if one entity observes malicious signatures in a new piece of malware, it can update the shared ledger so that others can automatically reject matching connections.

7.6 Table Summaries Within the Framework

Although the user prefers paragraphs, the following table structures certain aspects of the proposed framework succinctly. It is integrated here rather than appended at the end to aid clarity.

Table 1. Key components in the proposed blockchain-enhanced ICS security framework

Component	Purpose	Integration Method
Blockchain Layer	Maintains distributed and tamper-evident records of ICS events	Consortium nodes deployed at each participating entity
Smart Contracts	Enforces security policies, operational thresholds, and identity checks	Embedded code that validates each incoming command or data transaction
Secure Gateways	Convert ICS protocol data into blockchain transactions	Middleware bridging field devices with the ledger
Identity Registry	Stores cryptographic credentials for operators and devices	On-chain registry with real-time revocation
Incident Response	Allows automated and manual reactions to suspected breaches	Smart contracts trigger containment, alerts, or forensic record labeling

(Note: The numbering above refers to this table within Section 7.6, separate from the newly added tables in other locations.)

8. REGULATORY AND POLICY CONSIDERATIONS

8.1 The Evolving Legislative Context

Regulatory agencies have begun examining how distributed ledger solutions intersect with energy market rules, privacy mandates, and reliability standards (Saeed & Gull, 2024). Certain jurisdictions have developed specialized policies for blockchain-based energy trading, but these often address financial and transactional elements without deeply focusing on ICS security. Therefore, organizations seeking to deploy blockchain in ICS must interpret or request clarifications from regulators regarding compliance obligations.

Global regulatory variation adds further complexity (DOE et al., 2022). Where one country might approve the use of blockchain-based logs as valid evidence for compliance reporting, another may question the immutability principle if it conflicts with privacy laws stipulating user data deletion. Consequently, multi-national consortium blockchains must adopt flexible governance structures that segment data and align with each region's legal frameworks.

8.2 Balancing Transparency with Confidentiality

Regulators often emphasize transparency in critical infrastructure operations. They want to see thorough logs and traceability for system changes. Blockchain aligns well with this demand by enabling distributed audit trails. However, regulators and organizations alike worry about inadvertently exposing sensitive operational or business data (Freier, 2024). In a public blockchain, any participant could scrutinize transaction details. Even in a consortium model, unscrupulous insiders at one utility might glean confidential information about another utility's ICS.

Hence, the design must restrict read access to authorized nodes only. Transaction payloads may be encrypted, with the ledger storing only cryptographic hashes or references. This approach supplies immutability and traceability while limiting raw data exposure. Blockchain security does not automatically guarantee data privacy, so specialized cryptographic mechanisms, such as zero-knowledge proofs, may be necessary for certain transactions (Choo et al., 2020).

8.3 Liability and Legal Enforcement

Questions arise about who holds liability if a blockchain-based ICS security system fails to prevent an APT or inadvertently blocks legitimate operations. Conventional IT systems typically have well-defined ownership and software vendor obligations. A multi-stakeholder blockchain can complicate accountability since multiple entities share node operation and code validation responsibilities (Borkovcová et al., 2022).

Energy regulators and legal experts sometimes insist on binding legal agreements or service-level agreements among consortium members. These documents outline each party's duties for node security, data handling, and incident response. Smart contracts themselves might also specify financial penalties or dispute arbitration processes if a member fails to maintain required security measures. Although the technology offers decentralized trust, real-world liability must still be addressed in structured, legally recognized frameworks.

8.4 Policy as a Catalyst for Collaborative Security

When regulators acknowledge the potential of blockchain in securing ICS, they can encourage or mandate pilot programs. These initiatives then demonstrate feasibility and reveal operational constraints, enabling iterative improvements (Nai Fovino et al., 2021). Policymakers can also foster cross-sector intelligence sharing, urging energy operators, cybersecurity vendors, and other industrial participants to share data on new threats.

Financial incentives, including grants or tax benefits, can stimulate the adoption of blockchain-based cybersecurity solutions. Governments may see this as a strategic move if it helps deter crippling cyber incidents. Over time, successful implementations can shape future standards and best practices, establishing a baseline for decentralized trust solutions across critical infrastructure.

9. CASE STUDIES OF BLOCKCHAIN INTEGRATION IN ENERGY

9.1 Peer-to-Peer Microgrids and Security Implications

Several pilot projects in Europe and North America have tested microgrids where local prosumers can trade surplus energy via blockchain. Although the primary impetus is transactional efficiency and cost reduction, participants report improved resilience against common cyber fraud tactics (Andoni et al., 2019). In typical setups, each household or business uses a small meter or gateway that registers transactions on a permissioned blockchain. Attempts to falsify meter data are harder to conceal since the ledger provides a consensus-based validation mechanism.

Additionally, these experiments provide insights into how a limited set of nodes can coordinate consensus rapidly. Because microgrid transactions are generally smaller in volume than national markets, blockchain overhead remains manageable. Some communities have begun extending the model to incorporate limited ICS monitoring, where system status data is also hashed to the blockchain, serving as an immutable reference for performance audits (Bhavana et al., 2024).

9.2 ICS Monitoring in Renewable Energy Parks

Wind farms and solar arrays rely on ICS to optimize turbine pitch, panel angle, or battery storage levels. In some instances, operators integrate blockchain-based logs for performance and maintenance records. Each sensor reading, or aggregated set of readings, is cryptographically hashed and anchored to a permissioned ledger (Chibi et al., 2024). This method ensures that no single party can retrospectively manipulate performance data to mask equipment failures or inflate energy generation metrics.

In one reported case, the blockchain record was used to settle insurance claims after a storm damaged several turbines. The insurer requested verification that the turbines were operating below recommended wind-speed thresholds at the time of failure. The hashed logs offered a reliable record that had been corroborated by multiple nodes, including those operated by the insurer. This example emphasizes that blockchain use extends beyond direct cybersecurity advantages, producing tangible benefits in auditing and trust across different stakeholders.

9.3 Cross-Border Energy Exchanges Using Blockchain

Regions that share borders often exchange electricity to balance loads and manage surpluses. Traditional cross-border trading can be slow, reliant on a central clearinghouse, and prone to discrepancies. Some operators have begun experimenting with blockchain to automate settlement, employing smart contracts that convert metered flows into near-real-time settlements (Freier, 2024). This process reduces the time lag between delivery and payment, improving liquidity for energy producers and importers.

Security aspects arise when validating the data that triggers settlements. If an attacker injects false meter readings, they might tilt the balance of payments or cause blackouts. In a blockchain-based arrangement, each country's utility runs validating nodes. The resulting shared ledger ensures that no single participant can unilaterally alter flow data. Smart contracts can also incorporate ICS checks to confirm that actual device readings match what is reported for settlement. This robust alignment between operational data and financial transactions deters fraudulent manipulation.

9.4 Lessons Learned and Ongoing Limitations

Early experiences underline the importance of careful system design and user training. Some projects confronted integration difficulties between ICS protocols and blockchain gateways, leading to communication bottlenecks or data format inconsistencies (Kirkman et al., 2024). Scalability and latency remain technical considerations, particularly when dealing with high-volume real-time data streams. In addition, the privacy of sensitive operational data remains a top concern, necessitating partial or encrypted on-chain storage (Borkovcová et al., 2022).

The cost of running blockchain nodes can also deter smaller energy companies, especially if the system demands specialized hardware or large-scale computing resources. Nonetheless, consortia that share infrastructure and adopt efficient consensus algorithms have reported cost benefits over time. These implementations highlight how blockchain can serve as one of multiple layers in a comprehensive cybersecurity strategy, rather than a singular solution that displaces all existing measures.

10. FUTURE HORIZONS IN BLOCKCHAIN-BASED APT DEFENSE

10.1 Zero-Trust Integration

Zero-trust architectures assume that no user or device is inherently trustworthy. They enforce continuous authentication, least-privilege access, and micro-segmentation. Linking blockchain solutions with a zero-trust philosophy can create layered defenses. Blockchain provides distributed evidence of device and user states, while zero-trust ensures that each request is validated against contextual and historical patterns (DOE et al., 2022). This synergy can hamper APT operators who rely on continuous lateral movement, since they must re-authenticate at every stage, facing heightened chances of detection.

Below is a conceptual depiction of how zero-trust principles can work alongside a blockchain identity layer. Each action from a user or device is verified and recorded on-chain, while micro-segmentation further reduces the ability for attackers to move laterally.

Figure 3. Layered zero-trust architecture with blockchain

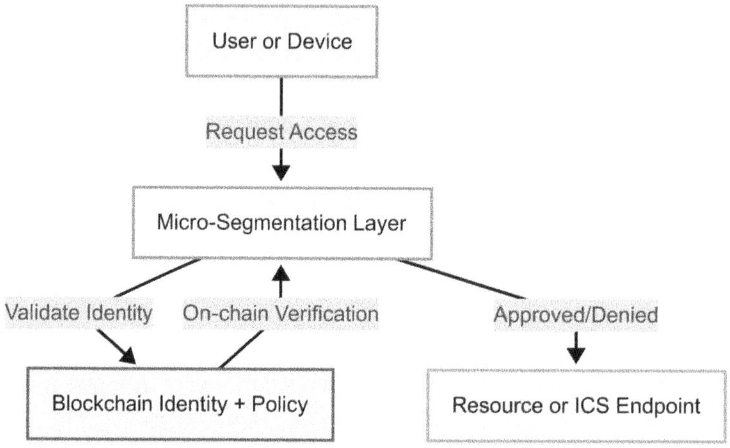

(The figure shows that every access request from a user or device goes through a micro-segmentation layer, which queries blockchain-based policies and identities. The blockchain verifies credentials and usage history, returning an approve/deny decision.)

10.2 Post-Quantum Blockchain Security

Quantum computing threatens classical cryptographic systems by potentially cracking widely used algorithms. Blockchain solutions deployed today may be vulnerable in a future scenario where an advanced adversary wields quantum capabilities (Choo et al., 2020). Energy providers seeking long-term security might evaluate post-quantum cryptography approaches, including lattice-based key exchanges or hash-based signatures. This forward-looking design ensures that ICS logs and identity credentials remain confidential and valid even when quantum technologies mature.

As a preview of how blockchain might evolve in tandem with new cryptographic methods, the following diagram illustrates a post-quantum–ready blockchain module for an ICS environment.

Figure 4. Example post-quantum-ready blockchain implementation for ICS

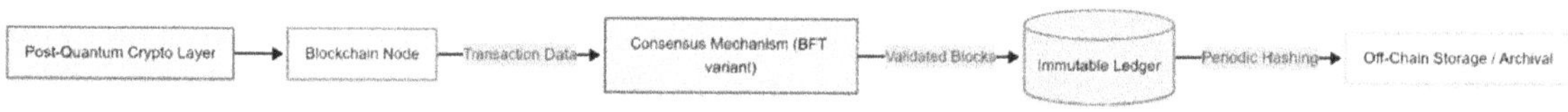

(The diagram shows how a post-quantum cryptographic layer can sit atop the blockchain node's communication protocols, ensuring that even if adversaries possess quantum decryption capabilities, the ICS transactions and logs remain secure.)

10.3 AI-Driven Threat Intelligence and Automated Defense

As the volume of ICS data grows, machine learning tools can scan network traffic, logs, and system behavior to identify anomalies indicative of APT infiltration. Blockchain can facilitate the secure sharing of threat intelligence across multiple entities (Prasada & Prasad, 2024). Each node can publish newly discovered indicators of compromise (IOCs) to the shared ledger, allowing others to update detection rules. Smart contracts could automatically adjust firewall settings or revoke access for suspicious IP addresses, compressing the time from threat detection to mitigation.

Still, caution is warranted. AI-generated classifications can produce false positives. If a blockchain-based defense system automatically blocks a critical control command, it might inadvertently endanger safety or compromise system availability. Therefore, a hybrid approach that employs AI for detection and relies on structured, transparent governance for final enforcement is often seen as prudent (Chibi et al., 2024).

10.4 Fostering Cross-Sector and Global Collaboration

APT defense in the energy sector goes beyond individual utilities. Attackers frequently target supply chains, third-party service providers, and adjacent industries. Blockchain consortia can accommodate multiple sectors, enabling secure data exchange regarding vulnerabilities and infiltration methods (Dragos, Inc., 2025). The sector's stability depends on organizations cooperating rather than competing in cybersecurity. Government agencies, private utilities, equipment manufacturers, and cybersecurity specialists must collaborate to share threat intelligence, standardize best practices, and coordinate incident response strategies.

International alliances can also emerge through frameworks that respect regional regulations while still providing a global ledger of known threats. Diplomatic agreements might formalize the data to be shared across borders. Over time, robust information exchange networks can form, reducing each participant's vulnerability to zero-day or emerging APT tactics. Blockchain underpins this collaboration, offering a neutral, tamper-evident repository of shared intelligence (Nai Fovino et al., 2021).

11. EXPANDED ANALYSIS OF IMPLEMENTATION CHALLENGES

11.1 Technical Integration Obstacles

The integration of blockchain and ICS technologies can face hurdles. Many ICS devices operate on real-time operating systems not designed to accommodate blockchain nodes. Even the gateway approach, where external hardware or software aggregates and signs data before committing it to the ledger, requires robust security to prevent tampering before data enters the blockchain (Krause et al., 2021).

Network bandwidth and latency also become concerns, especially in geographically distributed grids. Some sub-stations are located in remote areas with limited connectivity. If the consensus algorithm demands frequent communication among nodes, it may create performance bottlenecks. Developers might adopt asynchronous or partial consensus protocols that collect local transactions and batch them for sporadic updates. These methods reduce overhead at the cost of real-time finality.

A further technical challenge arises in the synergy between blockchain-based security rules and the ICS's operational technology (OT) environment. OT has stringent timing and reliability demands. The complexity of modern blockchains means that major changes in governance or code updates might take time to finalize. ICS processes cannot always pause or adapt swiftly to these updates. Thorough testing is needed to ensure that blockchain modifications do not impede critical operations.

Below is a summary of several common technical and organizational challenges encountered when integrating blockchain into ICS for APT mitigation, alongside potential solutions to address these issues.

Table 3. Challenges and proposed solutions for blockchain-integrated ICS

Challenge	Description	Proposed Solution
Real-Time Constraints in ICS	ICS processes demand rapid responses and minimal latency	Use lightweight or PBFT-like consensus; batch transactions; optimize gateways
Limited Connectivity at Remote Sites	Substations often have poor network infrastructure	Implement local data buffering; allow asynchronous updates to the ledger
Legacy Device Incompatibility	Older PLCs and RTUs lack blockchain-ready interfaces	Introduce secure gateway/middleware that translates ICS protocols to ledger transactions
Organizational Resistance to Change	ICS operators fear disruption of stable processes	Begin with non-critical pilots; provide clear ROI and training to stakeholders
Managing Large Volumes of Sensor Data	Excessive on-chain storage is impractical and costly	Store hashed summaries on-chain; maintain full logs in secure off-chain databases

11.2 Organizational and Cultural Barriers

Organizations often exhibit reluctance to adopt new technologies that appear to complicate well-established operations. ICS engineers, who prioritize stability and availability, may be skeptical of introducing blockchain layers and might require thorough explanations of how these do not disrupt real-time process controls (Ulghar et al., 2021). Leadership buy-in might depend on clear cost-benefit analyses and pilot results demonstrating minimal operational risk.

Additionally, an APT defense posture that relies on distributed validation calls for new skill sets among staff. Professionals with knowledge of both ICS engineering and blockchain architecture are rare. Training or hiring can pose logistical challenges. Cross-departmental collaboration between IT security teams, ICS engineers, and legal experts is required. This collaboration can be difficult if these groups rarely interact under normal circumstances. Overcoming organizational silos is thus crucial to implementing a cohesive security strategy.

11.3 Standardization and Interoperability

Given the global nature of energy markets, multiple organizations have begun forming specialized groups to define standards for blockchain-based systems. However, there is no single global standard for ICS security using blockchain. Ad hoc approaches often emerge in pilot projects, resulting in solutions that lack portability. Different blockchains use varied consensus protocols, data models, and smart contract languages, leading to interoperability limitations (Borkovcová et al., 2022).

If a utility invests significantly in developing a private blockchain solution, but the national regulator or a regional partner requires using a different blockchain protocol, integration complexities arise. Converters and bridging layers might be needed to link different ledgers, but these bridging solutions can introduce new vulnerabilities. Some industry consortia are exploring open interfaces and shared schemas that unify data representation, but standardization remains a work in progress (Freier, 2024).

11.4 Legal Ambiguities for Evidence and Compliance

One advantage of blockchain is that it can serve as a tamper-evident record, potentially strengthening the evidentiary weight of ICS event logs in legal contexts. However, some courts or regulatory agencies have yet to formalize guidelines on whether blockchain-based records automatically qualify as legally binding evidence. Attorneys may question the authenticity of digital signatures or the chain of custody for a ledger that spans multiple jurisdictions (Jaikaran, 2025).

Various privacy laws can also require partial redaction or erasure of personal data, which is in tension with blockchain's immutability principle. Solutions might revolve around storing only hashed references on-chain. Yet, data subjects or regulators might challenge whether hashing truly fulfills the right to be forgotten if the data can still be indirectly referenced and potentially reversed with enough computing power. Energy operators must navigate these novel legal intricacies in parallel with adopting the technology.

12. RECOMMENDATIONS AND BEST PRACTICES FOR PRACTITIONERS

The following paragraphs distill best practices for designing and implementing blockchain systems that mitigate APTs in energy ICS, presented in a narrative style rather than bullet points.

It is advisable for organizations to begin with pilot deployments focused on less critical data streams before integrating blockchain into mission-critical controls. For instance, operators might first anchor system logs or sensor metadata on a private ledger to improve auditability. This approach allows teams to refine processes, address latency or bandwidth constraints, and train personnel without risking major disruptions. Once successful pilot results are established, the scope can expand to more complex ICS processes, such as automated setpoint adjustments or cross-border energy trading settlements (Andoni et al., 2019).

Choosing the appropriate consensus mechanism is another important step. Energy ICS often necessitates near-real-time data validation, so protocols with high throughput and low latency, such as PBFT or proof-of-authority variants, are frequently more suitable than resource-intensive Proof of Work. Matching the consensus with operational requirements ensures that the blockchain does not become a bottleneck or introduce operational risks (Choo et al., 2020).

Governance frameworks should be carefully delineated in consortium agreements. This includes setting forth responsibilities for node hosting, system updates, and handling disputes or forks. Formal structures help maintain consistency in network membership and reduce confusion during threat incidents, such as the identification and removal of compromised nodes (Borkovcová et al., 2022). Similarly, organizations must develop standardized procedures for patching both ICS devices and the blockchain application layer, clarifying how changes are tested and approved.

Identity management remains central to mitigating APTs. In practice, robust key management solutions, such as hardware security modules, can safeguard private keys. Human users and ICS devices should each have distinct credentials, and the system must facilitate speedy revocation when suspicious activity arises (Ulghar et al., 2021). Periodic reviews of credentials and role-based access policies will further ensure that only authorized entities can issue commands or view sensitive data.

Privacy regulations and compliance mandates also require careful handling of personal or market-sensitive data. Organizations might store only transaction hashes or encrypted payloads on-chain while retaining complete records in protected off-chain databases. Access to these databases should be restricted, and cryptographic proofs should be validated through the ledger. This model meets the twin objectives of tamper-evident logging and privacy law adherence (Freier, 2024).

Finally, integrated threat intelligence exchange can significantly enhance the detection and response to APT campaigns. Consortium participants benefit from real-time alerts on newly detected infiltration tactics. Smart contracts can automate partial or full revocation of compromised identities across the network, shrinking the window of opportunity for attackers (Prasada & Prasad, 2024). This coordinated strategy embodies a collective defense paradigm, aligning with the recognition that resilient energy infrastructure is a shared responsibility among all stakeholders.

13. CONCLUDING DISCUSSION

The energy sector's digital evolution brings remarkable efficiencies but also invites relentless advanced threats. APT actors exploit ICS vulnerabilities to access and control critical infrastructures, as demonstrated by documented cases such as Stuxnet, Industroyer, and other significant intrusions. Traditional protective measures, although valuable, often prove insufficient against stealthy adversaries possessing considerable resources and expertise. The inherent difficulty in securing legacy ICS, combined with complex regulatory environments, provides malicious actors with numerous attack vectors.

Blockchain technology offers a path toward resilience. By establishing decentralized trust, immutability, and tamper-evident records, blockchain can limit the typical maneuverability of APT operations. Attackers seeking to erase evidence of intrusion or to inject false data will find these tasks more difficult when facing distributed consensus and encrypted identities stored across multiple nodes. Smart contracts further automate the enforcement of security policies, verifying that ICS commands comply with authorized parameters. The combination of these mechanisms can deter intrusion attempts, accelerate detection, and aid in forensic analysis (Chibi et al., 2024).

Nonetheless, blockchain integration in ICS is not a trivial endeavor. It requires reconciling real-time operational demands with consensus protocols, bridging legacy devices through secure gateways, and creating robust governance models that incorporate regulatory norms, liability considerations, and multi-stakeholder trust. Pilot implementations in microgrids, renewable parks, and cross-border trading highlight that blockchain can reduce fraud, improve auditability, and bolster ICS security in certain contexts, though scalability and privacy issues continue to demand attention (Andoni et al., 2019; Borkovcová et al., 2022).

Moving forward, the synergy between blockchain and advanced security paradigms such as zero-trust or post-quantum cryptography can further elevate ICS defenses. Multi-party collaboration remains central, as energy operators, technology vendors, regulators, and threat intelligence specialists must align to address rapidly evolving APT tactics. In the broader context, governments might support research, development, and standardization efforts, encouraging the formation of global frameworks that accommodate region-specific mandates while promoting consistent security practices.

Although blockchain does not constitute a comprehensive or sole solution, it can strengthen the layers of defense that energy infrastructures require. Its capacity to share data securely, enforce consistent logs, and automate essential checks through smart contracts is particularly relevant where ICS integrity is at stake. Incremental adoption, guided by rigorous testing, cross-disciplinary training, and robust governance, is likely the most prudent path. Over time, these decentralized architectures, when blended with conventional cybersecurity measures, can form a formidable bulwark against advanced persistent threats targeting one of society's most critical and interconnected sectors.

REFERENCES

Allied Market Research. (2022). *Advanced Persistent Threat (APT) market statistics, 2021–2030.* Retrieved from https://www.alliedmarketresearch.com

Alsayaydeh, J. A. J., & Gaid, A. S. A. (2025). Security of smart grid: Cybersecurity issues, potential cyberattacks, major incidents, and future directions. *Energies*, *18*(1), 141. DOI: 10.3390/en18010141

Andoni, M., Robu, V., Flynn, D., Abram, S., Geach, D., Jenkins, D., McCallum, P., & Peacock, A. (2019). Blockchain technology in the energy sector: A systematic review of challenges and opportunities. *Renewable & Sustainable Energy Reviews*, *100*, 143–174. DOI: 10.1016/j.rser.2018.10.014

Avertium. (2022, September 13). *The top 5 cyber threats in the energy sector.* Avertium Threat Report (blog).

Bhavana, G. B., Anand, R., Ramprabhakar, J., Meena, V. P., Jadoun, V. K., & Benedetto, F. (2024). Applications of blockchain technology in peer-to-peer energy markets and green hydrogen supply chains: A topical review. *Scientific Reports*, *14*(1), 21954. DOI: 10.1038/s41598-024-72642-2 PMID: 39304698

Borkovcová, A., Černá, M., & Sokolová, M. (2022). Blockchain in the energy sector—Systematic review. *Sustainability (Basel)*, *14*(22), 14793. DOI: 10.3390/su142214793

Chen, T. M., & Abu-Nimeh, S. (2011). Lessons from Stuxnet: A new era of cyber warfare. *Computer*, *44*(4), 91–93. DOI: 10.1109/MC.2011.115

Cherepanov, A., & Lipovsky, R. (2017). *Win32/Industroyer: A new threat for industrial control systems.* ESET White Paper. Retrieved from ESET.com

Chibi, N. T., Ait Oualhaj, O., Fassi Fihri, W., & El Ghazi, H. (2024). A novel approach based on machine learning, blockchain, and decision process for securing smart grid. *IEEE Access : Practical Innovations, Open Solutions*, *12*, 33190–33199. DOI: 10.1109/ACCESS.2024.3370239

Choo, K.-K. R., Dehghantanha, A., & Parizi, R. M. (Eds.). (2020). *Blockchain cybersecurity, trust and privacy.* Springer., DOI: 10.1007/978-3-030-38181-3

Department of Energy (DOE), & the Cybersecurity and Infrastructure Security Agency (CISA). National Security Agency (NSA); & Federal Bureau of Investigation (FBI). (2022). *APT cyber tools targeting ICS/SCADA devices (Joint Cybersecurity Advisory AA22-103A, April 2022).* U.S. Department of Energy.

Dragos, Inc. (2025). *8th Annual Year in Review – 2025 OT Cybersecurity Report.* Dragos. (Retrieved from dragos.com)

Freier, A. (2024). Blockchain in the energy sector: An advancing technology to tackle global climate change? ibidem Press (Columbia University Press).

Jaikaran, C. (2025). *Cybersecurity: Selected cyberattacks, 2012–2024 (CRS Report No. R46974).* Congressional Research Service.

Kirkman, S. S., Fulton, S., Hemmes, J., Garcia, C. M., & Wilson, J. C. (2024). A blockchain architecture to increase the resilience of industrial control systems from the effects of a ransomware attack: A proposal and initial results. *ACM Transactions on Cyber-Physical Systems, 8*(1), Article 9, 9:1–9:13. https://doi.org/DOI: 10.1145/3637553

Krause, T., Ernst, R., Klaer, B., Hacker, I., & Henze, M. (2021). Cybersecurity in power grids: Challenges and opportunities. *Sensors (Basel), 21*(18), 6225. DOI: 10.3390/s21186225 PMID: 34577432

Lee, R. M., Assante, M. J., & Conway, T. (2016). *Analysis of the cyber attack on the Ukrainian power grid.* Electricity Information Sharing and Analysis Center (E-ISAC) and SANS ICS. (March 2016 report).

Meng, X., & Zhu, L. (2024). Augmenting cybersecurity in smart urban energy systems through IoT and blockchain technology within the Digital Twin framework. *Sustainable Cities and Society*, *106*, 105336. DOI: 10.1016/j.scs.2024.105336

Nai Fovino, I., Andreadou, N., Geneiatakis, D., Giuliani, R., Kounelis, I., Lucas, A., & Steri, G. (2021). *Blockchain in the energy sector – WP3: Use cases identification and analysis (EUR 30782 EN).* Publications Office of the European Union., DOI: 10.2760/061600

Padmanaban, S., Dhanaraj, R. K., Holm-Nielsen, J. B., Krishnamoorthi, S., & Balusamy, B. (Eds.). (2022). *Blockchain-based systems for the modern energy grid.* Academic Press.

Prasada, P., & Prasad, D. S. (2024). Blockchain-enhanced machine learning for robust detection of APT injection attacks in cyber-physical systems. *International Journal of Computational and Experimental Science and Engineering*, *10*(4). Advance online publication. DOI: 10.22399/ijcesen.539

Saeed, S., Gull, H., Aldossary, M. M., Altamimi, A. F., Alshahrani, M. S., Saqib, M., Zafar Iqbal, S., & Almuhaideb, A. M. (2024). Digital transformation in energy sector: Cybersecurity challenges and implications. *Information (Basel), 15*(12), 764. DOI: 10.3390/info15120764

Ulghar, I., Jahankhani, H., & Kendzierskyj, S. (2021). Blockchain capabilities in defending advanced persistent threats using correlation technique and hidden Markov models (HMM). In Jahankhani, H., Kendzierskyj, S., & Akhgar, B. (Eds.), *Information Security Technologies for Controlling Pandemics* (pp. 21–64). Springer., DOI: 10.1007/978-3-030-72120-6_2

Chapter 16
Blockchain Applications for the Energy and Utilities Industry:
Future Trends and Emerging Opputunities

Anees Fathima
https://orcid.org/0009-0004-5021-5971
HKBK College of Engineering, India

Sufia Banu
https://orcid.org/0000-0002-4596-3340
HKBK College of Engineering, India

Noor Ayesha
https://orcid.org/0009-0003-5711-7656
HKBK College of Engineering, India

Salim Sharieff
https://orcid.org/0000-0003-2087-7689
HKBK College of Engineering, India

Zahira Tabassum
https://orcid.org/0000-0002-4912-0228
HKBK College of Engineering, India

Mohammed Anees
HKBK College of Engineering, India

ABSTRACT

Blockchain has evolved from supporting cryptocurrencies to transforming industries like finance, healthcare, and supply chain management. Emerging trends focus on scalability with Layer 2 solutions, sharding, and cross-chain interoperability. Sustainability efforts include transitioning to Proof-of-Stake, carbon-neutral blockchains, and renewable energy. AI integration enables decentralized models, secure data sharing, and AI-driven smart contracts. Governments explore CBDCs, while privacy technologies like Zero-Knowledge Proofs enhance security. Challenges remain in regulation, security risks, and adoption, but ongoing innovations are driving blockchain's widespread acceptance.

INTRODUCTION

Blockchain technology is a revolutionary innovation that has transformed the way digital transactions and data management are conductedBlockchain is a decentralized and distributed digital ledger that records transactions across multiple computers, ensuring secuity, transparency, and immutability..

DOI: 10.4018/979-8-3373-2439-5.ch016

Unlike traditional centralized databases, blockchain operates on a peer-to-peer (P2P) network, where each participant has a copy of the ledger, ensuring data integrity and security(Nakamoto, 2008).

Emerging technologies are constantly being developed to address societal sustainability and development concerns, including cloud computing, social media, the Internet of Things (IoT), big data analytics, ambient computing, the semantic web, and web 3.0(W. Mougayar, 2016; D. Tapscott & A. Tapscott, 2016). However, there remains a need for continuous innovation to resolve existing issues and the challenges faced by authorities in providing services. Among the latest advancements, Blockchain stands out as a game-changing technology. To comprehend the essence of Blockchain, an understanding of the concepts of data, IoT, and big data is crucial. Blockchain is a sophisticated and advanced form of digital transformation that provides security, auditability, and error-free functionality for various types of processes or transactions. Blockchain operates as a decentralized and distributed digital ledger, securely recording transactions across multiple computers. By eliminating the need for a central authority, it enhances transparency, security, and immutability, ensuring that data remains tamper-proof and verifiable across the network. It is a chain of blocks, each containing a record of multiple transactions or activities. These blocks are interconnected and, once combined, cannot be altered, making Blockchain secure and transparent. This technology does not depend on a central authority for authentication, making it resistant to tampering and fraud. Blockchain was first introduced in 2008 as a way to track transactions in a new digital currency called Bitcoin. Since then, the application of Blockchain technology and cryptocurrencies has grown, leading to the development of various Blockchain platforms and decentralized applications for several industries, including finance, governance, supply chain management, transportation, and more.

In the digital ecosystem, Blockchain can serve several different roles, depending on the specific application. One of the leading roles is to provide a safe and transparent way to store and manage data. By implementing a decentralized database, multiple parties can collaborate and share data without relying on a central authority or a single point of control, which increases trust and reduces the risk of fraud or tampering. Another crucial role of Blockchain in a digital ecosystem is to simplify the transfer of value among different groups. For instance, from the cryptocurrency perspective, Blockchain can expedite the transfer of digital assets between consumers. Employing Blockchain to track transactions makes it feasible to create a secure and transparent system for exchanging value between parties. Moreover, Blockchain can also be used to build smart contracts, which are self-executing contracts with the provisions of the agreement between the buyer and seller being promptly transcribed into lines of code. This can assist in automating and modernizing processes, making it easier to operate the business in a digital ecosystem.

Blockchain technology has numerous potential applications in various industries, including the financial sector, where it has the ability to transform traditional financial services through peer-to-peer transactions, increased transparency and security, and reduced reliance on intermediaries. It can also drive the development of innovative financial products and services, such as decentralized exchanges and stablecoins. In the realm of supply chain management, it has the potential to enhance transparency and efficiency by allowing the tracking of goods and materials throughout the supply chain process, improving traceability and reducing the risk of fraud and counterfeiting. In terms of identity management, it can be used to create decentralized identity systems that give individuals control over their personal information, helping to prevent identity theft. In the healthcare industry, it can be utilized to securely store and manage healthcare records, improving the efficiency of the healthcare system by enabling the sharing of information between different healthcare providers. For efficient governance, it has the potential to enhance the transparency and efficiency of government operations by providing a secure means of storing and managing records and documents, potentially reducing bureaucracy and improving public trust in government. In voting systems, it can be used to create secure and transparent voting systems that are less susceptible to fraud and tampering. It can also be applied in the real estate industry to create efficient record-keeping systems for real estate transactions. Blockchain technology has the potential to improve the transparency and efficiency of government operations by offering a secure method for storing and managing records and documents. This can help reduce bureaucracy and build public trust in governmental processes. In voting systems, blockchain can be implemented to develop secure and transparent voting platforms that are less vulnerable to fraud and tampering. Additionally, the technology can be applied in the real estate sector to streamline record-keeping for property transactions, enhancing accuracy and reducing administrative inefficiencies.

FUNDAMENTAL ROLES OF BLOCKCHAIN TECHNOLOGY

Blockchain technology has emerged as a promising solution with numerous potential benefits for a wide range of industries and applications. Scholars and stakeholders have proposed and implemented various advantages, applications, and uses of Blockchain technology. However, to provide long-term value and sustained growth, it is essential to identify the key characteristics of Blockchain technology that will shape future trends. Building upon a detailed review of the existing literature, this research proposes five fundamental roles of Blockchain technology that will drive future trends, as illustrated in **Fig 1**. The identification of these roles is based on the significance and vitality of the implementation system, as well as the ultimate purpose of Blockchain technology(F. H. Sharin, M, 2023).

Figure 1. Blockchain roles towards future application

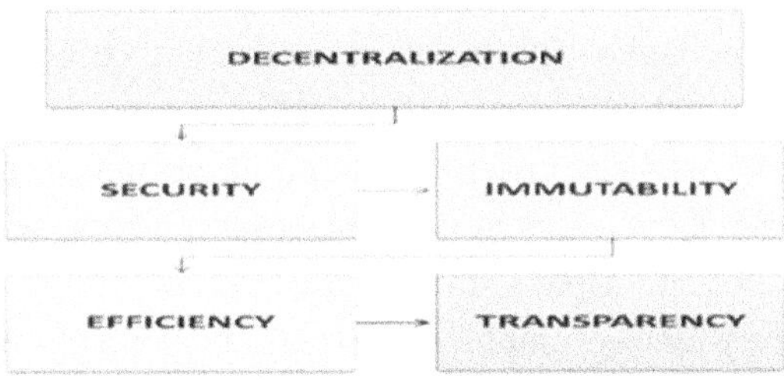

Decentralization

Decentralization, a core characteristic of blockchain technology, is anticipated to remain a key factor shaping its future development and applications.. By distributing power and control across a network of nodes or participants instead of relying on a single central authority, decentralization offers several advantages, including increased security, transparency, accessibility, and efficiency. Decentralization also plays a key role in the peer-to-peer (P2P) network, where each node or participant connects directly with others in the network, enabling the distribution of power and decision-making among network participants. Additionally, decentralization has the potential to revolutionize traditional financial systems and intermediaries through the use of decentralized finance (Defi), cryptocurrency networks, and smart contract platforms that use blockchain technology to securely and transparently track and verify transactions and data. The future of decentralization includes several potential trends, such as consensus management, which employs various approaches, including proof of work (PoW), proof of stake (PoS), or delegated proof of stake (DPoS) to validate transactions and activities on blockchain technologies. Furthermore, the emergence of decentralized autonomous organizations (DAOs) through decentralized governance is also paving the way for future trends. Decentralization has the potential to increase security and privacy, transparency and accountability, resiliency and reliability, and accessibility and inclusivity. Future examples of decentralization may include the widespread adoption of decentralized financial systems and platforms, such as decentralized exchanges (DEXs), and the adoption of decentralized identity systems and platforms.

Security

Blockchain technology has experienced significant attention for its ability to provide robust security through the use of cryptography to protect and secure data stored on the Blockchain. By utilizing cryptography, blockchain technology helps to prevent data breaches and unauthorized access to sensitive information. A significant application of blockchain technology in strengthening security is the development of decentralized identity systems, which leverage blockchain to securely store and safeguard personal information, reducing the risk of identity fraud. Additionally, blockchain can enhance security across various sectors, including facilitating secure financial transactions, ensuring safe data storage and management, establishing transparent and traceable supply chains, supporting secure voting platforms, maintaining reliable real estate transaction records, protecting intellectual property, managing healthcare records securely, and promoting efficient and transparent government operations. These security benefits will continue to develop and boost in the coming years as blockchain technology gains more widespread adoption. To further enhance the security of data and assets on blockchain technology, the use of zero-knowledge proofs, which enable one party to prove the possession of certain information without revealing the actual information, will become increasingly important to make blockchain transactions more difficult to manipulate Blockchain technology not only ensures secure record-keeping but also provides substantial benefits across multiple sectors. Industries like supply chain management, financial services, and intellectual property management can utilize blockchain to improve transparency, maintain data integrity, and mitigate fraud risks.

Immutability

The idea of immutability in the context of blockchain refers to the property of being unable to be modified or altered. This is a crucial aspect of a blockchain system, as it ensures that information and events stored in it cannot be tampered with or modified in the past. Once data is stored on the distributed ledger, it becomes very difficult to change or remove, making it more trustworthy and reliable. For instance, blockchain-based voting systems can enhance the reliability of elections by ensuring that results are accurate and cannot be manipulated.

The permanent nature of the distributed ledger is made possible through encryption techniques such as mining and digitally signed documents, which enable the chaining of transactions in an anonymous and irreversible manner. Thus, any attempts to alter data stored within the Blockchain will be detected.

This feature of blockchain technology can create a continuous and unchangeable record of events, which can improve accountability and transparency in organizational structures and processes, in addition to offering a secure record-keeping method. Various domains, including supply chain management, financial services, and intellectual property, could benefit from this.

Immutability plays a prominent role in blockchain technology by providing a secure and tamper-proof record-keeping system. Blockchain-based systems can be used to store and manage financial records, supply chain data, healthcare records, real estate transactions, and intellectual property records. The immutability of blockchain technology can also potentially lead to the development of new applications and technologies that take advantage of this property. For example, blockchain-based systems can create secure and transparent voting systems, facilitate the tracking and verification of supply chain events, or provide a decentralized platform for intellectual property management.

Moreover, the immutability of blockchain technology has the potential to disrupt traditional systems and intermediaries that rely on centralized and mutable record-keeping systems. For instance, the use of decentralized financial systems could disrupt traditional banking and payment processing systems, while blockchain-based supply chain systems could disrupt traditional supply chain intermediaries. The immutability of blockchain technology offers a promising opportunity for enhancing security and reliability in various domains and may pave the way for disruptive innovations in the future.

Efficiency

Blockchain technology has the potential to revolutionize various industries by reducing costs and streamlining complex operations. For instance, in the corporate sector, blockchain-based contract systems can simplify and speed up operations, resulting in reduced need for human involvement and middlemen. The technology can enhance productivity by producing a secure, transparent, and unchangeable record of every transaction and interaction, reducing the need for manual data entry and verification, and minimizing the threat of fraud. By deploying digital contracts that self-execute based on pre-determined conditions encoded in the blockchain, blockchain technology can automate this procedure and decrease human intervention.

Moreover, the blockchain's ability to establish peer-to-peer relationships and connections can eliminate the need for intermediaries, saving time and money. In supply chain management, blockchain-based systems can enhance accountability by providing a secure and visible trail of item flow, promoting efficiency through tracking, and reducing the likelihood of errors and delays. Blockchain technology can also improve data sharing and collaboration among different parties, enhancing communication and cooperation in the decision-making process.

Future scenarios for blockchain technology's productivity improvements include the automation and streamlining of financial transactions using intelligent contracts, secure tracking of the flow of items in supply chains, and enabling peer-to-peer conversations and transactions across different sectors. The technology can also enhance data sharing and collaboration in various industries such as government agencies, healthcare, and the food industry by establishing secure and comprehensive accounts of information flow. Overall, blockchain technology's potential to improve productivity and streamline operations is vast and can result in significant benefits for various industries.

Transparency

Blockchain technology provides a decentralized and distributed ledger system that allows for transparent and verifiable tracking of transactions and data by all parties involved in the network. The secure and immutable record of transactions and events on the Blockchain makes it difficult for a single party to manipulate or alter the data. This technology offers a high level of transparency, depending on the design and implementation of the Blockchain as well as the trust and participation among the parties involved. In the supply chain management, for example, it enables all partipants of the supply chain to track and verify the flow of commodities, while providing a safe, accessible record for the transit of resources and other objects documented on the distributed ledger.

Blockchain-based technologies have a broad range of applications in various industries and activities. For instance, it can enable money transfers in an open and traceable manner, lowering the likelihood of deception and improper behaviour. In healthcare, it can be employed for transparently store and maintain information in a secure way, enabling patients to access, verify and share their data with authorized medical personnel. In governance, it can be employed for to create decentralized governance platforms that allow for transparent and verifiable decision-making processes, such as through the use of a decentralized voting system.

The transparency offered by blockchain technology can help address future problems by enhancing trust and accountability among parties involved in the network, preventing fraud and corruption, improving decision-making, and increasing efficiency. In urban planning, for instance, increased transparency in urban mobility systems can help improve trust and accountability among the parties involved, such as transportation authorities, operators, and passengers. It can also help to prevent fraud and misconduct, as any discrepancies or anomalies in the data captured on the Blockchain would be easily detectable. Moreover, increased transparency can improve decision-making by enabling all parties to obtain and examine precise and trustworthy data. This can help identify inefficiencies and improvement opportunities in urban mobility systems, leading to more effective and sustainable transportation options.

BLOCKCHAIN FUNDAMENTALS

Blockchain technology is built upon a collection of fundamental principles that ensure its security, decentralization, and reliability. This chapter explores the core components that define blockchain functionality, including cryptographic security, consensus mechanisms, smart contracts, decentralization, and its evolution from Bitcoin to Web3.

Cryptographic Security

Cryptographic security is the foundation of blockchain, ensuring data integrity, confidentiality, and authenticity. It is achieved through hash functions, encryption techniques, and digital signatures.

Hashing in Blockchain

Blockchain make use of cryptographic hash functions like SHA-256 (Secure Hash Algorithm 256-bit) to convert data into a fixed-length hash. This ensures immutability, as even the smallest change in input generates a completely different hash.

Example of SHA-256 Hashing:

- Input: "Blockchain Security"
- Hash Output: 7c7f5a65d5f76b9f72e9e34dc9e3e92dc

A chain that guards against tampering is formed by each block containing the hash of the one before it.

Public and Private Key Cryptography

Asymmetric cryptography is used by blockchain to ensure safe transactions.

- Public Key – Shared with others for encryption.
- Private Key – Kept secret and used for decryption or signing transactions.
- Transactions are signed with a private key and verified with a public key, ensuring authenticity and security.

Digital Signatures

Digital signatures use Elliptic Curve Digital Signature Algorithm (ECDSA) to authenticate transactions. This prevents fraud and ensures that only the owner of a private key can authorize a transaction.

CONSENSUS MECHANISMS

Consensus mechanisms procedures that guarantee consensus among distributed nodes on the blockchain, preventing fraud and ensuring network security as shown in **Fig 2**.

Figure 2. Consensus mechanism

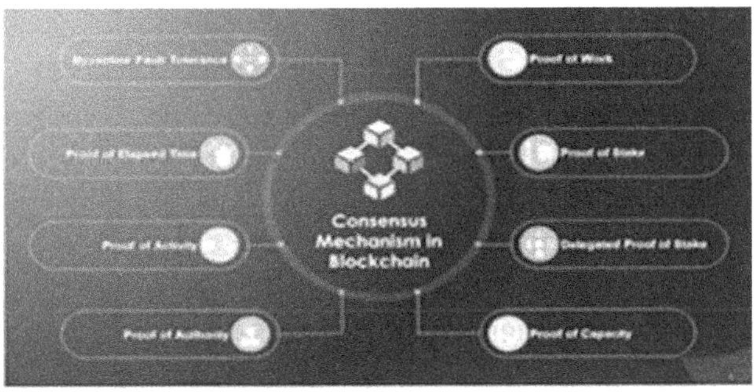

Proof of Work (PoW)

- Used in Bitcoin and early blockchains.
- Miners solve complex mathematical puzzles to validate transactions.
- Requires high computational power and energy.

Proof of Stake (PoS)

- Used in Ethereum 2.0 and other modern blockchains.
- Validators are chosen based on the number of coins they stake.
- More energy-efficient than PoW.

Delegated Proof of Stake (DPoS)

- Used in EOS and Tron blockchains.
- Stakeholders vote for trusted delegates to validate transactions.
- Processing transactions more quickly compared to PoW.

SMART CONTRACTS

Self-executing agreements are known as smart contracts,that automate transactions and processes without intermediaries.

Features of Smart Contracts

- Automation – Executes transactions based on predefined conditions.
- Transparency – Stored on the blockchain for public verification.
- Security – Encrypted and immutable, reducing risks of tampering.

Smart Contract Use Cases

- Finance – Automated lending, DeFi protocols, and token swaps.
- Supply Chain – Transparent product tracking and automated payments.
- Healthcare – Secure medical record sharing and insurance claims.

Smart contracts enhance efficiency and reduce costs by removing third-party involvement.

DECENTRALIZATION AND PEER-TO-PEER NETWORKS

With blockchain technology, there's no need for authority. Participants can transact directly in a peer-to-peer network.

How P2P Networks Work

- Every node (computer) in the network maintains a copy of the blockchain.
- Nodes communicate and validate transactions collectively.
- The blockchain can neither be controlled nor changed by any single person.

Benefits of Decentralization

1. Security – No single point of failure reduces the risk of attacks.
2. Transparency – Transactions are publicly recorded and verifiable.
3. Censorship Resistance – No government or organization can control transactions.

Decentralization is a key principle that makes blockchain trustless, secure, and resilient against failures or cyberattacks.

BLOCKCHAIN EVOLUTION: FROM BITCOIN TO WEB3

From the days of Bitcoin to the current use in DApps, NFTs, and even Web3, blockchain technology has undergone massive evolution.

Bitcoin (Blockchain 1.0) – The Birth of Cryptocurrency

Bitcoin, launched in 2009, introduced the first decentralized digital currency based on blockchain. It focused solely on financial transactions, offering a secure and transparent alternative to traditional banking.

Ethereum and Smart Contracts (Blockchain 2.0)

Ethereum, launched in 2015, expanded blockchain's capabilities by introducing smart contracts and decentralized applications (DApps). It enabled:

- ICO (Initial Coin Offerings) – New projects raising funds via blockchain.
- Decentralized Finance (DeFi) – Lending, borrowing, and trading without banks.

NFTs and Layer 2 Scaling (Blockchain 3.0)

Blockchain 3.0 focuses on scalability, efficiency, and new applications such as:

- NFTs (Non-Fungible Tokens) – Unique digital assets for art, gaming, and music.
- Layer 2 Solutions (e.g., Polygon, Lightning Network) – Faster and cheaper transactions.
- Interoperability (Polkadot, Cosmos) – Cross-chain communication between blockchains.

Web3 – The Future of Decentralization

Web3 aims to decentralize the internet, giving users control over their data and digital identity. Key Web3 applications include:

- Decentralized Storage (IPFS, Filecoin) – Replacing centralized cloud storage.
- Metaverse and Virtual Worlds (Decentraland, The Sandbox) – Blockchain-powered virtual environments.
- Decentralized Autonomous Organizations (DAOs) – Community-driven governance without central authorities.

The transition from Blockchain 1.0 (Bitcoin) to Web3 marks a paradigm shift towards a decentralized, user-controlled digital ecosystem that has the potential to redefine finance, governance, and digital interactions.

Table 1. Blockchain Network Types

Blockchain Type	Description	Access Control	Use Cases	Examples
Public Blockchain	A decentralized, open network where anyone can participate, validate transactions, and view data.	No restrictions (permissionless).	Cryptocurrencies, decentralized finance (DeFi), voting systems.	Bitcoin, Ethereum, Solana.
Private Blockchain	A controlled network where only authorized participants can validate and access data.	Restricted (permissioned).	Enterprise applications, supply chain management, internal record-keeping.	Hyperledger Fabric, R3 Corda.
Consortium Blockchain	A semi-decentralized network managed by multiple organizations instead of a single entity.	Partially restricted (permissioned by multiple entities).	Banking, interbank settlements, supply chain tracking.	Quorum, Hyperledger Fabric (in consortium mode).
Hybrid Blockchain	A combination of public and private blockchain features, allowing selective transparency and privacy.	Partially restricted (permission-based with public verification).	Healthcare, government services, enterprise applications.	XinFin (XDC), IBM Food Trust.

CURRENT APPLICATIONS OF BLOCKCHAIN

The transformation of blockchain technology started with cryptocurrencies and now, every industry exploits its application. Due to its decentralized, transparent, and secure nature, blockchain has found applications in the financial, healthcare, government, IoT, and even supply chain sectors. As mid-level disruption continues to affect these industries, so do the removal of intermediaries, increasement of security, and data integrity assurance. This chapter dives into the remarkable effects of blockchain on digital transactions and robust data management.

Blockchain In Finance And Cryptocurrencies

The financial sector has been one of the earliest adopters of blockchain technology, primarily driven by the growth of cryptocurrencies like Bitcoin and Ethereum. Beyond cryptocurrencies, blockchain is being widely used in financial services for cross-border payments, digital identities, and decentralized finance (DeFi). DeFi platforms use smart contracts to provide lending, borrowing, and trading services without intermediaries, ensuring greater financial inclusion. Moreover, central banks are exploring Central Bank Digital Currencies (CBDCs), which are government-issued digital currencies based on blockchain. Additionally, blockchain improves fraud detection and compliance through transparent and immutable transaction records. By providing real-time settlement and enhancing security, blockchain is revolutionizing financial exchange on a global scale.

Blockchain In Supply Chain Management

Blockchain has emerged as a game-changer in supply chain management, addressing critical challenges such as transparency, traceability, and efficiency shown in **Fig 3**. Traditional supply chains rely on multiple intermediaries, making it difficult to track products from the source to the consumer. Blockchain enables real-time tracking of goods by storing every transaction in an immutable ledger, ensuring authenticity and preventing fraud. Companies like IBM and Walmart have integrated blockchain-based supply chain solutions to monitor food safety, ensuring that products meet quality standards. Additionally, blockchain reduces paperwork by digitizing transactions, leading to faster and more cost-effective logistics management. The technology also combats counterfeit goods by verifying product origins and ensuring ethical sourcing. For example, luxury brands use blockchain to certify the authenticity of high-end products, while pharmaceutical companies track drugs to prevent counterfeit medicines from entering the market. By improving efficiency and trust, blockchain is revolutionizing the way supply chains operate.

Figure 3. Blockchain-enabled supply chain management

Blockchain In Healthcare And Medical Records

Blockchain technology is transforming the healthcare industry by enhancing security, patient data management, and interoperability shown in **Fig 4.** One of the biggest challenges in healthcare is maintaining accurate and secure patient records across multiple healthcare providers. Blockchain enables decentralized storage of medical records, ensuring that only authorized parties can access patient data. This improves data privacy while allowing seamless access for healthcare providers, reducing medical errors and improving patient care. Additionally, blockchain helps in verifying the authenticity of medical supplies, reducing fraud in pharmaceutical supply chains. Smart contracts are also being used for automated insurance claims processing, minimizing delays and fraud. Hospitals and research institutions are leveraging blockchain for clinical trials Maintaining data integrity and avoiding manipulation of research results. By enhancing transparency, security, and efficiency, blockchain is addressing critical issues in healthcare, ultimately leading to better patient outcomes.

Figure 4. Blockchain applications in healthcare

Blockchain In Government And Public Services

Governments worldwide are investigating the use of blockchain technology to enhance transparency, efficiency, and security in public services. A prominent application of blockchain in governance is digital identity management, offering a more secure and reliable alternative to traditional identity systems. Conventional identity management methods often face challenges such as fraud, data breaches, and operational inefficiencies, which blockchain can mitigate through its decentralized and tamper-resistant nature. **Fig 5** shows how a smart contract works. Additionally, blockchain is being used for secure voting systems, ensuring election integrity by preventing tampering and making vote counting more transparent. Another significant application is land registry and property transactions. Blockchain simplifies property transactions by maintaining an immutable record of ownership, reducing disputes and fraud. Governments are also using blockchain to enhance public sector financial management, ensuring transparency in the allocation and distribution of public funds. By leveraging blockchain, public services are becoming more efficient, accountable, and resistant to corruption(G. Vonitsanos et al., 2021).

Figure 5. Smart contracts and their working mechanism

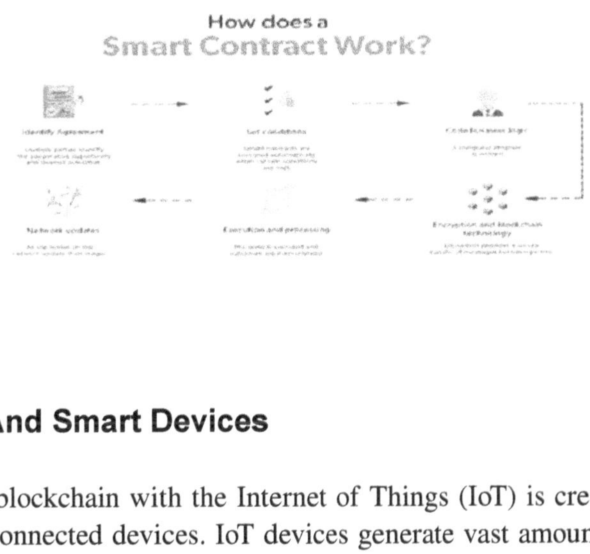

Blockchain In IoT And Smart Devices

The combination of blockchain with the Internet of Things (IoT) is creating a secure and decentralized framework for connected devices. IoT devices generate vast amounts of data, and traditional centralized systems struggle with security and scalability issues. Blockchain provides a decentralized ledger that ensures secure and tamper-proof communication between IoT devices, reducing risks such as data breaches and cyber-attacks. Industries such as smart homes, healthcare, and supply chains are leveraging blockchain to enhance IoT security and automation. For instance, blockchain enables smart contracts for automated transactions between IoT devices, such as self-executing agreements for energy trading between smart grids. It also enhances security in autonomous vehicles by ensuring real-time authentication of vehicle components. In industrial IoT, blockchain helps monitor equipment maintenance records, reducing downtime and improving efficiency. By combining blockchain with IoT, businesses can achieve greater automation, security, and transparency in smart device ecosystems.

Table 2. Applications of Blockchain in Various Sectors

Sector	Application of Blockchain	Benefits	Examples
Finance & Banking	Secure and transparent transactions, DeFi, cross-border payments	Faster transactions, reduced fraud, cost efficiency	Bitcoin, Ethereum, Ripple (XRP)
Supply Chain	Real-time tracking of goods, reducing fraud and errors	Transparency, efficiency, and traceability	IBM Food Trust, VeChain
Healthcare	Secure medical records, interoperability, and patient privacy	Data security, improved patient care	MedicalChain, MedRec
Government	Digital identity, voting systems, public record management	Transparency, security, and efficiency	Estonia e-Residency, Voatz
Energy & Utilities	Peer-to-peer energy trading, renewable energy tracking	Decentralization, efficiency, sustainability	Power Ledger, WePower
Real Estate	Smart contracts for property transactions	Faster transactions, reduced fraud	Propy, RealT
Education	Tamper-proof certification, digital diplomas	Prevents fraud, easy verification	MIT Digital Diplomas, Blockcerts
IoT & Smart Devices	Secure communication between IoT devices	Enhanced security, automation	IOTA, Waltonchain
Entertainment & Media	Copyright protection, NFT-based digital assets	Prevents piracy, enables direct monetization	Ethereum NFTs, Audius
Legal & Contracts	Smart contracts for automation of legal agreements	Reduces disputes, improves efficiency	OpenLaw, Accord Project

EMERGING TRENDS IN BLOCKCHAIN TECHNOLOGY

Blockchain technology is continuously evolving to address scalability, interoperability, privacy, and integration with emerging technologies like AI and IoT. As its adoption grows across industries, researchers and developers are innovating to enhance blockchain's efficiency, security, and usability. This chapter explores the latest advancements in blockchain technology, including scalability solutions, interoperability between different blockchain networks, integration with artificial intelligence and IoT, privacy-enhancing techniques, and decentralized identity solutions. These trends are shaping the future of blockchain, making it more adaptable for real-world applications.

Table 3. Blockchain-Based Energy Trading Platform

Platform Name	Description	Key Features	Use Cases
Power Ledger	Peer-to-peer (P2P) energy trading platform	Decentralized trading, real-time monitoring, transparency	Residential solar energy trading
WePower	Blockchain-based renewable energy marketplace	Tokenization of energy, direct P2P trading	Green energy financing & trading
LO3 Energy	Localized energy trading using blockchain	Smart meters, real-time energy pricing	Brooklyn Microgrid
The Energy Web Chain	Open-source blockchain for energy sector	Decentralized energy apps, scalability	Energy asset management, grid balancing
Grid+	Blockchain-based wholesale electricity provider	AI-driven energy optimization, lower costs	Smart energy consumption management
SunContract	Renewable energy marketplace	Direct consumer-producer trading, smart contracts	Solar and wind energy trading
Electron	Blockchain for energy flexibility markets	Grid optimization, energy trading	Grid balancing, energy efficiency

Scalability Solutions (Sharding, Layer-2)

One of the major challenges in blockchain technology is scalability, as traditional blockchains struggle to process a high volume of transactions. Bitcoin and Ethereum, the two largest blockchain networks, have faced congestion issues due to limited transaction throughput. To address this, researchers have introduced scalability approches such as sharding and Layer-2 solutions shown in **Fig 6**.

Sharding is a method that involves dividing a blockchain network into smaller, more manageable partitions known as shards. Each shard operates independently, processing its own transactions and executing smart contracts, thereby enhancing the network's scalability and efficiency. By distributing the workload across multiple shards, blockchain networks can process transactions in parallel, significantly increasing scalability. Ethereum 2.0 is one of the most notable implementations of sharding.

Layer-2 solutions operate on top of the main blockchain (Layer-1) to handle transactions off-chain, reducing congestion. Technologies like Lightning Network (for Bitcoin) and Rollups (for Ethereum) allow transactions to be processed off-chain and later settled on the main chain, improving speed and reducing costs. Layer-2 solutions are critical for enabling high-throughput applications such as decentralized finance (DeFi) and gaming.

Figure 6. Scalability solutions in blockchain

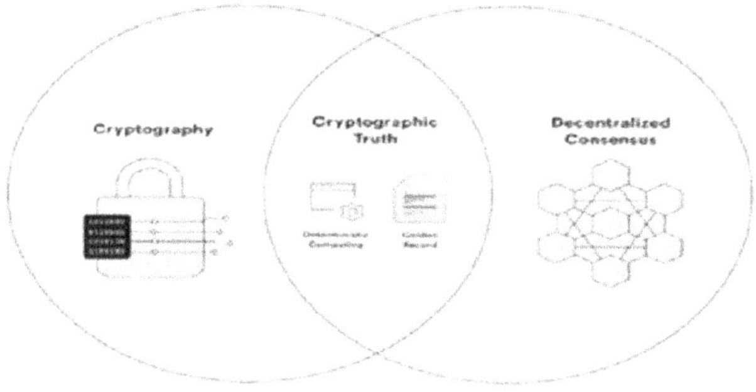

Interoperability Between Blockchains

The blockchain ecosystem consists of multiple networks, each with its own protocols and standards. However, the lack of interoperability between these networks limits their ability to communicate and share data. Interoperability solutions aim to bridge this gap by enabling seamless interactions between different blockchains.

Cross-chain communication protocols, such as Polkadot, Cosmos, and Chainlink, facilitate interoperability by allowing different blockchains to exchange information and assets securely. These protocols use relay chains and inter-blockchain communication (IBC) mechanisms to connect independent networks.

Interoperability is essential for creating a connected blockchain ecosystem where users can transfer assets across multiple platforms without intermediaries. For example, wrapped tokens, such as Wrapped Bitcoin (WBTC) on Ethereum, allow Bitcoin to be used in DeFi applications. Improved blockchain interoperability will lead to enhanced liquidity, efficiency, and collaboration across industries.

Integration With AI And IoT

The convergence of blockchain with artificial intelligence (AI) and the Internet of Things (IoT) is creating innovative applications that enhance automation, security, and efficiency. AI and blockchain complement each other, with blockchain providing data integrity and transparency, while AI enables automation and predictive analytics.

In IoT, blockchain enhances security by creating a decentralized network where devices can communicate securely without a central authority. This is crucial for smart cities, autonomous vehicles, and industrial automation, where billions of connected devices require secure and efficient data exchanges. Blockchain prevents unauthorized access, reduces cyber threats, and ensures tamper-proof records for IoT devices(T. N. Dinh & M. T. Thai, 2018).

AI-powered blockchain solutions are also transforming industries like finance, healthcare, and supply chain management. AI algorithms analyze blockchain data for fraud detection, automated decision-making, and personalized services. For example, AI-driven smart contracts can automatically execute agreements based on predefined conditions, minimizing reliance on intermediaries.

The integration of AI, IoT, and blockchain **Fig 7** is expected to create a decentralized intelligence ecosystem, where secure data-sharing and automation redefine business processes.

Figure 7. Blockchain and IoT integration model

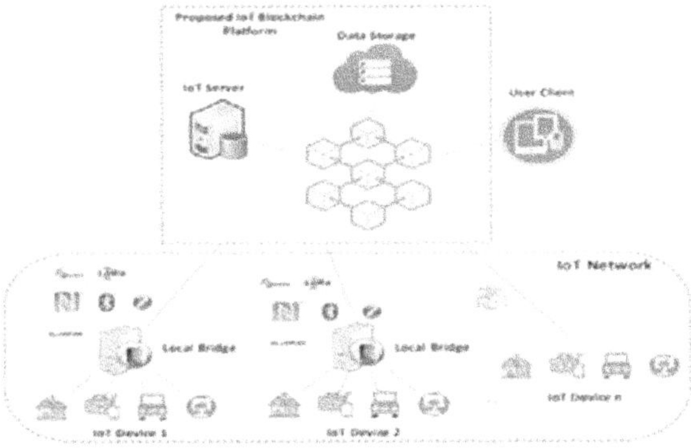

Privacy-Enhancing Technologies (Zero-Knowledge Proofs)

Privacy remains a significant concern in blockchain technology, particularly in public blockchains where transaction details are visible to all participants. While transparency is often considered a core advantage of blockchain, various applications in sectors like finance, healthcare, and identity management demand stricter data confidentiality.

To address this challenge, Privacy-Enhancing Technologies (PETs) have been developed, with Zero-Knowledge Proofs (ZKPs) being one of the most notable solutions as shown in **Fig 8**. ZKPs allow one party to prove the truthfulness of a statement without disclosing any underlying data, thereby ensuring privacy while maintaining trust.

A well-known implementation of ZKPs is zk-SNARKs (Zero-Knowledge Succinct Non-Interactive Argument of Knowledge). This cryptographic protocol provides secure verification without requiring direct interaction between the verifier and the prover. zk-SNARKs are commonly used in privacy-focused cryptocurrencies like Zcash, allowing users to conduct confidential transactions without revealing sensitive information on the blockchain.

By incorporating technologies like ZKPs, blockchain networks can offer both transparency and privacy, striking a balance that meets the security and confidentiality needs of modern applications.

Other PETs include Mimblewimble, homomorphic encryption, and multi-party computation, which improve privacy while maintaining blockchain security. Enterprise blockchain solutions, such as Hyperledger Fabric and Quorum, are adopting privacy-preserving techniques to enable confidential business transactions.

With growing concerns over data security and compliance with regulations like GDPR, privacy-enhancing technologies are becoming essential for mainstream blockchain adoption.

Figure 8. Zero-knowledge proofs for privacy in blockchain

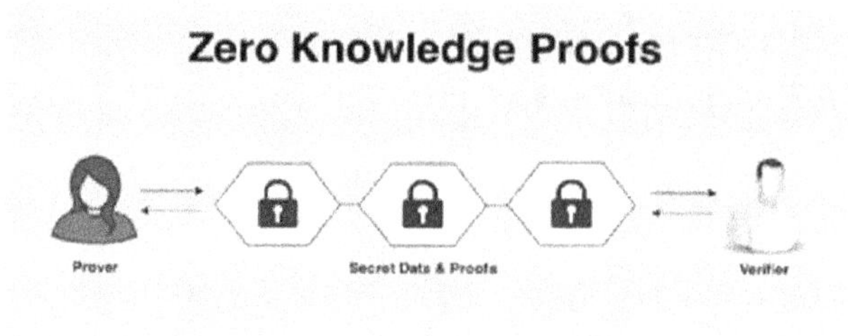

Blockchain For Decentralized Identity And Web3

Decentralized identity (DID) is a revolutionary concept that gives individuals control over their digital identities, terminating the need for centralized identity providers. Traditional identity management systems depend on third parties, making them vulnerable to data breaches, identity theft, and privacy violations.

Decentralized identity solutions empower users to establish self-sovereign identities (SSI), where personal data is stored on a blockchain and accessed only with user consent.

Organizations such as Microsoft (ION on Bitcoin) and Sovrin are developing DID solutions that enhance security and user control.

Decentralized identity plays a key role in Web3, the next evolution of the internet, where users own and control their data. Web3 applications, built on blockchain, promote decentralization, transparency, and data sovereignty, reducing dependence on tech giants and centralized platforms. Use cases of blockchain-based identity include secure login mechanisms, KYC verification, and credential management for banking, healthcare, and education. By eliminating third-party data storage, decentralized identity enhances privacy, reduces fraud, and empowers users in the digital ecosystem(G. Zyskind et al., 2015).

BLOCKCHAIN IN THE ENERGY AND UTILITIES INDUSTRY

The energy and utilities industry is undergoing a transformation driven by technological advancements, sustainability concerns, and the increasing demand for efficient energy management. Blockchain technology is emerging as a key enabler of this transformation, offering solutions for decentralized energy trading, grid modernization, distributed energy resources (DERs), and enhanced transparency. By leveraging blockchain, the energy sector can reduce inefficiencies, enhance security, and facilitate peer-to-peer (P2P) energy trading. This chapter explores the challenges in the energy sector and how blockchain is addressing them, along with its applications in energy trading and grid modernization.

CHALLENGES IN THE ENERGY SECTOR AND BLOCKCHAIN SOLUTIONS

The global energy sector faces several challenges, including centralized control, lack of transparency, inefficiencies in distribution, and increasing demand for renewable energy integration. The traditional energy system relies on large, centralized utilities that generate and distribute power, leading to monopoly control, high operational costs, and vulnerability to cyber threats. Additionally, the shift toward renewable energy sources, such as solar and wind, requires a decentralized approach to efficiently manage energy distribution and consumption.

Blockchain technology provides decentralized, tamper-proof, and transparent solutions to address these challenges. It enables:

- Decentralization: Eliminates the need for intermediaries, allowing direct transactions between energy producers and consumers.
- Transparency and Security: Immutable ledger technology ensures that energy transactions are recorded securely, preventing fraud and unauthorized alterations.
- Smart Contracts: Automate processes such as energy billing, payments, and compliance with regulations, reducing administrative costs and human errors.
- Real-Time Tracking: Blockchain enables real-time tracking of energy consumption and production, optimizing energy distribution and reducing waste.

By integrating blockchain, energy companies and consumers can create a more resilient, efficient, and sustainable energy ecosystem.

Table 4. Real-World Blockchain Implementations

Industry	Implementation	Key Features	Example Projects/Companies
Finance & Banking	Cross-border payments, fraud prevention	Decentralized ledgers, smart contracts, transparency	Ripple (XRP), JPM Coin, Stellar
Supply Chain	Product tracking, anti-counterfeiting	Immutable records, real-time tracking, transparency	IBM Food Trust, VeChain
Healthcare	Secure patient records, drug traceability	Privacy-focused, tamper-proof data storage	MediBloc, BurstIQ
Government & Public Services	Digital identities, voting systems	Secure authentication, reduced fraud	Estonia's e-Residency, Voatz
Energy & Utilities	P2P energy trading, grid management	Decentralized energy markets, smart metering	Power Ledger, Energy Web Chain
Real Estate	Tokenized property transactions	Smart contracts, reduced paperwork	Propy, RealT
IoT & Smart Devices	Device authentication, data integrity	Secure machine-to-machine transactions	IOTA, Helium

BLOCKCHAIN-BASED ENERGY TRADING AND P2P NETWORKS

Peer-to-peer (P2P) energy trading is a revolutionary model enabled by blockchain, allowing individuals and businesses to buy and sell excess energy directly without relying on centralized utility providers. This concept is particularly beneficial for prosumers—consumers who also produce energy through solar panels, wind turbines, or other renewable sources refer **Fig 9**.

How Blockchain-Based Energy Trading Works:

- Energy Generation: Prosumers generate excess electricity using renewable sources.
- Smart Meters: These IoT-enabled devices record energy production and consumption data on a blockchain.
- Decentralized Marketplace: A blockchain-based platform connects buyers and sellers, facilitating energy trade through smart contracts.
- Automated Transactions: Smart contracts execute trades when predefined conditions are met, ensuring secure and transparent transactions.
- Settlement and Payments: Transactions are settled instantly using cryptocurrencies or digital tokens, reducing delays and transaction fees.

Real-World Implementations:

- Brooklyn Microgrid (USA): One of the first blockchain-based P2P energy trading projects where residents trade solar energy directly.
- Power Ledger (Australia): A blockchain platform enabling real-time energy trading between households, businesses, and utilities.
- LO3 Energy: A startup focusing on decentralized energy markets using blockchain technology.

Blockchain-based energy trading reduces dependence on central utilities, lowers electricity costs, promotes renewable energy adoption, and improves energy resilience in communities.

Figure 9. Blockchain-Based P2P Energy Trading Model

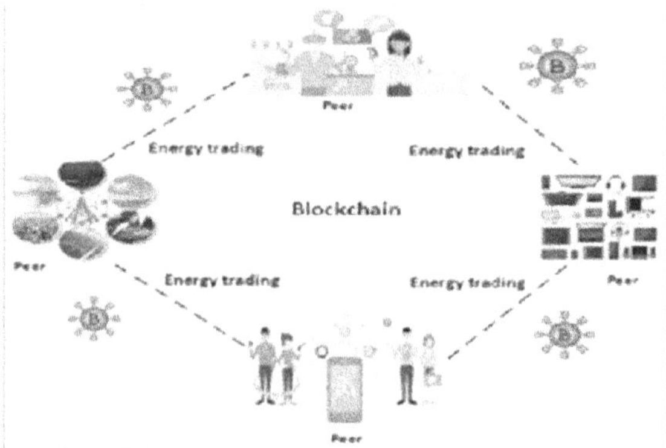

GRID MODERNIZATION AND DISTRIBUTED ENERGY RESOURCES (DERS)

The traditional energy grid is aging and struggling to cope with growing energy demands, renewable energy integration, and cybersecurity threats. Grid modernization is essential to create a more reliable, flexible, and efficient energy system refer **Fig 10**. Blockchain plays a fundamental role in modernizing the grid by enabling real-time energy tracking, automation, and decentralized control.

Key Areas of Grid Modernization with Blockchain:

Distributed Energy Resource (DER) Management

DERs, such as solar panels, wind farms, battery storage systems, and electric vehicles (EVs), are increasingly being integrated into the grid. However, managing these distributed resources effectively requires advanced coordination. Blockchain enables:

- Secure data sharing between DERs and utilities.
- Automated load balancing to optimize energy supply and demand.
- Dynamic pricing for energy storage and distribution.

Demand Response and Smart Grid Automation

Blockchain enables real-time energy demand-response mechanisms, where consumers adjust their energy usage based on price signals and grid conditions. Smart contracts automatically execute transactions when demand fluctuates, ensuring a stable and cost-effective energy supply.

Cybersecurity and Fraud Prevention

The energy grid is a prime target for cyberattacks, which can disrupt electricity supply and cause financial losses. Blockchain enhances security and resilience by:

- Decentralizing data storage, reducing the risk of single points of failure.
- Preventing unauthorized data manipulation through an immutable ledger.
- Ensuring identity verification for energy transactions.

Real-World Implementations:

- IBM & Energy Web Foundation: Developing blockchain-based solutions for decentralized grid management.
- Siemens & LO3 Energy: Collaborating on blockchain-based smart grids to optimize energy distribution.
- TenneT & Sonnen: Using blockchain for grid stabilization and balancing renewable energy fluctuations.

Figure 10. Grid modernization using blockchain

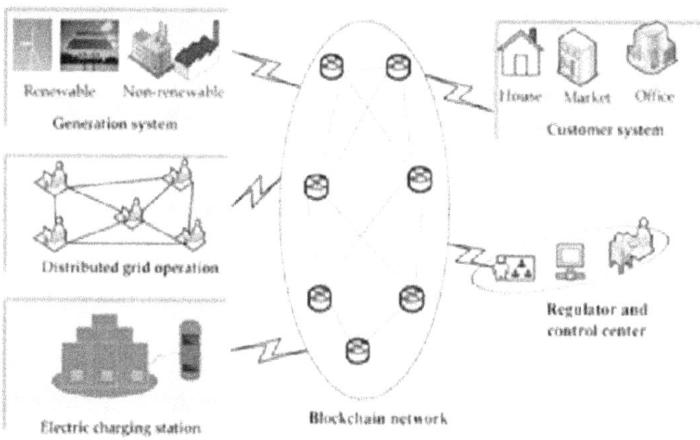

Blockchain For Microgrids And Energy Storage

Microgrids are localized energy networks that operate either independently or in conjunction with the main power grid. They are designed to provide energy resilience, efficiency, and sustainability, particularly in areas where centralized grids are unreliable or unavailable. Blockchain technology enhances the operation of microgrids by facilitating decentralized energy management, real-time monitoring, and secure peer-to-peer (P2P) energy transactions.

The function of Blockchain in Microgrids

1. Decentralized Energy Management

Traditional energy grids rely on centralized control, where large utility companies manage electricity generation and distribution. Microgrids, however, promote distributed energy generation through sources like solar panels, wind turbines, and battery storage. Blockchain activates decentralized energy management by allowing microgrid participants to track, verify, and trade energy seamlessly without depending on intermediaries.

2. Peer-to-Peer (P2P) Energy Trading

Blockchain-based P2P trading allows microgrid participants—such as households, businesses, and community organizations—to directly buy and sell electricity. Through smart contracts, blockchain enables:

- Automated settlements based on energy supply and demand.
- Transparent and tamper-proof transaction records.
- Cost-effective and real-time energy trading without third-party fees.

Energy Storage Optimization

Energy storage is a critical component of microgrids, as renewable sources like solar and wind generate electricity intermittently. Blockchain helps optimize battery storage management by:

- Tracking energy production and consumption in real-time.
- Predicting energy demand using AI-integrated smart contracts.
- Enabling dynamic pricing for stored energy based on market fluctuations.
- Brooklyn Microgrid (USA): Uses blockchain to enable P2P solar energy trading.

Real-World Implementations

- Power Ledger (Australia): Implements blockchain for decentralized energy storage and microgrid solutions.
- EDF & Electron (UK): Developing blockchain-powered grid flexibility markets.

Blockchain's ability to enhance efficiency, security, and autonomy in microgrids and energy storage positions it as a key technology in the transition toward a decentralized, resilient, and sustainable energy future.

Sustainability & Supply Chain Traceability

The global energy industry faces growing pressure to reduce carbon footprints, improve sustainability, and increase supply chain transparency. Blockchain gives a secure, immutable, and decentralized framework for tracking the entire lifecycle of energy production, distribution, and consumption, ensuring higher accountability and sustainability.

The function of Blockchain in Sustainable Energy

Renewable Energy Certification & Carbon Credit Trading

Governments and organizations use Renewable Energy Certificates (RECs) and carbon credits to incentivize sustainable practices. Blockchain enables:

- Tamper-proof certification: Ensuring RECs are authentic, verifiable, and fraud-resistant.
- Automated carbon credit trading: Smart contracts facilitate transparent and efficient transactions in carbon markets.
- Real-time emission tracking: Businesses can monitor and reduce their carbon footprint with blockchain-based audits.

Sustainable Supply Chain Traceability

The energy sector involves complex supply chains for raw materials, manufacturing, and distribution. Blockchain enhances sustainability efforts by:

- Tracking the origin of raw materials (e.g., lithium for batteries).
- Monitoring ethical and environmental compliance in the supply chain.

- Reducing fraud and improving confidence among stakeholders through immutable records.

Waste Reduction and Circular Economy

Blockchain facilitates the tracking and recycling of energy-related components, such as:

- Battery lifecycle management: Monitoring battery production, usage, and recycling.
- Solar panel disposal and recycling: Ensuring responsible waste management.
- Grid component tracking: Preventing counterfeit or low-quality materials from entering the market.

Real-World Implementations

- IBM Blockchain for Sustainability: Tracks supply chain emissions and sustainability metrics.
- Energy Web Chain: A blockchain platform for renewable energy certification.
- VeChain: Uses blockchain for carbon footprint tracking in energy supply chains.

By integrating blockchain, the energy sector can achieve greater transparency, regulatory compliance, and environmental sustainability, accelerating the transition to greener energy solutions.

REGULATORY CONSIDERATIONS AND GOVERNANCE FRAMEWORKS

As blockchain adoption in the energy sector grows, regulatory challenges and governance frameworks play a crucial role in ensuring compliance, security, and fair market practices. Governments and regulatory bodies must address concerns related to data privacy, energy trading laws, taxation, and smart contract enforceability.

Key Regulatory Considerations

Compliance with Energy Regulations

Many countries have strict regulations governing energy trading, grid connectivity, and carbon reporting. Blockchain-based solutions must align with:

- Electricity market rules to ensure fair competition.
- Data protection laws (e.g., GDPR) for user privacy in energy transactions.
- Renewable energy incentives that comply with government policies.

Smart Contract Legality & Dispute Resolution

Smart contracts enable automated energy trading, payments, and compliance enforcement. However, their legal recognition varies across jurisdictions. Challenges include:

- Contract enforceability: Courts may not yet recognize blockchain-based agreements.
- Regulatory approval: Some countries require licensed platforms for blockchain energy trading.

- Dispute resolution mechanisms: Governance frameworks must define how conflicts are settled in P2P transactions.

Cybersecurity & Data Privacy Concerns

Energy grids and microgrids face cybersecurity threats, such as hacking, data manipulation, and identity theft. Regulatory frameworks must:

- Ensure blockchain network security through encryption standards.
- Prevent unauthorized data access while maintaining transaction transparency.
- Mandate cybersecurity audits for blockchain-based energy solutions.

Global Governance Efforts

- European Union (EU): The EU promotes blockchain for energy efficiency under its Clean Energy Package and Digital Finance Strategy.
- United States (DOE & FERC): Researching blockchain's role in decentralized energy trading while ensuring compliance with federal energy laws.
- China & India: Exploring blockchain for smart grid management and renewable energy trading, with government oversight.

Future Regulatory Trends

As blockchain technology evolves, global energy regulations will need to adapt. Governments and industry leaders are working toward:

- Standardized blockchain protocols for energy trading.
- Cross-border blockchain regulations to enable international energy transactions.
- Hybrid governance models, combining decentralized energy management with centralized regulatory oversight.

CASE STUDIES AND REAL-WORLD IMPLEMENTATIONS

Blockchain technology has witnessed wide adoption across industries, with real-world implementations demonstrating its potential to enhance security, transparency, and efficiency. This chapter explores successful case studies in the energy and utilities sector, finance, and healthcare, along with key lessons learned and adoption barriers.

Successful Blockchain Projects In Energy And Utilities

The energy and utilities industry is rapidly integrating blockchain technology to streamline operations, enable decentralized energy trading, and enhance grid security. The following case studies highlight some of the most successful blockchain-based energy initiatives.

Brooklyn Microgrid (USA) Overview:

- A peer-to-peer (P2P) energy trading platform using blockchain.
- Enables local consumers to buy solar energy from their neighbors.

  Key Benefits:
- Reduces reliance on centralized grids.
- Encourages sustainable energy consumption.
- Uses smart contracts for automated transactions.

Power Ledger (Australia) Overview:

- A blockchain-based platform enabling real-time energy trading.
- Supports solar energy tracking and trading between consumers and businesses.

  Key Benefits:
- Facilitates transparent and cost-effective energy transactions.
- Provides audit trails for energy usage and sustainability tracking.
- Promotes renewable energy adoption through incentive mechanisms.

Energy Web Chain (Global) Overview:

- A blockchain infrastructure tailored for the energy sector.
- Used by companies like Shell, Engie, and PG&E for decentralized energy markets.

  Key Benefits:
- Enables efficient energy management and smart grid operations.
- Reduces administrative costs by automating compliance.
- Enhances cybersecurity for critical energy infrastructure.

These case studies demonstrate how blockchain is revolutionizing the energy sector by reducing inefficiencies, promoting sustainability, and enabling decentralized energy markets.

BLOCKCHAIN ADOPTION IN FINANCE AND HEALTHCARE

Beyond energy, blockchain is making a significant impact on finance and healthcare, solving long-standing issues such as fraud, inefficiencies, and data security.

Blockchain in Finance

The financial sector has been an early adopter of blockchain technology, primarily due to its ability to reduce transaction costs, eliminate intermediaries, and enhance security.

Ripple (Cross-Border Payments)

- Ripple Net uses blockchain to enable fast, low-cost cross-border transactions.

- Financial institutions like Santander and American Express use Ripple for real-time payments.

Impact:

- Reduces transaction time from days to seconds.
- Enhances transparency and fraud prevention in international payments.

JPM Coin (JP Morgan's Blockchain Initiative)

- A digital currency launched by JP Morgan for instantaneous payments between institutional clients.
- Improves liquidity management and cross-border transactions.

Impact:

- Reduces reliance on traditional banking systems.
- Enhances operational efficiency with automated settlements.

Blockchain in Healthcare

Healthcare systems face challenges related to data security, interoperability, and counterfeit drugs. Blockchain provides secure and decentralized solutions to address these issues.

MedRec (Healthcare Data Management)

- A blockchain-based system for patient health records.
- Developed by MIT Media Lab to enable secure and interoperable medical data.

Impact:

- Enhances data security and patient privacy.
- Enables seamless access to medical records across hospitals.

PharmaLedger (Drug Supply Chain Management)

- A European blockchain initiative for tracking pharmaceutical supply chains.
- Ensures drug authenticity and eliminates counterfeit medicines.

Impact:

- Improves drug traceability from manufacturer to patient.
- Reduces fraud and counterfeit drugs in the market.

These cases highlight blockchain's role in revolutionizing finance and healthcare, providing enhanced security, efficiency, and trust in critical operations.

LESSONS LEARNED AND ADOPTION BARRIERS

While blockchain adoption has seen success across industries, several challenges and barriers continue to hinder widespread implementation.

Lessons Learned from Blockchain Adoption

Interoperability is Key

- Many blockchain platforms operate in silos, making it difficult for different systems to communicate.

- Solution: Development of interoperable blockchain protocols (e.g., Polkadot, Cosmos).

Regulatory Clarity is Needed

- Governments are still developing regulations around blockchain and cryptocurrencies.
- Solution: Clearer legal frameworks and compliance standards will drive broader adoption.

Scalability Challenges Exist

- Blockchain networks often struggle with high transaction volumes (e.g., Ethereum congestion).
- Solution: Adoption of Layer-2 solutions like Lightning Network and Optimistic Rollups.

Security and Privacy Concerns

- While blockchain is secure, smart contracts can have vulnerabilities (e.g., DAO hack).
- Solution: Enhanced smart contract auditing and encryption mechanisms.

Adoption Barriers

1. High Implementation Costs
    - Setting up blockchain solutions requires significant investment in infrastructure and development.
    - Many small and medium enterprises (SMEs) find blockchain adoption expensive.
2. Resistance to Change
    - Traditional industries are often hesitant to replace legacy systems with blockchain solutions.
    - Education and training are needed to help organizations understand the benefits.
3. Energy Consumption Concerns
    - Blockchain networks like Bitcoin and Ethereum (before ETH 2.0) have faced criticism for high energy consumption.
    - Newer solutions like Proof of Stake (PoS) and energy-efficient blockchains are being developed.

Future Strategies for Overcoming Barriers

- Adoption of hybrid blockchain models (public + private blockchains).
- Integration with existing enterprise systems to minimize disruptions.
- Global regulatory cooperation to establish uniform blockchain governance.

FUTURE OUTLOOK AND OPPORTUNITIES

Blockchain's Role In Web3 And Defi

Web3 represents the next phase of the internet, focusing on decentralization, user control, and blockchain-based applications. Unlike traditional Web2, which is dominated by centralized entities, Web3 aims to create an ecosystem where users have ownership over their data and digital assets. Blockchain is the backbone of Web3, enabling trustless interactions, decentralized identity management, and transparent governance models(A. G. Gad et al., 2025; V. Buterin, 2014).

Decentralized Finance (DeFi) is most impactful blockchain applications within Web3. DeFi platforms remove intermediaries from financial transactions, allowing peer-to-peer lending, borrowing, trading, and asset management. Smart contracts play a important role in DeFi by automating transactions and ensuring trustless execution. However, scalability, security vulnerabilities, and regulatory concerns remain a greater concern that need to be addressed for mass adoption. Future developments in Layer-2 solutions, cross-chain interoperability, and improved smart contract security are anticipated to drive the next wave of DeFi innovation.

The Impact Of Quantum Computing On Blockchain

Quantum computing is a rapidly emerging technology that has the potential to revolutionize various industries, including blockchain. While blockchain relies on cryptographic security mechanisms such as SHA-256 and elliptic curve cryptography (ECC), quantum computers could potentially break these encryption methods using algorithms like Shor's algorithm. This poses a significant threat to the immutability and security of blockchain networks. o mitigate this risk, researchers are investigating post-quantum cryptography, which focuses on creating cryptographic algorithms that can withstand attacks from quantum computers. Several blockchain projects and institutions are investing in quantum-resistant protocols to future-proof their networks. The integration of quantum-safe encryption, lattice-based cryptography, and hybrid cryptographic models will be essential to ensuring blockchain's long-term security. Governments and organizations are also working on standards for quantum-resistant blockchain networks to prepare for the post-quantum era.

Policy And Regulatory Trends

As blockchain adoption grows, governments and regulatory bodies worldwide are focusing on establishing legal frameworks to govern its use. The lack of standardized regulations has been a major challenge, creating uncertainty for businesses and investors. However, several countries have made progress in defining clear guidelines for blockchain-based applications, particularly in finance, supply chain, and data privacy.

One of the critical areas of focus is the regulation of cryptocurrencies and DeFi platforms. Authorities are working to implement policies that ensure consumer protection, prevent illicit activities, and promote financial stability. Central Bank Digital Currencies (CBDCs) are also gaining attention, with governments exploring blockchain-based digital currencies as an alternative to traditional fiat money.

In addition to financial regulations, data privacy laws such as the General Data Protection Regulation (GDPR) and California Consumer Privacy Act (CCPA) have implications for blockchain applications. Developers are exploring privacy-enhancing technologies such as Zero- Knowledge Proofs (ZKPs) and confidential transactions to comply with regulatory requirements while maintaining the benefits of decentralization.

Next-Generation Blockchain Applications

Beyond finance and traditional use cases, the future of blockchain technology lies in next- generation applications that integrate with emerging technologies like artificial intelligence (AI), the Internet of Things (IoT), and edge computing. Some of the important areas include:

1. **Blockchain for Healthcare** – Secure patient records, clinical trials, and pharmaceutical supply chain management.
2. **Blockchain for Identity Management** – Self-sovereign identities (SSIs) that empower individuals to control their digital identities.
3. **Blockchain in AI and Machine Learning** – Decentralized AI models that improve transparency and data integrity.
4. **Blockchain for IoT Security** – Safe communication and authentication for IoT devices in smart cities, industrial automation, and home networks.
5. **Blockchain in Supply Chain and Sustainability** – Tracking the origin and authenticity of products, ensuring ethical sourcing, and reducing fraud in global trade.

As blockchain continues to evolve, research and development efforts will focus on improving scalability, reducing energy consumption, and enhancing user accessibility. The emergence of decentralized autonomous organizations (DAOs) and blockchain-based governance models will redefine decision-making processes in various industries.

Challenges And Ethical Considerations

A. **Consensus** Algorithms Smart city prerequisites have remarkably different requisitions that must be taken into account when designing them. Hence, these diverse requirements of the blockchain technology into a smart city can be considered a potential open research field. Regarding consensus algorithm, it must be noted that it supplies with regulations for coming to agreement between a few nodes included in a blockchain network. The implementation of consensus algorithms is characterized by several factors, like energy consumption or node identity (T. Ahram,2017).

B. **Cost** In order for alternating the existing framework, particularly when considering an infrastructure, it is important to identify numerous costs, like time and money. Hence, we need to make beyond any doubt that this novel technology does not only create financial advantages and meet the necessities of supervision, but also continuously experiences challenges from the inside organization which exists at a specific time period.

C. **High-performance** Computing Memories An exceptional raise in smart IoT devices is anticipated within the future of smart cities. The allowance of blockchain based services to this kind of devices postures a critical challenge of high storage necessities. This drives to the fact that each device in a blockchain network should maintain a total set of transactions and such type of high storage necessity limits the scalability of the blockchain network. Scalability is a fundamental prerequisite of smart cities and hence, it is of the utmost importance to utilize high-performance memory with high storage necessity for empowering the scalable operation of blockchain-enabled smart cities. Moreover, storage can be implemented outside the blockchain network in case the high-performance computing memories are not located at a blockchain node. This kind of storage can be enabled with use of high-performance computing memory although this can pose additional robustness and security challenges. Blockchains can utilize off-chain decentralized storage and file systems such as Inter Planetary File System (IPFS) and Swarm, rather than centralized storage. However, before uploading to IPFS, data can be encrypted despite the fact that this includes more encryption-decryption delay. Finally, the sharing of encryption-decryption keys in a decentralized way is another issue to be taken into consideration.

D. **Identity** When considering public/open blockchains, anyone has the ability to observe every transaction while every device can be identified with use of its public address. Regarding cryptocurrency-based smart city applications, the privacy issue can be moderated by creating a novel expendable address for each new payment. As of now, the identity of the users in smart city services is given with use of digital identity management systems, which are granted by central authorities. More to the point, Decentralized ID (DID) and Self-sovereign Identity (SSI) empower users to completely control their digital identity without an intermediate centralized third-party and this permits users to control how their personal information are public. Regarding IoT services, blockchain-based DID and SSI can be employed for initially identifying, as well as for the authentication of users in a decentralized way. However, there arise some issues, i.e. a user may lose the private key and therefore, formulating secure recovery mechanisms for DID and SSI could be a critical issue. Pioneer privacy strategies can be exploited for particular information sharing in mutual and anonymous multi-party transactions in a privacy-preserving way in terms of different smart city services .

E. **Latency** Latency indicates the processing time of each transaction while throughput represents a maximum number of transactions within a specific pre-defined time period. Both latency and throughput limitations significantly influence the scalability of smart cities. In a blockchain-enabled smart city, latency occurs because of the computation at decentralized nodes, as except computation at local nodes, data communication between nodes is further added to the total value of latency. Latency due to propagation delay causes forking and thus, for avoiding this forking impact, the propagation delay must be minimized. Following the forking taking place, the operation of block production restarts and this process continues until the update of the block occurs without forking.

F. **Scalability** Scalability in blockchain refers to the system's ability to maintain efficient performance and ensure **Quality of Service (QoS)** as the number of devices and transactions increases, particularly in smart environments like smart cities. While blockchain's key features — **decentralization, fault tolerance, and security** — offer significant advantages, they also introduce challenges in scaling the network.

A conventional blockchain network requires each participating node to store an ever-growing ledger of transactions and engage in the validation process. This inherently limits scalability since the decentralized structure demands considerable computational and storage resources.

With the anticipated growth of smart IoT devices, projected to reach 64 billion by 2025, this surge presents significant challenges for designing scalable blockchain-based smart city infrastructures. The Proof of Work (PoW) consensus algorithm, commonly used in blockchains like Bitcoin, ensures network security by requiring miners to solve complex mathematical problems. However, this process is computationally intensive and significantly limits the network's throughput, reducing its ability to process large numbers of transactions per second (TPS).

To address these scalability issues, researchers have proposed PoW-based parallel mining as a solution. This approach involves dividing the mining tasks among multiple nodes to perform computations in parallel, improving the network's transaction capacity while maintaining security. Additionally, other consensus mechanisms like Proof of Stake (PoS) and Sharding are also being explored to further enhance blockchain scalability, making it more adaptable for large-scale applications in smart city ecosystems.

G. **Sustainability Sustainability** in the context of blockchain-based smart cities involves designing systems that minimize the consumption of natural resources while maintaining efficient operations. However, the rapid growth of **smart IoT devices** within smart city environments significantly increases **energy consumption**. Additionally, traditional **consensus algorithms** like **Proof of Work (PoW)**, which require extensive computational power, further contribute to the energy burden.

To tackle these challenges, several strategies can be implemented:

1. **Energy-Efficient Planning**: Incorporating sustainable infrastructure and optimizing energy consumption through intelligent planning can reduce the environmental impact.
2. **Renewable Energy Sources**: Powering blockchain nodes using renewable energy such as solar, wind, or hydropower can significantly reduce carbon footprints.
3. **Sustainable Consensus Algorithms**: Algorithms like Proof of Stake (PoS), Delegated Proof of Stake (DPoS), and Proof of Authority (PoA) consume far less energy than PoW, making them more eco-friendly alternatives.
4. **Efficient Hardware**: Utilizing low-power hardware like Application-Specific Integrated Circuits (ASICs) can optimize computational efficiency while reducing energy consumption.
5. **Green Communication Networks**: Establishing low-energy communication systems for data transmission in smart cities can further contribute to sustainability.

For **long-term sustainability** from economic, environmental, and social perspectives, it is essential to adopt less energy-intensive frameworks. By integrating energy-efficient technologies and renewable resources, blockchain-enabled smart city services can operate in a more environmentally conscious manner, ensuring the responsible use of natural resources.

CONCLUSION

Blockchain technology has emerged as a groundbreaking innovation, transforming industries by enhancing security, transparency, and efficiency. Originally developed for cryptocurrencies, it is now widely applied in finance, healthcare, supply chain management, and energy. By enabling decentralized and trustless transactions, blockchain eliminates intermediaries, reduces fraud, and ensures data integrity. Its core principles—cryptographic security, consensus mechanisms, and smart contracts—form the foundation for its growing adoption across various sectors. Despite its vast potential, blockchain faces several challenges that hinder widespread adoption. Scalability remains a major concern, as traditional networks struggle with transaction speed and efficiency. Regulatory uncertainty and the lack of standardized legal frameworks create obstacles for businesses and governments looking to implement blockchain solutions. Additionally, interoperability between different blockchain networks, security vulnerabilities in smart contracts, and the high energy consumption of proof-of-work (PoW) systems pose significant barriers to growth. Looking ahead, continuous advancements are expected to address these challenges and unlock blockchain's full potential. The adoption of Central Bank Digital Currencies (CBDCs), interoperability solutions, quantum-resistant cryptography, and Decentralized Autonomous Organizations (DAOs) will drive blockchain's evolution. Energy-efficient consensus mechanisms, such as proof-of-stake (PoS), are making blockchain more sustainable. With ongoing innovation and supportive regulatory policies, blockchain will continue to revolutionize digital transactions, governance, and data management, shaping a more decentralized and secure future.

Blockchain technology has rapidly evolved from its initial application as the backbone of cryptocurrencies to a transformative force across various sectors. As industries and governments increasingly explore its potential, the future holds numerous opportunities and challenges. This chapter has outlined emerging trends and possible trajectories that blockchain technology may follow, highlighting its disruptive potential and the innovations driving its growth.

1. Expansion Beyond Cryptocurrencies

The use of blockchain technology has spread to industries including healthcare, finance, supply chain management, and energy, even if Bitcoin and other cryptocurrencies are still essential to its identity. Blockchain is the perfect answer for data verification, safe transactions, and trustless ecosystems because of its transparency, immutability, and decentralisation features. We can anticipate that blockchain use cases will continue to diversify in the upcoming years, especially in fields that require strong security and transparency.

2. Decentralized Finance (DeFi) and Financial Innovation

One of the biggest developments in blockchain technology is DeFi, which provides financial services without the need for conventional middlemen. Financial systems are becoming more inclusive and accessible as a result of the transformation of lending, borrowing, and trading brought about by smart contracts and decentralised applications (DApps). DeFi systems will probably grow in popularity and security as authorities set more precise rules. Financial inclusion will also be improved by the emergence of hybrid models that combine blockchain-based innovations with conventional finance.

3. Non-Fungible Tokens (NFTs) and Digital Ownership

With the introduction of provable digital ownership, NFTs have upended sectors including gaming, entertainment, and the arts. In addition to digital art, NFTs are being used in digital identity verification, real estate, and intellectual property management. NFTs will be incorporated into the metaverse and other virtual worlds in the future, resulting in new economic models powered by digital assets enabled by blockchain technology.

4. Central Bank Digital Currencies (CBDCs) and Financial Systems

In an effort to modernise financial institutions, central banks throughout the world are investigating the use of CBDCs. The traceability and transparency of blockchain technology can boost financial inclusion, lower transaction costs, and increase the effectiveness of monetary policy. However, the effectiveness of CBDCs will depend on their structure and management. Governments and blockchain developers must work together to guarantee financial stability, security, and privacy.

5. Sustainability and Green Blockchain Solutions

Energy-efficient consensus techniques like proof-of-stake (PoS) and proof-of-authority (PoA) have been developed in response to environmental concerns about blockchain's energy usage, especially in proof-of-work (PoW) systems. Furthermore, blockchain applications in renewable energy management and carbon credit tracking support environmental sustainability. Blockchain protocols will be further optimised in the future to reduce their carbon footprint.

6. Interoperability and Cross-Chain Solutions

The necessity of interoperability has been brought to light by the growth of blockchain systems. In order to facilitate smooth data and asset transfers between various blockchains, cross-chain solutions are starting to appear. Leading interoperability frameworks include Polkadot, Cosmos, and Chainlink protocols. Improved blockchain cooperation will open up new possibilities for multi-chain ecosystems and decentralized application

7. Blockchain and Artificial Intelligence (AI)

By improving automation, transparency, and data security, the combination of blockchain with AI has the potential to completely transform a number of sectors. Blockchain can guarantee data privacy while offering reliable datasets for AI training. In exchange, by anticipating network congestion and stopping fraudulent activity, AI systems can optimize blockchain networks. Blockchain applications will become more intelligent and effective as a result of this mutually beneficial interaction.

8. Regulation and Governance

Clarity in regulations will be essential to building stability and trust as blockchain adoption increases. The significance of balanced regulation that promotes innovation while safeguarding consumers and preserving financial stability is becoming more widely acknowledged by governments. Furthermore, decentralized autonomous organizations (DAOs) provide innovative governance structures that encourage accountability and transparency by enabling community involvement in decision-making.

9. Quantum-Resistant Blockchain

Current blockchain cryptography algorithms could be threatened by the development of quantum computing. Quantum-resistant cryptography is being actively developed by researchers to protect block-chain networks. Blockchain systems will remain resilient and secure if they are future-proofed against quantum attacks.

10. The Road Ahead

Web 3.0, the decentralized internet, is expected to be built on top of blockchain technology. Block-chain will provide people more control over their online identities and experiences by giving data back to individuals and lowering dependency on centralized organizations. A more just and effective global economy will also result from blockchain's contributions to safe voting procedures, open supply chains, and cross-border payments.

In summary, blockchain technology has a promising and wide-ranging future. Blockchain will continue to unlock its potential as regulatory frameworks and technological breakthroughs progress. Working together, developers, businesses, legislators, and researchers will be essential to overcoming obstacles and achieving blockchain's full potential. A new era of decentralised and transparent solutions will be ushered in by innovation,

REFERENCES

Ahram, T., Sargolzaei, A., Sargolzaei, S., Daniels, J., & Amaba, B. (2017, June). Blockchain technology innovations. In 2017 IEEE technology & engineering management conference (TEMSCON) (pp. 137-141). IEEE.

Buterin, V. (2014). "A Next-Generation Smart Contract and Decentralized Application Platform." *IEEE Blockchain Conference*

Gad, A. G., Mosa, D. T., Abualigah, L., & Abohany, A. A. (2025). Emerging Trends in Blockchain Technology and Applications: A Review and Outlook. *IEEE Access : Practical Innovations, Open Solutions.*

IEEE Blockchain Initiative. (2024). Blockchain Trends and Future Prospects. *IEEE Access : Practical Innovations, Open Solutions.*

Mougayar, W. (2016). *The Business Blockchain: Promise, Practice, and Application of the Next Internet Technology*. IEEE Transactions on Emerging Technologies.

Nakamoto, S. (2008). *Bitcoin: A Peer-to-Peer Electronic Cash System*. IEEE Transactions on Blockchain Technology.

Sharin, F. H., Hernandez, M. S., & Sentosa, I. (2023, May). Future trends of blockchain technology in the technological fields. In 2023 3rd International Conference on Advance Computing and Innovative Technologies in Engineering (ICACITE) (pp. 1307-1313). IEEE.

Swan, M. (2015). "Blockchain: Blueprint for a New Economy." *IEEE Blockchain Symposium*

Tapscott, D., & Tapscott, A. (2016). Blockchain Revolution: How the Technology Behind Bitcoin is Changing Money, Business, and the World. *IEEE Access : Practical Innovations, Open Solutions.*

Vonitsanos, G., Panagiotakopoulos, T., Kanavos, A., Maragoudakis, M., & Mylonas, P. (2021, November). Issues and challenges associated with blockchain in smart cities. In 2021 16th International Workshop on Semantic and Social Media Adaptation & Personalization (SMAP) (pp. 1-5). IEEE.

Wood, G. (2015). Ethereum: A Secure Decentralized Generalized Transaction Ledger. *IEEE Access : Practical Innovations, Open Solutions.*

Zyskind, G., Nathan, O., & Pentland, A. (2015). *Decentralizing Privacy: Using Blockchain to Protect Personal Data*. IEEE Transactions on Information Security.

Chapter 17

Overview of Blockchain Technology:

Fundamental Concepts, Key Features, and the Potential Impact of Blockchain on Various Industries

Zahira Tabasssum
https://orcid.org/0000-0002-4912-0228
HKBK College of Engineering, India

Sufia Banu
https://orcid.org/0000-0002-4596-3340
HKBK College of Engineering, India

Anees Fathima
HKBK College of Engineering, India

Noor Ayesha
https://orcid.org/0009-0003-5711-7656
HKBK College of Engineering, India

K. Nadeem Pasha
HKBK College of Engineering, India

Syed Zain Ul Abidin
HKBK College of Engineering, India

ABSTRACT

Blockchain technology, which was first created for the virtual currency Bitcoin, has expanded into a ground-breaking tool that has the potential to completely transform a number of different industries. Fundamentally, blockchain is a distributed, decentralized ledger that safely logs transactions over a network of computers, guaranteeing immutability, security, and transparency. Beyond crypto currencies, this technology provides cutting-edge solutions for transactions and data management in a variety of industries. This chapter offers a thorough overview of blockchain technology, going over its basic ideas, salient characteristics, advantages, and drawbacks. It investigates practical uses in sectors like energy, supply chain management, healthcare, and finance. Blockchain improves efficiency, lowers operating costs, and builds trust by doing away with middlemen. . This chapter serves as a valuable resource for understanding block chain's impact and future prospects in the digital era.

DOI: 10.4018/979-8-3373-2439-5.ch017

INTRODUCTION

Blockchain technology has evolved as a cornerstone of modern digital infrastructure, offering decentralization, security, and transparency across industries. Blockchain technology originated to support Bitcoin operations but it has now penetrated major industry sectors that include finance, healthcare, supply chain operations and energy management. Its unalterable trust-free transaction system functions as a main operational instrument to resolve security challenges and fight fraud together with operational constraints.

The worldwide blockchain market reached 17.57 billion in 2023 and research indicates it will grow with a 67.4% annual rate from 2023 through 2030 according to Grand View Research (2023). The fast-rising adoption patterns demonstrate why both public and private sectors continue integrating this technology into their practices. The technology transitioned from a specialized innovation towards becoming a mass-distributed digital solution when major institutions including IBM and JPMorgan and Tesla utilized blockchain to secure transactions besides the 420 million blockchain users across the globe. The digital world today welcomes blockchain technology as its revolutionary manifestation able to reshape various industries. The technology behind Bitcoin appeared in 2008 as a Bitcoin foundation before extending its reach beyond digital currencies for applications across finance, healthcare, supply chains and numerous other business fields. Blockchain functions as a distributed decentralized transaction system which provides secure cooperation between transparent and tamper-resistant transactions. Blockchain eliminates the need for the assurance mechanisms found in traditional centralized systems by leveraging smart contracts, distributed transparency, and cryptographic processes to achieve consensus. The solution offers advantages to businesses by providing improved security together with efficiency and automation features. Businesses and governmental institutions increasingly recognize blockchain potential because of its clear consequences. The operational models in various industries have strong potential for substantial redefinition because blockchain allows enhanced optimization of supply chains as well as medical record authenticity assurance and financial transaction management improvement. Widely accepted blockchain implementation requires resolving three main adoption barriers that include scaling difficulties and regulatory and power requirements. The study traces blockchain origins from its cryptocurrency roots to its adoption within general technology applications. Insurance Business UK provides a detailed look at blockchain fundamentals composed of nodes and blocks along with chains and cryptographic methodologies for safeguarding its non-modifiable framework against fraudulent behavior. The evaluation examines how POW and PoS consensus methods protect blockchain networks and minimize their power consumption issues. The special qualities of blockchain transform it into an effective industrial instrument for the present time. The digital transformation takes place mainly through blockchain technology because it boosts ancient business practices with five fundamental capabilities: decentralization, transparency, fraud resistance, automation and efficiency. Current research on blockchain includes critical innovations together with an analysis of current barriers along with emerging trends. The research investigates blockchain influence on supply chain management as well as government operations and real estate deals and financial sectors and healthcare services. The article explores blockchain applications in energy utilities in detail where it demonstrates how grid modernization meets sustainability goals through peer-to-peer (P2P) energy trading systems. This chapter provides a comprehensive overview of blockchain technology, how it works, and its potential impact on various industries. It also explores the challenges hindering its adoption and the future trends that could further enhance its capabilities. By understanding block chain's significance, businesses and individuals can better prepare for a future driven by decentralized innovation. Figure 1 (Vanmathi, 2024) shows the application of Block chain in various sectors and Figure 2(Vanmathi,2024) shows the application of block chain in IoT.

Figure 1. Applications of Block Chain

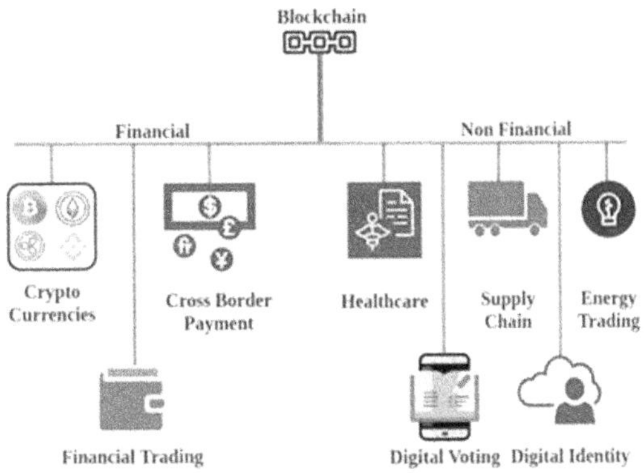

Figure 2. Applications of Blockchain in IoT

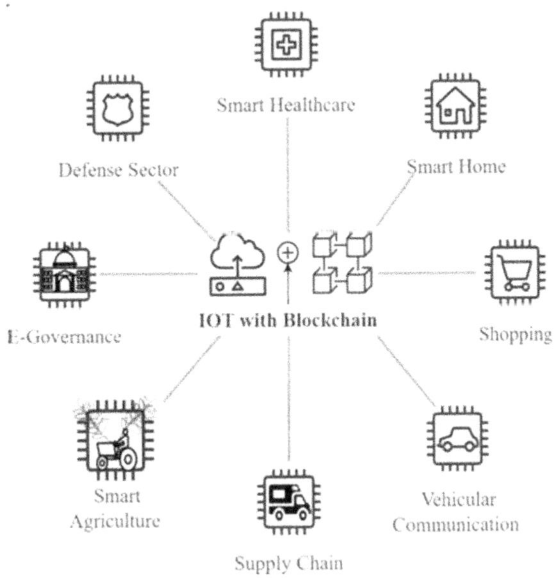

EVOLUTION AND FUNDAMENTALS OF BLOCKCHAIN TECHNOLOGY

The Birth of Blockchain and Its Introduction to the World through Bitcoin.

The conceptual foundation of blockchain technology was laid in 1991 by Stuart Haber and W. Scott Stornetta, who developed a cryptographically secured chain of blocks to timestamp and verify digital documents. However, it was in 2008 that an individual or group under the pseudonym Satoshi Nakamoto introduced blockchain as the underlying technology for Bitcoin, the world's first decentralized cryptocurrency.

The first Bitcoin block, known as the Genesis Block (Block 0), was mined on January 3, 2009, marking the inception of blockchain as a trustless, peer-to-peer financial system. The innovation of Bitcoin lay in its use of a Proof of Work (PoW) consensus mechanism, allowing transactions to be validated without the need for a centralized authority, such as a bank or government.

Milestones in Bitcoin and Blockchain Adoption

Blockchain technology gained prominence through Bitcoin's adoption and increasing market value, leading to widespread recognition. Some key milestones include:

· **2010:** The first-ever real-world Bitcoin transaction occurred when Laszlo Hanyecz paid 10,000BTC for two pizzas, equivalent to over $300 million today.
· **2011-2013:** The emergence of alternative cryptocurrencies (Litecoin, Ripple, and Ethereum) demonstrated block chain's versatility.
· **2015:** The launch of Ethereum, introducing smart contracts, significantly expanded block chain's application beyond financial transactions.
· **2017:**Bitcoin'sprice surged to $20,000, drawing institutional interest.
· **2020-2021:** Major corporations (Tesla, PayPal, and Visa) adopted Bitcoin, while El Salvador became the first country to recognize Bitcoin as legal tender.

As of 2024, Bitcoin has over 420 million users worldwide, and its blockchain processes more than 350,000 transactions daily (Statista, 2023).

Blockchain is defined as a decentralized, distributed ledger technology(DLT) that records transactions across multiple computers in a secure, immutable, and transparent manner. Each transaction is stored in a block, which is linked to previous blocks' using cryptographic hashing, forming a chain ofblocks—hence the term "blockchain."

Key characteristics that define blockchain include:

Decentralization – Unlike traditional centralized databases, blockchain operates on a peer-to-peer network, eliminating the need for a central authority.

Immutability – Once data is recorded on a blockchain, it cannot be altered or deleted, ensuring data integrity.

Transparency–All transactions are publicly visible (in public block chains) or accessible to authorized participants (in private blockchains).

Security – Security occurs through cryptographic algorithm encryption that defends against fraud and cyber threats.

Consensus Mechanisms – To confirm transactions, blockchain relies on consensus methods such as Practical Byzantine Fault Tolerance (PBFT), Proof of Work (PoW), and Proof of Stake (PoS).

Blockchain technology divides into four fundamental categories which distinguish between network openness and management structures together with their applications. The classification of blockchain system divides into four distinct types which are Public Blockchains, Private Blockchains, Hybrid Blockchains and Consortium Blockchains.

The open network of **Public Blockchains** welcomes all participants wishing to join their decentralized operations. All nodes can participate in blockchain networks which maintain distributed ledgers and permit user contribution for validating transactions while safeguarding the network. Bitcoin and Ethereum along with other examples make up popular public blockchains that let users execute decentralized financial deals and smart contracts while avoiding central administrative control. Open-source transparency together with decentralized control leads to high security on these blockchains yet they remain resource-intensive which results in slow process times because of proof-of-work or proof-of-stake consensus methods.

A single organization or entity manages and controls **Private Blockchains** since these networks operate within their clearly defined parameters. Private Blockchains achieve higher efficiency by restricting access to approved users and thus serve businesses that need secure and private operations. Internal business processes such as supply chain management and financial operations and data sharing among approved groups use these blockchains. The business sector relies on Hyper ledger Fabric as their private blockchain system because it offers enhanced operational visibility alongside efficient management of who accesses data and makes governance decisions.

Hybrid Blockchains unite aspects from public and private blockchains to create a versatile blockchain solution. Hybrid blockchains distribute their ledger data into two categories whereby specific portions become available to public viewing while other parts stay guarded under restricted access permissions. Businesses managing confidential data together with external stakeholders should use this model because it provides transparency alongside privacy protection. A well-known example of a hybrid blockchain is IBM Food Trust, which enhances supply chain transparency by allowing consumers to verify product origins while maintaining confidentiality for business participants.

Consortium Blockchains are a collaborative approach where multiple organization's govern and share the blockchain network. Unlike public blockchains, which are open to everyone, or private blockchains, controlled by a single entity, consortium blockchains are managed by a group of predefined members. These blockchains are commonly used in industries where different organization's need to work together while maintaining a secure and trusted environment.R3 Corda is a prime example of a consortium blockchain, widely used in the financial sector to streamline transactions and improve efficiency among multiple financial institutions.

Each type of blockchain serves distinct purposes and is adopted based on an organization's needs for security, scalability, decentralization, and transparency. Figure 3(crypto explorer) illustrates the classification of these blockchain types, highlighting their unique characteristics and use cases.

Figure 3. Types of Blockchain

EVOLUTION BEYOND CRYPTOCURRENCIES – SMART CONTRACTS, DEFI, AND ENTERPRISE SOLUTIONS

The evolution of blockchain technology has expanded beyond its initial use in cryptocurrencies, paving the way for Smart Contracts, Decentralized Finance (DeFi), and Enterprise Solutions. While Bitcoin demonstrated the feasibility of decentralized financial transactions, its limitations in scalability and programmability led to the development of more versatile blockchain frameworks. In 2015, Ethereum, created by Vitalik Buterin, introduced smart contracts, fundamentally transforming how blockchain networks function.

Smart Contracts – Self-Executing Agreements without Intermediaries

Smart contracts are self-executing digital agreements stored on the blockchain that automatically enforce terms once predefined conditions are met. These contracts eliminate the need for intermediaries such as banks, notaries, or brokers, streamlining processes across various industries:

- **Finance**: Enables automated lending, insurance claims processing, and investment management without manual intervention.
- **Supply Chain**: Enhances transparency by ensuring real-time tracking of goods and automating supplier payments.
- **Healthcare**: Secures patient data through encrypted medical records, improving privacy and interoperability.

As of 2023, over 45 million smart contracts have been deployed on the Ethereum blockchain, making it the leading platform for decentralized applications (dApps) (Ethereum Foundation, 2023). The adoption of smart contracts continues to grow as businesses and developers recognize their potential for improving efficiency, security, and automation in digital transactions.

Decentralized Finance (DeFi) – Redefining Financial Systems

Decentralized Finance (DeFi) is transforming traditional financial systems by leveraging blockchain technology to offer trustless, transparent, and permission less financial services. Unlike traditional banking, DeFi eliminates intermediaries such as banks and brokers, allowing users to interact directly with financial applications.

Some of the key DeFi services include:

Lending and Borrowing: Platforms like Aave and Compound enable users to lend or borrow crypto assets without relying on traditional financial institutions. Smart contracts automatically manage loan terms, collateral, and interest rates, reducing the risk of fraud and improving efficiency.

Decentralized Exchanges (DEXs): Platforms like Uniswap and Sushi Swap allow users to trade digital assets directly with one another, bypassing centralized exchanges. These peer-to-peer trading systems use liquidity pools instead of order books, improving accessibility and reducing trading fees. The growth of DeFi has been remarkable. By 2024, the total value locked (TVL) in DeFi protocols exceeded $60 billion, reflecting a 250% increase since 2020 (DeFi Llama, 2024). This surge highlights the increasing trust and adoption of DeFi applications, positioning them as a viable alternative to traditional financial services.

Enterprise Blockchain – Adoption by Multinational Corporations

Large enterprises across various industries are increasingly adopting blockchain technology to enhance security, traceability, and operational efficiency. Unlike public blockchains used for cryptocurrencies, enterprise blockchains are often private or consortium-based, enabling businesses to maintain control while benefiting from decentralization and transparency.

Several multinational corporations have successfully integrated blockchain into their operations:

- **IBM & Walmart**: Implemented blockchain-based food traceability through the IBM Food Trust network, ensuring greater food safety and reducing contamination risks by tracking products from farm to shelf.
- **Maersk & Trade Lens**: Developed Trade Lens, a blockchain-powered shipping logistics platform, improving transparency and reducing inefficiencies in global trade by enabling real-time tracking of shipments.
- **JPMorgan & Onyx**: Launched JPM Coin, a blockchain-based digital currency that facilitates real-time cross-border payments, significantly reducing settlement times from days to seconds and improving liquidity management.

The adoption of blockchain in enterprise solutions is growing rapidly. According to a 2023 Deloitte survey, 81 of the world's top 100 public companies are investing in blockchain-based solutions, demonstrating the technology's increasing relevance in industries such as finance, supply chain, and healthcare.

As enterprises continue to explore blockchain's potential, its role in enhancing security, reducing costs, and improving transparency is expected to expand further.

SIGNIFICANCE OF BLOCK CHAIN IN TODAY'S DIGITAL ECONOMY

Blockchain technology has evolved far beyond its initial role in cryptocurrencies, emerging as a critical infrastructure for the modern digital economy. Its core benefits—Decentralization, security, and transparency—are driving widespread adoption across industries. As businesses and governments seek efficient, fraud-resistant, and cost-effective solutions, blockchain's influence continues to grow.

Decentralization – Eliminating the Need for Central Authorities

Traditional financial and data management systems rely on centralized entities such as banks, governments, and corporations to facilitate transactions. However, this dependence increases costs, introduces inefficiencies, and creates single points of failure. Blockchain's peer-to-peer validation mechanism eliminates intermediaries, streamlining operations while enhancing security and accessibility.

According to (PwC, 2023), the global blockchain market is projected to reach $1.4 trillion by 2030, largely driven by enterprise adoption in sectors like finance, healthcare, and supply chain management.

Security and Fraud Prevention – A Tamper-Proof System

Blockchain leverages cryptographic hashing (such as SHA-256 in Bitcoin) to ensure that once data is recorded, it cannot be altered or tampered with. This immutability makes blockchain highly resistant to cyber attacks, fraud, and unauthorized modifications.

With cybercrime projected to cost $10.5 trillion annually by 2025 (Cybersecurity Ventures, 2023), blockchain-based security solutions are becoming increasingly crucial for protecting sensitive data and financial transactions.

Transparency and Efficiency – Borderless, Real-Time Transactions

Traditional cross-border payments using networks like SWIFT typically take 2-5 days due to multiple intermediaries and verification processes. In contrast, blockchain-based payment networks, such as Ripple's XRP Ledger, can settle transactions in just 3-5 seconds, significantly improving financial efficiency.

The World Economic Forum (2023) estimates that blockchain will reduce global transaction costs by over $1 trillion annually by 2030, revolutionizing global trade, remittances, and business operations.

Blockchain as a Foundation for Future Innovation

Since Bitcoin's inception in 2009, blockchain has evolved into a foundational technology that is transforming industries, including finance, governance, supply chains, and energy management. With continued advancements and integration with emerging technologies such as AI, IoT, and quantum computing, block chain's impact is expected to expand even further. It is no longer just an innovation—it has become a core pillar of the digital economy, driving efficiency, security, and trust across industries worldwide.

BREAKING DOWN THE BLOCKS: WORKING OF BLOCKCHAIN

Blockchain technology is a distributed ledger system that allows secure, transparent, and tamper-resistant transactions without the need for a central authority. It is the backbone of cryptocurrencies and has found applications in finance, healthcare, supply chain, energy, and more. According to a report by Markets and Markets (2023), the global blockchain market is expected to grow from $7.4 billion in 2022 to $94.0 billion by 2027, at a CAGR of 66.2%. The increasing adoption of blockchain is driven by its ability to enhance security, efficiency, and trust in digital transactions.

To understand how blockchain functions, it is essential to explore its core structure, consensus mechanisms, smart contracts, and cryptographic security. Fig 4 represents Distributed Ledger System.

Figure 4. Distributed Ledger System

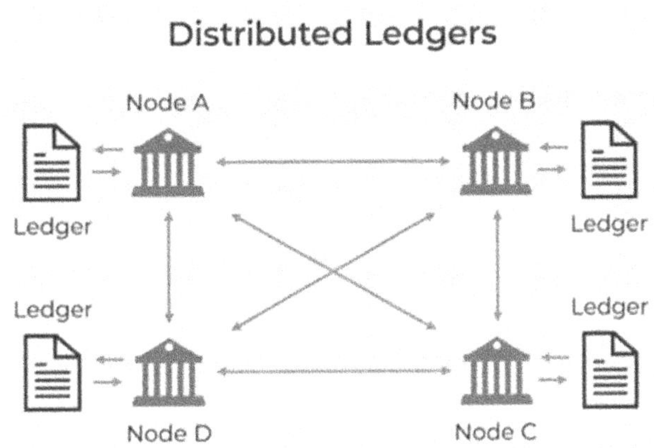

Anatomy of a Blockchain: Blocks, Chains, and Nodes

A blockchain is a distributed and decentralized database that maintains a continuously growing list of records, known as blocks, which are securely linked together in a chain. This structure ensures security, transparency, and immutability of data across the network. Each participant in the blockchain network operates a node, which is responsible for storing, validating, and propagating transactions. The decentralized nature of blockchain removes the need for a central authority, making the system trustless and resilient to attacks.

Each block in a blockchain consists of three main components: the header, which contains metadata such as the previous block's hash, a timestamp, and the Merkle root; the transaction data, which includes a list of validated transactions recorded in the block; and the nonce, a randomly generated number used in the Proof of Work (PoW) consensus mechanism to create a valid block hash. The linkage between blocks is secured using cryptographic hashing, ensuring that even a small change in one block would alter all subsequent blocks, making tampering nearly impossible.

To structure transaction data efficiently, blockchain systems use a Merkle tree, a hierarchical structure that hashes transactions recursively, ultimately producing a Merkle root—a single hash that serves as a fingerprint for all transactions in a block. The Merkle root is essential for improving blockchain performance and security. Instead of requiring verification of all transactions in a block, only a few hashes need to be checked, making the process highly efficient. Moreover, if even a single transaction is altered, its hash changes, affecting the entire Merkle tree (Hui Li, 2022) thereby ensuring that any tampering is immediately detectable. Merkle trees as shown in Figure 5 also contribute to blockchain scalability, allowing networks like Bitcoin and Ethereum to verify transactions without storing the entire dataset.

Through the combination of blocks, chains, nodes, cryptographic hashing, and Merkle trees, blockchain technology ensures a secure, transparent, and scalable digital infrastructure. These fundamental principles have made blockchain an essential component of modern applications, from cryptocurrencies to enterprise solutions, reinforcing trust in decentralized digital transactions.

Figure 5. Merkle tree

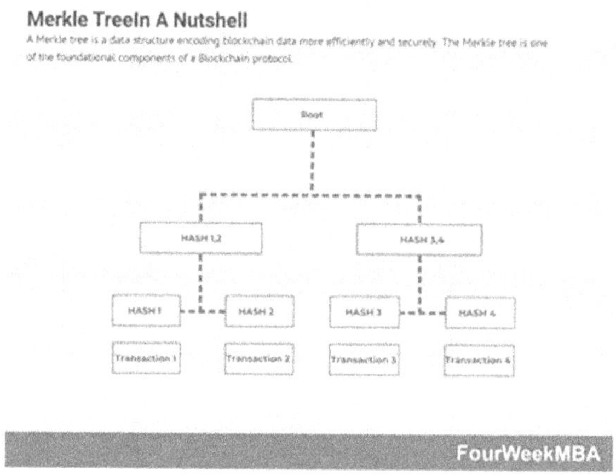

Blocks to Chain Formation

In a blockchain, each block is linked to the previous block through a unique cryptographic hash, forming a continuous, immutable chain. The cryptographic connection between blocks creates a protection mechanism that detects any past-block modifications which results in mutated following blocks thus rendering tampering attempts impossible. Any modification attempt on a previous block's transaction will cause its hash value to change thus rendering the whole blockchain invalid. The data immutability along with integrity protection which this feature provides is essential for financial transactions because it stops fraud and data tampering and double-spending attempts.

ROLE OF NODES IN BLOCKCHAIN NETWORKS

A node operates as a computer which takes part in blockchain networks with duties to store complete ledger information and execute transaction verification tasks. The network bases its decentralization security and consensus capabilities on these nodes that perform essential functions. The different network nodes in blockchain systems serve distinct operational purposes which together produce stable trustworthy systems.

Types of Blockchain Nodes

Each blockchain network consists of separate computer devices known as nodes that maintain the distributed ledger and validate and deprotect the network. The blockchain depends on nodes which perform particular functions to protect its integrity by maintaining consensus through decentralized operations. The major types of blockchain nodes include:

1. **Full Nodes**
   - Store a complete copy of the blockchain ledger.
   - Validate transactions and blocks independently.
   - Participate in consensus mechanisms and help enforce network rules.
   - Example: Bitcoin and Ethereum full nodes ensure that all transactions follow protocol rules.
2. **Light Nodes (SPV Nodes - Simplified Payment Verification)**
   - Store only blocks headers, not the full blockchain.
   - Rely on full nodes for transaction verification.
   - Commonly used in mobile wallets and low-storage devices.
3. **Mining Nodes**
   - Compete to solve complex cryptographic puzzles in Proof of Work (PoW) blockchains.
   - Contribute computing power to validate and add new blocks to the chain.
   - Example: Bitcoin miners use mining nodes to secure the network and earn rewards.
4. **Validator Nodes**
   - Used in Proof of Stake (PoS) and Delegated Proof of Stake (DPoS) networks.
   - Stake cryptocurrency to gain the right to validate transactions and propose new blocks.
   - Example: Ethereum 2.0 uses validator nodes in its PoS consensus mechanism.
5. **Masternodes**
   - Perform specialized functions such as governance voting, private transactions, and instant transactions.
   - Require staking a significant amount of cryptocurrency as collateral.
   - Example: Dash blockchain utilizes master nodes to enable instant transactions and privacy features.
6. **Archive Nodes**
   - Store the entire history of the blockchain, including past states and transactions.
   - Essential for historical data retrieval and auditing.
   - Commonly used by blockchain explorers and data analytics services.
7. **Relay Nodes**
   - Facilitate communication between nodes without participating in consensus.

- Help distribute transaction data and blocks across the network efficiently.

Blockchain nodes are fundamental to the operation of decentralized networks. Whether they are validating transactions, mining new blocks, or ensuring efficient data distribution, each type of node plays a specific role in maintaining the blockchain's security, transparency, and efficiency. The diverse functions of nodes allow blockchain networks to operate in a scalable, decentralized, and tamper-proof manner.

Figure 6. Types of Blockchain Nodes

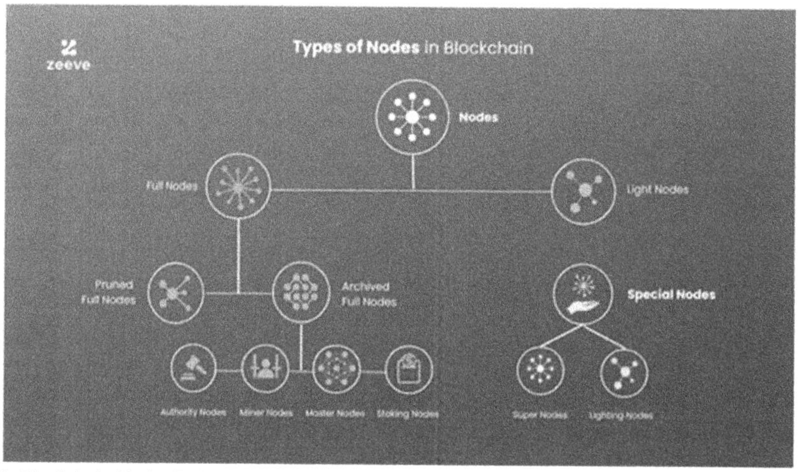

(In Depth Guide To Blockchain Nodes Training Ppt, n.d.).

THE POWER OF CONSENSUS: PROOF OF WORK, PROOF OF STAKE, AND BEYOND

Consensus mechanisms are the protocols that ensure all nodes in a blockchain network agree on the validity of transactions. They play a crucial role in maintaining the security, integrity, and decentralization of blockchain systems. Without effective consensus, blockchains could face issues such as fraud, double-spending, and network splits. Various consensus mechanisms have been developed, each with its advantages and trade-offs.

Proof of Work (PoW) – The Computational Backbone of Bitcoin

Proof of Work (PoW) is the first and most widely used consensus mechanism, originally pioneered by Bitcoin. It relies on a process called mining, where participants, known as miners, compete to solve complex mathematical puzzles to add new blocks to the blockchain.

Process: Miners use computational power to find a hash value that meets the network's difficulty conditions. Once a valid hash is found, the block is added to the blockchain, and the miner is rewarded with cryptocurrency.

Security: PoW is highly secure and resistant to attacks, as altering a block would require re-mining all subsequent blocks, which is computationally impractical.

Energy Consumption: PoW requires massive computational resources. Bitcoin's PoW network consumes approximately 120 terawatt-hours (TWh) per year, an amount comparable to the energy consumption of a small country (Cambridge Centre for Alternative Finance, 2023).

Due to its high energy consumption and environmental impact, many blockchain networks are shifting towards alternative consensus mechanisms that offer greater efficiency and scalability.

Figure 7. Proof of Work

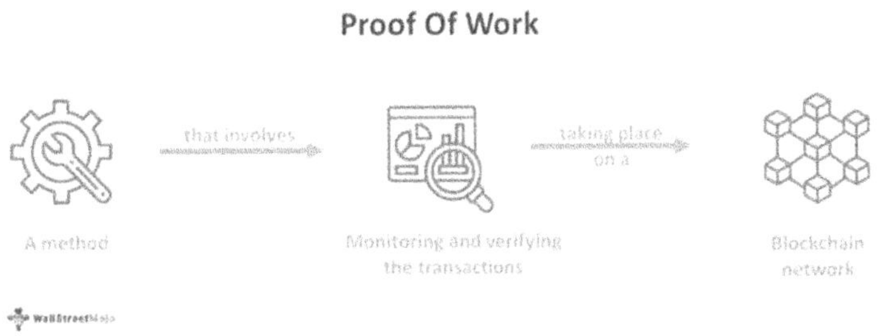

(John et al., 2023)

Proof of Stake (PoS) – An Energy-Efficient Alternative

Proof of Stake (PoS) was introduced as a **sustainable alternative** to Proof of Work (PoW), significantly reducing energy consumption while maintaining network security. Instead of miners competing to solve complex puzzles, PoS requires participants to **stake (lock up) tokens** as collateral to validate transactions.

Process: Validators are randomly selected to propose and validate blocks, with selection probability based on the number of tokens staked. The higher the stake, the greater the chances of being chosen.

Energy Efficiency: PoS is far more energy-efficient than PoW, consuming 99.95% less energy (Ethereum Foundation, 2022), making it an environmentally friendly alternative.

Adoption: Ethereum, the second-largest blockchain network, transitioned from PoW to PoS in 2022, reducing its energy consumption dramatically and enhancing scalability.

Figure 8. Proof of Stake Flow

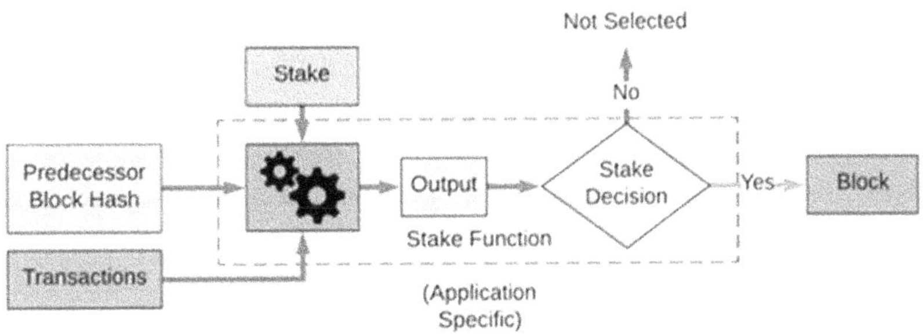

Other Consensus Mechanisms

Beyond PoW and PoS, several other consensus mechanisms have been developed to enhance speed, efficiency, and decentralization:

Delegated Proof of Stake (DPoS): A small number of elected validators are chosen by token holders to confirm transactions, making DPoS faster and more scalable than PoS. Used in networks like EOS and TRON.

Proof of Authority (PoA): Transactions are validated by pre-approved, trusted nodes rather than an open network of participants. PoA is commonly used in private and enterprise blockchains where control and efficiency are prioritized.

Proof of History (PoH): A consensus method used by Solana, PoH improves scalability by time stamping transactions before validation, allowing for extremely high transaction speeds.

As blockchain technology evolves, these consensus mechanisms continue to be refined, enabling greater scalability, security, and energy efficiency across decentralized networks.

SMART CONTRACTS – THE SELF-EXECUTING AGREEMENTS CHANGING INDUSTRIES

Smart contracts are self-executing programs stored on the blockchain that automatically enforce agreements once predefined conditions are met. By eliminating the need for intermediaries such as banks, notaries, and brokers, smart contracts reduce costs, increase efficiency, and enhance security across various industries.

Working of Smart Contracts

The execution of a smart contract follows a structured process.Figure 9 represents the working of smart contract.

Identify Agreement – Individuals or businesses define the rules of the contract and the expected outcomes.

Set Conditions – The contract specifies clear conditions (e.g., "If payment is received, then the product will be delivered").

Code Business Logic – A computer program encodes the logic of the contract, ensuring automatic execution when conditions are met.

Encryption & Blockchain Storage – The contract is securely stored on a blockchain, making it tamper-proof and immutable.

Execution & Processing – When conditions are met, the smart contract automatically executes actions such as transferring funds or granting access.

Network Updates – All nodes in the blockchain validate and update their records, ensuring transparency and trust in the system.

Figure 9. Working of Smart Contract

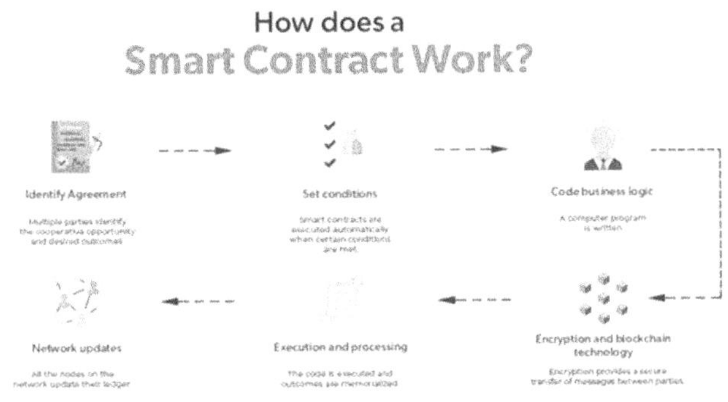

Example: Buying a Car Using a Smart Contract

- **Step 1**: The buyer and the seller create a smart contract stating: *"If 10, 00,000/- is paid, then car ownership is transferred."*
- **Step 2**: You send the money to the smart contract.
- **Step 3**: The contract verifies the payment and automatically transfers ownership to your name.
- **Step 4**: The blockchain records the transaction, and all parties receive an updated record.

Applications of Smart Contracts

Smart contracts are widely used across industries, transforming traditional processes:

- **Finance**: Enables automated lending, insurance claims, and decentralized exchanges without intermediaries.
- **Supply Chain**: Ensures real-time tracking, verification, and automated supplier payments.
- **Healthcare**: Protects patient records, ensures secure data sharing, and automates insurance claims processing.

As the leading platform for smart contracts, Ethereum hosts over 45 million deployed contracts (Ethereum Foundation, 2023), making it the dominant blockchain for decentralized applications (dApps).

CRYPTOGRAPHIC SECURITY: THE FORTRESS BEHIND BLOCKCHAIN'S IMMUTABILITY

Blockchain's security is fundamentally built on cryptographic techniques, ensuring data integrity, authenticity, and immutability. By combining cryptographic hashing and asymmetric encryption, blockchain prevents unauthorized modifications, making it one of the most secure digital technologies.

Blockchain security using cryptography

Blockchain employs two key cryptographic techniques to protect transactions:

Hash Functions (SHA-256) – Each transaction is processed using a cryptographic hash function, generating a fixed-length value unique to the input data. Any modification to the data results in a completely different hash, preventing tampering. Bitcoin, for example, uses SHA-256 for secure transaction hashing.

Public-Private Key Cryptography – Transactions are secured through asymmetric encryption, where each user has a public key (shared with others) and a private key (kept secret). This ensures that only the rightful owner can authorize a transaction, preventing unauthorized access.

Immutability and Data Integrity

Once data is recorded on the blockchain, it cannot be altered or deleted, ensuring permanent and tamper-proof records. This property makes blockchain particularly valuable for industries that require high data integrity, such as financial transactions, digital identity management, and secure record-keeping.

- Bitcoin's blockchain has never been hacked, showcasing the robustness of cryptographic security.
- Distributed ledger technology (DLT) ensures that all copies of the blockchain remain synchronized and protected from data breaches.

Necessity of Blockchain Security

Blockchain technology relies on multiple layers of security and decentralization to maintain trust, security, and transparency.

- Blocks, chains, and nodes work together to create an immutable ledger resistant to tampering.
- Consensus mechanisms like Proof of Work (PoW) and Proof of Stake (PoS) ensure decentralized validation of transactions.
- Smart contracts automate agreements, removing intermediaries and reducing fraud risks.
- Cryptographic security safeguards blockchain data from cyber threats, ensuring authenticity and integrity.

As blockchain gains traction, emerging innovations in scalability, security, and sustainability will continue to expand its applications across finance, supply chain management, healthcare, and beyond. No longer just a technological experiment, blockchain has become the foundation for a safer and more decentralized digital future.

THE DNA OF BLOCKCHAIN

Blockchain technology is widely regarded as a disruptive innovation due to its ability to facilitate decentralized, secure, and transparent transactions without reliance on intermediaries. Since its inception, blockchain has transformed from a niche technology supporting cryptocurrencies to a fundamental pillar of modern digital infrastructure.

According to Markets (2023), the global blockchain market is projected to grow from $7.4 billion in 2022 to $94.0 billion by 2027, representing a 66.2% compound annual growth rate (CAGR). This rapid expansion is driven by the growing demand for decentralized solutions, enhanced security, and transparency across various industries, including finance, healthcare, supply chain, and energy management.

KEY CHARACTERISTICS THAT MAKE BLOCKCHAIN REVOLUTIONARY

Blockchain shows revolutionary properties because of four main distinctive qualities that set it apart from traditional centralized systems.

1. Through decentralization the system cuts out the requirement for intermediates.

The blockchain system functions differently from traditional financial and data management programs because it relies on widespread nodes to operate as opposed to centralized banks. The system operates without centralized control so no one entity can govern it which builds security features and preserves autonomy.

Advantages of a Decentralized Network

The decentralized nature of blockchain technology provides essential advantages including security protection through elimination of central control and the establishment of enhanced system stability alongside limitation of censorship attempts. Basic centralized networks become disrupted completely when a single failure occurs through cyber attacks or technical problems. The distributed structure of blockchain keeps the network operational when nodes experience failures so it demonstrates high resistance to failures. One significant benefit obtained through decentralization is resistance against censorship. The structure of centralized systems enables authorities to exercise management of service access through control and possible complete access disruption. Users can make transactions within Bitcoin and Ethereum blockchain networks without worrying about an authoritarian body disrupting their activity. The value of decentralized networks is high for regions under financial restrictions and oppressive regulations since they conduct operations autonomously.

Security functions as one leading benefit that decentralized networks deliver to users. Blockchain uses a distributed network system which functions better than centralized databases because hackers cannot easily breach it. The decentralized network can operate normally when an attack occurs on one node because this attack does not compromise the integrity of other network components thus protecting against DDoS attacks and data breaches and malicious tampering. Real applications of decentralization networks show their usefulness in operational environments. More than 17,000 full nodes worldwide currently operate Bitcoin (BTC) (Bitnodes, 2023) to guarantee that nobody possesses control of the network and sustain its decentralized features. ETH demonstrated the change to Proof of Stake (PoS) which reduced mining demands without compromising decentralized network components. The storage solution known as File coin allows users to establish a decentralized network that provides cloud-storage services to challenge traditional providers such as Google Drive and AWS. The real-world examples demonstrate decentralization operates as a practical innovation shaping multiple worldwide industries.

2. Transparency – Ensuring trust through Public Ledgers/ Every Transaction, Verifiable by All.

The public or permissioned ledger system registers Blockchain transactions which allows users to verify and permanently store information and maintain data integrity. The public or permissioned ledger system creates trust alongside improved accountability while reducing fraud which becomes essential for transparent industries such as finance and healthcare and governance.

The central concept of blockchain technology rests on transparency which delivers dependable security measures and accountability for many industrial operations. The main blockchain feature of transparent open-ledger management stands out strongly in public networks such as Bitcoin and Ethereum which permit real-time transaction visibility to all users. A decentralized financial system develops through this approach which disconnects users from requiring bank operators or regulatory organizations for trust-based operations. Blockchain transparency achieves its most important characteristic through unalterable data. The blockchain network stores transactions with permanent accuracy because entries become un-alterable and undeletable when recorded. Blockchain operates as a dependable permanent registration system which diminishes the possibility of fraud and stops both data tampering and information loss.

Transaction transparency remains absent in conventional financial networks because traditional methods permit double spending and accounting fraud as well as other dishonest activities. Since blockchain data remains invulnerable to unauthorized changes it stops criminals from changing financial records thus decreasing the chances of fraudulent activities. Companies across the globe together with public authorities execute blockchain-based auditing platforms as their main tool to detect financial irregularities and curtail fraudulent activities. Blockchain transparency operates in two directions by illuminating industry operations through multiple sectors. IBM Food Trust and similar companies in supply chain management employ blockchain systems to track down food sources while preventing fraud and ensuring consumer safety through packaging monitoring throughout the supply chain process. Healthcare organizations use blockchain medical records systems which stop unauthorized modifications to enhance data fidelity and patient security. Additionally, governments are leveraging blockchain for e-governance solutions; for instance, Estonia has successfully implemented blockchain-based voting systems to enhance electoral security and transparency. The examples demonstrate that blockchain technology brings digital transparency while maintaining an unalterable record which transforms businesses to establish trust while boosting system accountability in modern digital environments.

3. Security – Cryptographic Protection Against Tampering

Blockchain operates with built-in tamper-resistant properties which makes it an extremely secure digital platform. Blockchain implements cutting-edge cryptographic defenses to protect data integrity and authenticate users and secure transactions which make its platform resist attacks effectively.

Cryptographic Security in Blockchain

Blockchain security implements cryptographic tools for securing deals and blocking unauthorized entry. The Bitcoin system relies on SHA-256 as its hash function to achieve data protection. Each transaction becomes a specific hash value of fixed length which creates complete new hash values when modified thus making data changes practically impossible. The security feature of public-private key cryptography stands among the most important elements of blockchain systems. Blockchain network users possess two essential elements: public keys serve as account numbers yet private keys function as transaction authorization tools. The asymmetric encryption mechanism enables the private key holder to be the sole party who can approve transactions because no other user can access protected assets.

Through Distributed Ledger Technology (DLT) data security improves because it spreads information across thousands of nodes located throughout the world. Blockchain operates securely when some nodes become compromised because network operations continue due to its decentralized structure and therefore proves highly resistant to cyber threats.

FRAUD RESISTANCE IN BLOCKCHAIN SYSTEMS

Blockchain stands out for its main benefit which enables fraud prevention alongside unauthorized changes. Users can attempt double spending in traditional digital payment systems because these platforms allow users to duplicate their digital assets therefore creating double-spend issues. The system uses consensus mechanisms to verify and confirm every transaction which subsequently gets recorded permanently thus eliminating fraud possibilities.

The resistance of blockchain against cyberattacks remains high. The manipulation of blockchain transactions requires hackers to seize control of at least 51% of the blockchain nodes yet this is impossible due to the extensive global presence of thousands of blockchain nodes in networks like Bitcoin and Ethereum. Blockchain protects identity security in addition to its preventive role against financial fraud. Traditional authentication technology stores its data base at one central location but hackers continuously seek out these centralized targets. User authentication managed through blockchain technology becomes secure because it employs decentralized encrypted record systems which minimize identity theft chances as well as data breach vulnerabilities.

Business operations rely heavily on blockchain due to its exceptional security features which defend digital deals while safeguarding information and stopping cyber scams across numerous business sectors.

a. Smart Contracts based on automation systems enforce agreements according to predefined conditions.

Blockchain-based programs called smart contracts contain self-execute functions that run through computer code stored on blockchain networks. The system executes transactions without human intervention by performing release functions whenever predefined criteria activate removing requirements for legal and banking institutions and notaries. The removal of human workflows and secondary actors through smart contracts leads to decreased expenses together with heightened operational speed and better system security.

Benefits of Smart Contracts

Speed and Automation

Standard contractual procedures need multiple weeks and multiple days for administrative staff to finalize paperwork while awaiting authorization which requires multiple intermediaries to process the documents. The execution of smart contracts via predefined conditions enables instant operation when requirements trigger their activation thus enabling quick transactions and time reductions.

Cost Reduction

Smart contracts reduce costs by cutting out conventional middleman entities which also eliminates fees for legal service and administrative procedures and banking expenses. The reduction in financial processing fees that smart contracts deliver results in more efficient and cheaper transactions.

Trust and Security

Use of blockchain technology for smart contracts provides the advantage of creating unalterable tamper-proof records. After deployment the contract becomes immutable because it maintains complete transparency and avoids disputes. The trustless system enables agreement parties to execute deals using this method without requiring external entities for enforcement.

Industries Benefiting from Smart Contracts

Finance

Blockchain-based agreements enable automated financial lending operations which eliminate the necessity of traditional financial institutions for issuing loans and processing payments. The automatic processing of insurance claims operates when all necessary verification conditions for accidents have been successfully met through blockchain-based agreements. Decentralized payment processing functions without banking intermediaries thus enables fast and affordable transaction that also provide security between peers. The financial industry benefits from these new technologies which deliver improved speed and dedication alongside complete business openness.

Healthcare

Smart contracts improve health care security together with operational efficiency by providing secure blockchain-based methods to handle medical records. The healthcare provider system maintains secure access to patient records because this structure stops unauthorized changes while protecting patient information and record integrity. Through blockchain automation the insurance industry decreases administrative load which improves both insurance claim processing speed and enhances automated patient verification processes. The adoption of blockchain technology simplifies healthcare procedures along with decreasing fraud opportunities while fortifying patient services.

Real Estate

Real estate operations benefit from smart contracts which create immediate house transfers through automated systems that work without brokerage services between buyers and sellers. The blockchain automatically transfers ownership funds when predefined payment verification conditions become satisfied in the system. The system bans traditional paperwork while lowering expenses and fighting unfair activities because it maintains impenetrable secure property documents. The implementation of blockchain technology enhances both transparency and efficiency which makes real estate deals proceed more securely as faster transactions while making them accessible to users.

b. Scalability – Expanding Blockchain's Reach and Performance

Blockchain early networks experienced performance limitations yet developers implemented Layer 2 technologies together with new consensus protocols to enhance the system's ability to handle numerous transactions effectively.

The Scalability: Current Limitations and Emerging Solutions

The growth of blockchain networks causes both delays in transaction execution and enhanced payment charges because of increasing transaction quantities. Bitcoin handles 7 TPS although Visa reaches 24,000 TPS as its processing network shows and Ethereum can initially do 15-30 TPS before Ethereum 2.0. Recent development of Ethereum 2.0 improved its operational scalability positively. Multiple solutions have been developed to overcome these emerging issues. The Lightning Network together with Rollups maintain separate processing networks outside the main blockchain to reduce its transactional load. Blockchain sharding produces multiple network chains with partitioned data that boosts processing speed alongside mixed blockchain frameworks which unite public and private blockchain systems for better operational performance. The enhancements developed for blockchain technology help it continue as a working solution for general adoption efforts. The innovative qualities of decentralization, transparency, security and automation, scalability have evolved business operations into what will dominate digital transaction and enterprise solutions in the future.

BLOCKCHAIN CREATES WIDE-RANGING INDUSTRY CHANGES BY TRANSFORMING THE REQUIRED ASPECTS OF THE WORLD

The industrial revolution earned its name through blockchain because this technology provides secure environments alongside transparent systems with automated functionalities and decentralized operations. Blockchains began as cryptocurrency tools before producers discovered its wide applications across financial institutions and healthcare and supply chains in addition to government entities and real estates and energy providers. Organizations implement blockchain systems because they want to remove middle parties while keeping fraud at bay and boosting operational processes. According to PwC's 2023 Blockchain Survey, blockchain has the potential to contribute $1.76 trillion to the global economy by 2030. The financial sector is expected to gain the most, with over 60% of banks actively exploring blockchain applications, while the health care industry is integrating blockchain to ensure secure medical records and data interoperability.

This section explores how blockchain is transforming finance and banking through decentralized finance (DeFi) and digital currencies, as well as its applications in healthcare, enabling data security, interoperability, and efficient patient management.

Finance & Banking: Disrupting Traditional Finance with DeFi and Digital Currencies

The finance and banking sector has been among the earliest adopters of blockchain. Decentralized Finance (DeFi), cryptocurrencies, and central bank digital currencies (CBDCs) are revolutionizing traditional financial systems by eliminating intermediaries, reducing transaction costs, and enhancing transparency.

The Rise of Decentralized Finance (DeFi)

Decentralized finance (DeFi) is a blockchain-based financial system that eliminates central authorities such as banks and financial institutions. It enables peer-to-peer (P2P) lending, trading, and asset management through smart contracts, reducing costs and improving efficiency.

Key Features of DeFi

Eliminates the Need for Banks as Intermediaries: Users can lend, borrow, and earn interest without traditional financial institutions.

Permission less Transactions: Anyone with an internet connection can access DeFi services

Transparency and Security: Transactions are recorded on the blockchain, ensuring fraud prevention.

Lower Costs: No banking fees or third-party commissions.

Figure 10 represents features of DeFi (Haleem, 2021) and Figure 11 represents Benefits of DeFi (Haleem, 2021)

Figure 10. Features of DeFi

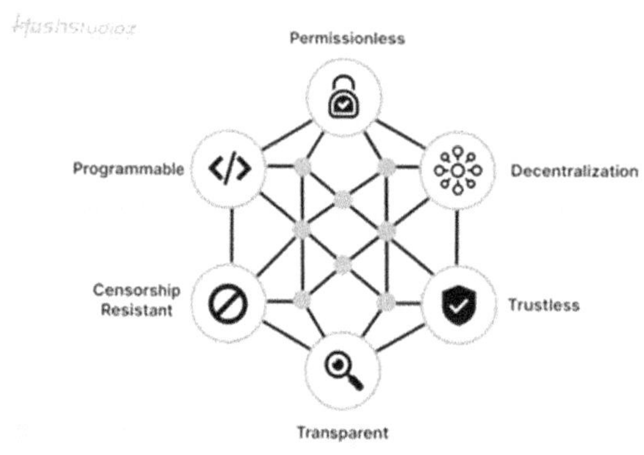

Figure 11. Benefits of DeFi

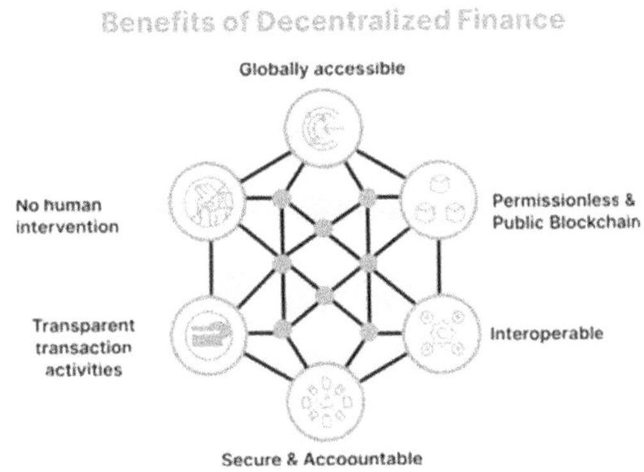

Growth of DeFi

The growth of decentralized finance (DeFi) has been remarkable over the past few years, revolutionizing traditional financial systems through blockchain technology. The total value locked (TVL) in DeFi protocols surged from $1 billion in 2019 to $80 billion in 2023, demonstrating the sector's rapid expansion and increasing adoption (DeFi Llama, 2023). The platform Ethereum leads DeFi application operations through its lead position in the blockchain marketplace with more than 60% market share thanks to its secure programming abilities combined with its safety standards and developer network ecosystem. The increase of DeFi usage can be attributed heavily to Uniswap, Aave, and MakerDAO because these platforms facilitate decentralized exchanges, lending and borrowing platforms and stablecoin protocols. The future of DeFi looks promising as it develops more with Layer 2 innovation and real-world digital assets and solutions for intersystem communication. Regulatory procedures together with security issues will function as fundamental determinants for DeFi's upcoming development path.

Real-World DeFi Applications

DeFi delivers numerous functional platforms that modernize the way people access traditional banking services.

The **decentralized exchange platforms** Uniswap along with Sushi Swap let their users execute digital asset trades without requiring centralized management. Through automated market-making (AMM) mechanisms both platforms achieve liquidity solutions as well as decrease users' dependence on traditional order books.

The **lending and borrowing** feature of Aave and Compound allows users to deposit their assets for interest generation and obtain loans by offering collateral assets. Financial services become more accessible because this distributed system does away with banks and their mandatory credit requirements.

Yield Farming presents one of the most popular routes for investors to get passive income through the process of depositing assets into liquidity pools. DeFi protocols enable users to provide liquidity through these pools to receive various rewards that include governance tokens and additional assets.

Users can experience the revolutionary power of DeFi in establishing a free and accessible financial system through these available applications.

Cryptocurrencies and Digital Assets

Numerous sectors and financial markets have witnessed the rapid rise of cryptocurrency patented systems as alternative digital money structures. The digital assets Bitcoin and Ethereum have transformed global financial operations through their decentralized transaction systems which provide secure borderless money transfers. The number of cryptocurrency users worldwide has increased massively in recent years to surpass 420 million in 2023 (Statista, 2023).Bitcoin continues to dominate the market, maintaining its position as the leading cryptocurrency with a market capitalization exceeding $800 billion. Meanwhile, Ethereum has solidified its role as the backbone of decentralized applications, driving innovation in smart contracts, non-fungible tokens (NFTs), and decentralized finance (DeFi). With Ethereum powering over 80% of NFT and DeFi transactions, its influence in the blockchain ecosystem remains unparalleled. As the digital asset space continues to evolve, key trends such as institutional adoption, regulatory developments, and advancements in blockchain scalability are expected to shape the future of cryptocurrencies.

Central Bank Digital Currencies (CBDCs)

Governments and central banks worldwide are actively exploring Central Bank Digital Currencies (CBDCs) as a means to provide secure, government-backed digital alternatives to traditional fiat money. Unlike decentralized cryptocurrencies, CBDCs are issued and regulated by central authorities, offering the benefits of digital transactions while maintaining monetary policy control. As of 2024, over 130 countries are engaged in research, development, or pilot programs for CBDCs (Atlantic Council, 2024), reflecting the growing interest in digital currency adoption. China has been at the forefront of this innovation with its Digital Yuan, which has already processed over $14 billion in transactions since its launch. Similarly, the European Central Bank (ECB) has announced plans to introduce a Digital Euro by 2026, aiming to modernize financial systems and enhance cross-border transactions. CBDCs develop momentum because they show potential to alter worldwide banking systems yet they must face challenges regarding data privacy together with security issues and regulatory oversight needs.

Blockchain in Banking and Financial Institutions

Traditional banks together with financial institutions adopt blockchain technology to improve efficiency metrics and security practices and regulatory compliance purposes. The most significant transformative blockchain use happens in cross-border payments because it reduces both transaction fees and processing duration between borders. The World Bank (2023) explains how blockchain stands as a solution which reduces bank transfer fees between 40-60% below traditional banking methods.

Through XRP Ledger Ripple provides transactions that finish within 3-5 seconds but SWIFT might require 2-5 days for international transfers. The fast processing coupled with low expenses makes blockchain an innovative force in international money transfers and trade payment processes.

Computer networks based on blockchain technology have transformed both fraud detection mechanisms and Know Your Customer (KYC) compliance systems. Blockchain deployment of decentralized digital identities results in eliminated identity fraud while offering stronger security features in payments processing. Two major international financial institutions including HSBC and JPMorgan continue using blockchain technology to track fraud instantly thereby providing better visibility while minimizing fraudulent activities. A shift in regulatory standards will propel acceleration in bank usage of blockchain which will advance transaction protection and automation of compliance requirements and financial services accessibility.

Case Study:

JPMorgan's Onyx and Blockchain-Based Payments

Onyx by JP Morgan represents a blockchain platform that operates as a financial transaction revolution. This platform offers secure swift cross-border payments as well as tokenized assets alongside DeFi solutions for institutional clients at cost-effective rates.

In 2020 JPMorgan Chase introduced the Onyx system for enhancing blockchain-based processing of payments which included wholesale transactions and interbank settlements. Traditional business transactions managed by SWIFT require 2-5 days to complete procedures while using multiple banks and causing substantial payment expenses. The use of JPM Coin within Onyx gives financial institutions immediate settlement abilities which lowers operational expenses and boosts operational effectiveness. JPM Coin operates on a permissioned block chain, cutting settlement times from days to seconds and lowering transfer costs by 30-50%. Another key innovation, Link, allows banks to share financial data securely, improving compliance and fraud detection. Over 400 financial institutions have joined Link, enhancing transparency and transaction security. JPMorgan also introduced blockchain-based repo transactions, enabling near-instant settlement and improved liquidity.

The impact of Onyx is substantial, enabling faster payments, lower costs, enhanced security, and regulatory efficiency. Its success has prompted Citibank, HSBC, and others to explore similar blockchain-based solutions. As Central Bank Digital Currencies (CBDCs) gain traction, blockchain- powered banking is expected to become a global standard by 2030.

JPMorgan's Onyx has set a precedent for modernizing financial transactions,demonstrating the potential of block chain in banking. By eliminating intermediaries and increasing transaction speed, Onyx is reshaping the future of cross-border payments with efficiency, security, and cost savings.

Healthcare: Securing Patient Data and Enabling Interoperability

Healthcare systems worldwide face significant challenges in ensuring data security, interoperability, and transparency. Medical records are often fragmented across different providers, making it difficult for patients and health care professionals to access vital health information in real time. Cybersecurity threats also pose a significant risk, with the healthcare sector being one of the primary targets for data breaches. Blockchain technology has the potential to address these issues by providing a decentralized, immutable, and transparent system for managing patient data, clinical trials, and medical supply chains. Figure 12 represents Features of healthcare in Blockchain domain (Sara Technologies Inc., n.d.).

Figure 12. Features of healthcare in Blockchain domain

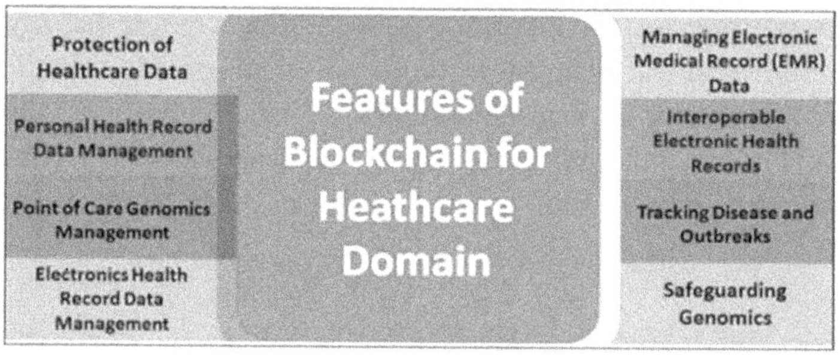

The lack of interoperability in healthcare systems creates inefficiencies and increases administrative costs. Hospitals, insurance companies, and other medical institutions rely on different databases and data management systems, making it difficult to exchange information seamlessly. Blockchain allows for a unified and decentralized health record system that enables real-time access to accurate medical histories.

A patient's complete medical record can be securely stored on a block chain network and accessed by authorized healthcare providers, reducing the risk of medical errors and improving treatment outcomes. Security is one of the biggest concerns in healthcare, with data breaches exposing millions of patient records each year. In 2023 alone, more than 40 million healthcare records were compromised in cyber attacks, according to the HIPAA Journal.

Traditional healthcare databases are vulnerable to hacking due to centralized storage systems. Blockchains cryptographic security and distributed ledger system mitigate this risk by ensuring that data cannot be altered or deleted without proper authorization. Every transaction recorded on a blockchain is time-stamped and immutable, reducing the possibility of fraud or unauthorized access.

Clinical trials and drug development processes also benefit from blockchain implementation. The pharmaceutical industry struggles with issues such as data manipulation, lack of transparency, and counterfeit drugs. Block chain can enhance the integrity of clinical trials by securely recording all research data, ensuring that results are accurate and verifiable.

This prevents fraudulent manipulation of trial outcomes and promotes trust among researchers and regulators. Furthermore, blockchain-based solutions can track drugs from production to distribution, ensuring the authenticity of pharmaceutical products and eliminating counterfeit medications. Figure 13 represents features of healthcare in Blockchain domain (Hassan, 2022).

Figure 13. Health care sphere with Blockchain Aspects

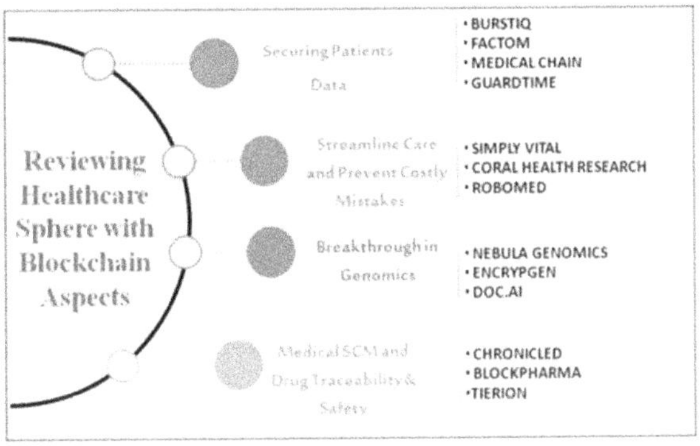

Blockchain is also transforming medical billing and insurance claims processing. Fraudulent claimsandincorrectbillingcontributetobillionsofdollarsinlossesannually.Blockchain'ssmartcontract functionality can automate insurance claims processing, reducing paperwork and administrative costs. When a patient receives treatment, the transaction is recorded on the blockchain, and the smart contract automatically initiates the claim process, ensuring quick and accurate reimbursements.

Several healthcare organizations and technology firms are already leveraging blockchain to improve data security and interoperability. Companies like IBM Watson Health and MediBloc are using blockchain networks to facilitate secure data sharing and patient-controlled health records. Estonia has successfully integrated blockchain into its national healthcare system, ensuring that patient records are securely accessible across different healthcare providers.

Figure14 represents how **blockchain technology** is used in **health care workflows**, ensuring secure, transparent, and efficient data management. Following a simple breakdown of the flow:

Figure 14. Integrated work-flow process of blockchain technology for healthcare culture.

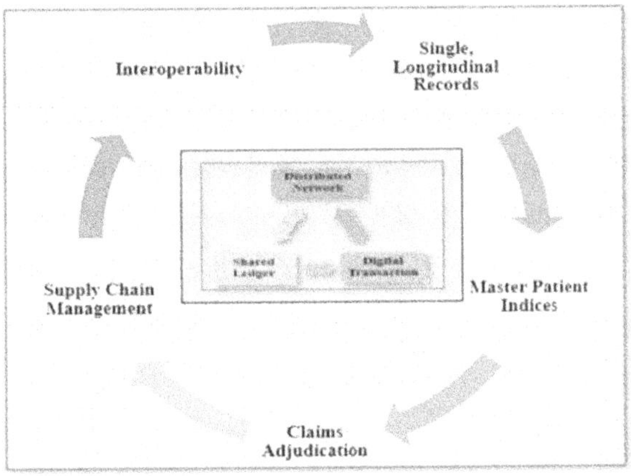

Core Components (Center Box)

· **Shared Ledger**→A common record of transactions that all authorized users can access.
· **Distributed Networks**→ Data is stored across multiple locations, increasing security and reducing risks.
· **Digital Transactions**→ Medical records, insurance claims, and other health care data are securely processed through blockchain.

Surrounding Processes (Arrows& Labels)

Interoperability (Blue Arrow) →Ensuring different health care systems can communicate and share data efficiently.

Single Longitudinal Records (Green Arrow)→Keeping a complete history of a patient's health data in one place.

Master Patient Index (Orange Arrow) →Creating a unique identifier for each patient to prevent duplication.

Claims Adjudication (Purple Arrow) →Automating the insurance claims process to reduce fraud and improve speed.

Supply Chain Management (Yellow Arrow) →Tracking the movement of medical supplies and drugs to prevent counterfeiting.

How It Works Together

· When a patient interacts with the health care system (e.g., getting gates t or treatment), a **digital transaction** is created.
· This transaction is recorded in the **shared ledger** and distributed across a **network**.
· Other health care services (like insurance, pharmacies, or hospitals) can access and verify the data in real time.
· This system helps improve **efficiency, security, and trust** in healthcare.

As blockchain adoption in healthcare continues to grow, it will play a crucial role in securing patient data, improving interoperability, and ensuring transparency in clinical trials and medical transactions. The transition to blockchain-powered healthcare systems will require collaboration between regulators, healthcare providers, and technology developers to create secure, scalable, and efficient solutions.

GOVERNMENT & PUBLIC SERVICES: DIGITAL IDENTITY, VOTING, AND GOVERNANCE INNOVATION

Governments worldwide are exploring blockchain to improve public services, enhance transparency, and reduce corruption. The ability of blockchain to provide a secure and immutable ledger makes it an ideal solution for identity management, voting systems, and governance applications. Identity management is a fundamental aspect of government services, but traditional identity verification systems often suffer from inefficiencies, security vulnerabilities, and lack of interoperability. Blockchain-based digital identity systems offer a secure and tamper-proof solution where individual scan store and manage their identity credentials. Governments can issue blockchain-based digital IDs that citizens can use for various services, including banking, healthcare, and online transactions. Figure 15 illustrates a **Blockchain-Based E-Voting System**, which ensures secure, transparent, and tamper-proof elections. Following is a breakdown of the process:

Figure 15. Blockchain-based e-voting system architecture

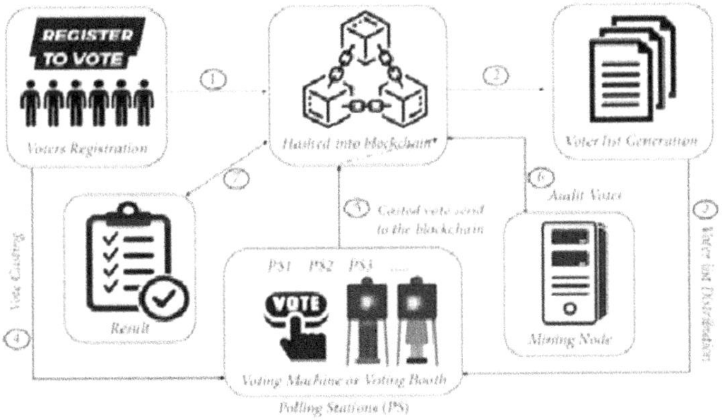

Blockchain in the Energy and Utilities Industry

Blockchain technology is transforming the energy and utilities sector by introducing decentralized, secure, and transparent solutions for energy trading, grid management, and sustainability tracking. Traditional energy systems rely on centralized authorities, making them prone to inefficiencies, high operational costs, and vulnerability to fraud. The integration of blockchain in energy infrastructure enables peer-to-peer (P2P) energy trading, smart grid management, and carbon credit tracking, ensuring a more sustainable, efficient, and transparent energy ecosystem. Figure 16 represents schematic diagram of an energy community using blockchain technology (Delardas, 2023).

Figure 16. Schematic diagram of an energy community using blockchain technology

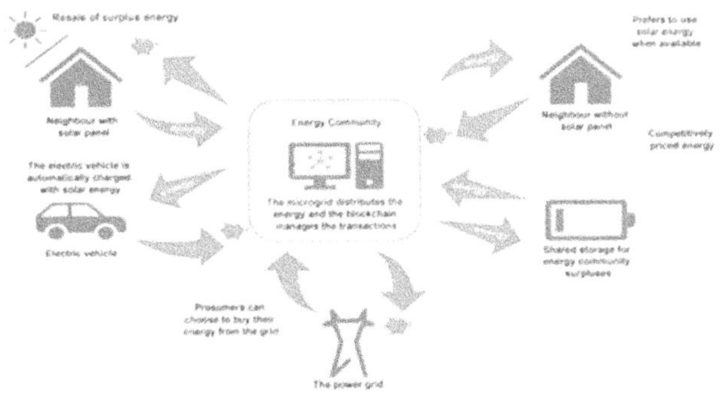

According to a Market report, the global blockchain in energy market was valued at $919 million in 2022 and is projected to reach $6.3 billion by 2027, growing at a CAGR of 45.3%. This rapid growth reflects the increasing adoption of blockchain-powered energy solutions by both governments and private enterprises. The transition to decentralized energy systems is crucial as the demand for renewable energy sources continues to rise. Traditional energy grids struggle with integrating decentralized renewable energy sources, such as solar and wind power, due to inconsistent energy generation and inefficient distribution mechanisms. Blockchain facilitates real-time tracking and automation of energy transactions, ensuring efficient energy distribution and seamless integration of renewable sources. Additionally, carbon credit tracking and sustainability initiatives are becoming more transparent with blockchain-based tracking systems. Traditional carbon credit markets suffer from fraud and double counting issues, but block chain's immutable ledger ensures transparency and accountability in the process. The World Economic Forum estimates that blockchain-based carbon credit systems could reduce fraudulent carbon credit claims by over 50%, enhancing trust in sustainability initiatives. In this section, two key case studies—Power Ledger and Brooklyn Microgrid—demonstrate the real-world impact of blockchain in the energy and utilities sector. These projects highlight how blockchain is enabling peer-to-peer energy trading, decentralized grid management, and enhanced energy efficiency.

CASE STUDY 1: POWER LEDGER – PIONEERING BLOCKCHAIN-BASED PEER-TO PEER ENERGY TRADING

Power Ledger, an Australia-based blockchain platform, is revolutionizing peer-to-peer (P2P) energy trading by enabling consumers to buy and sell excess solar energy directly within their communities. Traditional energy markets rely on centralized utilities, which often result in high transaction costs, energy wastage, and lack of consumer control over energy prices. Power Ledger's blockchain-based system eliminates intermediaries, allowing consumers to trade electricity transparently and efficiently.

Power Ledger Working

Power Ledger utilizes a blockchain-based trading platform that allows households and businesses to sell excess solar energy directly to other users in their local network. Smart contracts facilitate real-time energy transactions, ensuring automatic execution and fair pricing.

Key components of Power Ledger's system include:

- P2P Energy Trading: Consumers with solar panels can sell surplus energy to others without involving a traditional utility company.
- Smart Contracts: Transactions are automated; ensuring energy is bought and sold instantly without intermediaries.
- Transparent Pricing: Buyers and sellers negotiate energy prices, fostering a competitive and cost effective energy market.
- Decentralized Energy Ownership: Power Ledger empowers consumers to become "prosumers", both producing and consuming energy within the network. Figure 17(Kumar,2023) represents P2P trading platforms with Blockchain as an alternative space for energy trading and data exchange.

Figure 17. P2P trading platforms with Blockchain as an alternative space for energy trading and data exchange.

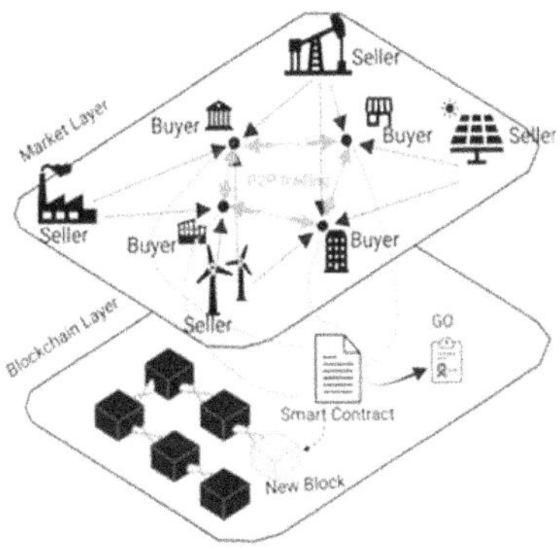

Figure 17 represents a blockchain-based peer-to-peer (P2P) energy trading system using the Power Ledger platform. Following is a simple breakdown process:

1. Market Layer (Top Layer)

Buyers & Sellers: Homes and businesses act as energy producers (sellers) or consumers (buyers). P2P Trading: Sellers can directly trade surplus energy with buyers within the local network.

2. Blockchain Layer (Bottom Layer)

Smart Contracts: Automated contracts handle transactions securely without intermediaries. New Block Creation: Every transaction is recorded in a blockchain, ensuring transparency, security, and tamper-proof records.

Key Benefits

Impact of Power Ledger: Since its launch, Power Ledger has partnered with multiple governments and organizations worldwide to implement blockchain-based energy trading solutions. Some key milestones include:

- Australia: Deployed in Western Australia, enabling thousands of households to trade solar energy.
- Japan: Partnered with KEPCO (Kansai Electric Power Company) to test blockchain-based energy trading.
- Thailand: Collaborated with BCPG to develop Southeast Asia's first blockchain-based energy marketplace.
- India: Partnered with Tata Power-DDL to enable localized energy trading in New Delhi. According to Power Ledger's internal reports, blockchain-based energy trading has reduced consumer electricity costs by up to 40% compared to traditional grid-based energy purchases. Furthermore, P2P

trading models could potentially save global energy consumers over $10 billion annually by 2030 by reducing energy transmission losses and optimizing local energy consumption.

Challenges and Future Prospects

While Power Ledger's platform has demonstrated significant potential, challenges remain. Regulatory hurdles, integration with existing energy grids, and scalability issues must be addressed to ensure wider adoption of blockchain-based energy trading. Despite these challenges, Power Ledger continues to expand, with increasing support from governments and energy companies worldwide. With global electricity demand rising and renewable energy adoption increasing, Power Ledger's blockchain-based model presents a scalable and sustainable solution for the future of energy trading and distribution.

CASE STUDY 2: BROOKLYN MICROGRID – A DECENTRALIZED ENERGY TRADING MODEL

Brooklyn Microgrid (BMG), developed by LO3 Energy, is one of the earliest blockchain-based energy projects aimed at creating localized, community-driven energy markets. Based in Brooklyn, New York, BMG enables residents to trade locally generated renewable energy within their neighborhood.

Brooklyn Microgrid Working

Brooklyn Microgrid operates on a decentralized energy model that allows consumers with solar panels to sell excess energy directly to their neighbors through a blockchain-based platform.

The project leverages:

- Smart Meters: These devices track real-time energy production and consumption.
- Blockchain Ledger: Securely records transactions and verifies ownership of energy.
- P2P Energy Marketplace: Residents buy and sell energy within their community, ensuring fair pricing and sustainability.
- Grid Resilience: Reduces reliance on centralized energy providers, making local energy networks more resilient.

Figure 18(delardas, 2023) shows that the residents of Brooklyn have created their micro-grid of green energy, facilitated by blockchain technology.

Figure 18. Brooklyn Microgrid, facilitated by blockchain technology.

Impact of Brooklyn Microgrid: Since its inception, Brooklyn Microgrid has demonstrated the feasibility of blockchain-based energy markets by providing a working model for decentralized energy trading. Some of its key achievements include: • Decentralized Energy Ownership: Enabled hundreds of households to participate in P2P energy trading. • Lower Energy Costs: Reduced energy bills by an average of 30% for participating consumers. • Increased Renewable Energy Utilization: Encouraged more households to adopt solar energy systems, reducing dependence on fossil fuels. • Grid Independence: Enhanced energy security by reducing reliance on large-scale power grids, which are vulnerable to failures and cyber attacks.

Challenges and Future Developments: While Brooklyn Microgrid has shown great potential, its expansion is limited by regulatory and policy constraints. In many regions, energy trading laws still require transactions to go through centralized utilities, making full-scale deployment of blockchain-based energy markets challenging. To address these barriers, Brooklyn Microgrid is collaborating with regulators and policymakers to develop new frameworks that enable blockchain-driven energy trading on a broader scale. Additionally, advances in blockchain scalability, smart grid technology, and AI-driven energy management could further enhance the efficiency and adoption of decentralized energy models. According to Bloomberg NEF, blockchain-based energy trading platforms could handle over 100 terawatt hours (TWh) of electricity transactions annually by 2030, significantly reshaping the global energy market.

Challenges in the Energy Sector

1. Grid Modernization and Infrastructure Limitations

The global energy grid is aging, with many existing power infrastructures built for centralized energy distribution. The transition to decentralized energy generation—which includes solar panels, wind farms, and distributed energy resources (DERs)—presents a significant challenge for grid operators. Traditional grids were not designed to accommodate bidirectional energy flows, leading to inefficiencies in managing energy distribution, demand fluctuations, and energy losses. According to the International Energy Agency (IEA), global electricity demand is expected to increase by 50% by 2050, necessitating a $14 trillion investment in grid modernization. The challenge lies in ensuring that energy grids can effectively handle decentralized energy sources while maintaining stability, efficiency, and security. Figure 19 illustrates Smart grid decentralized blockchain architecture with advanced wireless communication and data processing as a reliable and traceable energy trading and certification mechanism (Kumar, 2023). A local aggregator facilitates transactions between renewable energy sources, consumers, and storage systems, ensuring real-time validation and peer-to-peer trading using energy coins. Blockchain technology guarantees tamper-proof records, automation, and transparency, eliminating intermediaries and reducing transaction costs. Smart contracts enforce agreements autonomously, while wireless energy transfer enhances grid efficiency by enabling direct energy exchange. This architecture enhances sustainability, promotes energy democratization, and ensures trustworthy certification mechanisms, making the energy market resilient, scalable, and future-ready.

2. Renewable Energy Integration and Storage Management

The growing shift towards renewable energy sources such as solar and wind brings intermittency challenges—meaning that energy production fluctuates based on environmental conditions. Without proper energy storage solutions, excess energy generated during peak production periods is wasted, and energy shortages occur when production is low. The U.S. Department of Energy reports that nearly 20% of renewable energy in the U.S. is lost due to inefficient storage and grid limitations. Traditional energy grids lack the flexibility to manage real-time energy balancing, making the integration of renewable energy difficult. Blockchain facilitates renewable energy integration by enabling automated energy trading, decentralized storage management, and predictive analytics through smart contracts. However, challenges remain in scaling blockchain-based energy storage solutions and ensuring cost-effective deployment across different regions.

3. Transparency and Accountability in Energy Transactions

The energy industry has long struggled with transparency issues, including fraud in carbon credit trading, lack of real-time energy tracking, and limited consumer control over energy pricing. Inaccurate energy reporting and manipulation in carbon credit markets lead to billions of dollars in financial losses annually. According to a World Economic Forum report, over 30% of carbon credits in global markets are either double-counted or misrepresented, undermining efforts to combat climate change. Additionally, traditional energy billing systems often lack transparency, making it difficult for consumers to verify the accuracy of their charges. Blockchain solves these challenges by providing a decentralized, immutable ledger that records every energy transaction, ensuring transparency, traceability, and fraud prevention. However, the adoption of blockchain in energy tracking requires industry-wide standardization, regulatory compliance, and integration with smart meter technology.

Overcoming Challenges and Paving the Future of Blockchain in Energy and Beyond

Blockchain technology has revolutionized multiple industries, including finance, healthcare, supply chain, real estate, and energy. The energy sector, in particular, has witnessed significant interest in decentralized energy trading, grid modernization, and sustainability tracking. However, despite its transformative potential, blockchain adoption faces several challenges, including scalability, regulatory compliance, integration with legacy systems, and energy consumption concerns. Overcoming these barriers is critical to realizing a fully decentralized and transparent energy infrastructure powered by blockchain.

Challenges Hindering Blockchain Adoption

1. Scalability Issues
2. Energy Consumption
3. Computational power
4. Regulatory Uncertainty
5. Interoperability Between Blockchains
6. Security and Privacy Concerns
7. Lack of Skilled Talent
8. High Initial Investment
9. Public Perception and Trust
10. Legal and Compliance Issues
11. Adoption of Legacy Systems

CONCLUSION

Blockchain technology has emerged as a transformative force across multiple industries, offering decentralization, transparency, security, and automation. Originally developed to support Bitcoin, it has since evolved into a revolutionary tool in finance, healthcare, supply chain management, real estate, and the energy sector. Its ability to record immutable transactions on a distributed ledger has helped address critical challenges such as fraud, inefficiencies, lack of interoperability, and energy transparency.

In finance, blockchain has revolutionized cross-border payments, decentralized finance (DeFi), and digital assets, enabling faster and more cost-efficient transactions. According to the World Economic Forum, blockchain could reduce banking infrastructure costs by up to 30% and lower cross-border remittance fees by 40–60%. Similarly, in healthcare, it enhances secure patient data management and interoperability, minimizing the risks of data breaches and medical fraud.

The supply chain industry has leveraged blockchain to improve product traceability, prevent counterfeiting, and optimize logistics. The World Economic Forum estimates that counterfeit goods account for 3.3% of global trade, costing businesses hundreds of billions of dollars annually. Blockchain-based supply chain management can significantly reduce fraud and improve consumer trust.

Governments are also embracing blockchain to enhance digital identity verification, voting systems, and public services, ensuring greater transparency and fraud prevention. One of the most promising applications of blockchain is in the energy and utilities sector, where it facilitates decentralized energy trading, grid modernization, and carbon credit tracking. The global blockchain in energy market is projected to grow from $919 million in 2022 to $6.3 billion by 2027, at a CAGR of 45.3%, reflecting its rising adoption. Case studies such as Power Ledger and Brooklyn Microgrid showcase the benefits of peer-to-peer (P2P) energy trading, enabling consumers to buy and sell renewable energy directly, reducing reliance on centralized energy providers.

Despite these advancements, blockchain adoption still faces challenges, including scalability limitations, regulatory hurdles, energy consumption concerns, and integration with existing infrastructure. Traditional blockchain networks like Bitcoin and Ethereum struggle with transaction bottlenecks, with Bitcoin processing only seven transactions per second (TPS) compared to Visa's 24,000 TPS. However, innovations such as Layer-2 scaling solutions, Proof of Stake (PoS), and hybrid blockchain models are being developed to enhance efficiency and sustainability. Ethereum's transition to PoS in 2022 reduced its energy consumption by 99.95%, making blockchain more environmentally viable.

Regulatory concerns also present significant challenges, as many energy laws were not designed for decentralized trading and carbon credit tracking. Governments and policymakers must establish clear regulatory frameworks that support blockchain innovation while ensuring compliance with consumer protection and financial laws.

Looking ahead, blockchain is expected to become a foundational technology across industries. AI-driven smart contracts, interoperability between blockchain platforms, and government-backed blockchain initiatives will accelerate adoption. The future of blockchain in energy, finance, and public services will depend on collaborative efforts among governments, businesses, and technology developers to create a secure, scalable, and sustainable blockchain ecosystem.

REFERENCES

Alhomdy, S., Thabit, F., Abdulrazzak, F. A. H., Haldorai, A., & Jagtap, S. (2021). The role of cloud computing technology: A savior to fight the lockdown in COVID 19 crisis, the benefits, characteristics and applications. *International Journal of Intelligent Networks*, *2*, 166–174. DOI: 10.1016/j.ijin.2021.08.001

(2022). Anwar ul Hassan, C., Hammad, M., Iqbal, J., Hussain, S., Ullah, S. S., AlSalman, H., ... & Arif, M. (2022). A Liquid Democracy Enabled Blockchain-Based Electronic Voting System. *Scientific Programming*, (1), 1383007.

de l'Énergie, E. (n.d.). *Blockchain: Uses and applications in the energy sector*. https://www.encyclopedie-energie.org/en/blockchain-uses-applications-energy-sector/

Delardas, O., & Giannos, P. (2022). Towards energy transition: Use of blockchain in renewable certificates to support sustainability commitments. *Sustainability (Basel)*, *15*(1), 258. DOI: 10.3390/su15010258

HashStudioz. (n.d.). *What is DeFi? A beginner's guide to decentralized finance*. https://www.hashstudioz.com/blog/what-is-defi-a-beginners-guide-to-decentralized-finance/

Li, H., Zou, L., Jiang, T., Liao, D., Zhang, M., & Chen, X. (2022). A Master-slave Multi-chain Vehicle Networking Architecture.

Phemex. (n.d.). *Real estate blockchain and crypto*. https://phemex.com/academy/real-estate-blockchain-and-crypto

Rosli, A., Hassan, S., Omar, M. H., Sajat, M. S., Attarbashi, Z. S., & Zaini, K. M. (2024, November). Blockchain Applications and Management: A Multidisciplinary Analysis. In 2024 7th International Conference on Internet Applications, Protocols, and Services (NETAPPS) (pp. 1-8). IEEE.

Sara Technologies Inc. (n.d.). *Blockchain in logistics management*. https://www.saratechnologies.com/logistics-blockchain-management

SlideTeam. (n.d.). *In-depth guide to blockchain nodes training [PowerPoint slides]*. https://www.slideteam.net/in-depth-guide-to-blockchain-nodes-training-ppt.html

Solunke, H., & Bhaladhare, P. (2024, July). An In-Depth Investigation of Machine Learning Powered Blockchain Approached for Privacy Preservation. In *2024 International Conference on Emerging Trends in Networks and Computer Communications (ETNCC)* (pp. 1-5). IEEE. Available: https://ieeexplore.ieee.org/abstract/document/10767529/

Srinivas, B., & Rao, K. N. (2024, December). Cloud-Block: An Integrated Cloud and Blockchain using Secure Cloud Computing. In *2024 3rd International Conference on Automation, Computing and Renewable Systems (ICACRS)* (pp. 651-657). IEEE. https://www.cryptoexplorer.cc/

Vanmathi, C., Farouk, A., Alhammad, S. M., Bhattacharya, S., & Kasyapa, M. S. (2024). The Role of Blockchain in Transforming Industries Beyond Finance. *IEEE Access : Practical Innovations, Open Solutions*.

WallStreetMojo. (n.d.). *Proof of work*. https://www.wallstreetmojo.com/proof-of-work/

Compilation of References

Abdollahi, A., Sadeghvaziri, F., & Rejeb, A. (2023). Exploring the role of blockchain technology in value creation: A multiple case study approach. *Quality & Quantity*, *57*(1), 427–451. DOI: 10.1007/s11135-022-01348-2

Adhikari, P., Lehner, C., & Sing, L. J. (2021). Transaction Cost Reductions in Renewable Energy Financing: A Blockchain Model. *Financial Innovation*, *7*(2).

Aggarwal, R., Li, D., & Zhang, K. (2021). AI-Driven Risk Assessment in Energy Token Investments. *Expert Systems with Applications*, *168*, 114232.

Ahlstrom, M., He, K., & Meng, J. (2021). *Transaction Costs and Energy Project Financing* (Vol. 144). Energy Policy.

Ahmad, T., Zhang, D., Huang, C., Zhang, H., Dai, N., Song, Y., & Chen, H. (2021). Artificial intelligence in sustainable energy industry: Status quo, challenges and opportunities. *Journal of Cleaner Production*, *289*, 125834. DOI: 10.1016/j.jclepro.2021.125834

Ahmad, T., Zhu, H., Zhang, D., Tariq, R., Bassam, A., Ullah, F., AlGhamdi, A. S., & Alshamrani, S. S. (2022). Energetic systems and artificial intelligence: Applications of Industry 4.0. *Energy Reports*, *8*, 334–361. DOI: 10.1016/j.egyr.2021.11.256

Ahmed, N., Lee, D., & Park, S. (2020). Rapid Disaster Response and the Role of Blockchain-based Grids. *IEEE Transactions on Engineering Management*, *67*(3), 725–736.

Ahram, T., Sargolzaei, A., Sargolzaei, S., Daniels, J., & Amaba, B. (2017, June). Blockchain technology innovations. In 2017 IEEE technology & engineering management conference (TEMSCON) (pp. 137-141). IEEE.

Aitzhan, N. Z., & Svetinovic, D. (2018). Security and privacy in decentralized energy trading through multi-signatures, blockchain and anonymous messaging streams. *IEEE Transactions on Dependable and Secure Computing*, *15*(5), 840–852. DOI: 10.1109/TDSC.2016.2616861

Al Shareef, A. M., Seçkiner, S., Eid, B., & Abumeteir, H. (2024). Integration of blockchain with artificial intelligence technologies in the energy sector: A systematic review. *Frontiers in Energy Research*, *12*, 1377950. DOI: 10.3389/fenrg.2024.1377950

Al-Abri, T., Onen, A., Al-Abri, R., Hossen, A., Al-Hinai, A., Jung, J., & Ustun, T. S. (2022). Review on energy application using blockchain technology with an introduction to the pricing infrastructure. *IEEE Access : Practical Innovations, Open Solutions*, *10*, 80119–80137. DOI: 10.1109/ACCESS.2022.3194161

Alam, R., Jeong, S., & Li, C. (2022). Renewable Energy Certification in Blockchain Networks: A Practical Review. *IEEE Transactions on Engineering Management*, *69*(4), 1235–1244.

Alam, T., Goodman, J., & Doh, Y. (2022). Decentralized Ledgers and Smart Contracts in Energy Markets: A Comprehensive Study. *Energy Informatics*, *5*(2), 22–34.

Aleisa, M. A. (2025). Blockchain-enabled zero trust architecture for privacy-preserving cybersecurity in IoT environments. *IEEE Access : Practical Innovations, Open Solutions*, *13*, 18660–18676. Advance online publication. DOI: 10.1109/ACCESS.2025.3529309

Al-Fatlawi, H. H., Khayoon, Q. M., Albhadly, A. J., & Ibrahim, A. K. (2024). AI-enhanced demand response strategies in smart grids: Toward a sustainable energy future. *Journal of Electrical Systems*, *20*(7s), 1054–1064. DOI: 10.52783/jes.3535

Alhomdy, S., Thabit, F., Abdulrazzak, F. A. H., Haldorai, A., & Jagtap, S. (2021). The role of cloud computing technology: A savior to fight the lockdown in COVID 19 crisis, the benefits, characteristics and applications. *International Journal of Intelligent Networks*, *2*, 166–174. DOI: 10.1016/j.ijin.2021.08.001

Ali, M. S., Dolui, K., & Antonelli, F. (2018). IoT data privacy via blockchains and IPFS. In Proceedings of the 7th International Conference on the Internet of Things (pp. 14:1-14:7). ACM.

Ali, M., Prakash, K., Hossain, M. A., & Pota, H. R. (2021). Intelligent energy management: Evolving developments, current challenges, and research directions for a sustainable future. *Journal of Cleaner Production*, *314*, 127904. DOI: 10.1016/j.jclepro.2021.127904

Alketbi, A., Nasir, Q., & Talib, M. A. (2018). Blockchain for government services: Benefits and challenges. *IEEE Access : Practical Innovations, Open Solutions*, *7*, 43426–43433.

Alladi, T., Chamola, V., Rodrigues, J. J. P. C., & Kozlov, S. (2019). Blockchain in smart grids: A review on different use cases. *Sensors (Basel)*, *19*(22), 4862. DOI: 10.3390/s19224862 PMID: 31717262

Allen, C. (2016). The path to self-sovereign identity. Retrieved from https://www.lifewithalacrity.com/2016/04/the-path-to-self-sovereign-identity.html

Allen, R., Berg, T., & Reuben, P. (2020). Regulating Security Tokens: Legal Challenges and Market Dynamics. *Journal of Financial Regulation*, *6*(1).

Allied Market Research. (2022). *Advanced Persistent Threat (APT) market statistics, 2021–2030*. Retrieved from https://www.alliedmarketresearch.com

Almarri, S., & Aljughaiman, A. (2024). Blockchain technology for IoT security and trust: A comprehensive SLR. *Sustainability (Basel)*, *16*(23), 10177. DOI: 10.3390/su162310177

Alrammal, M., Abu-Amara, F., Ismail, Z., & Nadeem, M. (2024). Blockchain technology for sustainable management of electricity and water consumption. In *Proceedings of the International Conference on Recent Advances in Science and Engineering (RAiSE 2023)* (Engineering Proceedings Vol. 59, No. 1, Article 223). MDPI. https://doi.org/DOI: 10.3390/engproc2023059223

Al-Saif, N., Ahmad, R., Salah, K., Yaqoob, I., Jayaraman, R., & Omar, M. (2021). Blockchain for electric vehicles energy trading: Requirements, opportunities, and challenges. *IEEE Access : Practical Innovations, Open Solutions*, *9*, 156947–156961. DOI: 10.1109/ACCESS.2021.3130095

Alsayaydeh, J. A. J., & Gaid, A. S. A. (2025). Security of smart grid: Cybersecurity issues, potential cyberattacks, major incidents, and future directions. *Energies*, *18*(1), 141. DOI: 10.3390/en18010141

Altin, N., & Eyimaya, S. E. (2023). Artificial intelligence applications for energy management in microgrid. In *Proceedings of the 2023 11th International Conference on Smart Grid (icSmartGrid)* (pp. 1–6). IEEE. DOI: 10.1109/icSmartGrid58556.2023.10170860

Ambhika, C., Gayathri, S., Anish, T. P., & Sheena, B. G. (2024, August). Enhancing Predictive Modeling in High Dimensional Data Using Hybrid Feature Selection. In *2024 5th International Conference on Electronics and Sustainable Communication Systems (ICESC)* (pp. 873-879). IEEE. DOI: 10.1109/ICESC60852.2024.10690153

Andersen, B. (2021). Blockchain Regulation for the Energy Sector: An Evolving Landscape. *Energy Policy*, *154*, 112294.

Anderson, R., & Moore, T. (2006). The economics of information security. *Science*, *314*(5799), 610–613. DOI: 10.1126/science.1130992 PMID: 17068253

Andoni, M., Robu, V., Flynn, D., Abram, S., Geach, D., Jenkins, D., & Peacock, A. (2019). Blockchain Technology in the Energy Sector: A Systematic Review of Challenges and Opportunities. *Renewable & Sustainable Energy Reviews*, *100*, 143–174. DOI: 10.1016/j.rser.2018.10.014

Androulaki, E., Barger, A., Bortnikov, V., Cachin, C., Christidis, K., De Caro, A., & Muralidharan, S. (2018). Hyperledger Fabric: A distributed operating system for permissioned blockchains. In *Proceedings of the 13th EuroSys Conference* (pp. 1-15). ACM. DOI: 10.1145/3190508.3190538

Antonopoulos, A. M. (2014). *Mastering Bitcoin: Unlocking digital cryptocurrencies*. O'Reilly Media, Inc.

Asgari, M., & Nemati, M. (2022). Application of distributed ledger platforms in smart water systems—A literature review. *Frontiers in Water*, *4*, 848686. Advance online publication. DOI: 10.3389/frwa.2022.848686

Aswini, J., Sudha, K., Ganesh, G., Subramanian, S., & Ghinea, G. (2024). AI-powered parallel computing architecture and its applications. In *The convergence of self-sustaining systems with AI and IoT* (pp. 23–39). IGI Global. DOI: 10.4018/979-8-3693-1702-0.ch002

Atzei, N., Bartoletti, M., & Cimoli, T. "A Survey of Attacks on Ethereum Smart Contracts," in *Proc. 6th Intl. Conf. on Principles of Security and Trust*, 2017, pp. 164–186. DOI: 10.1007/978-3-662-54455-6_8

Avertium. (2022, September 13). *The top 5 cyber threats in the energy sector.* Avertium Threat Report (blog).

Azad, M. A., Abdullah, S., Arshad, J., Lallie, H., & Ahmed, Y. (2024). Verify and trust: A multidimensional survey of zero-trust security in the age of IoT. *Internet of Things : Engineering Cyber Physical Human Systems*, *27*, 101227. DOI: 10.1016/j.iot.2024.101227

Aziz, A. F. A., Khalid, S. N., Mustafa, M. W., & Aliyu, N. (2019). Blockchain technology for renewable energy: Principles, applications and prospects. *IOP Conference Series. Earth and Environmental Science*, *268*, 012113.

Azouvi, S., Maller, M., & Meiklejohn, S. (2019). Egalitarian society or benevolent dictatorship: The state of cryptocurrency governance. In *Financial Cryptography and Data Security* (pp. 127–143). Springer. DOI: 10.1007/978-3-662-58820-8_10

Badidi, E. (2022). Edge AI and blockchain for smart sustainable cities: Promise and potential. *Sustainability (Basel)*, *14*(13), 7609. DOI: 10.3390/su14137609

Baker, J., & Chen, M. (2021). IoT-Enabled Microgrid Coordination: A Review of Methods and Technologies. *Energy Informatics*, *4*(1).

Basuki, I., & Rosiyadi, D. (2019). Joint transaction-image steganography for high-capacity covert communication. Proc. 2019 International Conference on Computer, Control, Informatics and its Applications (IC3INA). IEEE Press.

Batool, S., & Channi, H. K. (2024). Role of blockchain IoT in smart building for distributed renewable power. In *Advances in Computational Intelligence and Robotics (Chapter 4, pp. 63–80)*. IGI Global. https://doi.org/DOI: 10.4018/979-8-3693-1586-6.ch004

Belotti, M., Božić, N., Pujolle, G., & Secci, S. (2019). A vademecum on blockchain technologies: When, which, and how. *IEEE Communications Surveys and Tutorials*, *21*(4), 3796–3838. DOI: 10.1109/COMST.2019.2928178

Benhamou, G., Patel, N., & Chen, T. G. (2022). Blockchain Scalability in Multi-Stakeholder Energy Trading Platforms. *Energy Reports*, *8*(3), 1120–1131.

Ben-Sasson, E., Chiesa, A., Garman, C., Green, M., Miers, I., Tromer, E., & Virza, M. (2014). Zerocash: Decentralized anonymous payments from Bitcoin. In *IEEE Symposium on Security and Privacy* (pp. 459-474). IEEE.

Bhargavan, K., Delignat-Lavaud, A., Fournet, C., Gollamudi, A., Gonthier, G., Kobeissi, N., & Zanella-Béguelin, S. (2016). Formal verification of smart contracts: Short paper. In *Proceedings of the 2016 ACM Workshop on Programming Languages and Analysis for Security* (pp. 91-96). ACM. DOI: 10.1145/2993600.2993611

Bhavana, G. B., Anand, R., Ramprabhakar, J., Meena, V. P., Jadoun, V. K., & Benedetto, F. (2024). Applications of blockchain technology in peer-to-peer energy markets and green hydrogen supply chains: A topical review. *Scientific Reports*, *14*(1), 21954. DOI: 10.1038/s41598-024-72642-2 PMID: 39304698

Bhushan, B., Khamparia, A., Sagayam, K. M., Gupta, D., Chowdhary, C. L., & Joshi, G. P. (2020). Blockchain for smart cities: A review of architectures, integration trends, and future research directions. *Sustainable Cities and Society*, *61*, 102360. DOI: 10.1016/j.scs.2020.102360

Bicer, C., Murturi, I., Donta, P. K., & Dustdar, S. (2023). Blockchain-based zero trust on the edge. In *Proceedings of the 2023 International Conference on Computational Science and Computational Intelligence (CSCI)* (pp. 1006–1013). IEEE. https://doi.org/DOI: 10.1109/CSCI62032.2023.00167

Binyamin, S., & Ben Slama, S. (2025). IntelliGrid AI: A blockchain and deep-learning framework for optimized home energy management with V2H and H2V integration. *AI, 6*(2), 34. DOI: 10.3390/ai6020034

Bishaw, F.Faisal Ghazi Bishaw. (2024). Artificial intelligence applications in renewable energy systems integration: A review. *Journal of Electrical Systems, 20*(3), 566–582. DOI: 10.52783/jes.2983

BloombergNEF (BNEF). (2022). *Energy Transition Investment Trends*. Available: https://about.bnef.com/

Borkovcová, A., Černá, M., & Sokolová, M. (2022). Blockchain in the energy sector—Systematic review. *Sustainability (Basel), 14*(22), 14793. DOI: 10.3390/su142214793

Boucher, P. (2020). *Blockchain for the environment: Opportunities & challenges* (EPRS Study PE 648.791). European Parliamentary Research Service.

Bouckaert, S., Pales, A., Mekhilef, S., McGlade, C., Remme, U., & Wanner, B.. (2021). *Net Zero by 2050: A roadmap for the global energy sector*. International Energy Agency.

Bouri, E., Molnár, C., & Roubaud, A. (2019). Cryptocurrency Scams and Investor Due Diligence. *Finance Research Letters, 28*, 251–257.

Brayton, K., & O'Neill, M. (2020). Infrastructure Hardening and Resilience: A Comprehensive Review. *Resilient Systems Quarterly, 2*(4), 55–66.

Bustamante, P., Cai, M., Zambrano, R., & Weiss, M. (2022). Government by code? Blockchain applications to public sector governance. *Frontiers in Blockchain, 5*, Article 869665.

Buterin, V. (2014). "A Next-Generation Smart Contract and Decentralized Application Platform," *Ethereum White Paper*. https://ethereum.org/en/whitepaper/

Buterin, V. (2014). "A Next-Generation Smart Contract and Decentralized Application Platform." *IEEE Blockchain Conference*

Campbell, R. J. (2019). *Bitcoin, blockchain, and the energy sector* (CRS Report R45863). Congressional Research Service.

Cao, , Yin, H., Gao, F., Zhang, Z., Khoussainov, B., Xu, S., & Zhu, L. (2022). Chain-based covert data embedding schemes in blockchain. *IEEE Internet of Things Journal, 9*(16), 14699–14707. DOI: 10.1109/JIOT.2020.3040389

Carney, M. (2019). "The Future of Money," Speech at the Bank of England. https://www.bankofengland.co.uk/

Casino, F., Dasaklis, T. K., & Patsakis, C. (2019). A systematic literature review of blockchain-based applications: Current status, classification and open issues. *Telecommunication Systems, 71*(1), 53–74.

Cauich, E., Gomez, C. R., & Watanabe, R. (2005). Data hiding in identification and offset IP fields. *Proceedings of the 5th International School and Symposium*, 247-261. DOI: 10.1007/11533962_11

Cengiz, F. (2023). Blockchain governance and governance via blockchain: Decentralized utopia or centralized dystopia? *Policy Design and Practice, 6*(4), 446–464. DOI: 10.1080/25741292.2023.2247203

Chang, Y., Ng, J., & He, C. (2020). Blockchain-Enhanced Transparency in Supply Chains: A Systematic Review. *Computer Standards & Interfaces, 70*, 103430.

Chauhan, A., Li, Y., & Zhou, X. (2022). A Survey on Consensus Mechanisms for IoT-Blockchain Integration. *Computer Networks, 203*, 108623.

Chen, G., Xu, B., Lu, M., & Chen, N. S. (2018). Exploring blockchain technology and its potential applications for education. *Smart Learning Environments, 5*(1), 1–10. DOI: 10.1186/s40561-017-0050-x

Chen, J., Yang, D., & Liu, K. (2023). Emerging Threats and Resilience Measures for Next-Generation Grids. *IEEE Transactions on Smart Grid, 14*(1), 101–112.

Chen, K., & Bellavitis, F. (2020). Blockchain Disruption and Decentralized Finance: The Future of Banking. *Financial Innovation, 6*(8).

Chen, S., Ping, J., Yan, Z., Li, J., & Huang, Z. (2022). Blockchain in energy systems: Values, opportunities, and limitations. *Frontiers in Energy, 16*(1), 9–18. DOI: 10.1007/s11708-022-0818-8

Chen, T. M., & Abu-Nimeh, S. (2011). Lessons from Stuxnet: A new era of cyber warfare. *Computer, 44*(4), 91–93. DOI: 10.1109/MC.2011.115

Chen, T. S., & Fan, L. (2020). Public Access to Green Projects Through Tokenized Infrastructure. *Environmental Innovation and Societal Transitions, 37*, 156–166.

Cherepanov, A., & Lipovsky, R. (2017). *Win32/Industroyer: A new threat for industrial control systems.* ESET White Paper. Retrieved from ESET.com

Chibi, N. T., Ait Oualhaj, O., Fassi Fihri, W., & El Ghazi, H. (2024). A novel approach based on machine learning, blockchain, and decision process for securing smart grid. *IEEE Access : Practical Innovations, Open Solutions, 12*, 33190–33199. DOI: 10.1109/ACCESS.2024.3370239

Choo, K.-K. R., Dehghantanha, A., & Parizi, R. M. (Eds.). (2020). *Blockchain cybersecurity, trust and privacy.* Springer., DOI: 10.1007/978-3-030-38181-3

Christidis, K., & Devetsikiotis, M. (2016). Blockchains and smart contracts for the Internet of Things. *IEEE Access : Practical Innovations, Open Solutions, 4*, 2292–2303. DOI: 10.1109/ACCESS.2016.2566339

Clack, D., Bakshi, V., & Braine, L. "Smart Contract Templates: Legal and Design Perspectives," *arXiv preprint* arXiv:1612.04496, 2017.

Climate Bonds Initiative (CBI). (2020). *Green Bond Market Summary.* https://www.climatebonds.net/

Cloud Security Alliance. (2024). *Zero Trust Guidance for Critical Infrastructure.* Cloud Security Alliance.

Coli, P., Pflueger, C., & Campbell, T. (2021). *Blockchain uses for microfinance institutions in the water and sanitation sector: Pilot study.* Inter-American Development Bank.

Craver, S. (1998). On public-key steganography in the presence of an active warden. *International Workshop on Information Hiding*, 355-368. DOI: 10.1007/3-540-49380-8_25

Crosby, M., Pattanayak, P., Verma, S., & Kalyanaraman, V. (2016). Blockchain technology: Beyond bitcoin. *Applied Innovation, 2*, 6–10.

Cybersecurity and Infrastructure Security Agency (CISA). (2023). *Zero Trust Maturity Model Version 2.0*. U.S. Department of Homeland Security.

Czarnecki, B., & Jochem, P. (2020). Blockchain-Based Certification of Green Electricity Origin: Potentials and Pitfalls. *Applied Energy*, *277*, 115573.

Dao, T., Sharma, V., & Fu, G. (2022). Decentralized Governance for Investment Funds Using Blockchain: A Conceptual Framework. *IEEE Access : Practical Innovations, Open Solutions*, *10*, 55622–55633.

de l'Énergie, E. (n.d.). *Blockchain: Uses and applications in the energy sector*. https://www.encyclopedie-energie.org/en/blockchain-uses-applications-energy-sector/

Deb, H. (2021). Cyber Resilience in Modern Power Systems. *Energy Security Journal*, *5*(2), 21–32.

Delardas, O., & Giannos, P. (2022). Towards energy transition: Use of blockchain in renewable certificates to support sustainability commitments. *Sustainability (Basel)*, *15*(1), 258. DOI: 10.3390/su15010258

Dellermann, D., Ebel, P., Söllner, M., & Leimeister, J. M. (2019). Hybrid intelligence. *Business & Information Systems Engineering*, *61*(5), 637–643. DOI: 10.1007/s12599-019-00595-2

Department of Energy (DOE), & the Cybersecurity and Infrastructure Security Agency (CISA). National Security Agency (NSA); & Federal Bureau of Investigation (FBI). (2022). *APT cyber tools targeting ICS/SCADA devices (Joint Cybersecurity Advisory AA22-103A, April 2022)*. U.S. Department of Energy.

Dhillon, V., Metcalf, D., & Hooper, M. (2017). *Blockchain-enabled applications: Understand the blockchain ecosystem and how to make it work for you*. Apress. DOI: 10.1007/978-1-4842-3081-7

Di Silvestre, M. L., Gallo, P., Guerrero, J. M., Musca, R., Sanseverino, E. R., Sciumè, G., & Zizzo, G. (2018). Blockchain for power systems: Current trends and future applications. *Renewable & Sustainable Energy Reviews*, *113*, 109268.

Di Silvestre, M., Gallo, L., & Riva Sanseverino, E. (2020). Blockchain for Energy Communities: A Comprehensive Review on Technologies and Challenges. *Renewable & Sustainable Energy Reviews*, *133*, 110301.

Diffie, W., & Hellman, M. E. (1976). New directions in cryptography. *IEEE Transactions on Information Theory*, *22*(6), 644–654. DOI: 10.1109/TIT.1976.1055638

Ding, . (2019). Survey on key issues in networks covert channel. *Journal of Software*, *30*(8), 2470–2490.

Dorfleitner, G., Muck, F., & Scheckenbach, I. (2021). Blockchain Applications for Climate Protection: A Global Empirical Investigation. *Renewable & Sustainable Energy Reviews*, *149*, 111378. DOI: 10.1016/j.rser.2021.111378

Dorri, A., Kanhere, S. S., Jurdak, R., & Gauravaram, P. (2017). Blockchain for IoT security and privacy: The case study of a smart home. In *IEEE International Conference on Pervasive Computing and Communications Workshops* (pp. 618-623). IEEE. DOI: 10.1109/PERCOMW.2017.7917634

Dragos, Inc. (2025). *8th Annual Year in Review – 2025 OT Cybersecurity Report*. Dragos. (Retrieved from dragos.com)

Drescher, D. (2017). *Blockchain Basics: A Non-Technical Introduction in 25 Steps*. Apress. DOI: 10.1007/978-1-4842-2604-9

Ekanem, U., & Kim, Y. S. (2023). Empirical study of factors influencing adoption of blockchain technology converged with e-government in Nigerian public sector. *International Journal of Contents*, *19*(1), 67–81. DOI: 10.5392/IJoC.2023.19.1.067

El-Atawy, A., Duan, Q., & Al-Shaer, E. (2017). A novel class of robust covert channels using out-of-order packets. *IEEE Transactions on Dependable and Secure Computing*, *14*(2), 116–129. DOI: 10.1109/TDSC.2015.2443779

ElHusseini, H., Assi, C., Moussa, B., Attallah, R., & Ghrayeb, A. (2020). Blockchain, AI and smart grids: The three musketeers to a decentralized EV charging infrastructure. *IEEE Internet of Things Magazine*, *3*(2), 24–29. DOI: 10.1109/IOTM.0001.1900081

Elkind, P. (2018). Blockchain and Disaster Recovery: A Blueprint for Rapid Response. *International Journal of Disaster Resilience*, *5*(1), 22–33.

EnergyFunders. (2021). https://www.energyfunders.com/

Epishkina, , & Kogos, . (2015). Protection from binary and multi-symbol packet length covert channels. *Proceedings of the 8th International Conference on Security of Information and Networks*, 196-202.

Esfandi, S., Tayebi, S., Byrne, J., Taminiau, J., Giyahchi, G., & Alavi, S. A. (2024). Smart cities and urban energy planning: An advanced review of promises and challenges. *Smart Cities*, *7*(1), 414–444. DOI: 10.3390/smartcities7010016

EU Blockchain Observatory and Forum. (2022). *Blockchain applications in the energy sector.* European Commission report.

European Commission – EU Blockchain Observatory. (2019). *Blockchain applications in the energy sector*. European Commission.

European Commission. (2022). *Markets in Crypto-Assets (MiCA)*. https://ec.europa.eu/

Evens, M., Ercoli, P., & Arteconi, A. (2023). Blockchain-enabled microgrids: Toward peer-to-peer energy trading and flexible demand management. *Energies*, *16*(18), 6741. DOI: 10.3390/en16186741

Fernández-Caramés, T. M., & Fraga-Lamas, P. (2018). A review on the use of blockchain for the Internet of Things. *IEEE Access : Practical Innovations, Open Solutions*, *6*, 32979–33001. DOI: 10.1109/ACCESS.2018.2842685

Ferrag, M. A., & Maglaras, L. (2020). DeepCoin: A novel deep learning and blockchain-based energy exchange framework for smart grids. *IEEE Transactions on Engineering Management*, *67*(4), 1285–1297. DOI: 10.1109/TEM.2019.2922936

Financial Conduct Authority (FCA). (2021). *Regulatory Sandbox Report*. https://www.fca.org.uk/

Fisch, C. (2019). Initial Coin Offerings (ICOs) for Financing Energy Ventures. *Journal of Business Venturing*, *34*(1), 1–22. DOI: 10.1016/j.jbusvent.2018.09.007

Fischer, M. J., Lynch, N. A., & Paterson, M. S. (1985). Impossibility of distributed consensus with one faulty process. *Journal of the Association for Computing Machinery, 32*(2), 374–382. DOI: 10.1145/3149.214121

Fox, R. (2021). *Critical Challenges in Power Grid Modernization*. EnergyTech Press.

Francisco, K., & Swanson, D. (2018). The Supply Chain Has No Clothes: Technology Adoption of Blockchain for Supply Chain Transparency. *Logistics (Basel), 2*(1), 2. DOI: 10.3390/logistics2010002

Freier, A. (2024). Blockchain in the energy sector: An advancing technology to tackle global climate change? ibidem Press (Columbia University Press).

Fu, S., Tan, Y., & Xu, Z. (2023). Blockchain-based renewable energy certificate trade for a low-carbon community of active energy agents. *Sustainability (Basel), 15*(23), 16300. DOI: 10.3390/su152316300

Gad, A. G., Mosa, D. T., Abualigah, L., & Abohany, A. A. (2025). Emerging Trends in Blockchain Technology and Applications: A Review and Outlook. *IEEE Access : Practical Innovations, Open Solutions.*

Gadekallu, T. R., Huynh-The, T., Wang, W., Yenduri, G., Ranaweera, P., Pham, Q. V., . . . Liyanage, M. (2022). Blockchain for the metaverse: A review. *arXiv preprint arXiv:2203.09738.*

Gai, K., She, Y., Zhu, L., Choo, K. K. R., & Wan, Z. (2023). A blockchain-based access control scheme for zero trust cross-organizational data sharing. *ACM Transactions on Internet Technology, 23*(3), 1–25. DOI: 10.1145/3511899

Gai, K., Wu, Y., Zhu, L., Qiu, M., & Choo, K. K. R. (2019). Privacy-preserving energy trading using consortium blockchain in smart grid. *IEEE Transactions on Industrial Informatics, 15*(6), 3548–3558. DOI: 10.1109/TII.2019.2893433

Gao, P., & Singh, A. (2022). Blockchain-Enhanced Traceability for Renewable Energy Certificates. *Applied Energy, 311*, 118664.

Gao, , Zhu, L., Gai, K., Zhang, C., & Liu, S. (2020). Achieving a covert channel over an open blockchain network. *IEEE Network, 34*(2), 6–13. DOI: 10.1109/MNET.001.1900225

Ghisetti and F. Quatraro. (2020). *Green Technologies and Financial Opportunities: Investigating the Link* (Vol. 169). Ecological Economics.

Giffifin, . (2020). Covert messaging through TCP timestamps. *Proceedings of 2002 International Conference on Privacy Enhancing Technologies*, 194-208.

Girling, C. G. (1987). Covert channels in LAN's. *IEEE Transactions on Software Engineering, 13*(2), 292–296. DOI: 10.1109/TSE.1987.233153

Goodman, J., & Alam, M. (2021). Examining the Energy Consumption of Consensus Algorithms in Blockchain Systems. *IEEE Systems Journal, 15*(4), 5395–5406.

Govea, J., Gaibor-Naranjo, W., & Villegas-Ch, W. (2024). Securing critical infrastructure with blockchain technology: An approach to cyber-resilience. *Computers, 13*(5), 122. DOI: 10.3390/computers13050122

Grebović, M., Popović, T., & Sinđić Grebović, R. (2023). Blockchain Technology for Weather Data Management. In Proc. 22nd Int. Symposium INFOTEH-JAHORINA (INFOTEH 2023) (pp. 1–6). IEEE. DOI: 10.1109/INFOTEH57020.2023.10094082

Guo, , Shi, L., Xu, M., & Yin, H. (2021). MRCC: A practical covert channel over Monero with provable security. *IEEE Access : Practical Innovations, Open Solutions*, *9*, 31816–31825. DOI: 10.1109/ACCESS.2021.3060285

Hahn, A., Singh, R., Liu, C.-C., & Chen, S. (2017). Smart contract-based campus demonstration of decentralized transactive energy auctions. In *2017 IEEE Power & Energy Society Innovative Smart Grid Technologies Conference (ISGT)* (pp. 1–5). IEEE. https://doi.org/DOI: 10.1109/ISGT.2017.8086092

Hajian Berenjestanaki, M., Barzegar, H. R., El Ioini, N., & Pahl, C. (2023). Blockchain-based e-voting systems: A technology review. *Electronics (Basel)*, *13*(1), 17. DOI: 10.3390/electronics13010017

Hajian, A., & Chang, H.-C. (2023). A blockchain-based smart grid to build resilience through zero-trust cybersecurity. In Ayyub, B. M., Wegrzyn, K., & Shah, V. P. (Eds.), *Handbook of Smart Energy Systems* (pp. 1–19). Springer., DOI: 10.1007/978-3-030-97940-9_161

Handel, T. G., & Sandford, M. T.II. (1996). Hiding data in the OSI network model," *International Workshop on Information Hiding*, 23-38. DOI: 10.1007/3-540-61996-8_29

Hansen, T., & Rodriguez, E. "Decentralized Ledgers in Critical Energy Infrastructure: Opportunities and Pitfalls," in *Proc. IEEE Intl. Conf. on Future Power Systems*, 2019, pp. 11–18.

Han, Y., Meng, J., & Luo, Z. (2024). Multi-agent deep reinforcement learning for blockchain-based energy trading in decentralized electric vehicle charger-sharing networks. *Electronics (Basel)*, *13*(21), 4235. DOI: 10.3390/electronics13214235

Hasan, H., Latif, M., & Javaid, N. (2021). Security and Privacy in Blockchain-Based Energy Trading: A Comprehensive Review. *IEEE Access : Practical Innovations, Open Solutions*, *9*, 55337–55357.

Hasan, M. K., Alkhalifah, A., Islam, S., Ali, R., & Beg, M. O. (2022). Blockchain technology on smart grid, energy trading, and big data: Security issues, challenges, and recommendations. *Wireless Communications and Mobile Computing*, *2022*, 9065768. DOI: 10.1155/2022/9065768

HashStudioz. (n.d.). *What is DeFi? A beginner's guide to decentralized finance*. https://www.hashstudioz.com/blog/what-is-defi-a-beginners-guide-to-decentralized-finance/

Henry, W., Miller, B., & Maewall, S. (2020). *Blockchain and IoT for wastewater management* (White paper). Deloitte Consulting LLP.

Holmes, L., Bond, S., & Kimmel, D. (2021). Challenges of Blockchain Adoption in Energy Markets. *Computer Standards & Interfaces*, *73*, 103469.

Hou, J., Li, G., & Zhang, M. (2020). Tokenization of Solar Assets for Fractional Ownership. *Energy Reports*, *6*, 421–432.

Hua, W., Chen, Y., Qadrdan, M., Jiang, J., Sun, H., & Wu, J. (2022). Applications of blockchain and artificial intelligence technologies for enabling prosumers in smart grids: A review. *Renewable & Sustainable Energy Reviews, 161*, 112308. DOI: 10.1016/j.rser.2022.112308

Huh, S., Cho, S., & Kim, S. (2017). Managing IoT devices using blockchain platform. In *International Conference on Advanced Communication Technology* (pp. 464-467). IEEE.

Hussain, A., Bui, V., & Kim, H. (2019). Microgrids as a Resilient Power System Technology for Community Empowerment: A Review. *Renewable & Sustainable Energy Reviews, 101*, 154–164.

IEEE Blockchain Initiative. (2024). Blockchain Trends and Future Prospects. *IEEE Access : Practical Innovations, Open Solutions.*

IEEE Standards Association. (2024). *IEEE Std 3219-2023: Standard for Blockchain-Based Zero-Trust Framework for the Internet of Things (IoT).* IEEE.

International Energy Agency (IEA). (2021). *World Energy Outlook 2021.* IEA Publications.

International Energy Agency (IEA). *World Energy Investment2021.* Online: https://www.iea.org/

International Renewable Energy Agency (IRENA). (2020). *Innovation landscape brief: Peer-to-peer electricity trading.* IRENA.

Irazoqui, G., Eisenbarth, T., & Sunar, B. (2016). Cross processor cache attacks. *Proceedings of the 11th ACM on Asia Conference on Computer and Communications Security*, 353-364.

Jaikaran, C. (2025). *Cybersecurity: Selected cyberattacks, 2012–2024 (CRS Report No. R46974).* Congressional Research Service.

Jha, N., Prashar, D., Khalaf, O., Alotaibi, Y., Alsufyani, A., & Alghamdi, S. (2021). Blockchain-Based Crop Insurance: A Decentralized Insurance System for Modernization of Indian Farmers. *Sustainability (Basel), 13*(16), 8921. DOI: 10.3390/su13168921

Jithish, J., Alangot, B., Mahalingam, N., & Yeo, K. S. (2023). Distributed anomaly detection in smart grids: A federated learning-based approach. *IEEE Access : Practical Innovations, Open Solutions, 11*, 7157–7179. DOI: 10.1109/ACCESS.2023.3237554

Judijanto, L. (2023). Implementation of blockchain technology in improving transparency of public services: A case study on government service delivery in Indonesia. *West Science Information System and Technology, 1*(2), 63–71. DOI: 10.58812/wsist.v1i02.477

Kademeteme, E., & Bvuma, S. (2023). Using blockchain technology to improve the integrity and transparency of procurement processes between SMMEs and government: A systematic literature review. *Journal of The British Blockchain Association, 6*(1), 1–13. DOI: 10.31585/jbba-7-1-(1)2024

Kang, J., Yu, R., Huang, X., Maharjan, S., Zhang, Y., & Hossain, E. (2017). Enabling localized peer-to-peer electricity trading among plug-in hybrid electric vehicles using consortium blockchains. *IEEE Transactions on Industrial Informatics, 13*(6), 3154–3164. DOI: 10.1109/TII.2017.2709784

Kassem, M. (2023). Blockchain and digital governance: Decentralization of decision making policy. *The Review of Policy Research, 42*(1), 45–67.

Kaushik, A. K. (2019). *The development of smart water markets using blockchain technology* (Working Paper). New America.

Khalifa, F., & Marzouk, M. (2024). Integrated blockchain and digital twin framework for sustainable building energy management. *Journal of Industrial Information Integration*. Advance online publication. DOI: 10.1016/j.jii.2024.100747

Khan, A. A., Laghari, A. A., Rashid, M., Li, H., Javed, A. R., & Gadekallu, T. R. (2023). Artificial intelligence and blockchain technology for secure smart grid and power distribution automation: A state-of-the-art review. *Sustainable Energy Technologies and Assessments*, *57*, 103282. DOI: 10.1016/j.seta.2023.103282

Khang, A., Chowdhury, S., & Sharma, S. (Eds.). (2022). *The data-driven blockchain ecosystem: fundamentals, applications, and emerging technologies*. DOI: 10.1201/9781003269281

Khan, M. A., & Salah, K. (2018). IoT security: Review, blockchain solutions, and open challenges. *Future Generation Computer Systems*, *82*, 395–411. DOI: 10.1016/j.future.2017.11.022

Khan, M., Li, G., Wu, Y., & Ghosh, S. (2021). Blockchain-Envisioned Peer-to-Peer Energy Trading Mechanisms: A Systematic Review. *IEEE Access : Practical Innovations, Open Solutions*, *9*, 16571–16594.

Khan, S., Shael, M., Majdalawieh, M., Nizamuddin, N., & Nicho, M. (2022). Blockchain for governments: The case of the Dubai Government. *Sustainability (Basel)*, *14*(11), 6576. DOI: 10.3390/su14116576

Khaqqi, K. N., Sikorski, J. J., Hadinoto, K., & Kraft, M. (2018). Incorporating seller/buyer reputation-based system in blockchain-enabled emission trading application. *Applied Energy*, *209*, 8–19. DOI: 10.1016/j.apenergy.2017.10.070

Khatoon, A., Verma, P., Southernwood, J., Massey, B., & Corcoran, P. (2019). Blockchain in energy efficiency: Potential applications and benefits. *Energies*, *12*(17), 3317. DOI: 10.3390/en12173317

Kindervag, J. (2010). *No More Chewy Center: Introducing the Zero Trust Model of Information Security*. Forrester Research.

Kirkman, S. S., Fulton, S., Hemmes, J., Garcia, C., & Wilson, J. C. (2024). A blockchain architecture to increase the resilience of industrial control systems from the effects of a ransomware attack: A proposal and initial results. *ACM Transactions on Cyber-Physical Systems*, *8*(1), 1–13. DOI: 10.1145/3637553

Kirli, D., Couraud, B., Robu, V., Salgado-Bravo, M., Norbu, S., Andoni, M., Antonopoulos, I., Negrete-Pincetic, M., Flynn, D., & Kiprakis, A. (2022). Smart contracts in energy systems: A systematic review of fundamental approaches and implementations. *Renewable & Sustainable Energy Reviews*, *158*, 112013. DOI: 10.1016/j.rser.2021.112013

Kloppenburg, S., & Boekelo, M. (2019). Digital Platforms and the Future of Energy Provisioning. *Energy Research & Social Science*, *49*, 68–73. DOI: 10.1016/j.erss.2018.10.016

Knapp, C., Novak, E., & Rossi, M. (2021). Scaling Blockchains in Energy Systems: Layered Designs and Sidechain Approaches. *IEEE Access : Practical Innovations, Open Solutions*, *9*, 118954–118965.

Kolahan, A., Maadi, S. R., Teymouri, Z., & Schenone, C. (2021). Blockchain-based solution for energy demand-side management of residential buildings. *Sustainable Cities and Society*, *75*, 103316. DOI: 10.1016/j.scs.2021.103316

KPMG. (2021). *Sustainability-Linked Finance: Guidance Document*. https://kpmg.com/

Krause, T., Ernst, R., Klaer, B., Hacker, I., & Henze, M. (2021). Cybersecurity in power grids: Challenges and opportunities. *Sensors (Basel)*, *21*(18), 6225. DOI: 10.3390/s21186225 PMID: 34577432

Kshetri, N. (2017). Blockchain's roles in strengthening cybersecurity and protecting privacy. *Telecommunications Policy*, *41*(10), 1027–1038. DOI: 10.1016/j.telpol.2017.09.003

Kumar, P. R., Meenakshi, S., Shalini, S., Devi, S. R., & Boopathi, S. (2023). Soil Quality Prediction in Context Learning Approaches Using Deep Learning and Blockchain for Smart Agriculture. In *Effective AI, Blockchain, and E-Governance Applications for Knowledge Discovery and Management* (pp. 1-26). IGI Global. DOI: 10.4018/978-1-6684-9151-5.ch001

Kumar, N. M., Chand, A. A., Malvoni, M., Prasad, K. A., Mamun, K. A., Islam, F. R., & Chopra, S. S. (2020). Distributed energy resources and the application of AI, IoT, and blockchain in smart grids. *Energies*, *13*(21), 5739. DOI: 10.3390/en13215739

Kumar, T. W. K., & Gupta, S. (2021). DeFi in the Energy Industry: Potential and Challenges. *IEEE Access : Practical Innovations, Open Solutions*, *9*, 138765–138778.

Kuzior, A., Sira, M., & Brożek, P. (2022). Using Blockchain and Artificial Intelligence in Energy Management as a Tool to Achieve Energy Efficiency. *Virtual Economics*, *5*(3), 69–90. DOI: 10.34021/ve.2022.05.03(4)

Kuzlu, M., Pipattanasomporn, M., & Rahman, S. (2014). Communication network requirements for major smart grid applications in HAN, NAN and WAN. *Computer Networks*, *67*, 74–88. DOI: 10.1016/j.comnet.2014.03.029

Kwok, J. S. H., & Koh, L. H. (2019). Is blockchain technology a watershed for e-government? The case of Seoul's blockchain 2.0 initiatives. *Asia Pacific Journal of Public Administration*, *41*(1), 37–50.

Lakumarapu, S., Nithyanandhan, R., Bhargavi, V. S., Anish, T. P., & Nalini, M. (2024, August). Machine Learning Approaches for Liver Disease Prediction: A Comparative Analysis. In *2024 5th International Conference on Electronics and Sustainable Communication Systems (ICESC)* (pp. 159-164). IEEE. DOI: 10.1109/ICESC60852.2024.10689974

Lampson, B. W. (1973). A note on the confinement problem. *Communications of the ACM*, *16*(10), 613–615. DOI: 10.1145/362375.362389

Laroiya, C., Saxena, D., & Komalavalli, C. (2020). Applications of blockchain technology. In *Handbook of research on blockchain technology* (pp. 213–243). Academic press. DOI: 10.1016/B978-0-12-819816-2.00009-5

Lee, R. M., Assante, M. J., & Conway, T. (2016). *Analysis of the cyber attack on the Ukrainian power grid*. Electricity Information Sharing and Analysis Center (E-ISAC) and SANS ICS. (March 2016 report).

Lee, D., Pang, M., & Ryu, Y. (2021). *Fractional Ownership and Liquidity for Renewable Energy Assets* (Vol. 153). Energy Policy.

Lee, D., Ryu, S., & Jeong, D. (2022). Peer-to-Peer Energy Transactions in a Blockchain Environment: A Case Study of Solar Communities. *IEEE Transactions on Smart Grid*, *13*(5), 3671–3681.

Li, H., Zou, L., Jiang, T., Liao, D., Zhang, M., & Chen, X. (2022). A Master-slave Multi-chain Vehicle Networking Architecture.

Li, . (2021). *Block chain network covert communication method based on multi-node timestamp collusion: CN111245570B*. Patent.

Li, . (2022). Survey on key technologies of covert channel in ubiquitous network environment. *Journal of Communication*, *43*(4), 186–201.

Liang, G., Weller, S. R., Luo, F., Zhao, J., & Dong, Z. Y. (2018). Distributed blockchain-based data protection framework for modern power systems against cyber attacks. *IEEE Transactions on Smart Grid*, *10*(3), 3162–3173. DOI: 10.1109/TSG.2018.2819663

Liang, K., Shetty, S., & Tosh, D.. (2017). ProvChain: A Blockchain-Based Data Provenance Architecture in Cloud Environment with Enhanced Privacy and Availability. In *Proc. 17th IEEE/ACM Int. Symp. on Cluster, Cloud and Grid Computing (CCGRID 2017)* (pp. 468–477). IEEE. DOI: 10.1109/CCGRID.2017.8

Li, J., Herdem, M. S., Wen, S., & Kuzior, A. (2023). Methods and applications for Artificial Intelligence, Big Data, Internet of Things, and Blockchain in smart energy management. *Energy and AI*, *11*, 100208. DOI: 10.1016/j.egyai.2022.100208

Liu, . (2015). Last-level cache side-channel attacks are practical. *Proceedings of 2015 IEEE Symposium on Security and Privacy*, 605-622. DOI: 10.1109/SP.2015.43

Liu, . (2021). Research progress of blockchain in data security. Indian's. *Journal of Computers*, *44*(1), 1–27.

Liu, C., Tan, R., Wu, Y., Feng, Y., Jin, Z., Zhang, F., Liu, Y., & Liu, Q. (2024). Dissecting zero trust: Research landscape and its implementation in IoT. *Cybersecurity*, *7*(1), 20. Advance online publication. DOI: 10.1186/s42400-024-00212-0

Liu, Y., Lu, Q., Zhu, L., Paik, H. Y., & Staples, M. (2023). A systematic literature review on blockchain governance. *Journal of Systems and Software*, *197*, 111576. DOI: 10.1016/j.jss.2022.111576

Li, Y., Xie, J., Yang, J., Ren, J., & Zhai, N. (2022). Application of blockchain technology in water rights trading in the irrigation area under the Internet-of-Things environment. *Sustainability*, *14*(12), 7017. DOI: 10.3390/su14127017

Li, Y., Zhao, Y., Wu, L., & Zeng, Z. (2024). *Artificial intelligence enabled computational methods for smart grid forecast and dispatch*. Springer.

Li, Z., Kang, J., Yu, R., Ye, D., Deng, Q., & Zhang, Y. (2018). Consortium blockchain for secure energy trading in industrial Internet of Things. *IEEE Transactions on Industrial Informatics*, *14*(8), 36903700.

LO3 Energy. (2017). *Brooklyn Microgrid Project*. Retrieved from https://lo3energy.com/

Lucena, . (2004). Syntax and semantics-preserving application-layer protocol steganography. *International Workshop on Information Hiding*, 164-179. DOI: 10.1007/978-3-540-30114-1_12

Luo, . (2022). A novel covert communication method based on bitcoin transaction. *IEEE Transactions on Industrial Informatics*, *18*(4), 2830–2839. DOI: 10.1109/TII.2021.3100480

Lyu, J. S., & Cao, X. C. (2021). Covert communication technology based on bitcoin. *Journal of Cybersecurity*, *6*(2), 143–152.

Ma, C., Zhu, F., & Hu, B.. (2023). A Blockchain Solution for Remote Sensing Data Management Model. *Applied Sciences (Basel, Switzerland)*, *13*(17), 9609. DOI: 10.3390/app13179609

Mahula, S., Tan, E., Crompvoets, J., & Timmers, P. (2025). What motivates public sector organisations to use blockchain? [in press]. *International Journal of Public Sector Management*, *38*(1), 120–136. DOI: 10.1108/IJPSM-12-2023-0361

Mancarella, J., Chicco, G., & Pistono, F. (2020). Integrated Energy Systems: A View on Planning, Operation, and Market Design. *IEEE Power & Energy Magazine*, *15*(1), 38–55.

Mathews, J., & Tan, H. (2020). Circular Economy and Blockchain: Uniting Industrial Symbiosis. *Resources, Conservation and Recycling*, *156*, 104715.

Mazurczyk, W., & Szczypiorski, K. (2012). Evaluation of steganographic methods for oversized IP packets. *Telecommunication Systems*, *49*(2), 207–217. DOI: 10.1007/s11235-010-9362-7

Mazzucato, M., & Semieniuk, G. (2018). Financing Renewable Energy: Who is Financing What and Why it Matters. *Technological Forecasting and Social Change*, *127*, 8–22. DOI: 10.1016/j.techfore.2017.05.021

Mengelkamp, E., Gärttner, J., Rock, K., Kessler, S., Orsini, L., & Weinhardt, C. (2018). Designing microgrid energy markets: A case study with the Brooklyn Microgrid. *Applied Energy*, *210*, 870–880. DOI: 10.1016/j.apenergy.2017.06.054

Meng, X., & Zhu, L. (2024). Augmenting cybersecurity in smart urban energy systems through IoT and blockchain technology within the digital twin framework. *Sustainable Cities and Society*, *106*, 105336. DOI: 10.1016/j.scs.2024.105336

Meyers, G., van Oirsouw, C. C. K., Keymolen, E. L. O., & Goossens, J. (2023). After the announcement: An interdisciplinary analysis of blockchain development in governments. *Policy Design and Practice*, *6*(4), 433–447. DOI: 10.1080/25741292.2023.2272332

Michel, B., Yang, X., & Clark, R. (2020). Smart Contracts for Automated Demand Response in Microgrids. *IEEE Transactions on Industrial Informatics*, *16*(12), 7687–7696.

Millen, J. (1999). 20 years of covert channel modeling and analysis. *Proceedings of the 1999 IEEE Symposium on Security and Privacy*, 113-114. DOI: 10.1109/SECPRI.1999.766906

Morris, R. (2020, January 21). Using blockchain to manage Colorado's water supply. *ASME News*. Retrieved from https://www.asme.org

Morstyn, T., Farrell, N., Darby, S. J., & McCulloch, M. D. (2018). Using peer-to-peer energy-trading platforms to incentivize prosumers to form federated power plants. *Nature Energy*, *3*(2), 94–101. DOI: 10.1038/s41560-017-0075-y

Morstyn, T., Teytelboym, A., & McCulloch, M. (2019). Bilateral Contract Networks for Peer-to-Peer Energy Trading. *IEEE Transactions on Smart Grid*, *10*(2), 2026–2035. DOI: 10.1109/TSG.2017.2786668

Moudoud, H., Cherkaoui, S., & Khoukhi, L. (2019). An IoT Blockchain Architecture Using Oracles and Smart Contracts: The Use-Case of a Food Supply Chain. In *Proc. IEEE 30th Annu. Int. Symposium on Personal, Indoor and Mobile Radio Communications (PIMRC 2019)* (pp. 1–6). IEEE. DOI: 10.1109/PIMRC.2019.8904404

Mougayar, W. (2016). *The Business Blockchain: Promise, Practice, and Application of the Next Internet Technology*. IEEE Transactions on Emerging Technologies.

Muchene, D. N., Luli, K., & Shue, C. A. (2013). *Reporting insider threats via covert channels. Proceedings of 2013 IEEE Security and Privacy Workshops*. IEEE Press.

Mukherjee, D., Hussain, M., & Li, H. (2018). IoT-Integrated Blockchain Framework for Energy Data Accuracy. *IEEE Transactions on Engineering Management*, *67*(4).

Munsing, E., Mather, J., & Moura, S. (2017). Blockchain for a transactive energy system: An experimental case study. *Proceedings of the 1st IEEE International Conference on Energy Internet (ICEI)*, 85-90.

Mustafa, G., Rafiq, W., Jhamat, N., Arshad, Z., & Rana, F. A. (2024). Blockchain-based governance models in e-government: A comprehensive framework for legal, technical, ethical and security considerations. *International Journal of Law and Management*. (Advance online publication)

Nai Fovino, I., Andreadou, N., Geneiatakis, D., Giuliani, R., Kounelis, I., Lucas, A., & Soreli, G. (2021). *Blockchain in the energy sector: Use cases identification and analysis (EUR 30782 EN)*. Publications Office of the European Union.

Nai Fovino, I., Andreadou, N., Geneiatakis, D., Giuliani, R., Kounelis, I., Lucas, A., & Steri, G. (2021). *Blockchain in the energy sector – WP3: Use cases identification and analysis (EUR 30782 EN)*. Publications Office of the European Union., DOI: 10.2760/061600

Nai Fovino, I., Poursanidis, I., Soupionis, I., & Steri, G. (2021). Blockchain in the energy sector: WP3 – Use cases identification and analysis [EN]. Luxembourg: Publications Office of the European Union.]. *EUR*, •••, 30782.

Nakamoto, S. (2008). Bitcoin: A peer-to-peer electronic cash system.

Nakamoto, S. (2008). *Bitcoin: A Peer-to-Peer Electronic Cash System*. https://bitcoin.org/bitcoin.pdf

Nakamoto, S. (2008). Bitcoin: A peer-to-peer electronic cash system. Retrieved from https://bitcoin.org/bitcoin.pdf

Nakamoto, S. (2008). Bitcoin: A peer-to-peer electronic cash system. *Consulted*, *28*(1), 21260–21268.

Nakamoto, S. (2008). *Bitcoin: A Peer-to-Peer Electronic Cash System*. IEEE Transactions on Blockchain Technology.

Naqash, M. T., Syed, T. A., Alqahtani, S. S., Siddiqui, M. S., Alzahrani, A., & Nauman, M. (2023). A blockchain-based framework for efficient water management and leakage detection in urban areas. *Urban Science (Basel, Switzerland)*, *7*(4), 99. DOI: 10.3390/urbansci7040099

National Security Telecommunications Advisory Committee (NSTAC). (2022). *Report to the President on Zero Trust and Trusted Identity Management*. U.S. DHS.

Neufund. (2020). *Equity Tokens for Energy Startups*. https://neufund.org/

Nguyen, T. Q., Das, A. K., & Tran, L. T. (2019). Neo Smart Contract for Drought-Based Insurance. In *Proc. IEEE Canadian Conf. Electrical and Computer Engineering (CCECE 2019)* (pp. 1–4). IEEE.

Nie, S., Ren, J., Wu, R., Han, P., Han, Z., & Wan, W. (2025). Zero-trust access control mechanism based on blockchain and inner-product encryption in the Internet of Things in a 6G environment. *Sensors (Basel)*, *25*(2), 550. DOI: 10.3390/s25020550 PMID: 39860920

Novo, O. (2018). Blockchain meets IoT: An architecture for scalable access management in IoT. *IEEE Internet of Things Journal*, *5*(2), 1184–1195. DOI: 10.1109/JIOT.2018.2812239

Ortiz, P., Park, S., & Hauser, G. "Using Blockchain for Power Grid Event Logging and Forensic Analysis," in *Proc. IEEE PES General Meeting*, 2021, pp. 1–5.

Padmanaban, S., Dhanaraj, R. K., Holm-Nielsen, J. B., Krishnamoorthi, S., & Balusamy, B. (Eds.). (2022). *Blockchain-based systems for the modern energy grid*. Academic Press.

Park, J.-H., Park, S., & Youm, H. (2025). A proposal for a zero-trust-based multi-level security model and its security controls. *Applied Sciences (Basel, Switzerland)*, *15*(2), 785. DOI: 10.3390/app15020785

Park, L. W., Lee, S., & Chang, H. (2018). A sustainable home energy prosumer-chain methodology with energy tags over the blockchain. *Sustainability (Basel)*, *10*(3), 658. DOI: 10.3390/su10030658

Partala, J. (2018). Provably secure covert communication on blockchain. *Cryptography*, *2*(3), 18–35. DOI: 10.3390/cryptography2030018

Patuck, R., & Hernandez-Castro, J. (2013). Steganography using the extensible messaging and presence protocol (XMPP). arXiv Preprint, arXiv: 1310.0524.

Peng, Z., Guo, X., Zhen, F., & Li, J. (2021). A Blockchain-Based Airplane Meteorological Data Sharing Incentive System. In *Proc. 2021 IEEE Int. Conf. on Information Technology, Big Data and Artificial Intelligence (ICITBAI)*. IEEE.

Peterson, J., & Prahalad, C. K. (2020). Transparency in Energy Finance: A Frontier for Governance. *Finance and Stochastics*, *6*(2), 11–26.

Petitcolas, F. A. P., Anderson, R. J., & Kuhn, M. G. (1999). Information hiding-a survey. *Proceedings of the IEEE*, *87*(7), 1062–1078. DOI: 10.1109/5.771065

Phemex. (n.d.). *Real estate blockchain and crypto*. https://phemex.com/academy/real-estate-blockchain-and-crypto

Piccardo, G., Conti, L., & Martino, A. (2024). Blockchain technology and its potential to benefit public services provision: A short survey. *Future Internet*, *16*(8), 290. DOI: 10.3390/fi16080290

Poberezhna, A. (2018). Addressing water sustainability with blockchain technology and green finance. In A. Marke (Ed.), *Transforming Climate Finance and Green Investment with Blockchains* (pp. 189–196). Academic Press (Elsevier). DOI: 10.1016/B978-0-12-814447-3.00014-8

Polcumpally, A. T., Pandey, K. K., Kumar, A., & Samadhiya, A. (2024). Blockchain governance and trust: A multi-sector thematic systematic review and exploration of future research directions. *Heliyon*, *10*(12), e32975. DOI: 10.1016/j.heliyon.2024.e32975 PMID: 38984295

Poon, J., & Buterin, V. (2017). Plasma: Scalable autonomous smart contracts. White Paper, 1-47.

Pop, C., Cioara, T., Antal, M., Anghel, I., Salomie, I., & Bertoncini, M. (2018). Blockchain-based decentralized management of demand response programs in smart energy grids. *Sensors (Basel)*, *18*(1), 162. DOI: 10.3390/s18010162 PMID: 29315250

Pop, C., Cioara, T., Salomie, I., & Antal, M. (2018). Blockchain-Based Decentralized Energy Trading in Community Microgrids. *Sustainability*, *10*, 116.

Popov, S. (2018). The tangle. White Paper, 1-28.

Power Ledger. (2017). *White paper: An introduction to the Power Ledger platform.*

PowerLedger. (2021). *Peer-to-Peer Energy Trading Platform.* https://www.powerledger.io/

Prabha, D., Subramanian, R. S., Dinesh, M. G., & Girija, P. (2024). Sustainable Farming Through AI-Enabled Precision Agriculture. In *Artificial Intelligence for Precision Agriculture* (pp. 159–182). Auerbach Publications. DOI: 10.1201/9781003504900-8

Prakash, A., Bose, C., & Suh, B. (2022). Vulnerability Assessments of Electrical Grids Under Extreme Events. *Energy Studies Review*, *34*(3), 77–88.

Prasada, P., & Prasad, D. S. (2024). Blockchain-enhanced machine learning for robust detection of APT injection attacks in cyber-physical systems. *International Journal of Computational and Experimental Science and Engineering*, *10*(4). Advance online publication. DOI: 10.22399/ijcesen.539

Queiroz, M. M., Fosso Wamba, S., Wamba-Taguimdje, S. L., & Lu, Q. (2024). How emerging technologies can solve critical issues in organizational operations: An analysis of blockchain-driven projects in the public sector. *Government Information Quarterly*, *41*(1), 101912. DOI: 10.1016/j.giq.2024.101912

Rahman, M., Chowdhury, S., Shorfuzzaman, M., Hossain, M. K., & Hammoudeh, M. (2023). Peer-to-peer power energy trading in blockchain using efficient machine learning model. *Sustainability (Basel)*, *15*(18), 13640. DOI: 10.3390/su151813640

Raj, H., Saroiu, S., Wolman, A., Aigner, R., Cox, J., England, P., & Jaeger, T. (2016). fTPM: A software-only implementation of a TPM chip. In *25th USENIX Security Symposium* (pp. 841-856). USENIX Association.

Rathor, S. K., & Saxena, D. (2020). Energy management system for smart grid: An overview and key issues. *International Journal of Energy Research*, *44*(6), 4067–4109. DOI: 10.1002/er.4883

Reyna, A., Martín, C., Chen, J., Soler, E., & Díaz, M. (2018). On blockchain and its integration with IoT. Challenges and opportunities. *Future Generation Computer Systems, 88*, 173–190. DOI: 10.1016/j.future.2018.05.046

Rios, R., Onieva, J. A., & Lopez, J. (2012). *HIDE_DHCP: covert communications through network configuration messages*. Information Security and Privacy Research.

Rodríguez Furones, A., & Tejero Monzón, J. I. (2023). Blockchain applicability in the management of urban water supply and sanitation systems in Spain. *Journal of Environmental Management, 344*, 118480. DOI: 10.1016/j.jenvman.2023.118480 PMID: 37459816

Rodriguez Müller, A. P., Martin Bosch, J., & Tangi, L. (2024). An overview of the expected public values arising from blockchain adoption in the European public sector.* [forthcoming]. *International Journal of Public Sector Management, 37*(2), 240–257.

Rodríguez-Molina, J., Martínez-Núñez, M., Martínez, J. F., & Baena, W. (2020). Blockchain in Smart Grid: A Critical Review of Progress and Challenges. *Energies, 13*(4), 838.

Rose, S., Borchert, O., Mitchell, S., & Connelly, S. (2020). Zero trust architecture (SP 800–207). National Institute of Standards and Technology.

Rosli, A., Hassan, S., Omar, M. H., Sajat, M. S., Attarbashi, Z. S., & Zaini, K. M. (2024, November). Blockchain Applications and Management: A Multidisciplinary Analysis. In 2024 7th International Conference on Internet Applications, Protocols, and Services (NETAPPS) (pp. 1-8). IEEE.

Rowland, C. H. (1997). Covert channels in the TCP/IP protocol suite. *First Monday, 2*(5), 1. DOI: 10.5210/fm.v2i5.528

Saeed, M., Amin, R., Aftab, M., & Ahmed, N. (2022). Trust management technique using blockchain in smart building. In *Proceedings of the 2021 International Symposium on Networks, Computers and Communications (Engineering Proceedings Vol. 20, No. 1, p. 24)*. MDPI. https://doi.org/DOI: 10.3390/engproc2022020024

Saeed, S., Gull, H., Aldossary, M. M., Altamimi, A. F., Alshahrani, M. S., Saqib, M., Zafar Iqbal, S., & Almuhaideb, A. M. (2024). Digital transformation in energy sector: Cybersecurity challenges and implications. *Information (Basel), 15*(12), 764. DOI: 10.3390/info15120764

Salami, Y., & Hosseini, S. R. (2023). BSAMS: Blockchain-Based Secure Authentication Scheme in Meteorological Systems. *Nivar, 47*(120–121), 181–197.

Salem, M. J., Ndolu, F. H. E., Hidayatullah, D. E. R., & Sari, R. F. (2021). Developing NEO Smart Contract for Weather-Based Insurance. In *Proc. 4th Int. Seminar on Research of Information Technology and Intelligent Systems (ISRITI 2021)* (pp. 603–608). IEEE. DOI: 10.1109/ISRITI54043.2021.9702853

Sara Technologies Inc. (n.d.). *Blockchain in logistics management*. https://www.saratechnologies.com/logistics-blockchain-management

Satilmisoglu, T. K., Sermet, Y., Kurt, M., & Demir, I. (2024). Blockchain opportunities for water resources management: A comprehensive review. *Sustainability (Basel), 16*(6), 2403. DOI: 10.3390/su16062403

Savur, D., & Emekci, Ş. (2024). A review on smart building and blockchain technology. *Politeknik Dergisi*, *27*(1), 1–13. DOI: 10.2339/politeknik.1403561

Shahinzadeh, H., Zanjani, S., & Moradi, J. (2022). The transition toward merging big data analytics, IoT, and artificial intelligence with blockchain in transactive energy markets. In *Proceedings of the 2022 IEEE Global Energy Conference (GEC)*. IEEE. DOI: 10.1109/GEC55014.2022.9986604

Sharin, F. H., Hernandez, M. S., & Sentosa, I. (2023, May). Future trends of blockchain technology in the technological fields. In 2023 3rd International Conference on Advance Computing and Innovative Technologies in Engineering (ICACITE) (pp. 1307-1313). IEEE.

Shen, S., Luo, J., & Chan, M. (2020). *Blockchain-Based Energy Investment Funds: Architecture and Operational Models* (Vol. 30). Energy Strategy Reviews.

Shetty, S. S., Kamhoua, C. A., & Njilla, L. L. (Eds.). (2019). *Blockchain for Distributed Systems Security*. Wiley-IEEE Press. DOI: 10.1002/9781119519621

Sikorski, J. J., Haughton, J., & Kraft, M. (2017). Blockchain technology in the chemical industry: Machine-to-machine electricity market. *Applied Energy*, *195*, 234–246. DOI: 10.1016/j.apenergy.2017.03.039

Simmons, G. J. (1984). The prisoners' problem and the subliminal channel. *Proceedings of CRYPTO*, 51-67.

Singh, K., & Singha, N. (2024). Credit-based energy trading system using blockchain and machine learning. *The Journal of Supercomputing*, *80*(11), 1–22. DOI: 10.1007/s11227-024-06073-1

Singh, M., Rathee, G., & Kumar, P. (2020). Blockchain and AI amalgamation for energy cloud management. *Journal of Systems Architecture*, *110*, 101746.

SlideTeam. (n.d.). *In-depth guide to blockchain nodes training [PowerPoint slides]*. https://www.slideteam.net/in-depth-guide-to-blockchain-nodes-training-ppt.html

Sohail, H., Hassan, M., Elmagzoub, M. A., Rajab, A., Rajab, K., Ahmed, A., Shaikh, A., Ali, A., & Jamil, H. (2023). BBSF: Blockchain-Based Secure Weather Forecasting Information through Routing Protocol in VANET. *Sensors (Basel)*, *23*(11), 5259. DOI: 10.3390/s23115259 PMID: 37299987

Solunke, H., & Bhaladhare, P. (2024, July). An In-Depth Investigation of Machine Learning Powered Blockchain Approached for Privacy Preservation. In *2024 International Conference on Emerging Trends in Networks and Computer Communications (ETNCC)* (pp. 1-5). IEEE. Available: https://ieeexplore.ieee.org/abstract/document/10767529/

Song, S., & Peng, W. (2020). BLOCCE +: An improved blockchain-based covert communication approach. *Journal of Chongqing University of Technology*, *34*(9), 238–244.

Soto, P., Ramirez, D., & Alvarado, J. (2021). Smart Contracts for Renewable Energy Projects. *Energies*, *14*(3).

Sousa, M. J. (2023). Blockchain as a driver for transformations in the public sector. *Policy Design and Practice*, *6*(4), 415–432. DOI: 10.1080/25741292.2023.2267864

Sousa, S., Silvestre, J., & Vardasca, R. (2019). Blockchain-based peer-to-peer energy trading and billing platform for microgrids. *International Journal of Electrical Power & Energy Systems, 112*, 232–241.

Sovacool, B., Hook, A., & Martiskainen, M. (2021). Blockchain for Transparent Energy Markets. *Renewable Energy, 156*, 1022–1032.

Sreelakshmi, S., Babu, A. A., Lakshmipriya, C., Gracious, L. A., Nalini, M., & Subramanian, R. S. (2024, October). Enhancing Intrusion Detection Systems with Machine Learning. In *2024 2nd International Conference on Self Sustainable Artificial Intelligence Systems (ICSSAS)* (pp. 557-564). IEEE. DOI: 10.1109/ICSSAS64001.2024.10760341

Srinivas, B., & Rao, K. N. (2024, December). Cloud-Block: An Integrated Cloud and Blockchain using Secure Cloud Computing. In *2024 3rd International Conference on Automation, Computing and Renewable Systems (ICACRS)* (pp. 651-657). IEEE. https://www.cryptoexplorer.cc/

Staples, M., Chen, S., Falamaki, S., Ponomarev, A., Rimba, P., Tran, A. B., & Zhu, J. (2018). Risks and opportunities of blockchain technologies in the public sector. In *Proceedings of the 51st Hawaii International Conference on System Sciences (HICSS)* (pp. 5470–5479). IEEE.

StartEngine. (2022). *Equity Crowdfunding and Token Offerings*. https://www.startengine.com/

Steffen, B. (2021). *A Look into the Financing Barriers for Small-Scale Renewable Projects* (Vol. 282). Applied Energy.

Subha, S., Balakrishnan, C., & Anish, T. P. (2024, August). Optimizing Machine Learning Algorithms for Heart Disease Prediction in Healthcare: A Comparative Study. In 2024 5th International Conference on Electronics and Sustainable Communication Systems (ICESC) (pp. 866-872). IEEE. DOI: 10.1109/ICESC60852.2024.10689877

Sun, D., Li, L., & Shen, C. (2019). Physical Reinforcements for Resilient Distribution Networks. *IEEE Access : Practical Innovations, Open Solutions, 7*, 99532–99545.

SunExchange. (2021). https://thesunexchange.com/

Sun, L. (2024). Exploration of blockchain-based building energy management. *Smart Systems and Green Energy, 6*(1), 60–67. DOI: 10.23977/ssge.2024.060108

Sutriyan, H., Sunaryadi, A., & Sinambela, M. (2022). Blockchain-Based Multiple Server Database System Prototype on BMKG Automatic Weather Station (AWS) Center Architecture. In *Proc. 7th Int. Conf. on Informatics and Computing (ICIC 2022)* (pp. 1–6). IEEE. DOI: 10.1109/ICIC56845.2022.10006923

Swan, M. (2015). "Blockchain: Blueprint for a New Economy." *IEEE Blockchain Symposium*

Swan, M. (2015). *Blockchain: Blueprint for a New Economy*. O'Reilly Media.

Syed, N. F., Shah, S. W., Shaghaghi, A., Anwar, A., Baig, Z., & Doss, R. (2022). Zero Trust Architecture (ZTA): A comprehensive survey. *IEEE Access : Practical Innovations, Open Solutions, 10*, 57143–57179. DOI: 10.1109/ACCESS.2022.3174679

Tahir, . (2016). Sneak-Peek: high speed covert channels in data center networks," *Proceedings of the 35th Annual IEEE International Conference on Computer Communications*, 1-9. DOI: 10.1109/INFO-COM.2016.7524467

Tan, E. (2023). The missing piece: The link between blockchain and public policy design. *Policy Design and Practice*, *6*(4), 488–504. DOI: 10.1080/25741292.2023.2233160

Tanquattrini, R., Palmaccio, M., Turco, M., & Manzari, A. (2024). The contribution of blockchain technologies to anti-corruption practices: A systematic literature review. *Business Strategy and the Environment*, *33*(1), 4–18. DOI: 10.1002/bse.3327

Tapscott, D., & Tapscott, A. (2016). Blockchain Revolution: How the Technology Behind Bitcoin is Changing Money, Business, and the World. *IEEE Access : Practical Innovations, Open Solutions.*

Tapscott, D., & Tapscott, A. (2016). *Blockchain revolution: How the technology behind Bitcoin is changing money, business, and the world.* Penguin.

Tapscott, D., & Tapscott, A. (2018). *Blockchain Revolution: How the Technology Behind Bitcoin and Other Cryptocurrencies is Changing the World.* Portfolio.

Tarek, D., & Ismail, T. (2022, December). Integrating smart contracts with WDNs framework for energy management and secure transactions. In *Proceedings of the 10th International Japan-Africa Conference on Electronics, Communications and Computations (JAC-ECC 2022)* (pp. 1–6). IEEE. https://doi.org/ DOI: 10.1109/JAC-ECC56395.2022.10044027

Tariq, M. U. (2024). Application of blockchain and Internet of Things (IoT) in modern business. In *Future of customer engagement through marketing intelligence* (pp. 66–94). IGI Global. DOI: 10.4018/979-8-3693-2367-0.ch004

Teng, F., Zhang, Q., Wang, G., Liu, J., & Li, H. (2021). A comprehensive review of energy blockchain: Application scenarios and development trends. *International Journal of Energy Research*, *45*(12), 17546–17578. DOI: 10.1002/er.7109

Tian, F. (2016, June). An agri-food supply chain traceability system for China based on RFID & blockchain technology. In 2016 13th international conference on service systems and service management (ICSSSM) (pp. 1-6). IEEE.

Torres, M., Li, Y., & Brown, T. (2022). Redundancy Strategies for Grid Continuity During Weather Extremes. *Renewable & Sustainable Energy Reviews*, *142*, 110877.

Treiblmaier, H. (2018). The Impact of the Blockchain on Supply Chain Management: A Theory-Based Research Framework and a Call for Action. *Supply Chain Management*, *23*(6), 545–559. DOI: 10.1108/SCM-01-2018-0029

Treiblmaier, H., & Sillaber, M. (2020). The Convergence of Blockchain and Supply Chain in E-Commerce: Future Perspectives. *Electronic Commerce Research and Applications*, *39*, 100935.

Treleaven, J., Gendal Brown, R., & Yang, D. (2017). Blockchain Technologies in Asset Securitization. *The Computer Journal*, *60*(9), 1378–1397.

Truong, N. B., Sun, K., Lee, G. M., & Guo, Y. (2019). GDPR-compliant personal data management: A blockchain-based solution. *IEEE Transactions on Information Forensics and Security, 15*, 1746–1761. DOI: 10.1109/TIFS.2019.2948287

Tseng, Y., Chen, L., & Chong, B. (2018). Combatting Counterfeit Drugs: A Blockchain-Based Approach. *IEEE Internet of Things Journal, 5*(6), 4011–4019.

TXOne Networks. (2023). *Insights Into ICS/OT Cybersecurity 2022*. TXOne Networks.

U.S. Department of Energy. (2024). *AI for Energy: Opportunities for a modern grid and clean energy economy*. U.S. DOE Office of Critical and Emerging Technologies.

U.S. Dept. of Energy (DOE). (2019). Roadmap to Achieve Energy Delivery Systems Cybersecurity.

U.S. Securities and Exchange Commission (SEC). (2020). *Framework for 'Investment Contract' Analysis of Digital Assets*. https://www.sec.gov/

Ul Haq, F., Zhang, J., & Tohid, R. (2023). Smart Contract-Based Fault Isolation in Autonomous Grids. *Electronics (Basel), 12*(4), 897.

Ulghar, I., Jahankhani, H., & Kendzierskyj, S. (2021). Blockchain capabilities in defending advanced persistent threats using correlation technique and hidden Markov models (HMM). In Jahankhani, H., Kendzierskyj, S., & Akhgar, B. (Eds.), *Information Security Technologies for Controlling Pandemics* (pp. 21–64). Springer., DOI: 10.1007/978-3-030-72120-6_2

United Nations Environment Programme (UNEP). (2021, September 23). *In battle against climate crisis, don't overlook the blockchain* [Press release].

Vakil, S., Petro, T., & Goyne, S. (2018). *Blockchain: A true disruptor for the energy industry – Use cases and strategic questions.* Deloitte Insights White Paper. Deloitte.

Vanmathi, C., Farouk, A., Alhammad, S. M., Bhattacharya, S., & Kasyapa, M. S. (2024). The Role of Blockchain in Transforming Industries Beyond Finance. *IEEE Access : Practical Innovations, Open Solutions, 12*, 148845–148867. DOI: 10.1109/ACCESS.2024.3468611

Ventura, D., & King, J. (2020). Cascading Failures in Centralized Grids: Causes, Consequences, and Mitigations. *IEEE Power & Energy Magazine, 18*(5), 45–57.

Vionis, P., & Kotsilieris, T. (2024). The potential of blockchain technology and smart contracts in the energy sector: A review. *Applied Sciences (Basel, Switzerland), 14*(1), 253. DOI: 10.3390/app14010253

Vogel, R. H., & Li, Y. (2020). Real-Time GHG Footprint Recording Using Blockchain. *Sustainability, 12*(24).

Vonitsanos, G., Panagiotakopoulos, T., Kanavos, A., Maragoudakis, M., & Mylonas, P. (2021, November). Issues and challenges associated with blockchain in smart cities. In 2021 16th International Workshop on Semantic and Social Media Adaptation & Personalization (SMAP) (pp. 1-5). IEEE.

WallStreetMojo. (n.d.). *Proof of work*. https://www.wallstreetmojo.com/proof-of-work/

Wang, . (2020). Categorization of covert channels and its application in threat restriction techniques. *Journal of Software*, *31*(1), 228–245.

Wang, G. (2019). Blockchain Technology Helps the Development of Meteorological Informatization. *Journal of Business and Management Sciences*, *7*(3), 112–120. DOI: 10.12691/jbms-7-3-2

Wang, J. (2025). Multi-agent system based smart grid anomaly detection using blockchain machine learning model in mobile edge computing network. *Computers & Electrical Engineering*, *121*, 109825. DOI: 10.1016/j.compeleceng.2024.109825

Wang, Q., & Su, Z. (2020). Integrating blockchain technology into the energy sector—From theory of blockchain to research and application of energy blockchain. *Computer Science Review*, *37*, 100275. DOI: 10.1016/j.cosrev.2020.100275

Wang, T., Ching, C., & Qu, M. (2021). A Taxonomy of Cyber Threats in Energy Systems. *Journal of Power Security*, *9*(2), 31–47.

Wang, X., Zha, X., Ni, W., Liu, R. P., Guo, Y. J., Niu, X., & Zheng, K. (2019). Survey on blockchain for Internet of Things. *Computer Communications*, *136*, 10–29. DOI: 10.1016/j.comcom.2019.01.006

Wardrop, J., Zhang, B., & Zhan, T. (2019). Blockchain-Based Dividend Distribution in Energy Tokens. *Computer Standards & Interfaces*, *68*, 35–44.

Wardrop, S., Zhang, B., & Rau, R. (2015). *Rise of Crowdfunding in Energy Projects*. Cambridge Centre for Alternative Finance.

Weber and K. Neuhoff. (2020). Accelerating the Energy Transition: Funding Bottlenecks and Solutions. *Energy Journal*, *41*(5).

Wendzel, S., Kahler, B., & Rist, T. (2012). Covert channels and their prevention in building automation protocols: a prototype exemplified using BACnet. *Proceedings of 2012 IEEE International Conference on Green Computing and Communications*, 731-736. DOI: 10.1109/GreenCom.2012.120

Wendzel, , Zander, S., Fechner, B., & Herdin, C. (2015). Pattern-based survey and categorization of network covert channel techniques. *ACM Computing Surveys*, *47*(3), 50. DOI: 10.1145/2684195

WePower. (2020). *White Paper*. https://wepower.com/

Werbach, K. (2018). *The blockchain and the new architecture of trust*. MIT Press. DOI: 10.7551/mitpress/11449.001.0001

Wolf, M. (1989). *Covert channels in LAN protocols. Local Area Network Security*. Springer.

Wood, G. (2015). Ethereum: A Secure Decentralized Generalized Transaction Ledger. *IEEE Access : Practical Innovations, Open Solutions*.

World Bank. (2020). *Energy Sector Directions Paper*.

World Economic Forum. (2018). *Building Block(chain)s for a Better Planet*. Cologny, Switzerland: World Economic Forum.

World Economic Forum. (2018). *Building block(chain)s for a better planet*. WEF/PwC White Paper.

Wörner, D., & von Bomhard, T. (2018). When your sensor earns money: Exchanging data for cash with Bitcoin and blockchain technologies. *IEEE Sensors Applications Symposium (SAS)*, 1-6.

Wu, . (2015). POSTER: biTheft: stealing your secrets by bidirectional covert channel communication with zero-permission android application. *Proceedings of the 22nd ACM SIGSAC Conference on Computer and Communications Security*, 1690-1692. DOI: 10.1145/2810103.2810108

Xia, W., Chen, X., & Song, C. (2022). A framework of blockchain technology in intelligent water management. *Frontiers in Environmental Science*, *10*, 909606. DOI: 10.3389/fenvs.2022.909606

Xie, H., Wang, Y., Ding, Y., Yang, C., Liang, H., & Qin, B. (2024). Industrial wireless internet zero trust model: Zero trust meets dynamic federated learning with blockchain. *IEEE Wireless Communications*, *31*(2), 22–29. DOI: 10.1109/MWC.001.2300368

Xiong, Z., Zhang, Y., Niyato, D., Wang, P., & Han, Z. (2018). When mobile blockchain meets edge computing. *IEEE Communications Magazine*, *56*(8), 33–39. DOI: 10.1109/MCOM.2018.1701095

Xue, Q., Yu, Y., Cen, S., & Li, Z. (2025). Secure and transparent energy management using blockchain and machine learning-based anomaly detection: A case study of the Ausgrid dataset. *Computers & Industrial Engineering*, *171*, 111040.

Xu, L., & Alstone, P. (2019). Islandable Microgrids for Community Energy Resilience: A Holistic Perspective. *Energy Reports*, *5*, 121–130.

Xu, X., Weber, I., & Staples, M. (2018). *Architecture for blockchain applications*. Springer.

Xu, Z., Chen, J., & Wu, M. (2019). Blockchain for Supply Chain: A Literature Review on Current Developments and Future Directions. *The Journal of Supply Chain Management*, *24*(1), 56–68.

Yamada, T., & Huang, Q. (2020). Distributed Optimization for Microgrid Management. *IEEE Transactions on Power Systems*, *35*(2), 789–800.

Yang, X., Wang, T., Wen, X., & Cai, W. (2019). Pricing mechanism for data trading in IoT data marketplace based on blockchain. In ICC 2019-2019 IEEE International Conference on Communications (pp. 1-6). IEEE.

Yan, Z., Zhang, Q., Ma, X., & Huang, H. (2019). Environmental monitoring system based on blockchain. In *Proceedings of the 4th International Conference on Crowd Science and Engineering (ICCSE '19)* (pp. 114–118). ACM. DOI: 10.1145/3371238.3371245

Yao, A. C. (1982). Protocols for secure computations. In *23rd Annual Symposium on Foundations of Computer Science* (pp. 160-164). IEEE. DOI: 10.1109/SFCS.1982.38

Yildiz, Ö. (2019). Financing Renewable Energy via Citizen Participation: The Case of Germany. *Renewable Energy*, *68*, 677–685. DOI: 10.1016/j.renene.2014.02.038

Youssef, S. B. H., Rekhis, S., & Boudriga, N. (2019, April). A blockchain-based secure IoT solution for dam surveillance. In *Proceedings of the 2019 IEEE Wireless Communications and Networking Conference (WCNC)* (pp. 1–6). IEEE. https://doi.org/DOI: 10.1109/WCNC.2019.8885479

Zander, S., Armitage, G., & Branch, P. (2007). Covert channels and countermeasures in computer network protocols. *IEEE Communications Surveys and Tutorials*, *45*(12), 136–142. DOI: 10.1109/COMST.2007.4317620

Zhang, . (2020a). A covert communication method using special bitcoin addresses generated by vanitygen. *Computers, Materials & Continua*, *65*(1), 597–616. DOI: 10.32604/cmc.2020.011554

Zhang, . (2020b). *The research on covert communication model based on blockchain: a case study of Ethereum's whisper protocol.* Frontiers in Cyber Security.

Zhang, C., Wu, J., Zhou, Y., Cheng, M., & Long, C. (2018). Peer-to-peer energy trading in a microgrid. *Applied Energy*, *220*, 1–12. DOI: 10.1016/j.apenergy.2018.03.010

Zhang, J. (2023). Application of blockchain technology in water management and new energy economic development. *Water Science and Technology: Water Supply*, *23*(10), 4227–4241. DOI: 10.2166/ws.2023.258

Zhang, S., Yang, G., & Lee, D. (2022). *Milestone-Based Disbursement of Funds in Smart Contracts: Methods and Case Studies* (Vol. 46). Computer Law & Security Review.

Zhang, X., & Wen, Q. (2021). Blockchain-Based Energy Trading and Consensus Algorithms: Challenges and Future Directions. *IEEE Access : Practical Innovations, Open Solutions*, *9*, 76133–76146.

Zhang, Y., & Wen, J. (2017). The IoT electric business model: Using blockchain technology for the internet of things. *Peer-to-Peer Networking and Applications*, *10*(4), 983–994. DOI: 10.1007/s12083-016-0456-1

Zhang, , Zhang, Z., Jin, Z., Su, Y., & Wang, Z. (2021). An approach of covert communication based on the Ethereum whisper protocol in blockchain. *International Journal of Intelligent Systems*, *36*(2), 962–996. DOI: 10.1002/int.22327

Zhang, , Zhang, Z., Wang, W., Jin, Z., Su, Y., & Chen, H. (2022). Research on a covert communication model realized by using smart contracts in blockchain environment. *IEEE Systems Journal*, *16*(2), 2822–2833. DOI: 10.1109/JSYST.2021.3057333

Zhao, B., Li, K., & Thompson, J. (2021). Smart Contract Security in Energy Trading Platforms. *Future Generation Computer Systems*, *120*, 117–129.

Zhao, S., Zhu, S., Wu, Z., & Jiang, B. (2022). Cooperative energy dispatch of smart building clusters based on smart contracts. *International Journal of Electrical Power & Energy Systems*, *138*, 107896. DOI: 10.1016/j.ijepes.2021.107896

Zheng, X., Xiong, F., & Liu, Q. (2018). Programmable Finance and Cross-Border Transactions: A Blockchain Perspective. *Computer Standards & Interfaces*, *54*, 46–58.

Zheng, Z., Xie, S., Dai, H. N., Chen, X., & Wang, H. (2018). Blockchain challenges and opportunities: A survey. *International Journal of Web and Grid Services*, *14*(4), 352–375. DOI: 10.1504/IJWGS.2018.095647

Zheng, Z., Xie, S., Dai, H., Chen, X., & Wang, H. (2018). An Overview of Blockchain Technology: Architecture, Consensus, and Future Trends. [IEEE.]. *Proceedings. IEEE International Congress on Big Data*, *2017*, 557–564.

Zhou, Z., Wang, B., Guo, Y., & Zhang, Y. (2019). Blockchain and computational intelligence inspired incentive-compatible demand response in internet of electric vehicles. *IEEE Transactions on Emerging Topics in Computational Intelligence*, *3*(3), 205–216. DOI: 10.1109/TETCI.2018.2880693

Zou, . (2005). *The research on information hiding based on command sequence of FTP protocol*. Knowledge-Based Intelligent Information and Engineering Systems. DOI: 10.1007/11553939_151

Zyskind, G., Nathan, O., & Pentland, A. (2015). *Decentralizing Privacy: Using Blockchain to Protect Personal Data*. IEEE Transactions on Information Security.

Zyskind, G., Nathan, O., & Pentland, A. S. (2015). Decentralizing privacy: Using blockchain to protect personal data. In *IEEE Security and Privacy Workshops* (pp. 180–184). IEEE. DOI: 10.1109/SPW.2015.27

About the Contributors

Someah Alangari is currently an associate professor of Computer Science at Shaqra University, Saudi Arabia. Her research interests include Artificial Intelligence, Machine learning, Deep learning, Software Engineering, Blockchain and digital transformation. She looks forward to improving and applying scientific knowledge to real-world problems.

Mohammad Al-Khaldy is an Assistant professor of Business Intelligence and Data Analytics at University of Petra, Jordan. He completed his PhD in Atrificial Intelligence and Data Science for the University of Hull -United Kingdom - 20017. Alkhaldy's research interests include data analytics, machine learning, predictive analytics, NLP, and decision support systems. His research has been published extensively in academic journals and conferences. Alkhaldy taught a variety of courses in Artificial intelligence and business intelligence, including Data Mining, Business Analytics, Computer Programming, Intelligent business systems, Data Visualization, and machine learning. Alkhaldy is a member of Jordan Computers Society, and Arab Robotics & AI Association. He also a reviewer for several other academic conferences and journals.

Mohammad Alauthman (mohammad.alauthman@uop.edu.jo) is an Associate Professor in the Information Security Department at the Faculty of Information Technology, University of Petra, Amman, Jordan. His research interests include network security, intrusion detection systems, and applying AI techniques—such as machine learning and deep learning—for botnet and DDoS detection, spam filtering, IoT security, and network traffic classification. He earned his Ph.D. in Computer Science from Northumbria University, UK, in 2016. Dr. Alauthman has received multiple research grants supporting advanced AI-driven intrusion detection systems and international collaboration. He is also actively involved in Erasmus+ projects, including BITTCOIN-JO (technology transfer), RL4Eng (remote engineering labs), Pro-GREEN LABs (green competences in education), and COMMO (mobility and cooperation across Mediterranean and Balkan institutions), contributing to innovation, sustainability, and cross-border academic development.

Amjad Aldweesh is a computer associate professor interested in the Blockchain and Smart contracts technology as well as cyber security. Amjad has a Bachelor degree in computer science. He has a MSc degree in advanced computer science and security from the University of Manchester with distinction. Amjad is the second in the UK and the first in the middle east to have a PhD in the Blockchain and Smart contracts technology from Newcastle University.

Ammar Almomani Higher Colleges of Technology, Dept. of Information Science, Sharjah, UAE IT Dept., Al-Huson University College, Al-Balqa Applied University, Irbid, Jordan Email:

ammarnav6@bau.edu.jo Prof. Ammar Almomani is a seasoned academic, researcher, and consultant with 20+ years of experience in cybersecurity, AI, and computer science. He holds a Ph.D. in Advanced Internet Security from Universiti Sains Malaysia and has taught over 45 specialized courses. He serves in the IT departments at Al-Balqa Applied University (Jordan) and Higher Colleges of Technology (UAE), contributing to research, curriculum design, and program accreditation (ABET, AACSB). He holds 10 patents (Germany and India) and has published 165+ articles (104 in Scopus), earning global recognition, a UAE Golden Visa, and ranking among the Top 2% of Global Researchers (Stanford & Scopus, 2023–2024). Prof. Almomani also supervises MSc/Ph.D. students internationally, chairs conferences, and engages in industry collaborations. With 113,000+ ResearchGate readers.Prof. Almomani continues to make a significant impact in the fields of cybersecurity,AI and higher education fostering innovation .

Enrique Arias-Antúnez is a full professor at the University of Castilla-La Mancha (UCLM). His research interests include high-performance computing applied to science (climate change, meteorology, bioinformatics) and engineering (control and high-speed railways) with more than 25 years of experience. He has published more than 100 scientific publications including conferences, journals, etc.. Arias-Antúnez received his Ph.D. degree in computing engineering from the Technical University of Valencia and has been the supervisor of 4 PhD, and 2 PhD currently under development. Enrique has participated in International, Europea, National and Regional projects both with public and private funding.

B. Yamini is a faculty member at SRM Institute of Science and Technology (SRMIST), Kattankulathur, Tamil Nadu. She is actively involved in research, teaching, and scholarly publications in her field. With a strong academic background, she contributes to advancing knowledge and student development. Her dedication to research enriches the academic environment of SRMIST.

Kashinath Basu is a Senior Lecturer in Computer Science at Oxford Brookes University. He has over over 20 years of research and teaching experience in the field of computer science, especially focussing around networking and web technologies. His recent research includes application of AI/ML and big data technologies in the softwarisation of network infrastructure, cloud computing and cyber security and application of AI/Ml and big data in the domain of networks and cloud. Dr Basu regularly publishes in prestigeous journals, conferences and books and has been in editorial board, organising committees and chairs of several international scholastic events. He also serves in a number of IEEE and Internet Engineering Task Force committees on internation standards and has active research collaboration with a number of global technology companies.

Ala Hamarsheh is an associate professor at the Faculty of Engineering and Information Technology of the Arab American University of Jenin. Dr. Hamarsheh has obtained a PhD in engineering sciences from Vrije Universiteit Brussel (VUB)/Brussels-Belgium in 2012. He graduated in computer science at the Faculty of Science, Birzeit University, Palestine, in 2000. He obtained an MSc degree in computer science at the Kind Abdullah II School for IT, The University of Jordan, Jordan, in 2003. Dr. Hamarsheh has published numerous papers in international refereed journals and conferences.

Vinay Kumar Kasula has a Master's in Technology from Osmania University, followed by a second Master's from Northwest Missouri State University. His academic journey has been driven by a deep-rooted interest in research and advanced technologies. This passion led him to pursue and complete a Ph.D. at the University of the Cumberlands, where he graduated in 2024, focusing on areas including Blockchain, Artificial Intelligence, Deep Learning, Machine Learning, Data Analytics, and Cybersecurity. With over a decade of industry experience, currently as a Senior Systems Analyst, he has applied and expanded his expertise in these domains, continually exploring and integrating innovative solutions to meet evolving technological challenges. This blend of education, research, and practical experience fueled his commitment to advancements in these cutting-edge fields.

Bhargavi Konda works as a senior systems analyst in the human resources information systems department for a healthcare company. She mainly works with data analysis and generates reports for various departments in the organization. She has been one of the major contributors in maintaining employee records, supporting all HR departments, and especially working with payroll operations from her team. She earned her master's degree from Osmania University (2001-2004). She received the gold medal and Star Student of the Year awards for her outstanding performance. She was placed in an IT consulting company right after her master's, and she worked as an Oracle technical consultant from 2004 to 2012. In that period, she was awarded the Outstanding Trainer award for her commitment and dedication. She has experience implementing applications in the banking and manufacturing sectors. She was passionate about higher studies, so she started to pursue a Ph.D. in IT in 2019. She earned a Ph.D. degree in Information Technology in the year 2024 from the University of The Cumberlands. She is excited to work on many journals and publications in the future and wants to continue her research. She can be contacted at email: Bhargavikonda@ieee.org. Her ORCID is 0009-0001-4704-1526.

L. A. Anto Gracious is an accomplished academic professional in Information Technology and Computer Science, known for his dedication to education, technology, and research. He holds a B.Tech in IT from C.S.I Institute of Technology (2006), an MBA in Technology Management, and an M.Tech in IT from Francis Xavier Engineering College (2011). He began his career as a Programmer at Software Solutions Integrated Limited, later transitioning to academia in 2011 as an Assistant Professor at Marthandam College of Engineering and Technology. Over eight years, he excelled in teaching, led the Software Development Cell, and earned multiple awards for his contributions. He later served at St. Xavier's Catholic College of Engineering and currently holds a position at R.M.K. College of Engineering and Technology. A prolific researcher, Anto focuses on machine learning, wireless sensor networks, and healthcare technologies. His work is widely published, and he has received accolades such as the Best Paper Award and multiple Best Faculty Awards.

Dinesh MG is an accomplished researcher and faculty member at Easa College of Engineering and Technology, specializing in cybersecurity. With a strong academic background and numerous publications, he is dedicated to advancing research and education in his field.

Ali Al Maqousi is an Associate Professor of Information Security at the University of Petra in Jordan and serves as the Director of the Office of External Funding and Technology Transfer. He holds a Ph.D. in Computer Science (Computer Networks) from Oxford Brookes University, UK, and specializes in cybersecurity, computer networks, blockchain, and smart systems. With over 30 peer-reviewed publications, Dr. Al Maqousi has led and contributed to numerous research initiatives, including high-impact Erasmus+ projects. These include BITTCOIN-JO, which focuses on establishing Technology Transfer Offices in Jordan; RL4Eng, which develops remote and virtual labs for engineering education;

Pro-GREEN LABs, promoting sustainable development skills in architecture; and COMMO, a mobility program enhancing international academic exchange across the Mediterranean and Western Balkans. Dr. Al Maqousi contributed to the organization of four international conferences as Organization Chair (AEECT 2011, IT-DREPS 2013, AEECT 2013, and AEECT 2015) and two international conferences as General Co-Chair (JIEEEC 2017 & IT-DREPS 2017) and one international conference as a publication chair (ICCIAA 2025)

Siva Subramanian R is associated with the Department of Computer Science and Engineering at R.M.K. College of Engineering and Technology. As a corresponding author, he actively contributes to research and scholarly publications. He is committed to mentoring students and driving technical innovation in his field. His work strengthens the academic and research environment of the institution.

Akhila Reddy Yadulla holds an MS in Computer Science from Northwestern Polytechnic University in 2016, an MS in Information Technology Management from Campbellsville University in 2019, and a Ph.D. in Information Technology from the University of the Cumberlands in 2024. She is an accomplished Senior Software Engineer with over eight years of experience in Java full-stack development and expertise in the Software Development Life Cycle. Dr. Yadulla is proficient in Java, Spring Boot, Hibernate, RESTful APIs, Angular, React, Microservices, SQL/NoSQL databases, DevOps, and Agile methodologies. Her research focuses on Cybersecurity, Artificial Intelligence, Data Science, Big Data, Blockchain Technologies, Amazon Web Services, and Machine Learning. She combines deep technical knowledge, advanced problem-solving skills, and AI-driven innovative solutions in cutting-edge technologies. Notably, she led a team in developing a secure and scalable microservices architecture for significant telecommunication and banking platforms, resulting in a major increase in system performance. She can be contacted at email: Akhilareddyyadulla@ieee.org. Her ORCID is 0009-0005-6377-6265.

Index

Q

R

S

T

U

W

Z

www.ingramcontent.com/pod-product-compliance
Lightning Source LLC
LaVergne TN
LVHW062108300926
849161LV00010B/2395